DIGITISED OPTICAL SKY SURVEYS

ASTROPHYSICS AND
SPACE SCIENCE LIBRARY

A SERIES OF BOOKS ON THE RECENT DEVELOPMENTS
OF SPACE SCIENCE AND OF GENERAL GEOPHYSICS AND ASTROPHYSICS
PUBLISHED IN CONNECTION WITH THE JOURNAL
SPACE SCIENCE REVIEWS

PROCEEDINGS
VOLUME 174

DIGITISED OPTICAL
SKY SURVEYS

PROCEEDINGS OF THE CONFERENCE ON
'DIGITISED OPTICAL SKY SURVEYS',
HELD IN EDINBURGH, SCOTLAND, 18–21 JUNE 1991

Edited by

H.T. MACGILLIVRAY

and

E.B. THOMSON

Royal Observatory, Edinburgh, Scotland

Springer Science+Business Media, B.V.

Library of Congress Cataloging-in-Publication Data

Conference on "Digitised Optical Sky Surveys" (1991 : Edinburgh,
  Scotland)
    Digitised optical sky surveys : proceedings of the Conference on
  "Digitised Optical Sky Surveys" held in Edinburgh, Scotland, 18-21
  June 1991 / edited by H.T. MacGillivray & E.B. Thomson.
        p.   cm. -- (Astrophysics and space science library ; v. 174)
    Includes indexes.
    ISBN 978-94-010-5091-3   ISBN 978-94-011-2472-0 (eBook)
    DOI 10.1007/978-94-011-2472-0
    1. Astrometry--Data processing--Congresses. 2. Photometry,
  Astronomical--Data processing--Congresses.   I. MacGillivray, H. T.
  II. Thomson, E. B.   III. Title.   IV. Series.
  QB807.C66 1991
  522--dc20   92-2891

ISBN 978-94-010-5091-3

Printed on acid-free paper

TABLE OF CONTENTS

PREFACE

H.T. MacGillivray

Royal Observatory, Blackford Hill, Edinburgh EH9 3HJ.

The conference on 'Digitised Optical Sky Surveys' held at Heriot-Watt University Conference Centre in Edinburgh, Scotland, during 18–21st June 1991, was the second such meeting on the subject. The first was held in Geneva in May of 1989 and was co-organised by Professor Carlos Jaschek (then of the Strasbourg Stellar Data Centre) and the present author. The meetings were the original inspiration of Professor Jaschek, to whom must be given the credit for the idea of bringing together workers active in this very important area. The second meeting continued that original aim, with further goals of stimulating discussion of data processing techniques, of identifying the problems inherent in the extraction of the sheer volume of information contained on astronomical photographs, and of ensuring the unlocking of that information in a reliable manner, thereby providing the basis for a powerful legacy for future generations of astronomers.

This second meeting in Edinburgh was attended by 70 astronomers from all over the world, and involved persons with a wide range of skills: from photographers involved in the production of the photographic sky surveys themselves and reproduction for sky atlases, engineers involved in the construction of the photographic-plate digitisation devices and in using the latest state-of-the-art technology for digitisation of astronomical photographs, software specialists responsible for the development of sophisticated techniques for the processing of the data, to, finally, the astronomers who are actively involved in large-scale statistical investigations based on the digitised data or in large-scale optical identification programmes for identifying the optical counterparts to objects which emit in other regions of the electro-magnetic spectrum (e.g. X-ray, infra-red and radio).

I am grateful to all those who made the conference a success. Special mention must be made of the conference secretary, Mrs Anne Bryans, who throughout the period leading up to the conference was a never-ending mine of ideas and assistance. Anne was aided during the conference by Mrs Dorothy Skedd, whose help was also invaluable. My co-editor, Eve Thomson, has laboured long and hard with the written versions of the contributions, first getting them onto disk (often having to type them herself) and also putting them into a unified LaTeX format. Finally, I would like to thank the participants at the meeting, both for their willingness to contribute in a very constructive manner, and also for their eagerness to participate in post-talk discussions. This conference has been, for me at least, one of the most lively with regard to vigorous exchanges after presentations. In this respect, I also owe much to Professor Gart Westerhout, who throughout the four days of the meeting kept us old plate-digitisation campaigners on our toes with frequent references to the potential future sky surveys based on CCD devices.

This volume presents written versions of most of the contributions delivered at the meeting. It is divided into 5 parts:- Part I presents the current situation with regard to photographic sky surveys and calibration programmes; Part II presents the latest results regarding current and future digitisation programmes; Part III presents some recent developments with regard to software techniques applied to digitised scans; in Part IV is described latest results from large-scale statistical studies based on the digitised data, while in Part V is reported the results of recent optical identification work.

LIST OF PARTICIPANTS

Baruch, J.	Department of Electrical Engineering, University of Bradford, Bradford BD7 1DP, U.K.
Burch, T.	Tesa Metrology Ltd., P.O. Box 418, Halesfield 8, Telford, Shropshire TF7 4QN, U.K.
Beard, S.M.	Royal Observatory, Blackford Hill, Edinburgh EH9 3HJ, U.K.
Bienaymé, O.	Observatoire de Besançon, 41 bis Av. de l'Observatoire, F-25010 Besançon, France.
Blanchard, A.	DAEC, Observatoire de Meudon, F-92190 Meudon, France.
Bonnarel, F.	Observatoire de Strasbourg, 11 rue de l'Université, F-67000 Strasbourg, France.
Bridgeland, M.	Royal Greenwich Observatory, Madingley Road, Cambridge CB3 0EZ, U.K.
Broadbent, A.	Department of Physics, University of Durham, South Road, Durham DH2 3LE, U.K.
Broadhurst, T.	Royal Observatory, Blackford Hill, Edinburgh EH9 3HJ, U.K.
Cannon, R.D.	Anglo-Australian Observatory, P.O. Box 296, Epping, NSW 2121, Australia.
Clowes, R.G.	Royal Observatory, Blackford Hill, Edinburgh EH9 3HJ, U.K.
Collins, C.A.	Royal Observatory, Blackford Hill, Edinburgh EH9 3HJ, U.K.
Cormack, W.A.	Royal Observatory, Blackford Hill, Edinburgh EH9 3HJ, U.K.
Coutures, Ch.	DPhPe/SEPh, Cen Saclay, 91191 Gif/Yvette Cedex, France.
Crolla, A.	Department of Electrical Engineering, University of Bradford, Richmond Road, Bradford BD7 4DP, U.K.
Davenhall, A.C.	Royal Observatory, Blackford Hill, Edinburgh EH9 3HJ, U.K.

Deul, E.	Sterrewacht Leiden, P.O. Box 9513, 2300 RA Leiden, The Netherlands.
Dodd, R.J.	Carter Observatory, P.O. Box 2902, Wellington 1, New Zealand.
Doi, M.	Department of Astronomy, University of Tokyo, Bunkyo-ku, Tokyo 113, Japan.
Duerbeck, H.W.	Astronomisches Institut, Domagstraße 75, D-4400 Münster, Germany.
Gardiner, L.T.	Department of Astronomy, University of Edinburgh, Blackford Hill, Edinburgh EH9 3HJ, U.K.
Goldschmidt, C.A.	Department of Astronomy, University of Edinburgh, Blackford Hill, Edinburgh EH9 3HJ, U.K.
Goldsmith, C.	Department of Physics, University of York, 4700 Keele Street, North York, Ontario M3J 1P3, Canada.
Grosbøl, P.	European Southern Observatory, Karl-Schwarz-schild-Straße 2, D-8046 Garching bei München, Germany.
Guibert, J.	Institut National des Sciences de L'Univers, CAI/MAMA, Bat, Observatoire de Paris, F-75014 Paris, France.
Gursky, H.	Naval Research Laboratory, Code 4100, 4555 Overlook Ave. SW, Washington DC 20375, USA.
Hale-Sutton, D.	Department of Physics, University of Durham, South Road, Durham DH1 3LE, U.K.
Hambly, N.	Department of Astronomy, University of Leicester, University Road, Leicester LE1 7RH, U.K.
Hawkins, M.R.S.	Royal Observatory, Blackford Hill, Edinburgh EH9 3HJ, U.K.
Humphreys, R.	Department of Astronomy, University of Minnesota, 116 Church Street SE., Minneapolis MN 55455, USA.
Irwin, M.J.	Royal Greenwich Observatory, Madingley Road, Cambridge CB3 0HA, U.K.
Laidler, V.G.	Space Telescope Science Institute, 3700 San Martin Drive, Baltimore, Maryland 21218, USA.
Lasker, B.M.	Space Telescope Science Institute, 3700 San Martin Drive, Baltimore, Maryland 21218, USA.
Leggett, S.K.	U.S. Naval Observatory, Flagstaff Station, P.O. Box 1149, Flagstaff AZ 86002, USA.
MacGillivray, H.T.	Royal Observatory, Blackford Hill, Edinburgh EH9 3HJ, U.K.
McLean, B.	Space Telescope Science Institute, 3700 San Martin Drive, Baltimore MD 21218, USA.

Maddox, S.J.	Department of Astrophysics, University of Oxford, Keble Road, Oxford OX1 3RH, U.K.
Miller, L.	Royal Observatory, Blackford Hill, Edinburgh EH9 3HJ, U.K.
Monet, D.G.	U.S. Naval Observatory, Flagstaff Station, P.O. Box 1149, Flagstaff AZ 86002, USA.
Moniez, M.	Laboratoire de l'Accelerateur Lineaire, IN2P3-CNRS, Université de Paris-Sud, F-91405 Orsay Cedex, France.
Morgan, D.H.	Royal Observatory, Blackford Hill, Edinburgh EH9 3HJ, U.K.
Murdin, P.G.	Royal Observatory, Blackford Hill, Edinburgh EH9 3HJ, U.K.
Nichol, R.C.	Department of Astronomy, University of Edinburgh, Blackford Hill, Edinburgh EH9 3HJ, U.K.
Ochsenbein, F.	Observatoire de Strasbourg, 11 rue de l'Université, F-67000 Strasbourg, France.
Odewahn, S.C.	Dept. of Astronomy, University of Minnesota, 116 Church St. SE., Minneapolis MN 55455, USA.
Okamura, S.	Department of Astronomy, University of Tokyo, Bunkyo-ku, Tokyo 113, Japan.
Parker, Q.A.	Royal Observatory, Blackford Hill, Edinburgh EH9 3HJ, U.K.
Paterson, M.	Royal Observatory, Blackford Hill, Edinburgh EH9 3HJ, U.K.
Peacock, J.A.	Royal Observatory, Blackford Hill, Edinburgh EH9 3HJ, U.K.
Pennington, R.L.	Dept. of Astronomy, University of Minnesota, 116 Church St. SE., Minneapolis MN 55455, USA.
Phillipps, S.	Department of Physics, University of Wales College of Cardiff, P.O. Box 913, Cardiff CF1 3TH, U.K.
Quebatte, J.	European Southern Observatory, Karl-Schwarzschild-Straße 2, D-8046 Garching bei München, Germany.
Raychaudhury, S.	Institute of Astronomy, Madingley Road, Cambridge CB3 0HA, U.K.
Roche, P.	High Energy Astrophysics Group, Department of Physics, University of Southampton, Southampton SO9 5NH, U.K.
Romer, C.	Department of Astronomy, University of Edinburgh, Blackford Hill, Edinburgh EH9 3HJ, U.K.

Russell, K.S.	Anglo Australian Observatory, Private Bag, Coonabarabran NSW 2357, Australia.
Schachter, J.	Harvard-Smithsonian Center for Astrophysics, 60 Garden Street, Cambridge, MA 02138, USA.
Schuecker, P.	Wilhelm-Klemm-Straße 10, D-W 4400 Munster, Germany.
Seitter, W.	Astronomisches Institute, Domagstraße 75, D-4400 Münster, Germany.
Speikermann, G.	Astronomisches Institut der Universität Munster, Wilhelm-Klemm-Straße 10, D-W 4400 Münster, Germany.
Stobie, R.S.	Royal Observatory, Blackford Hill, Edinburgh EH9 3HJ, U.K.
Taff, L.	Space Telescope Science Institute, 3700 San Martin Drive, Baltimore MD 21218, USA.
Tritton, S.B.	Royal Observatory, Blackford Hill, Edinburgh EH9 3HJ, U.K.
Van Dessel, E.L.	Royal Belgian Observatory, Av. Circulaire 3, 1180 Brussels, Belgium.
De Vegt, Chr.	Hamburg Observatory, Gojeubergsweg 112, 2050 Hamburg 80, Germany.
Voges, W.	Max Plank Institut für Extraterrestriche Physik, Karl-Schwarz-schild-Straße, D-8046 Garching bei München, Germany.
Weir, N.	Division of Physics, Mathematics and Astronomy, California Institute of Technology, Pasadena CA 91125, USA.
Westerhout, G.	Department of the Navy, US Naval Observatory, 34th and Massachusetts Ave. NW, Washington DC 20392-5100, USA.
Winter, L.	Hamburg Observatory, Gojeubergsweg 112, 1050 Hamburg 80, Germany.
Wolstencroft, R.D.	Royal Observatory, Blackford Hill, Edinburgh EH9 3HJ, U.K.
Yamagata, T.	National Astronomical Observatory, Mitaka, Tokyo 181, Japan.
Yentis, D.	Naval Research Laboratory, Code 4100, 4555 Overlook Ave. SW., Washington DC 20375, USA.

Participants Appearing in the Conference Photograph

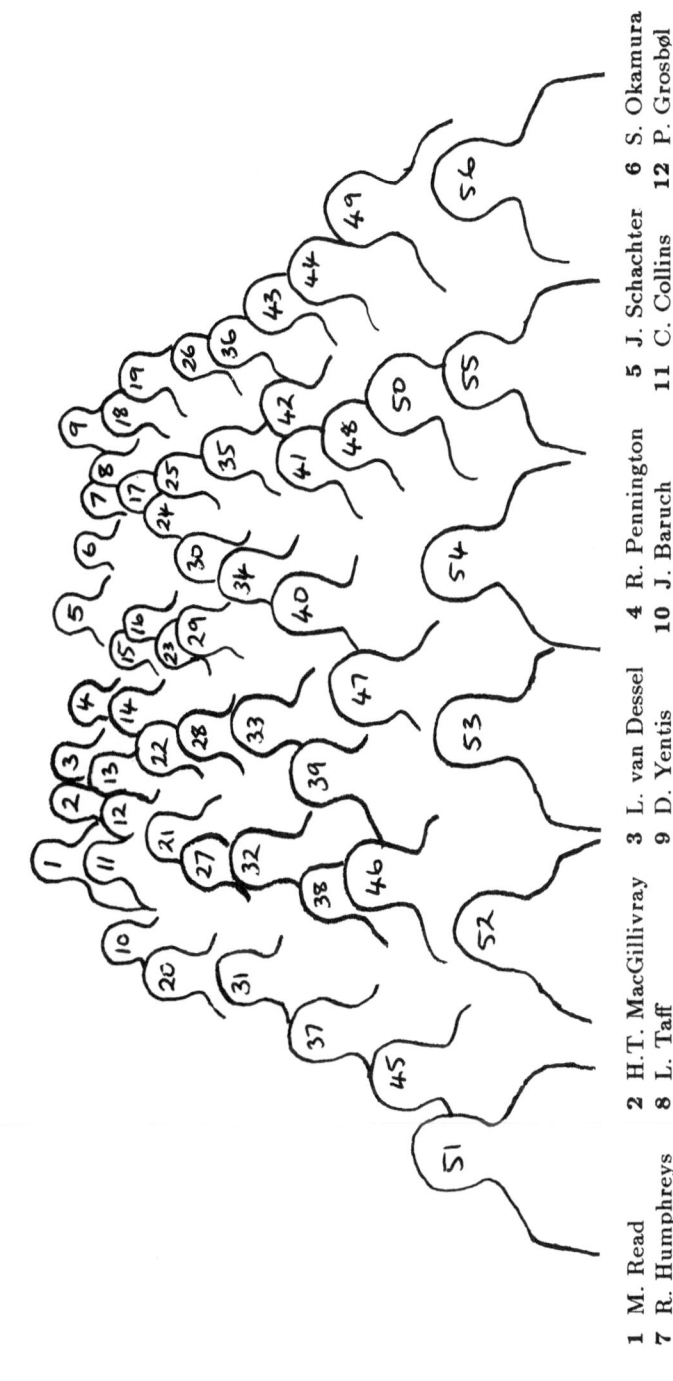

1 M. Read	2 H.T. MacGillivray	3 L. van Dessel	4 R. Pennington	5 J. Schachter	6 S. Okamura
7 R. Humphreys	8 L. Taff	9 D. Yentis	10 J. Baruch	11 C. Collins	12 P. Grosbøl
13 M. Hawkins	14 R. Clowes	15 S. Leggett	16 W. Seitter	17 S. Tritton	18 H. Duerbeck
19 R. Wolstencroft	20 N. Hambly	21 L. Gardiner	22 L. Miller	23 L. Winter	24 A. Blanchard
25 J. Peacock	26 S. Maddox	27 N. Weir	28 D. Monet	29 R. Bonnarel	30 J. Quebatte
31 B. McLean	32 F. Ochsenbein	33 G. Westerhout	34 E. Deul	35 J. Guibert	36 Q. Parker
37 S. Odewahn	38 C. Davenhall	39 P. Roche	40 R. Stobie	41 R. Dodd	42 T. Yamagata
43 S. Beard	44 C. Goldsmith	45 C. Coutures	46 G. Spiekermann	47 P. Schuecker	48 S. Phillipps
49 O. Bienaymé	50 H. Gursky	51 M. Moniez	52 V. Laidler	53 B. Lasker	54 D. Morgan
55 W. Cormack	56 K. Russell				

WELCOMING ADDRESS:
TWO THOUSAND YEARS OF OPTICAL SKY SURVEYS

P.G. Murdin

Royal Observatory, Blackford Hill, Edinburgh EH9 3HJ, U.K.

The first optical sky survey which we know about is a poem, *Phaenomena*, by Aratus (c. 350 BC). It sets out the shape of the constellation figures and the stars which form them. Aratus' survey is based on one now lost, by Eudoxus, who visited Alexandria for the purpose of examining the records of observations kept in the library and for consultations with the community of scholars there — these observations were originally made about 2000 BC, perhaps from Minos in Crete.

Aratus gave some interpretations of his survey in the form of navigational and weather lore; a more scientifically valuable interpretation was made by Hipparchus two centuries afterwards. He noticed that parts of the sky described by Aratus (the southern parts of the constellation Argo) were no longer visible above the horizon seen from Greece, and other parts that were visible had no described constellations. This was due to a change in tilt in the Earth's axis relative to the stars — Hipparchus thus discovered precession.

The story of the modern Sky Surveys, such as the SERC surveys, is almost exactly parallel to this first sky survey of the classical era. The UK Schmidt telescope and its survey observations were completed in the 1970s and 80s, and we can cast our first speaker Professor Vincent Reddish in the same rôle as a Minoan priest. Its records are housed in a library, not in Alexandria but in Edinburgh. Like Eudoxus, ROE astronomers such as Harvey MacGillivray visit the Plate Library to examine the records with the COSMOS machine.

Who parallels Aratus? There are several candidates, because numerous astronomers have access to the digital survey. I spoke to one such person yesterday, namely Herb Gursky, from whom we will hear later in this conference. He is interpreting the COSMOS scans for clusters of galaxies in connection with ROSAT soft X-ray observations. His catalogue of galaxy clusters will be a modern equivalent of the list of constellations, I am sure with more underlying physical reality. I don't know who the Hipparchus is who is going to make the most dramatic discoveries with the survey he is making — perhaps some graduate student whose name is not yet known.

H. T. MacGillivray and E. B. Thomson (eds.), Digitised Optical Sky Surveys 1–2.
© *1992 Kluwer Academic Publishers.*

This conference is a little about history, it is a good deal about the present and, I hope, it is a lot about the future. I am confident that the current and future sky surveys hold as dramatic discoveries as Hipparchus' discovery of precession.

I welcome you all to the conference and thank you all for making the effort, especially the extra effort made by those who came from furthest away, voyaging arduously, like Eudoxus, to consult with their colleagues in remote lands!

INTRODUCTION

V.C. Reddish

Ex–Director, Royal Observatory, Blackford Hill, Edinburgh EH9 3HJ, U.K.

When he asked me to give this introduction, Harvey MacGillivray described me as the father of the UK Schmidt Telescope and of COSMOS. I think he really meant the grandfather but was too tactful to say so! I hope you will allow me to reminisce a little on family matters; it was of course a wider family, with a growing number of aunts and uncles to teach the children how to behave; I shall mention some of them.

My first experience of an automatic measuring machine was 26 years ago, in 1965, with **GALAXY** — General Automatic Luminosity and **XY** measuring machine. It was conceived by Professor Brück and Peter Fellgett to measure the positions and magnitudes of star images on Schmidt Telescope photographs with an accuracy limited only by the signal/noise in the image, for the purpose of obtaining parallaxes and proper motions of large numbers of stars in the local region of the Galaxy. When Peter left, Professor Brück called me in and asked if I would have a look at the project to see if I could take it over. I found that we had a completed carriage system, massive, stable and very accurate, of the type then in use in nuclear physics measuring machines; but there was no control or data system, no design except for a magazine article, and no money left. My first child was left on the doorstep and found to have a body but no brains!

Just at that time the contractor — Ferranti — suffered the cancellation of a major defence contract and in the resulting reshuffle of people and resources within the firm, the GALAXY project gained a very experienced and determined electronic design engineer, Bruce Walker, to lead it. Bruce was a remarkable person. He took up sailing and in his small 5.5 metre yacht entered the Granton to Aberdeen singlehanded race. For those of you not familiar with the local geography, it is a fairly tough coastline. He won the race, of course. When being presented with the prize he pointed out that even if he had not come in first all the other competitors should have been disqualified because he was the only truly single-handed sailor in the race. Bruce had only one arm. I tell this story because it is typical of the skill, humour and determination which he brought to everything he did, including the GALAXY project. Determined people are not to everyone's liking but we got on fine together. I doubt that the GALAXY project would have been completed successfully but for the cancellation of that defence contract and the arrival of

3

H. T. MacGillivray and E. B. Thomson (eds.), Digitised Optical Sky Surveys 3–8.
© 1992 *Kluwer Academic Publishers.*

Bruce Walker. We soon produced a detailed specification for the control and data systems and some logical designs for the analogue and digital scanning systems, the necessary money was found, and in due course the machine was finished and worked to specification; but it was limited to 900 images per hour, and could only measure stars.

By this time Ferranti had sold out this part of their firm to a small Company, Bruce Walker had gone, George Walker and Phil Williams had come in, and a member of our Observatory staff, Neil Pratt, had written the control system and data reduction software. While all the problems of designing and building GALAXY were fresh in our minds, we got together to discuss the design of a next generation machine.

We already knew that the UK was to have a 1.2 metre Schmidt telescope in Australia (the UKST) and this of course changed the situation completely. Instead of the thousands of images on each photograph taken by our small telescope at Edinburgh we would have millions, and a large proportion of them would be galaxies. So we needed a machine to measure the positions, sizes, shapes, orientations and magnitudes of images, and to do so at a rate of a million per hour rather than a thousand. We produced a detailed specification of what the machine must do and sufficient ideas on how it would do it to be able to place a contract. The contract price was £63K. Neil Pratt wrote the basic machine control and data reduction software (as he did also for the RGO GALAXY machine) and our total in-house costs including all salaries and overheads and computing costs were £30K to a working machine.

The story of how COSMOS got its name has been described as a likely story, but it is true. We had called the machine GALAXY II to provide a link with a previous successful project (GALAXY had made the front page of THE TIMES, but that is another and fascinating story!) and because for financial reasons we had to use the same mechanical system; but it was our desire to find a more appropriate name by the time the machine was complete. Neil Pratt and I were discussing the problem in my office one day, and I remarked that what we really wanted was a name which would describe what the machine would do, so I wrote on my blackboard a list of the things it would measure — the Co-Ordinates, Size, Magnitude, Orientation, Shape; by chance, the order of the first letters spelled **COSMOS**.

With the Anglo-Australian 3.8 metre telescope (AAT) already under construction, the need for a large Schmidt telescope to provide a catalogue of objects to observe was a matter of urgency. The Science Research Council (SRC) decided that the quickest way would be to build a copy of the Palomar Schmidt with the minimum necessary changes. The Palomar Observatory generously gave them a complete set of copies of drawings.

When I was asked to take charge of the project, this policy did not seem to me to be a good one. A lot of changes had been made to the Palomar Schmidt after it was first built and there was no certainty that they were all on the drawings. Good though the telescope now was, it had some shortcomings, and a lot of technological advances had been made in twenty five years. My suggestion to the Secretary of the SRC that we should design a new telescope caused worry that I would offend both Palomar and the Council. I was due to visit Palomar to examine the telescope and use it for a few nights, a visit that confirmed my doubts, so I went to see their chief engineer, Bruce Rule, and explained my predicament to him. He listened patiently and then said "Vince, if you go to the Ford Motor Company and ask them to build you a Ford T they'll do it; it will cost a lot of money and won't be a good car by current standards. If I were you, I would build the best Schmidt telescope I could."

Thus comforted, I came back and wrote a very detailed performance requirement specification covering every aspect of the design and operation of the telescope's optical, mechanical and drive and control systems, which I believed could just be achieved with the best technology available at the time. The object, of course, was to obtain the best possible image quality with long exposures. Three competitive design studies were commissioned, and then we went out to competitive tender for the design and manufacture, and the contract was won by Grubb Parsons on all three criteria of price, time and technology. The invitation to tender stated that the telescope was to follow closely the design of the Palomar Schmidt with only those changes being made which were necessary to meet the Performance Requirement Specification. I and the contractor knew that in practice this involved a complete re-design. The only thing copied from the Palomar design was the shutter, and a modified version of their plateholder loading system was used.

The reasons for wanting the best possible images were, of course, to provide the most detail on the structure of galaxies and to reach the faintest limiting magnitude. To assist us we had three more advantages over earlier telescopes — an achromatic corrector plate, a very good photoelectric autoguider developed by Gordon Adam at ROE, and sensitised IIIa-J emulsions.

During my visit to Palomar, I had seen their system for baking emulsions to drive out the moisture which was inhibiting the sensitivity. It seemed odd that after keeping the plates in a freezer to prevent the growth of background fog we should take them out and put them in an oven. Certainly the sensitivity increased, but so did the background fog. It occurred to me that a better way would be to put the plates in a dry atmosphere so that the moisture would diffuse out of them; if the atmosphere was changed repeatedly the water vapour would be swept away. The cheapest and most easily available gas which would do the job without harming the emulsion was dry nitrogen. The more I thought about it the more convinced I became that it would work — the theories of diffusion and partial pressures could not be wrong — so when I went to the Hamburg conference on Schmidt telescopes I committed us to using the IIIa-J emulsion for the Southern Sky Survey. Years later, Chip Arp said to me "We all thought you were crazy, Vince, but you knew exactly what you were doing, didn't you?" Well, perhaps not exactly, but I hoped near enough, because I could see that they thought I was crazy! Experiments at Edinburgh proved that the idea worked and in due course our staff at Siding Spring created an extremely efficient system for sensitising with dry nitrogen and later also with hydrogen.

The result of all these technological advances was a substantial increase in limiting magnitude and hence in penetration into space. To this was added the objective prism to search for quasars and other emission line objects.

The telescope was designed, manufactured and constructed in two years, and worked to the tight specification laid down — a remarkable achievement by Grubb

Parsons which owed a lot to the late David Brown, and to Andrew Lawrie.

When I was asked to take charge of the UK Schmidt Telescope project, it was recognised that I would have to give up my own research for some years, and so the Science Research Council offered me a generous inducement — all the non-survey time on the telescope for my personal use until the survey was completed! What I would have done with all that time, even with COSMOS to help me, I cannot imagine. However, it gave me the opportunity to strike a different bargain. My first years on the staff of Edinburgh University doing research with a small telescope in a bad climate with a budget of £30 (yes, thirty pounds) per annum were still clear in my mind. So I said that I would take on the project providing that instead of the telescope being primarily a support telescope for the AAT, the UKST and COSMOS both became national facilities for the support of University research, while accepting that the sky surveys would take priority. This was agreed with the then Director of the astronomy division in the SRC, Jim Hosie, and with the Chairman of Council Brian Flowers. That was in 1970.

I returned from Australia in time to rescue the COSMOS project from the bankruptcy of the contractor. The hardware was complete but not assembled, and we were able to extricate it from the factory and get it up to the Observatory, but the equally crucial problem was that of the two vital engineers — George Walker and Phil Williams. Heriot-Watt University stepped in to rescue them for us and we were able to transfer the contract for completion to them. Completed it was, and also worked to specification.

Meanwhile, Russell Cannon had taken over the management of the UKST in Australia and the survey was rapidly making progress. High quality photographic laboratories were established at Edinburgh — the intervention of Bart J. Bok with a letter to the SRC (unknown to me until later) proving crucial in getting the necessary finance.

I will not attempt to detail the rather traumatic period we went through trying to get the staff and funds to progress from a basic working COSMOS machine to an effective national facility, but much of the credit must go to Harry Atkinson, who had succeeded Jim Hosie in the SRC; we sometimes disagreed on timescales but shared a common objective. The credit for making effective use of those resources belongs primarily to Bob Stobie, who took charge not only of COSMOS but of our other measuring machines and all the computer systems as well.

Soon the UKST and COSMOS were pouring out data to astronomers, in UK universities and round the world, data they had previously only dreamed about. The astronomical community owes a considerable debt to Russell Cannon and to Bob Stobie for the magnificent service they provided over those years, at the sacrifice of most of their own research. All sorts of things became possible; studies of the distribution of galaxies, not just of clustering and superclustering but the question of holes in a random distribution; the discovery of large numbers of quasars and their distributions in redshift and over the sky; the stellar luminosity function at the faint end, leading to the discovery of brown dwarfs; and so on. The results from recent studies will no doubt emerge during this conference.

In due course, the UKST and COSMOS were joined by the ESO Schmidt Telescope, the Automatic Plate Measuring machine (APM) at Cambridge, a refurbished

Palomar Schmidt Telescope, and other instruments around the world. We have seen a remarkable expansion in the means of gathering and analysing astronomical data, and it has been an exciting time for all of us who had a part to play in it. The most excellent relations developed between the institutions involved:- ROE, ESO, Palomar, the AAO, and others, and the most memorable aspect of these projects for me was the friendships which they produced worldwide.

Although I have spoken to you of my personal experiences, they doubtless reflect those of many who have been engaged in similar projects, and I hope this account will be regarded as merely a typical example of what is involved in creating the systems needed for digitised optical sky surveys.

Now the desire to go further with the latest technology is producing SuperCOS-MOS, with a very high speed and vast data handling capacity.

The Royal Observatory Edinburgh has been at the forefront of high speed measurement technology for more than twenty years. I think you will all agree with me that it has been a remarkable achievement, sustained through a succession of scientists and of industrial firms.

If the Science and Engineering Research Council were ever to find itself without the financial means to support an Institution with this record of success it would be a bad day not only for British, and indeed international, science but for British technology and British industry because they designed and built these world-leading machines.

It is worth emphasising that such developments result from a partnership between the scientists in the research institution, and the scientists and engineers in the manufacturing industry. That is a vital component; these things would not otherwise be developed. Market forces have been with us for many thousands of years but they are not the answer to everything; they are not a substitute for leadership in science.

COSMOS has been used not just for exploring the Universe but for research in industry — studies in pollution, in oceanography, geology and so on. Any study where large numbers of objects need to be counted and their distributions analysed; where they can be distinguished by size, shape, or colour and need to be sorted accordingly; where there is a need to search for the oddity, the one-in-a-million; that is work for COSMOS. The software developed at Edinburgh has extremely wide applications. The potential demand from industry for the use of machines such as COSMOS and SuperCOSMOS is so great that there is a strong case for creating an Institute of Industrial High Speed Measurement, funded by European Industry and with the technological help and support of the Royal Observatory Edinburgh and the Science and Engineering Research Council. What better place for it than here at the Heriot-Watt University's Industrial Research Park? This is a case where industry can benefit enormously from technology developed for the pursuit of fundamental science.

The remarkable changes which we are seeing in the relationships between countries around the World, from militarisation to civilisation, will release resources in money, and skills in science and engineering. Most, no doubt, will go into manufacturing industry, communications and transport; hopefully some will go into front-line developments such as SuperCOSMOS for science and industry; but some should surely go into those sciences which, like art and music, enrich our civilisation.

Ever since humankind got up off all fours and was able to gaze at the sky it has been fascinated by the Sun, Moon, planets and stars. The drawings on the walls of caves are evidence enough of that, and the popularity of television programmes such as 'The Sky at Night' shows that the fascination is as strong as ever. People want to know about the Universe in which they live. How big is it? How old is it? How did it begin? How will it end? We do not pursue this research just for ourselves, but to satisfy the curiosity which is a driving force in humankind.

So what is going to be done with SuperCOSMOS?

More comprehensive studies of the distributions of stars, galaxies and quasars will no doubt tell us much more about the structure and evolution of our Universe, but there is one matter to which I would give priority.

If we add a hundred photographs of the same field, our naive expectation is that we will gain about a factor ten in signal/noise and some 2.5 magnitudes in limiting magnitude. I dare say that some of you will have carried out experiments by adding several plates over a limited area and can add the voice of experience to expectation, but a gain there must be and it should become greater as techniques are perfected.

The technical difficulty of adding 100 photographs over the whole field of our large Schmidt plates is enormous, especially if it is to be done without significant loss of resolution, but the prize for success is also enormous.

Always we should seek to penetrate deeper into space, further back in time, if we are to find the answers to the questions all of us ask.

Whatever you do, I wish you good fortune, and thank you for listening to me.

Discussion

MacGillivray :

How do you see the future of photographic sky surveys and high-speed plate digitisation machines in the light of plans to use large-format CCD arrays to undertake digital sky surveys?

Reddish :

The photograph is a very efficient means of storing data. I think it depends on what improvements in sensitivity result from research into emulsions — whether or not they can approach that of the CCDs. If they can, the form of the permanent record of the photograph has major advantages.

DIGITISED OPTICAL SKY SURVEYS

Part One:

SKY SURVEYS AND CALIBRATION PROGRAMMES

CURRENT AND FUTURE PROGRAMMES WITH THE UK SCHMIDT TELESCOPE

D.H. Morgan [1], *S.B. Tritton* [1], *A. Savage* [2], *M. Hartley* [2] *and R.D. Cannon* [3]

[1] Royal Observatory, Blackford Hill, Edinburgh EH9 3HJ, U.K.
[2] UK Schmidt Telescope, Anglo-Australian Observatory, Siding Spring Observatory, Coonabarabran, NSW, Australia.
[3] Anglo-Australian Observatory, 167 Vimiera Road, Epping, NSW 2121, Australia.

Abstract

The status of the major sky surveys and sky atlases is reviewed. The new second epoch southern sky survey being taken with the UK 1.2m Schmidt Telescope (UKST) is described. Trends in the astronomical use of the UKST are presented, and an outline of possible future work is given.

1 Introduction

This conference is one of many demonstrations of the importance of wide field astronomy projects undertaken with large Schmidt Telescopes, and of sky surveys in particular; see also, for example, the proceedings of *IAU Colloquium*, No. 78, *'Astronomy with Schmidt-Type Telescopes'*. We will not repeat the details of the importance of surveys (see Cannon 1984) but will concentrate on the work of the UK Schmidt Telescope (UKST) in Australia which was operated by the Royal Observatory Edinburgh (ROE) until mid-1988 and, since then, by the Anglo-Australian Observatory (Morgan & Savage 1991). Morgan & Tritton (1988 – hereafter referred to as MT) reviewed the survey work, and some of that report will be updated here. This paper will also describe some of the major 'non-survey' projects which have been carried out on the UKST over the past two decades as well as those in progress now.

H. T. MacGillivray and E. B. Thomson (eds.), Digitised Optical Sky Surveys 11–22.
© 1992 *Kluwer Academic Publishers.*

2 Sky Surveys

There are at least fourteen major sky surveys which have been completed by, or are still in progress at, the three large Schmidt Telescopes (the ESO 1.0m in Chile, the Palomar Oschin 1.2m and the UKST). These surveys are summarized in Table 1. There are a few changes from a similar table presented by MT; these relate mostly to the dates associated with active surveys. Another change is the inclusion of magnitude limits for the POSS-II surveys (Reid et al. 1991). These limits appear brighter than the nominal values for the southern sky surveys; but this difference probably arises from the different ways used to define the limits. The topic of survey uniformity was discussed by MT.

The most significant difference between Table 1 and the version of MT is the inclusion of the AAO-R Survey. This survey was started in 1990 and is specified in Table 2. Briefly, it is a second epoch UKST survey which covers the same area of southern sky as the SERC-J Survey, but approximately 15 years later. However, the AAO-R survey is in red rather than blue light. The principal goal of the AAO-R Survey (or Second Epoch Survey – SES) is to obtain proper motions of many objects in the southern sky. The ESO-R Survey was, for the most part, taken more recently than the SERC-J Survey; 60% of the ESO-R plates were taken between 1984 and 1986 whereas the SERC-J Survey plates are concentrated in the years 1974-1979.

One of the driving forces behind the AAO-R Survey was the need of the Hubble Space Telescope to have accurate positional measurements for its guide stars at a recent epoch. Most of the SERC-J Survey plates were obtained in the mid-1970s, and since then a significant fraction of these moderately bright guide stars will have moved by more than the allowed 0.3 arcsec. The new survey will provide the necessary positions.

The second major driving force was the need to study southern hemisphere proper motions. The 15 year baseline between first and second epoch surveys is sufficient to discover large numbers of nearby high proper motion stars, as was done in the north by Luyten (1963) using plates with a 12 year baseline. With this new survey, we expect to increase the total sample size for the rarest types of nearby star, to push to fainter apparent and hence absolute magnitudes, and to find the best southern examples of various categories of low luminosity star for further study. However, it is worth noting that the numbers of very faint stars with high proper motion will not be large; most will lie within 25pc, so that apparent magnitude 20 will correspond to absolute magnitude >18. Thus most will be cool red dwarfs or cool degenerate 'white' dwarfs; the majority of nearby hot white dwarfs should already have been discovered. Luyten (1979) has estimated that the Schmidt surveys south of his POSS limit at -33° should yield about 500 new stars with proper motions in excess of 0.5 arcsec per annum and brighter than B=21. Of particular interest is the class of brown dwarf. Probst (1986) has estimated that the new POSS-II survey could yield some two dozen brown dwarfs by extending the magnitude limit to R=21, and a comparable number might turn up in a systematic southern search to similar limits. As well as yielding examples of rare types of star, the AAO-R Survey will be used to determine smaller proper motions for large

Table 1. Details of sky surveys completed or still in progress.

Survey	Col	Dec centres	N	Emul	Filter	$\Delta\lambda$ nm	Mag	Dates	Ref
POSS-I	Blue	$\geq -30°$	935	103a-O	-	350-500	21	1950-58	1
POSS-I	Red	$\geq -30°$	935	103a-E	pg†	620-670	20	1950-58	1
POSS-I	IR	$\geq 0°$*	80	IV-N	WR88a	770-900	19	1975-79	2
ESO-B	Blue	$\leq -20°$	606	IIa-O	GG385	385-500	21	1973-78	3
SERC-J	Blue	$\leq -20°$	606	IIIa-J	GG395	395-540	23	1974-87	4
ESO-R	Red	$\leq -20°$	606	IIIa-F	RG630	630-690	22	1978-90	5
SERC-EJ	Blue	$-15° - 0°$	288	IIIa-J	GG395	395-540	23	1979-	4
SERC-ER	Red	$-15° - 0°$	288	IIIa-F	OG590	590-690	22	1984-	4
SERC-I/SR	IR	$\leq 0°$*	163	IV-N	RG715	715-900	19	1978-85	6
SERC-I	IR	$\leq 0°$‡	731	IV-N	RG715	715-900	19	1980-	-
AAO-R	Red	$\leq -20°$	606	IIIa-F	OG590	590-690	22	1990-	7
POSS-II	Blue	$\geq 0°$	894	IIIa-J	GG385	385-540	22.5	1987-	8
POSS-II	Red	$\geq 0°$	894	IIIa-F	RG610	610-690	20.8	1987-	8
POSS-II	IR	$\geq 0°$	894	IV-N	RG9	730-900	19.5	1987-	8

* limited to galactic latitudes $|b^{II}| < 10°$; matching short exposure red plates are included in the Atlas. The SERC-I/SR Survey also includes the Magellanic Clouds.
‡ limited to galactic latitudes $|b^{II}| > 10°$ and excluding the Magellanic Clouds.
† number 2444 red Plexiglass filter.
Notes:
(1) The SERC- and AAO- surveys are taken with the UK 1.2m Schmidt Telescope, the ESO- surveys are taken with the ESO 1.0m Schmidt Telescope, and the POSS- surveys are taken with the Palomar 1.2m Schmidt Telescope, (2) N is the number of fields in the survey, (3) $\Delta\lambda$ is the nominal waveband for each survey but is not defined in a consistent manner, (4) mag is the nominal limited magnitude but again is not defined in a consistent manner, (5) the dates are those of the first and last original survey plates.

References:
(1) Minkowski & Abell 1963. (2) Hoessel et al. 1979. (3) West & Schuster 1982. (4) Cannon 1984. (5) West 1984. (6) Hartley & Dawe 1981. (7) This paper. (8) Reid et al. 1991.

numbers of normal stars and thereby give access to the field of galactic kinematics (see Murray 1986).

Some problems occur as a result of using red rather than blue plates as the second epoch database. The main one is that very red and, to a much lesser extent, very blue stars will be absent from either first or second epoch plates, so that the magnitude limit for these extreme stars is not as faint as it could be. This means that the search for brown dwarfs is unable to reach R=21 using SERC-J for the first epoch. However, ESO-R Survey plates can be used as the first epoch database

Table 2. The AAO-R Southern Sky Survey.

Component	Specification
Plate material	Hypersensitized Kodak IIIa-F emulsion
Filter	Schott OG590
Waveband	590-690nm
Exposure	Sky-limited exposures (nominal 60min)
Image and depth quality	Acceptance standards are the same as those used for the original plates of the SERC-J Survey
Cosmetic quality	Rejection of plates with emulsion flaws will be less severe than for the SERC-J Survey
Size of plate	356mm × 356mm
Area covered	6.5° × 6.5°
Field centres	The plates centres are spaced 5° apart between −90° and −20°, and match those used for the SERC-J survey
Number of fields	606
Start date	1990
Timescale	7 years
Scheduling	Plates will be taken in an order so as to keep the average baseline between the J-Survey and the AAO-R Survey as close as possible to 15 years
Priority fields	450 fields at high galactic latitude
Copies	Positives and three films will be made. The film copies are for the three contributing organizations

for this particular purpose which does not require such a long baseline, since brown dwarfs are expected to have proper motions of about of $0.3''/yr$. Another problem is that atmospheric chromatic effects will compromise the astrometric accuracy. However, the advantage is that colour information which is needed for full exploitation of the astrometry is obtained from just two sets of plates.

3 Survey Availability

Access to the data contained in the Sky Surveys can be achieved in more than one way.

3.1 Sky Atlases

Most astronomers have access to survey material through the Atlas sets which have been made from the original plates and sold worldwide. Three materials have been used: paper, glass and film. In general the paper copies do not give such faithful reproduction as the glass and film copies and have not been used for Atlas production for many years. Film copies are the best for visual inspection purposes and glass copies for machine measurement. Table 3 lists details of published Sky Atlases and estimates of future publications. It should be noted that there are no plans to reproduce the AAO-R Survey as an Atlas. The reason is that the ESO-R Atlas covers the same area of sky in more or less the same wavebands and is already well distributed throughout the world. However, glass positives have been and are being made by the AAO along with three films of each field. Thus, the possibility of making Atlas sets from the positives is retained should there be a demand. Similarly, the SERC-I Survey is not expected to be complete for a number of years and plans for its reproduction have not been set. The ROE Photolabs are now engaged in making copies of the SERC-EJ Atlas on glass and film, and are expecting to continue with copies of the SERC-ER Atlas, again on both glass and film.

Table 3. Published Sky Atlases (completed and planned).

Survey	Colour	Material	Publication period
POSS-I	Blue	Paper, glass	1959-63
POSS-I	Red	Paper, glass	1959-63
POSS-I	IR	Paper	1982-82
ESO-B	Blue	Film, glass	1972-75
SERC-J*	Blue	Film, glass	1975-88
ESO-R*	Red	Film, glass	1978-91
SERC-EJ	Blue	Glass	1984-92
SERC-EJ	Blue	Film	1991-93
SERC-ER	Red	Film	1993-96
SERC-ER	Red	Glass	1992-96
SERC I/SR	IR	Film	1982-85
SERC-I	IR	Film	?
POSS-II	Blue	Film, glass	1991-97
POSS-II	Red	Film, glass	1991-97
POSS-II	IR	Film, glass	1991-97

* The SERC-J and ESO-R Atlases are produced by ESO and are issued as the single publication 'The ESO/SERC Southern Sky Atlas'.
Dates in 1991 and beyond are estimates.

3.2 UKST Original Plates

Clearly, a set of original plates which forms the raw material for a Sky Atlas can also be used for digitization purposes. In the case of the UKST, the SERC-J and SERC-

EJ original plates have not been made available for measurement and digitization, and access to these surveys has been limited to the Atlas copies. However, UKST master original plates for the SERC-ER and SERC-I Surveys have been released for digitization because glass Atlas copying is not expected to start until the SERC-EJ glass Atlas is complete. It is intended that all the AAO-R Survey originals will be measured, probably independently by several measuring machines.

For many survey fields ($\sim$ 50%), there are other plates which were not quite good enough to be accepted as masters for the sky atlases. This plate collection is important because it provides a second measurement on many fields and therefore improved photometry.

3.3 Microspots

One of the most disappointing aspects of modern emulsions is their propensity to develop microspots. These microspots are small spots usually of a yellow or gold colour and commonly with a mirrored surface (Good & Gourlay 1984); they render the affected areas unusable for many purposes, and, since they often preferentially affect the calibration wedges, make the use of the whole plate difficult. Although certain processing techniques seem to inhibit microspot formation (Good 1988), many plates are affected. For the most recent statement about microspots on UKST plates see Tritton & Morgan (1991). Most atlas quality UKST originals have been bathed in selenium toner at the recommendation of Kodak, as have many other original plates. However, toning sometimes introduces drying marks of area $\sim 0.5 \text{cm}^2$ on the emulsion surface, many of which pose serious threats to the validity of machine measurements in those areas. Unfortunately, a large fraction of the ER Survey originals are affected by microspots and are not suitable for large area survey work. Since positives were made soon after the ER plates were taken, glass Atlas sets can be made without the need to repeat these hard-gained plates. Fortunately, microspots affect very few positives (see Tritton & Morgan 1991).

These facts argue strongly for the making of positives as soon as possible after the time of observation, the measurement of the original plates at an early stage, and the toning of them soon afterwards.

4 Large Projects

Much of the 'non-survey' work of the UKST and the UK measuring machines has been devoted to special projects which require large numbers of plates. These fall broadly into two categories: those which need one or two plates on a large number of fields, and those which need sets of plates on a small number of fields. The latter can consist of sets of plates in several colours, often in pairs to increase photometric accuracy, or large numbers of similar plates to allow the detection of variable objects (photometrically variable or moving objects such as asteroids). The UKST Plate Library at the ROE now has a large number of data sets consisting of 50 or more plates. The 16 largest sets are listed in Table 4. It is clear that these have been used for a wide range of astronomical projects, and that they will provide excellent databases for many other wide-field astronomy studies.

Table 4. Largest programmes with the UKST.

Programme	Plate material Colour	Exp	No of Fields	No of Plates	Dates Obtained
* Blue object survey	U B	short	178‡	398	1985-
Parallaxes	V(b•)	short	3	312	1974-84
* Variable objects	BₗRIU	lim	1	199	1978-
* Parallaxes+colours	V(ubⱼvri•)	short (lim)	3	181	1987-
Multicolour survey	UBₗVRI	med	13	172	1985-88
* SMC/LMC red variables	I	lim	3	159	1973-
Comet Halley	IF,BR	med	1	130	1985-86
Emission nebulae	Hα	various	83	127	1974-76
WR star searches	HeII,IF	various	33	114	1983-90
Asteroids	Bₗ	lim	⋆	109	1981
* Three colour photography	BVR	lim	22	104	1976-
HST guide stars	V	short	92	101	1986-88
RR Lyrae searches	V	short	5	97	1976-77
Multicolour	UBVRI	lim	6	90	1982-86
x-ray identifications	U Bₗ	short	39	80	1983-86
Gamma ray burster search	B R	short	7	76	1983-87
* Magellanic Cloud Atlas	UBₗVRIP	lim	12	36†	1978-
* Prism survey	Bₗ prism	lim	80‡	80†	1977-

* These are continuing programmes
• lower case letters refer to a minority of plates
† The total number of plates here refers to 'accepted survey plates'
‡ Number of fields taken to date **not** total number required
⋆ A moving block of 6-8 fields on the ecliptic
IF, HeII – unspecified and HeII(λ4686Å) interference filters.

Notes:
The data given refer to the majority of plates taken for the programme; some additional plates ($< 10\%$) may have been taken on different emulsions or fields or outside the quoted dates etc.
Programmes which mainly requested 'priority survey plates' (especially before the southern sky atlases were published) have been excluded from these statistics as they do not represent separate data sets.

The final two entries in Table 4 are out of place, but have been included because they are of relevance to this meeting. The first is what is known as 'the objective prism survey'. It consists of unfiltered, sky-limited IIIa-J plates taken on survey centres with the 2400Å/mm low-dispersion objective prism (Savage et al. 1985). Plates have usually been taken as a direct response to requests from astronomers for specific research purposes, but top quality plates have been copied and stored as survey master copies.

The second project is one to obtain sets of plates in several colours on the Large and Small Magellanic Clouds, and to copy these and market film sets for use as research material. The colours to be included are U, B_J, V, R (deep and short), and I, and low-dispersion objective prism plates will be available as well. Twelve fields will be covered. The films will be sold as sets of twelve in each of the individual colours. The I and SR plates were included as part of the SERC-I/SR Atlas but can soon be purchased separately. The IIIa-J plates will not be those used for the SERC-J Atlas, but will be plates taken since 1989. These plates will be more uniform than the SERC-J plates because the Atlas plates of the Magellanic Clouds were obtained before nitrogen flushing was introduced into the plateholder during exposure (see MT). Film sets for the deep red and I/SR plates will be produced first; the others will be made as the originals become available. Few U and V plates are yet available, but the other sets will be produced soon.

5 Past and Future Trends

The nature of the research programmes carried out on the UKST has changed steadily throughout the eighteen years of UKST operation. To investigate these trends we calculated the numbers of plates taken each year for projects devoted to the study of eight different classes of astronomical object. Figure 1 shows these numbers as percentages of the total numbers of plates taken for 'non-survey' programmes, for the years 1973-1990. This shows that work on stars has increased throughout the period whereas work on external galaxies has decreased steadily from a peak around 1979. Quasar work increased from 1975 to 1985 and has remained at the 10-15% level since. Studies of variable objects have not changed dramatically since 1978, though they do show a relatively large scatter, probably because sets of plates have often been taken in groups over periods of one or two months. Solar system studies appear to show a large scatter, but this is due to two major programmes in 1981 and 1986 (an asteroid search and Comet Halley studies); otherwise solar system studies are continuing to increase. The category 'identification' refers to the identification of unknown objects detected at other frequencies (e.g. radio and x-ray); these programmes are on the decrease, most probably because they can now be generally satisfied from the large digitised databases, such as the COSMOS database of the Southern Sky (see Yentis et al., these proceedings; Voges, these proceedings). Finally, there are the two smaller components, nebulae and star clusters. Nebulae were frequently observed at the start of telescope operation especially through narrow-band interference filters, while star clusters have received a recent surge of interest.

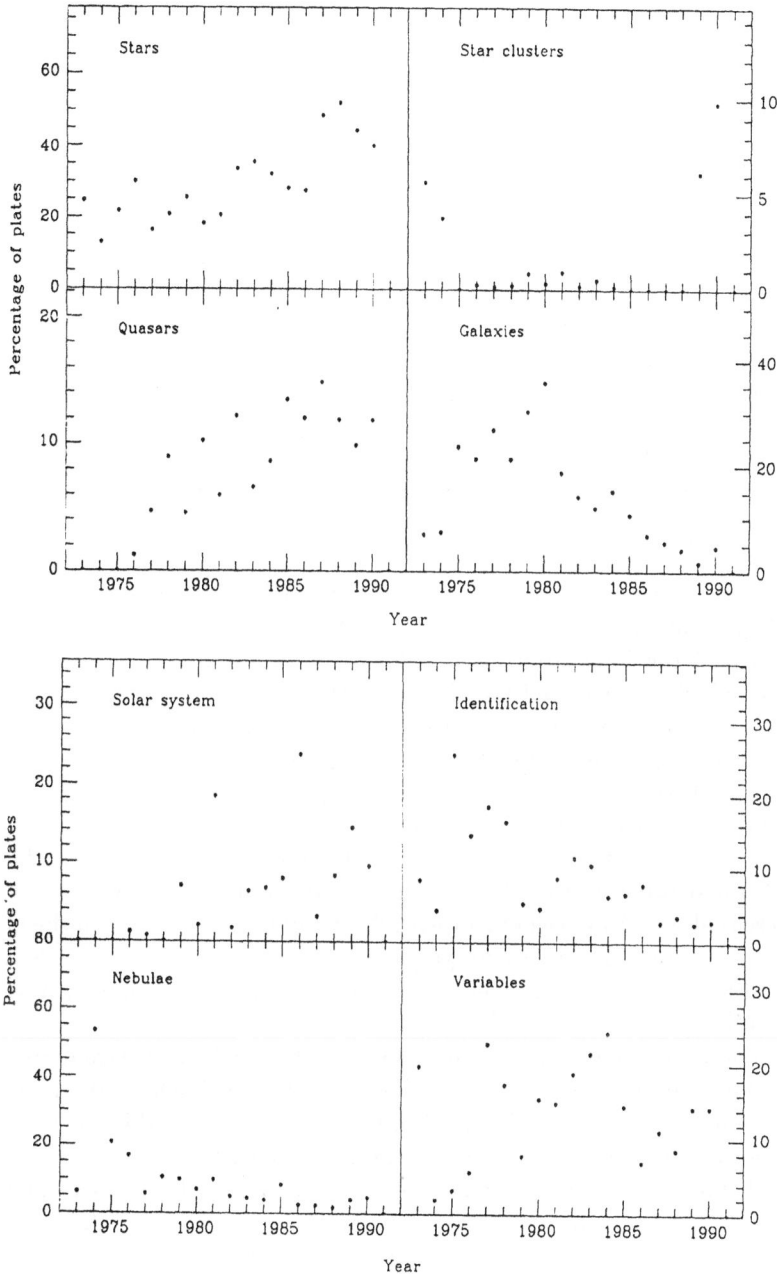

Fig. 1. Changes in the scientific usage of the UKST. The individual plots show the change, as a function of time, of the amount of UK Schmidt Telescope effort devoted to projects working on particular classes of object. The ordinate is the number of plates taken for each object class as a percentage of the total number of plates taken for all classes (Survey plates are excluded.)

Broadly speaking, these trends reflect the increasing availability of the sky survey material. For example, many of the projects on galaxies, which ten or more years ago required new plates, can now be carried out on survey material. Also, the increase in stellar and quasar work has accompanied the improved performances of the fast measuring machines. Much of this type of work requires several plates in several colours on selected fields and cannot be carried out on survey material. Consequently, we expect these trends to continue a little longer and then flatten out once all the surveys are issued. Another controlling influence on these statistics is the seeing requirement. The UKST still has a long survey programme ahead, so there will continue to be severe competition for prime dark time. However, there will still be plenty of opportunity for special programmes which can utilize poorer seeing conditions (e.g. 3-4 arcsec).

A future survey could be a second epoch survey in the B_J waveband. The main purpose for this would be the improved accuracy in proper motion measurements as a result of both the increased baseline between first and second epochs and the matching of the wavebands used. Other major projects could include detailed multi-colour surveys of selected areas and systematic surveys with the objective prisms.

A major influence on the future work of the UKST is the development of the FLAIR multi-object spectroscopic system (Watson et al. 1990) which uses optical fibres to feed the light from about 100 objects simultaneously into an efficient spectrograph fitted with a CCD detector. Whilst the details of this system are not relevant to these proceedings, it should be said that its use will take telescope time away from the photographic programmes. On the other hand, it can provide an ideal means of acquiring large numbers of spectra per field in follow-up spectroscopic programmes (e.g. determination of galaxy and quasar redshifts and the classification of objects selected through photographic techniques — see for example the articles by Parker and Broadbent et al., both in these proceedings).

There is also a new development with the use of film as detector in the telescope. This was found to be possible some years ago (Humphries & Morgan 1988) and with new emulsions available on film has now become a realistic possibility. It is described elsewhere in these proceedings by Russell et al.

The telescope is not the only source of material for wide-field astronomy projects. As the plate collection housed at ROE increases (now $\sim$ 15000 original plates), the number of projects which can be satisfied by existing plates within the Library rises and is now at the 50% level. It is the enormous scientific value of this collection which is behind the policy of providing plates on temporary loan to astronomers and encouraging borrowers to return the plates to the Library on the shortest possible timescale so that they can be available for other projects.

6   Conclusions

The role of the UKST is expected to continue into the next decade along similar lines to the past decade, with a large survey programme and many smaller research projects. FLAIR spectroscopy will absorb an increasing amount of telescope time;

and the number of projects which can be carried out on published survey material and on the 15000+ plates in the ROE Plate Library will increase.

References

Cannon, R.D., 1984. *'Astronomy with Schmidt-Type Telescopes'*, *IAU Colloquium* No. 78, p. 25, ed. M. Capaccioli, D. Reidel.

Good, A.R., 1988. *Astrophotography*, p28, ed S. Marx, Springer-Verlag.

Good, A.R., Gourlay, G., 1984. *Astronomical Photography*, p. 93, eds M.E. Sim, K. Ishida, Royal Observatory Edinburgh.

Hartley, M., Dawe, J.A., 1981. *Proc. Astr. Soc. Aust.*, **4**, 251.

Hoessel, J.G., Elias, J.H., Wade, R.A., Huchra, J.P., 1979. *Publ. Astron. Soc. Pacific*, **91**, 41.

Humphries, C.M., Morgan, D.H., 1988. *Astrophotography*, p. 234, ed. S. Marx, Springer-Verlag.

Luyten, W.L., 1963. *A Proper Motion Survey with the [Palomar] 48-inch Schmidt Telescope, paper I*. Minneapolis: U. of Minnesota.

Luyten, W.J., 1979. *The LHS Catalogue*, Minneapolis: U. of Minnesota.

Minkowski, R.L., Abell, G.O., 1963. *Stars and Stellar Systems, vol III*, p. 481, ed. K.Aa. Strand.

Morgan, D.H., Savage, A., 1991. *Proceedings of IAU Photographic Working Group*, ESO, Garching. In press.

Morgan, D.H., Tritton, S.B., 1988. *'Mapping the Sky'*, *IAU Symposium* No. 133, p. 349, eds S. Debarbat et al., D. Reidel. (MT)

Murray, C.A., 1986. *Mon. Not. R. Astron. Soc.*, **223**, 649.

Probst, R.G., 1986. *Astrophysics of Brown Dwarfs*, p. 22, eds Kafatos et al., Cambridge University Press.

Reid, I.N., Brewer, C., Brucato, R.J., McKinley, W.R., Maury, A., Mendenhall, D., Mould, J.R., Mueller, J. et al., 1991. *Publ. Astron. Soc. Pacific*, **103**, 661.

Savage, A, Waldron, J.D., Morgan, D.H., Tritton, S.B., Cannon, R.D., Dawe, J.A., Bruck, M.T., Beard, S.M., Palmer, J.B., 1985. *The UKST Objective Prisms: II. Catalogue of Objects and Technical Data*, Royal Observatory Edinburgh.

Tritton, S.B., Morgan, D.H., 1991. *Proceedings of IAU Photographic Working Group*, ESO, Garching. In press.

Watson, F.G., Oates, A.P., Gray, P.M., 1990. *Proc. S.P.I.E.*, **1235**, 736.

West, R.M., 1984. *'Astronomy with Schmidt-Type Telescope'*, *IAU Colloquium* No. 78, p. 13, ed. M. Capaccioli, D. Reidel.

West, R.M., Schuster, H.-E., 1982. *Astron. Astrophys. Suppl. Ser.*, **49**, 577.

Discussion

Gursky :

You discussed in detail plans to make available film copies of survey plates. Are there any plans to distribute the information in an electronic form; specifically, the pixel data resulting from plate scans?

MacGillivray :

You will see from **Dr. Baruch's** contribution that there are indeed plans to distribute and make generally available the thresholded scans of the sky survey material from COSMOS. We plan a major programme of distribution of the complete survey pixel data from SuperCOSMOS.

THE USE OF EASTMAN KODAK 4415 FILM IN THE UK SCHMIDT TELESCOPE

K.S. Russell [1], *D.F. Malin* [1], *A. Savage* [1], *M. Hartley* [1] and *Q.A. Parker* [2]

[1] Anglo-Australian Observatory, Private Bag, Coonabarabran, NSW 2357, Australia.
[2] Royal Observatory, Blackford Hill, Edinburgh EH9 3HJ, U.K.

1 Introduction

In this paper we report experiments with 4415 film in the UK Schmidt telescope which may enable us to use a much wider range of sensitized materials than was previously possible, and to exploit the superior imaging properties of modern materials that are only available on film.

The focal plane of the UK Schmidt telescope (UKST) is spherically curved to a radius of about 3 m, so the thin, 356 mm square glass plates normally used in the telescope are mechanically deformed to fit a spherical-section mandrel during exposure. Glass has traditionally been used because it immediately and fully recovers its flatness on release of the mechanical constraint without displacement of or damage to the photographic emulsion. Experiments using the UKST with film products for colour photography (Malin 1978a) had been limited to 203 × 254 mm sheets, which were stuck to a glass backing plate for exposure. Some buckling and de–focusing of the corners were noted with this technique.

Later, some tests were made using full size, thick Estar base Aerographic Duplicating film, the only large format material available to us at the time (Humphries & Morgan 1988). This film had been mechanically deformed to fit the curved mandrel during exposure. Although this material was never intended for telescope exposure, it showed that useful image quality could be maintained over most of the focal plane. Recently we have been able to obtain some samples of Eastman Kodak 4415 emulsion coated on large sheets of thick Estar base.

2 The Film

Kodak Technical Pan Film 4415 has a formulation very similar or identical to the more widely available Technical Pan 2415 but is coated onto a 7–mil (178 μm) thick Estar base.

H. T. MacGillivray and E. B. Thomson (eds.), Digitised Optical Sky Surveys 23–33.
© 1992 *Kluwer Academic Publishers.*

The technical characteristics of 2415 film are given in Kodak leaflet P–255 (Kodak 1982). It is described there as a panchromatic black-and-white negative film with extended red sensitivity that offers extremely fine grain and high resolution. The spectral sensitivity extends from the UV to about 650 nm with some useful sensitivity as far as 690 nm; however, the region of maximum sensitivity and the sensitivity distribution across the visible spectrum are not the same as the superficially similar IIIa-F plate. To date there is no 4415 equivalent of the blue–green sensitive IIIa-J.

The resolution is considerably higher than that of the normally–used IIIa emulsions. Kodak publication P–315 (Kodak 1987) quotes 200 lp/mm for IIIa-F against 320 lp/mm for Tech Pan (no developer specified) and the diffuse RMS granularity is lower by a factor of two, a very substantial improvement. The film may be processed in a number of ways to provide a wide contrast range, at the high end equalling and possibly surpassing that of the IIIa materials.

3 Exposure and Processing

While current hypering experiments are still in their early stages, we have been able to produce films with an astronomically useful speed, although we have not firmly established what the optimum sky background density should be. Test samples have already reached similar sky densities (0.8–1.00) to those achieved using conventional IIIa-J/IIIa-F emulsions on glass with comparable exposure times. When used without a filter the film appears to be fully exposed after about 20 minutes.

Comparison of the published characteristic curves of 4415 with the more familiar IIIa materials (Kodak 1987) shows that the newer emulsion has a less pronounced 'toe' to its characteristic curve which suggests that maximum contrast (and thus optimum output signal to noise) might be achieved at a lower density, enabling us to reduce exposure times. The lower sky density might have other practical benefits, enabling a higher signal to noise to be achieved on the current measuring machines.

4 Mounting

The UKST is not designed to accept emulsions coated on a thin, plastic base, so one of our main concerns before tests began was how we might clamp the film tightly against the curved mandrel without stretching or buckling it. We had earlier found that it might not be possible to retain perfect focus around the edge of the field.

We shall not go into detail here, except to say that as far as we can tell the new method of loading film produces exposures with images which appear to be in perfect focus across the entire field. The method is simple and reliable enough to be considered routine. We can furnish technical details to anyone who might be interested.

5 Astrometry

Published information (e.g. Kodak 1972) indicates that the dimensional stability of polyester film, while not as high as glass, is extremely good. However, these figures apply to film exposed flat without mechanical stress, so this observation may not apply in our case.

We have yet to carry out definitive tests on the actual astrometric properties of the film. Recent techniques used in the analysis of Schmidt plates, however, show quite clearly that we still do not fully understand the distortions present in glass plates and the data is best reduced over small areas at a time. This same technique should serve to minimise any additional problems we might encounter with film.

6 Photometry

The earliest systems of astronomical photometry were based on the spectral sensitivities of the photographic materials then available and these early photographic characteristics determined the standard BV photometric system for many years.

The introduction of the IIIa emulsions in the late 1960s marked a radical departure from these early spectral sensitisations and greatly complicated photographic photometry. What was to become the IIIa-J had initially a standard blue (i.e. O) response, but was given an extended sensitisation into the green in order to give astronomically useful speed. The awaited IIIa-D never appeared, and the IIIa-F has a red sensitizing that was unlike any previous red plate. Transformations of one passband to another were eventually evolved, but not without difficulty and loss of precision. The lack of a fine grain V band emulsion has long been evident.

With the 4415 emulsion, we are faced with the introduction of yet another spectral sensitization that does not match any of the previous emulsions. The relative spectral sensitivity curves of the IIIa-F and 2415 are shown in the Kodak literature. When plotted on a linear scale, one is immediately aware of the spectral non–uniformity of these materials, especially IIIa-F. Tech Pan film Type 2514 betrays its origins as a material designed to monitor solar flares, which emit strongly in the 656 nm line of Hα thus accounting for its excellent sensitivity in the 650 nm region.

7 Imaging Properties

7.1 Direct Comparison

To compare the properties of 4415 film with the more familiar IIIa materials we obtained a deep exposure (OR14310, exposure 70min) of a field in Virgo for which we had a corresponding deep IIIa-F plate (OR9911, exposure 60min). These exposures cover very similar passbands and were taken under similar conditions of seeing etc. Simple inspection of the film and plate on a light table confirms that they are very different. Examination with a hand magnifier immediately reveals that the 4415 material is almost grainless when compared to the IIIa-F. A comparison of identical regions of the two exposures is shown in Figs. 1a, b.

The improved resolution is immediately apparent, and though the difference in grain structure is obvious to the eye, it is quite difficult to record photographically without a microscope. Also, it is not immediately obvious that the film reveals

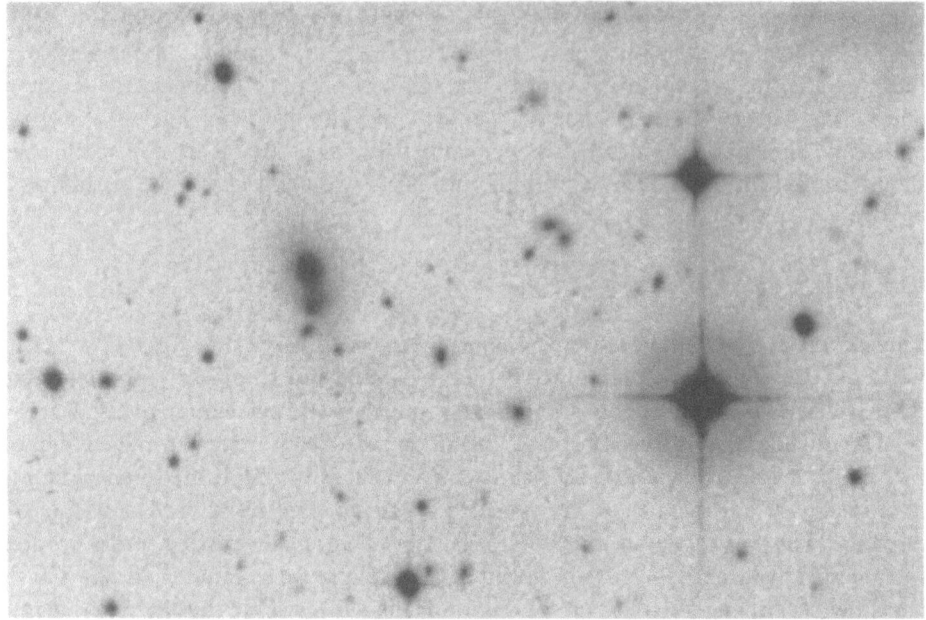

Fig. 1 a) Direct enlargement from the Virgo field — 4415 film.

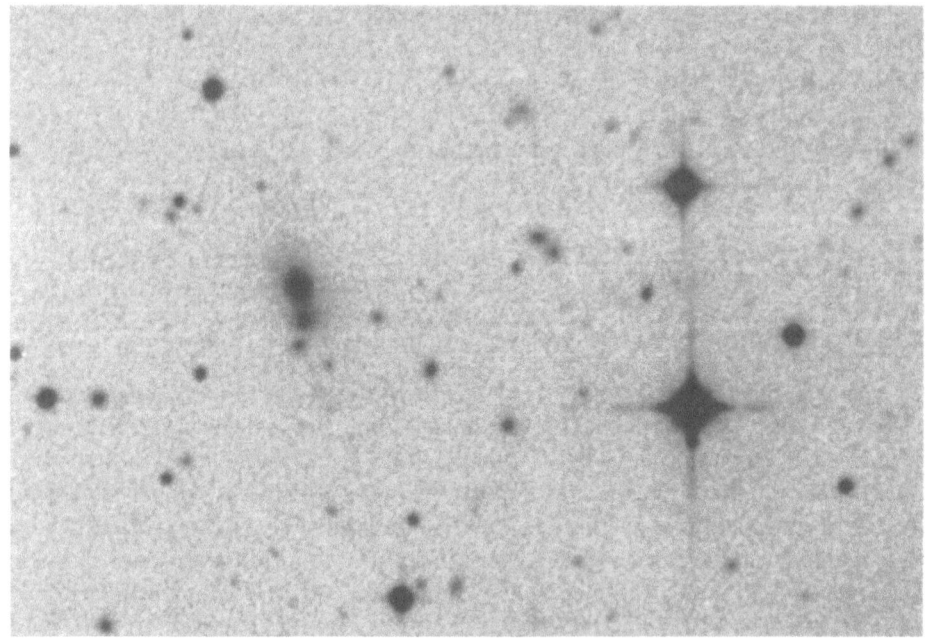

Fig. 1 b) Direct enlargement from the Virgo field — glass plate.

significantly deeper images. A series of simple qualitative tests was undertaken to see if these differences could be made more evident.

7.2 Photographic Amplification

This technique (Malin 1978b) is ideal for extracting the faintest images from a plate and for revealing non–uniformities. It also emphasises grain structure. The same area that appeared in Fig. 1 was used for this test (Figs. 2a, b) and now it is obvious that the 4415 has a much finer grain. Even more impressive is the much improved resolution, greatly simplifying star–galaxy separation. Note also the low surface brightness dwarf elliptical galaxy. The IIIa-F image hints that it has a nucleus. The 4415 film leaves no doubt. Thus the finer grain and higher resolution combine to give much improved information content and reveal for the first time that the image quality of the UKST has never been fully exploited by the IIIa emulsions.

As a further test, two well known objects with both stellar and extended faint features were examined with photographic amplification. The detection of faint extended features requires good uniformity and high contrast, but not necessarily high resolution, while the detection of faint stellar features requires fine grain and corresponding good spatial resolution.

The elliptical galaxy M 89 was found by Malin (1979) to have a faint jet–like structure and a shell. It is clear that both are better seen on the photo–amplified 4415 film than they are on the corresponding IIIa-F plate (Figs. 3a, b). Similarly, the galaxy M 87 has a well–documented family of faint globular clusters that are unresolved [see Huchra (1988) for a review]. Most of these are invisible on the IIIa-F plate but clearly seen on the 4415 film (Figs. 4a, b). The faintest have a corresponding blue magnitude of about 23 (Racine 1968). In these last examples, unsharp masking (Malin 1990) has been used on both originals to mask the diffuse image of M 89 itself.

It is clear from these simple tests that 4415 film has much better imaging qualities than IIIa-F and on this count alone must be considered as its replacement where very deep images of red objects are needed. Of course, because the night sky is relatively much brighter at those wavelengths where Type 4415 film is most sensitive, a blue–sensitive equivalent would also be very useful.

7.3 COSMOS Analysis

A film (OR14355, exposure 60 min) was selected for measurement on COSMOS. This was analysed to produce the plots (Fig. 5) often used in separating stellar images from those of galaxies. These plots were compared with similar data derived from a glass survey plate of comparable passband (Fig. 6).

The stellar images should be restricted to a fairly strict locus for ease of separation. It can be seen from the plots that in all four cases the stellar locus is much tighter on the film data than that obtained from the glass plate. Since a great deal of research effort is being directed towards large scale galaxy surveys, this result suggests that it should be possible to classify galaxies more accurately from exposures on 4415 film than from IIIa emulsions.

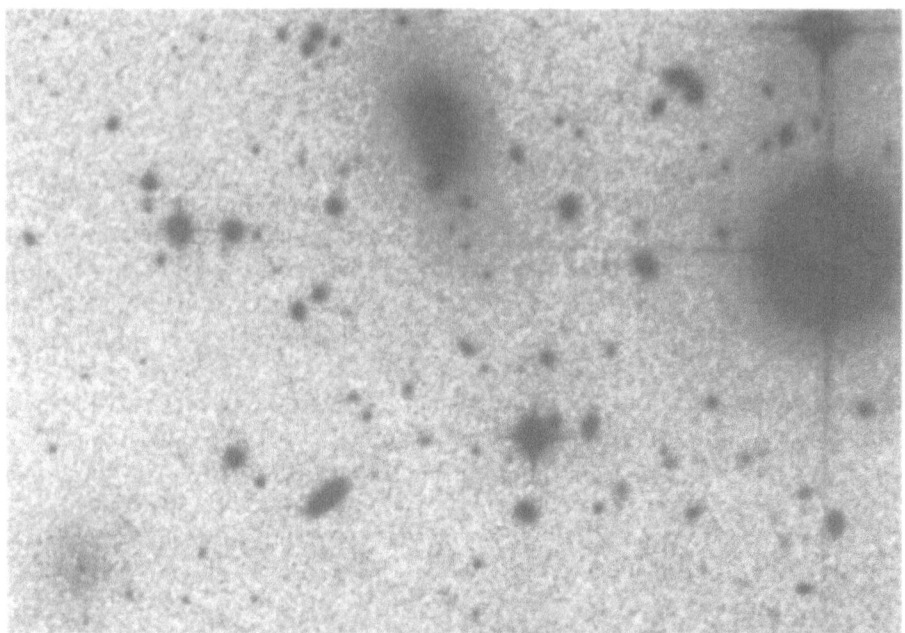

Fig. 2 a) Same area as in Fig. 1 but photo-amplified — 4415 film.

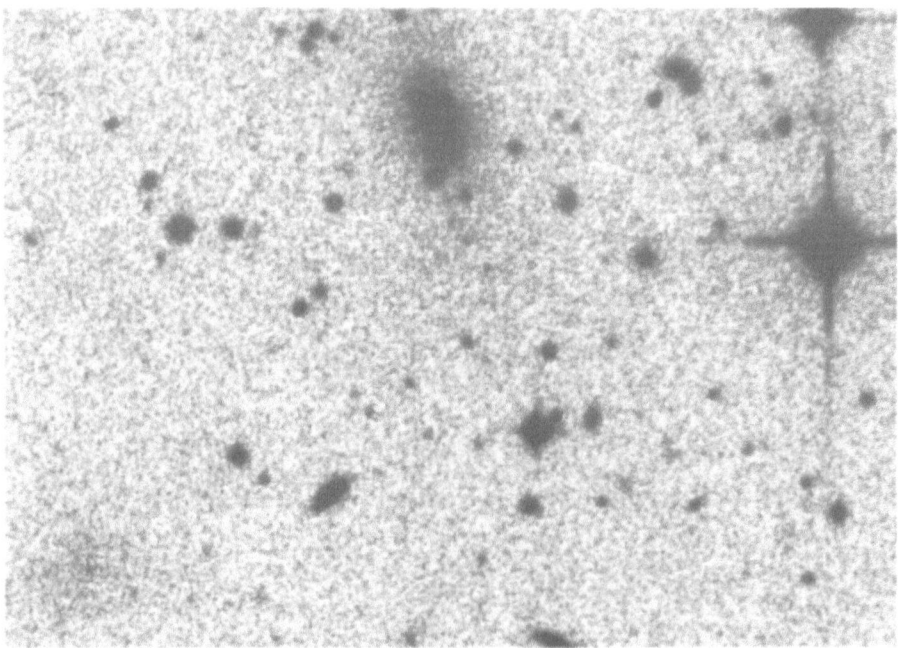

Fig. 2 b) Same area as in Fig. 1 but photo-amplified — glass plate.

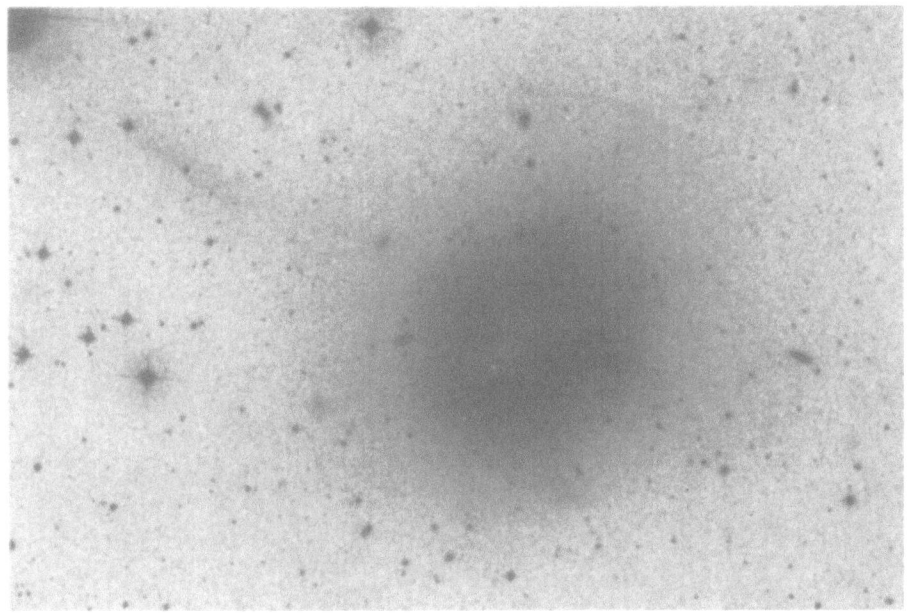

Fig. 3 a) M89 photo-amplified, showing jet and shell structure — 4415 film.

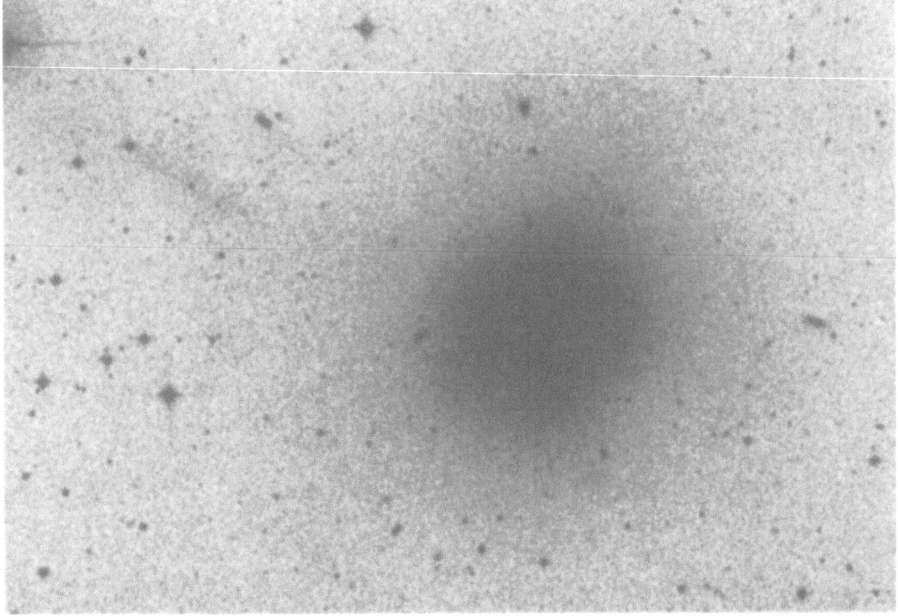

Fig. 3 b) M89 photo-amplified, showing jet and shell structure — glass plate.

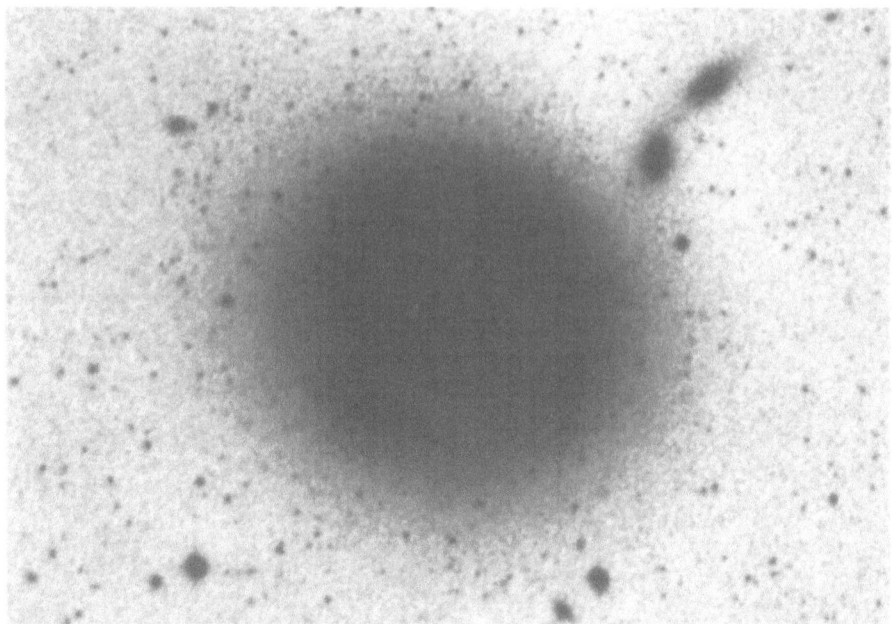

Fig. 4 a) M87 photo-amplified, showing globular clusters — 4415 film.

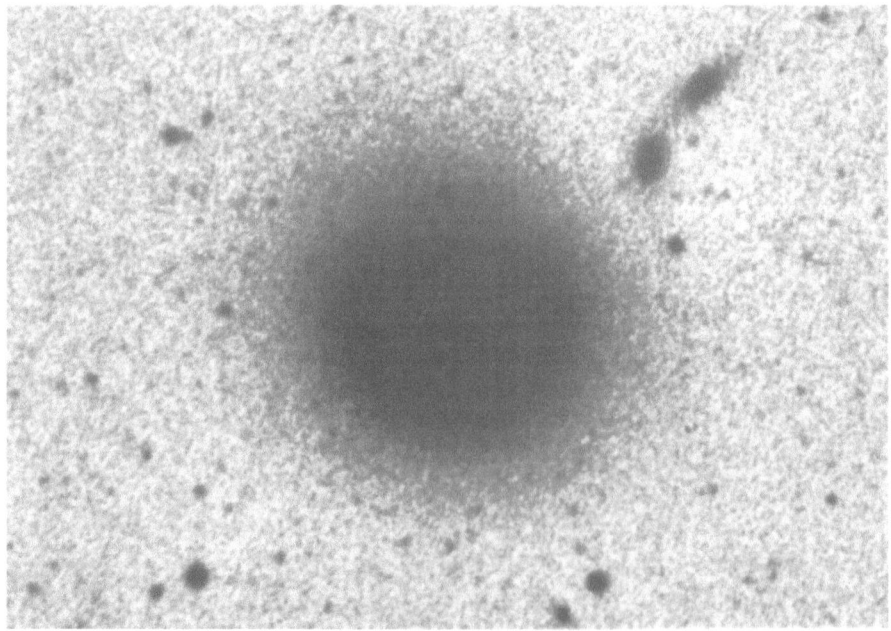

Fig. 4 b) M87 photo-amplified, showing globular clusters — glass plate.

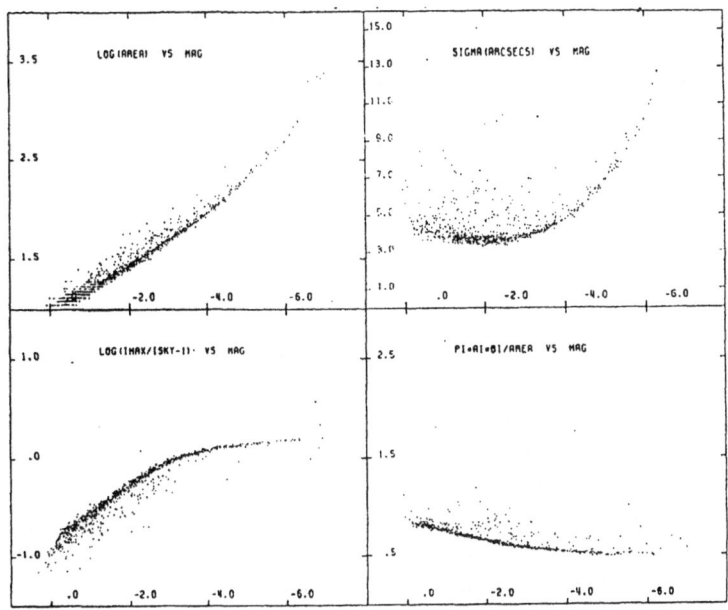

Fig. 5 COSMOS quality control information — film.

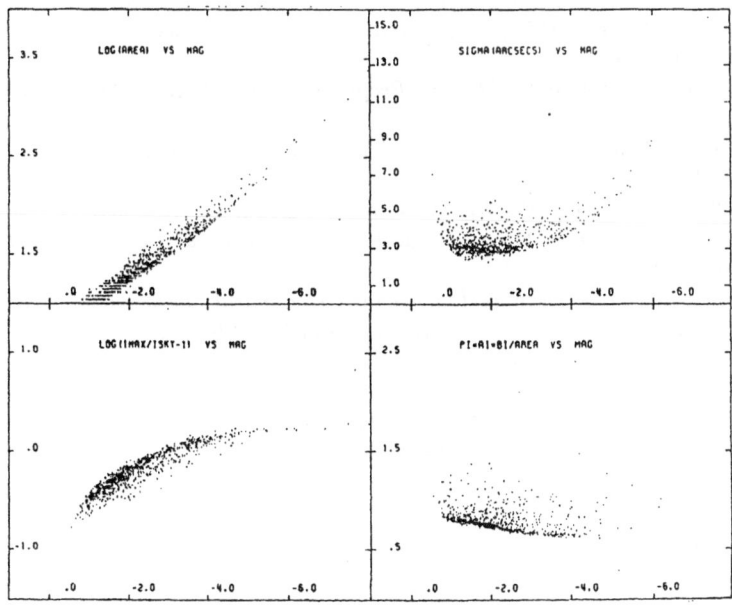

Fig. 6 COSMOS quality control information — glass plate.

8 Conclusions

Overall, the film–based 4415 emulsion appears to outperform IIIa-F on glass plates by about one stellar magnitude, but the increase in information content may be much greater due to the enhanced resolution and fine grain structure. Extended, low surface brightness structures are more visible on the 4415 film and image classification can be carried out more accurately than on corresponding IIIa plates.

The quality of these early results and the low cost of the film–based product both open up possibilities for extended or new programs of work at the UK Schmidt. We shall be carrying out an extensive series of tests of this material during the next six months. We also intend to investigate the possibility of obtaining the 4415 film with different sensitisations to allow more general use of film as a direct replacement for some current programs using IIIa-J and IIIa-F emulsions on glass.

It is mostly the remarkably fine grain and high resolution which has surprised us and we suspect that in order to fully extract the extra information inherent in this emulsion, automated measuring machines with extremely small scanning apertures and high levels of digitisation will be required. While current, or planned, measuring machines can probably achieve the level of digitisation required, it is not at all clear that their pixel size will be small enough to make best use of exposures on 4415 film. This is a question we intend to address during the tests mentioned above.

References

Huchra, J., 1988. In *The Harlow Shapley Symposium on Globular Cluster Systems in Galaxies*, p. 255-268, eds. Grindlay & Davis Philip, Kluwer, Dordrecht.

Humphries, C.M., Morgan, D.H., 1988. In *Astrophotography*, p. 234, Ed. S. Marx, Springer–Verlag, Heidelberg.

Kodak, 1972. Publication M–62 *Physical Properties of Kodak Aerial Films*, Eastman Kodak Co.

Kodak, 1976. Publication P–316A *Kodak Special Plates Type 127-04 and Type 127-05*, Eastman Kodak Co.

Kodak, 1982. Publication P–255 *Kodak Technical Pan Film 2415*, Eastman Kodak Co.

Kodak, 1987. Publication P–315 *Scientific Imaging with Kodak Films and Plates*, p. 7d, Eastman Kodak Co.

Malin, D.F., 1978a. In *'Modern Techniques in Astronomical Photography'*, Proceedings of *ESO Conference*, p. 235-237, eds. West and Heudier, ESO, Geneva.

Malin, D.F., 1978b. *Nature*, **276**, 591–593.

Malin, D.F., 1979. *Nature*, **277**, 279–280.

Malin, D.F., 1990. *Tech Bits*, Issue No. 1, 1990. Eastman Kodak Publication.

Racine, R., 1968. *J. Roy. Astr. Soc. Canada*, **62**, 367–376.

Discussion

MacGillivray :

Do you have a feeling for the optimum parameters at which the high-speed machines should be digitising these films?

Russell :

Early results from the film tests demonstrate that the IIIa emulsions are not fully exploiting the superb quality of the UKST optics, which are sub-arcsecond across the visible spectrum. We have some isolated exposures on glass with 12μm images using the IVN emulsion which confirm this performance. I suspect that in order to fully exploit the new high resolution films we will require pixel sizes of less than 10μm.

Parker :

In the slide of M89 there was a faint 'extra' feature apparent on the 4415 film not visible on the IIIa-F. However, there was also a similar faint feature visible on both images. Is this feature real?

Russell :

I can't really say but the similar feature which I did point out is also visible on a completely independent film.

THE CHINESE LARGE SCHMIDT TELESCOPE PROJECT

J.-S. Chen

Beijing Astronomical Observatory, Chinese Academy of Sciences, Beijing 100080, P.R. China.

Abstract

The idea of building the world's largest Schmidt telescope in China has been discussed for several years among the Chinese astronomical community. The basic point to motivate such a facility is the following: China, as a developing country, has only modest means for the development of astrophysics, including the capabilities of financial support, technical feasibility, and site quality. Chinese astronomers should determine a strategical direction in which they can make, for the coming decades, important contributions to the development of world astronomy. We also hope that such a facility would be a unique one, so that it would become a very real attraction for other astronomers of the world.

1 The Challenge

In the past two decades, rapid advances in astrophysical research have led to significant discoveries in astronomical fields at all space levels; interaction with different branches of physics has brought about new breakthroughs in the understanding of the origin and evolution of the stars, galaxies and the Universe; advanced new techniques, especially those of space science, have opened up a new era for observational astronomy, such that the age-long restriction imposed by the Earth's atmosphere is henceforth removed and the information carried by cosmic radiation of all wavelengths in the electromagnetic spectrum may now be utilized. Many countries, notably the developed countries with close international cooperation, pay great attention to the study of astrophysics. In the past ten years, a number of large astronomical instruments have been put into operation, and for the coming decade more giant instruments (including a ground based optical telescope of 10m class, large radio telescope systems, and advanced infrared, X-ray and optical space telescopes) are expected to come onto the scene.

H. T. MacGillivray and E. B. Thomson (eds.), Digitised Optical Sky Surveys 35–39.

2 Two Directions for the Development of Astronomy

We understand that there are two main approaches or directions for the future development of modern astronomy. One direction is to study the detailed properties of carefully selected objects by using as powerful facilities as possible, including the whole coverage of the electromagnetic spectrum and using a large, even huge, telescope to obtain high spatially-resolved images, high resolution and high signal-to-noise spectra. In order to achieve high spatial resolution in the optical and infrared band, the concept of a space observatory is under progress, while on the ground a sub-arcsecond site is being explored and established. Survey astronomy is also carried out with such facilities but is limited to small areas although down to faint limiting magnitudes. On the other hand, large-field sky surveys including imaging, multi-colour photometry and optical fibre redshift surveys have rapidly developed during the last decade, have produced large amounts of information for the study of large-scale structure and statistical properties of large samples, and have led to many important discoveries. Both directions intersect at the point where survey astronomy will serve to search out rare and interesting new objects for the large telescopes for detailed study, and in turn the results of detailed studies of individual objects by large telescopes often demand statistical investigation of large samples of objects of the same category with the help of survey astronomy.

3 Revolutionized Schmidt Astronomy

The Schmidt telescope is the widest-field telescope and is most suitable for large-scale and survey astronomy. The important rôle played by Schmidt astronomy has been well demonstrated by the scientific contribution from the existing Schmidt telescopes (as well as witnessed by these proceedings). During the last two decades, the breakthrough in handling large amounts of Schmidt data and the applications of new high-technology have brought Schmidt astronomy into a new era.

The field of view of a Schmidt telescope may reach $6° \times 6°$, about forty times larger than that of an optical reflecting telescope. So the wealth of information contained in a single Schmidt photographic plate is correspondingly large. Resorting to the development of high speed, high precision, automatic plate measuring machines and utilization of digital image processing systems, astronomers are able now to make full use of the information thus obtained. Complete digitized optical sky surveys are now being carried out efficiently. An order of magnitude improvement in the astrometric measurement of existing Schmidt telescope plates is now being achieved by the next generation of the COSMOS machine (SuperCOSMOS) developed by the Edinburgh group. The science coming from the whole sky, such as unprecedented quantities of faint object astrometry, is just being tapped.

With the rapid development of CCD techniques, such as large format (as big as 4096×4096 pixels), multi-gate read-out, low noise, and of the data acquisition techniques for large data bases with high speed and low cost (such as the transputer and the optical disc media), the possibility of using CCDs as the full Schmidt-field detector is now open. This means that digitized sky imaging and photometric surveys can be done with unprecedented speed and accuracy and down to fainter limiting magnitudes.

With the introduction of the techniques of fibre optics in astronomical observation, the rôle of the Schmidt telescope in astronomical research has become more important. Images of objects at different parts of the field can be brought together by optical-fibres for processing (e.g. to be aligned onto a single inlet of a spectrograph). With such techniques, it becomes possible to combine the advantages of the large fields of view given by the Schmidt telescope and the high sensitivity offered by modern detecting devices, so that a large number of object images may be treated all at the same time. This highly efficient way of observation is particularly powerful in the astrophysical study of the statistics of large samples of objects to a medium depth.

All these developments have revolutionized the whole of Schmidt astronomy, and we may still expect even greater and unpredicted successes to follow.

To give full play to the function of the Schmidt telescope will definitely lead to important advances in various fields of astronomical research, noticeably in the study of galactic and extragalactic astronomy, large scale structure of the Universe, etc. As an example, one may mention the discovery of large redshift ($Z > 3.5$) quasars. To find out the existence of such objects is of great importance to the extension of Man's knowledge to greater depths of the Universe. However, no progress had been made for nearly ten years in the search for them until recently, when a number of quasars with $Z > 4$ were discovered with the help of the U.K. Schmidt telescope and a fast measuring machine of superb performance.

4 Opportunity

In contrast to the great amount of support given to the development of a giant reflecting telescope (which often amounts to the order of 100 million US dollars per item), attention paid to exploring the capability of the Schmidt telescope is by far insufficient. Especially in the northern hemisphere, the 120/180 cm Schmidt telescope at Mt. Palomar (California, U.S.A.), which has played a dominating role in Schmidt astronomy, was built forty years ago. Though the contribution it has made to contemporary astronomy has been large, it would be difficult for the Palomar Schmidt telescope by itself to cope with the essential needs for the 1990s. Moreover, as it is not equipped with an objective prism, the important information offered by slitless spectra is excluded. Equally serious is the situation of city light pollution at the Palomar site, which causes the quality of observations to deteriorate. This means that there is an opportunity open to us to build the world's largest Schmidt telescope, with state-of-the-art technology and equipped with sophisticated equipment, at a site free from city light. This would thus be of strategic value for the development of Chinese astronomy and astrophysics.

5 Specification and Capabilities

The main contents of the proposed Chinese Schmidt telescope are as follows:

achromatic corrector plate	150cm
main mirror	220cm
focal length	375cm
apex angle of objective prism	44 arcmin

The telescope will be designed for direct photography, direct CCD imaging, slitless spectroscopy and the adoption of fibre optics and other new techniques.

The telescope should have the following capabilities:

1. deep sky atlas production;
2. multi-colour photometry;
3. direct CCD imaging and photometry;
4. slitless spectral survey work;
5. fibre optical astronomy;
6. large field astrometry;
7. narrow band survey work.

The telescope should be equipped with the most advanced instruments including: a high speed, high accuracy, fully automatic plate measuring machine; mosaic large area CCD camera; fibre optic coupler; and with plate library and facilities for handling huge amounts of digitized astronomical data.

6 Feasibility

The optics and mechanics involved in building this telescope are not much different from a two meter size reflector which has already been built by Chinese industry. We have a spare cervit blank with this size available. The Chinese optical factory is able to cast the glass for the corrector and prism plates. Technically, we need to master the skill of making an achromatic corrector plate. Therefore, we will be able to build the main body of the telescope based completely within domestic industry. The site requirements for a Schmidt type telescope, though, are strict (especially for the world's largest Schmidt), but are not as strict as those required for a large reflector, which normally requires sub-arcsecond seeing conditions. We can probably find a site of modest seeing condition (say 1.5 arcsecond seeing) and with high darkness inside China.

The high-technology part, including the plate measuring machine, large field mosaic CCD camera system, image processing system and fibre optic system and other items, which can not be provided from within China, need to be imported.

7 Present Status of the Proposal

Phase A studies are being carried out, which include the studies of the general layout and the design of the mechanical structure of the telescope, the optical design, the

60 cm size experiment to understand how to manufacture the achromatic corrector plate, and the study of guiding methods with the main telescope. Site testing has been carried out during the past 2 years. It has become clear that the existing optical site is not suitable for the world's largest Schmidt telescope. A new site survey is being prepared in the north-western part and southern part of China to identify a really dark site with reasonably good seeing.

THE DEEP 2-MICRON SURVEY OF THE SOUTHERN SKY

A. Blanchard [1] and N. Epchtein [2]

[1] Observatoire de Meudon, DAEC, F-92190 Meudon, France.
[2] Observatoire de Meudon, DESPA, F-92190 Meudon, France.

1 Introduction

Infrared Astronomy is one of the major potential sources of important progress in Astronomy. The IRAS satellite has allowed an all sky survey in the range 10-100μm. During the last recent years, the performance of 2-D infrared receptors at wavelengths $\geq$ 1μm has been greatly improved, and has become comparable to optical CCDs. This permits us to envisage a digital sky survey in the near infrared. The absence of the equivalent to Schmidt plates has left the near infrared domain almost unexplored. The purpose of DENIS (Deep Near Infrared Survey of the Southern Sky) is to provide an all-sky survey in the J (1.25 μm) and K (2.2 μm) bands with a pixel size of 3''. An optical band, I (0.9μm), with a better spatial resolution will be also included. The survey will be done on the 1 meter ESO telescope, with a camera using 256$\times$256 *Rockwell* arrays.

The limiting magnitude in the K-band should be 14.5. The importance of such surveys for infrared astronomy will be comparable to Schmidt surveys in the optical range.

2 Scientific Objectives

The advantages of observations in the near infrared are numerous and this survey will offer a remarkable data base for both stellar and extragalactic purposes. One of the most obvious advantages comes from the fact that the Galaxy is almost transparent at these wavelengths.

2.1 Stars

The stellar population of galaxies is dominated by low mass stars, and therefore the light in the K-band is a much better tracer of the bulk mass in the Galaxy.

41

H. T. MacGillivray and E. B. Thomson (eds.), Digitised Optical Sky Surveys 41–42.

This is not the case for previous surveys: in the blue, the luminosity is dominated by the young stellar population, while in the IRAS bands it includes dust emission. The expected total number of stars that will be detected in the survey is 10^8. The luminosity function of the low temperature stars will be improved. The sample will contain a substantial number of brown dwarfs ($\geq 10^5$), out of which more than 10^4 should be identifiable.

This survey will allow the deep probing of giant molecular clouds, and the low luminosity proto-stars that they contain. The stellar luminosity of these regions will be studied in detail. The IMF and SFE will be well sampled.

2.2 Galaxies

The interest for extragalactic purposes of the K band is considerable. First, as we mentioned above, the Galaxy is almost transparent at these wavelengths (the absorption at $|b| \approx 1°$ is less than one magnitude). This permits the consideration of a full-sky survey of galaxies, while optical surveys are usually limited at solid angle of the order of 1 sr because of absorption. In addition, the photometry will be based on digital data while optical surveys are based on Schmidt plates. There are several advantages over the IRAS survey. From the number counts in the K-band (Glazebrook et al. 1991), the total number of galaxies in the survey is expected to be of the order of 2×10^6. Furthermore, the K-band is a reliable tracer of the stellar mass content of the galaxies: the M/L in K is remarkably constant for different galaxy types (Struck-Marcell & Tinsley 1978; Jablonka & Arimoto 1991). In that sense the 2μm survey will provide a unique data base. The number counts at the bright end will be known with a much better accuracy (the presently available counts at K = 14.5 are uncertain by more than a factor of 2). The projected galaxy distribution could be investigated in great detail with such a sample. The galaxy correlation function could be determined free of any possible Galactic absorption. Because galaxy clustering is different for IRAS and optically-selected samples, it is critical to have a survey selected in the K band: the correlation function will be fully representative of the galaxy distribution. Obviously, this survey will provide a wealth of information on QSOs, AGN, radio-galaxies, etc., allowing the investigation of the properties of these objects in this domain. A spectroscopic follow-up of the galaxy sample will be necessary to fully exploit the survey.

The main problem is certainly the star/galaxy confusion. The $3''$ pixel in the K band does not allow their separation. The addition of an optical band with a smaller pixel size will allow the identification of extended sources. Even in that case, the star/galaxy separation might become a problem at low Galactic latitudes because of the confusion limitation.

References

Glazebrook, K., Peacock, J.A., Miller, L., Collins, C.A., 1991. *Proceedings of the Blois Conference: Physical Cosmology*, Editions Frontières, Gif-sur-Yvette, France.
Jablonka, P., Arimoto, N., 1991. Preprint.
Struck-Marcell, J., Tinsley, B.M., 1978. *Astrophys. J.*, **221**, 562.

NEXT GENERATION OPTICAL SKY SURVEYS

M.J. Irwin

Royal Greenwich Observatory, Madingley Road, Cambridge CB3 0EZ, U.K.

Abstract

By building a prime focus CCD array camera system capable of taking advantage of the full 40 arcmin diameter field of the 2.5m Isaac Newton telescope on La Palma, a dramatic advance in optical sky surveys would be possible. With such a system, a deep CCD-based sky survey in several optical passbands over several hundred square degrees could be achieved using an existing medium sized telescope at modest cost. The key technology and expertise in processing sky surveys already exist and the scientific returns would be enormous: ranging from Solar System studies, such as searches for primordial comets, out to searches for the most distant objects in the Universe, such as high redshift quasars and primordial galaxies.

1 Introduction

Astronomy is basically an observational science rather than an experimental one, and the development and advancement of the subject has relied heavily on surveys of the sky at optical wavelengths to expand our knowledge of the observable Universe. Surveys form a basic foundation of observational astronomy, and are needed to obtain quantitative statistical information on the distribution of objects in our own galaxy and the Universe, to discover radically new types of object and to select representative samples of certain types of objects, particularly the brightest examples, for further study with large telescopes. Reliable, quantitative surveys will form a vital constituent in the future success of any large telescope programme, just as the surveys based on photographic plates taken with the Palomar Schmidt and UK Schmidt have been vital to the success of the present generation of large telescopes. Although the importance of next generation survey work is self-evident, rather little practical work has been carried out to develop survey instruments beyond the

43

H. T. MacGillivray and E. B. Thomson (eds.), Digitised Optical Sky Surveys 43–52.

Palomar and UK Schmidt photographic surveys. Compare this to the effort that has been devoted to the development and construction of large telescopes. Moreover, the competitiveness of the very large telescope programmes, in fundamental areas of research, such as large scale structure and the formation and evolution of galaxies, strongly depends on the quality of the survey work. We cannot efficiently run 8-16m telescopes in the next decade if we do not have at the same time proper survey facilities with which to select the best objects to follow-up.

1.1 Current Situation of Sky Surveys

Despite the considerable advances of detector technology within astronomy, very little improvement has been made in surveys beyond those available in the 1950s when the Palomar Sky Survey was carried out. A photographic plate taken on a 1.2m Schmidt telescope saturates in about 1 hour but is only 1-2% efficient. Therefore our best sky surveys amount to no more that a 60 second glance of the Universe. The recent dramatic improvement in exploiting the existing survey material is due entirely to the advent of the fast automatic plate measuring machines. By routinely processing the information on a Schmidt plate in a matter of a few hours, these machines have opened up new areas of quantitative survey astronomy. However, the planned and current photographic surveys (UKST second epoch R and new Palomar Sky Survey) will not provide a significant improvement in this situation since they will essentially be simply extending what already exists. To anticipate one of the results of the next section of this paper, because of the low efficiency and relatively high grain noise of photographic plates, the intrinsic photometric errors are typically around $0.^m1$ whilst the astrometric errors are around 1μm and are unlikely to improve dramatically (though see Russell et al., these proceedings). In addition field errors of $0.^m25$ on photographic plates are not uncommon, exacerbating the problem of extracting reliable astronomical results. The current generation of measuring machines, including those nearing completion, fully exploit the present and planned photographic sky surveys. In order to make a significant advance (say a factor of 10) over current material, any new survey must be based on detectors that are essentially photon noise-limited, such as CCDs.

Up until recently, CCD arrays have been physically small and the pixel sizes not well matched to the plate scales available on potential wide–field survey telescopes. In addition there has been the problem of what to do with the vast amounts of data that a CCD sky survey would produce. These drawbacks now no longer limit the potential for CCD-based sky surveys — large format 2048 x 2048 CCDs are now available and still larger mosaics are being fabricated (e.g. Geary et al. 1991); whilst computing resources and storage media have advanced sufficiently to enable both full archiving and 'real-time' processing of the survey to take place.

1.2 Next Generation Optical Surveys

Although the existing photographic surveys cover the whole sky, there is no requirement that a fainter survey should do the same. Significant scientific progress would be made by simply initially doing a survey both 10 times fainter and a factor of 5 more accurate than the existing sky surveys, but only covering a few hundred to a thousand square degrees. By fully exploiting the red sensitivity of CCDs, particularly in the I and Z band, we would also be exploring new wavebands which as yet have received little survey attention. These advantages will be fundamental in both searching for rare types of object (e.g. surveys for objects with Lyman alpha emission at redshifts above $z = 6$ which cannot be done with present surveys), discovering new categories of object and in doing accurate large volume statistical surveys.

A CCD–based survey covering a thousand square degrees at high galactic latitude ($|b| > 30°$) and reaching an equivalent R magnitude of 24, would almost double the surveyed volume of the Universe and at the very least would generate a catalogue of more than several million galaxies and tens of thousands of quasars. It would permit an order of magnitude reduction in the errors compared to a photographic survey and it would reach several magnitudes fainter. The scientific case for such a survey instrument could be considered self-evident since it would become the basic foundation of our future astronomical programmes. The range of science possible encompasses subjects as widely disparate as: solar system studies, such as searches for primordial comets and satellites of the major planets; to searches for the most luminous and distant objects in the Universe, such as high redshift quasars and primordial galaxies; whilst high photometric accuracy, with well sampled data taken in good seeing, is vital for the full range of both star and galaxy studies.

2 Error Limitations of Photographic Plates

2.1 Random Errors

Random noise on photographic plates is dominated by the fluctuations in the spatial distribution of grains, rather than photon noise from the observed targets (e.g. Dainty & Shaw 1974). For a modern emulsion the noise in a spatial resolution element is ~0.1D, where D denotes optical density. In contrast the available signal range is bounded between sky (typically at 1D) and a level which depends on the measuring machine used (3D for a multi- element or flying spot scanner to 5D for a fixed optic single sample system such as a PDS). Consequently, the maximum achievable peak signal/rms noise level is only in the range 20 – 40. This severely limits the attainable precision with which measurements can be made. In order to investigate this, consider the minimum theoretical error attainable for measuring positions and intensities of images from 2D arrays of data as discussed by King (1983) and Irwin (1985). For Gaussian-shaped images in constant random noise, which is a reasonable approximation for faint images measured on photographic plates, the variance of the positional error is given by

$$var[\hat{x}] = \frac{\sigma_G^2}{2\eta} \cdot \frac{4\pi\sigma_G^2\sigma_N^2}{\eta} = \frac{2\sigma_N^2}{\pi I_P^2} \tag{1}$$

whilst the variance of the intensity error is given by

$$var[\frac{\hat{\eta}}{\eta}] = \frac{1}{\eta} \cdot \frac{2\pi\sigma_G^2\sigma_N^2}{\eta} = \frac{2\sigma_N^2}{\pi\sigma_G^2 I_P^2} \tag{2}$$

where η is the total image intensity in some arbitrary unit, σ_N is the random noise per pixel in the same units as η and σ_G is the radius at which the Gaussian is exp -1 of the peak height, I_P.

Note that the minimum attainable error for both intensity and positional estimates is inversely proportional to the available dynamic range of the recording medium, I_P/σ_N. Since the useful dynamic range of most plates is typically ~ 20, this severely limits the attainable precision with which images can be measured. In order to improve upon this, either emulsions of much finer grain size have to be used to reduce the grain noise dominated term, σ_N (e.g. Russell et al., these proceedings) or alternatively devices with much larger dynamic range such as CCDs need to be used.

For brighter stellar images, saturation and the shape of the outer wings of the profile become the dominating characteristics in determining the theoretical error limit. On a typical IIIa-J or IIIa-F UKST plate, $\sigma_N \approx 0.1$D for 7.5μm (1/2") pixel size and σ_G varies from 3 to 5 pixels (2"- $3\frac{1}{2}$" FWHM seeing). The peak intensity for the faintest detected images ranges upward from 1-2 σ_N. Stellar images start to saturate ~3 magnitudes above the plate limit, severely restricting the attainable astrometric accuracy with which brighter images can be measured. This is illustrated in Fig. 1 which shows the theoretical minimum possible error for measuring the intensities and positions of stars on survey quality sky-limited UKST IIIa-J and AAT IIIa-J plates. It is immediately obvious that the theoretical minimum error for measuring positions is never lower than about $0.3 - 0.4\mu$m and for the vast majority of images is significantly worse than this. A similar situation holds for the intensity errors, with the minimum attainable percentage error for a UKST IIIa-J plate being ~2% with an AAT plate potentially a factor of ~4 better. Again, for the majority of images on a UKST survey plate the minimum attainable percentage intensity error lies between 5-10%. Positional errors for galaxies would be correspondingly larger since the effective σ_G is larger and the error scales as σ_G^2.

Whilst it is relatively straightforward to come close to the minimum attainable error for faint stellar images, once images start to saturate the profile becomes dominated by the wings of the seeing profile and telescope-induced artifacts (e.g. diffraction spikes, reflection halos, coma etc.). For these brighter images it would be extremely impractical to routinely use a maximum likelihood method and the likely realistic error bound is probably at least a factor of 2 higher (i.e. $> 0.5\mu$m for positions and $> 5\%$ for intensities). The situation for plates taken for the old Palomar sky survey is even worse because of the much coarser grain size.

2.2 Systematic Errors

The photographic emulsion is an imperfect elastic medium subject to various abuses during preparation (baked/flushed), on the telescope (distorted), and development (swollen, washed, etc.) not to mention storage. It would indeed be surprising if it responded to this without systematic shifts over the surface. Indeed it has been shown, both theoretically and empirically, that radial distortions of up to 15 μm occur from simply deforming the plate to fit the focal plane of a Schmidt telescope.

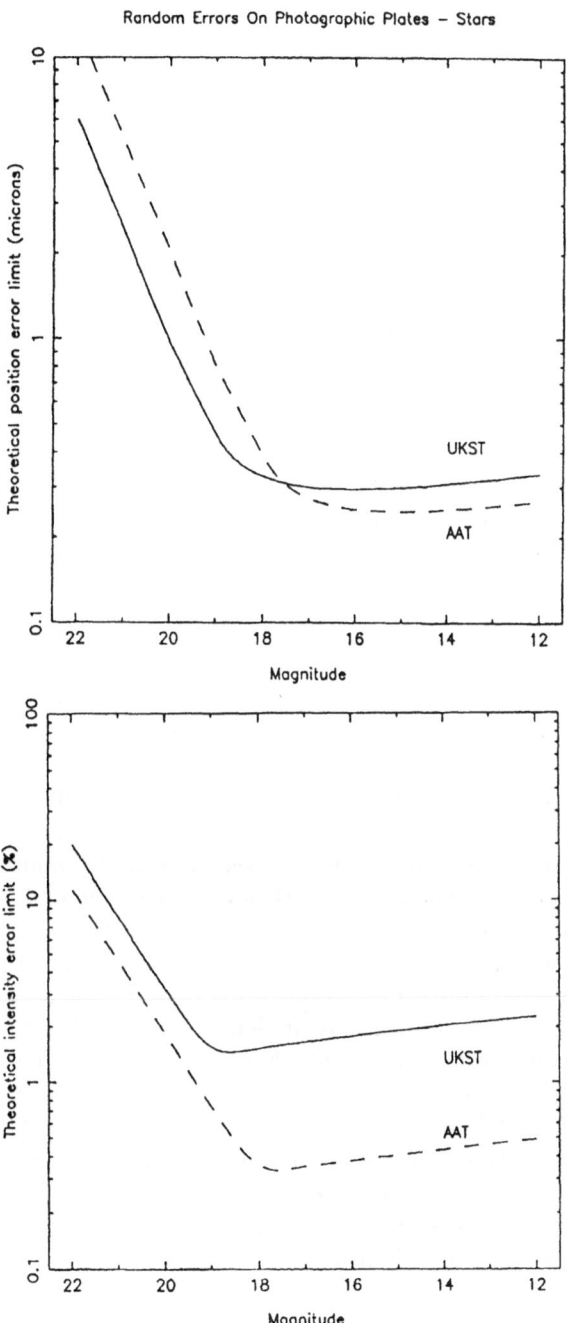

Fig. 1. Theoretical limits for attainable precision in measuring intensities and positions of stellar images on survey quality IIIa-J UKST and AAT plates.

The glass+emulsion is not a perfect elastic medium and it is impossible to predict to within a few microns what distortions will occur.

The telescope also introduces systematic distortions over the emulsion, both fixed pattern optical (e.g. coma) and dynamic (e.g. differential field rotation, differential refraction) up to several tens of μm in scale. Furthermore, temperature gradients in the plate of 1°C, at either telescope or measuring stage, also introduce distortions of a few μm across the plate, present particularly away from the plate centre. The combination of these factors results in a complex pattern of systematic non-linear emulsion distortions typically up to 10 μm in size with scale lengths ranging from 1-2 cm upwards. Fortunately, for virtually all astrometric projects, these systematic errors can be easily dealt with and removed. Any extra global systematic error up to a few μm introduced at the measuring stage is not a problem either, provided it is either fixed pattern or on sufficiently large scales (cm) such that it can be mapped and thereby removed.

3 Wide-Field CCD Camera

Any new wide-field survey instrument must be capable of covering a large area of sky with a small pixel scale, to take advantage of the best seeing conditions, and with a high quantum efficiency in order to provide accurate multi-passband photometry and astrometry to faint magnitudes. A CCD array camera system mounted at the prime focus of the 2.5m Isaac Newton Telescope (INT) on La Palma would not only satisfy these requirements but would enable a deep CCD-based sky survey to be carried out with an existing medium-sized telescope at modest cost. The combination of increased telescope aperture over the 1.2m Schmidt telescopes, the much better site (sub-arcsecond seeing, dark sky and clear nights), the use of high quantum efficiency digital linear detectors and recent developments in computer technology make this the ideal time to initiate a next generation sky survey. Table 1 compares the expected limiting magnitudes and photometric accuracy in several passbands for such a system with similar figures for the UKST. With realistic exposure times, typically 600s, it is feasible to survey several hundred square degrees of sky in multiple passbands to the desired accuracy and depth in a timescale of 1-2 years (assuming $\sim$ 25% of time on the INT was available for the survey). The proposed CCD camera would replace the current prime focus direct imaging system and would become one of the main detectors for the INT. By fitting a wide field CCD camera, the imaging capability of the INT would improve by a factor of 50 transforming the INT into a wide field imaging telescope. In the future the INT could then be oriented towards providing the first of the new generation of deep wide field quantitative surveys and therefore complement large telescopes, such as the William Herschel Telescope (WHT), rather than compete with them. The INT prime focus design philosophy is also compatible with the future possibility of a new wide field corrector (up to 2 degrees) for the INT which could give an additional factor of 10 in area available.

Table 1.1 Sample exposure times for INT.

Waveband	U	B	V	R	I
Sky Brightness (no moon)	22.0	22.7	21.8	20.9	19.9
(7 day old moon)	19.9	21.6	21.4	20.6	19.7
Exposure time (seconds)	900	900	600	600	600
Limiting magnitude (S/N)=5	24.0	25.0	25.0	24.5	24.0

Notes:
(1) Sky brightness estimates are from CTIO observers guide.
(2) S/N estimates assumes 1 arc second seeing and profile fitting.
(3) Limiting magnitudes are rounded to nearest 0.5 mag.

Table 1.2 Sample exposure times for UKST.

Waveband	U	B	V	R	I
Exposure time (minutes)	180	60	60	60	180
Limiting magnitude (S/N)=5	21.5	22.0	19.5	20.5	19.0

Notes:
(1) All exposures are sky limited (dark sky) including the I band.
(2) S/N estimates are based on APM measurements of UKST plates.

3.1 Camera Design and Prime Focus Modifications

Whilst the existing corrector on the INT is adequate, providing images 0.3 arcsec or better over the unvignetted field of diameter 40 arcmin for wavelengths in the range 3650 to 10140 Å, a new A&G unit, CCD-based autoguider and a filter wheel to take 125mm filters will be required. An advantage of a CCD-based autoguider is that it would be capable of routinely guiding on stars down to V $\sim$ 17.

The unvignetted field of the INT prime focus, with diameter 40 arcmin (0.35 deg^2), could be completely covered by an EEV-based CCD array of 4400 $\times$ 4400 pixels of size 22μm giving an image sampling of 0.55 arcsec in the 2.5m f/3.3 beam. Alternatively, a Ford CCD-based mosaic of 6144$\times$6144 pixels of size 15μm giving an image sampling of 0.37 arcsec could be used. Both of these pixel scale sizes are well matched to the median seeing conditions on La Palma, although in the best seeing conditions the Ford system would be preferable (see Table 2 for details). In practice not all of the 40 arcmin (97mm) field of view would be available for direct imaging due to the constraints of an autoguider assembly. CCD mosaics based on either 4 EEV 2186 × 1152 chips (22μm pixels) or 6 Ford 2048 × 2048 chips (15μm pixels) satisfy a number of mildly conflicting requirements such as collecting area, image scale, field of view and cost (see Fig. 2). Both systems would give an unvignetted area of $\sim$ 0.25 deg^2 a 50 fold increase of that currently available. This leaves an area of 0.06 deg^2 within which guide stars must be obtained. This is more than adequate, since the average number of stars brighter than V $\sim$ 17 at high galactic latitude available to the autoguider is $\gtrsim$ 20. Since digitised photographic data now exists for all the sky, both the guide stars and an optimum survey region could be preselected.

Table 2.1 CCD system based on EEV P88530 chips.

Pixel size (μm)	22.5
INT scale (arc seconds per pixel)	0.55
Pixel format	2186 × 1152
Physical size of imaging area per chip (mm)	49.2 × 25.9
(arc min)	20.2 × 10.7
Imaging area produced by 4 chips (arc min)	41.8 × 20.2
(deg^2)	0.24
Physical size of mosaic including 1mm gaps (mm)	106.7 × 49.2

Table 2.2 CCD system based on Ford chips.

Pixel size (μm)	15.0
INT scale (arc seconds per pixel)	0.37
Pixel format	2048 × 2048
Physical size of imaging area per chip (mm)	30.7 × 30.7
(arc min)	12.8 × 12.8
Imaging area produced by 6 chips (arc min)	38.4 × 25.6
(deg^2)	0.27
Physical size of mosaic including 1mm gaps (mm)	94.1 × 62.4

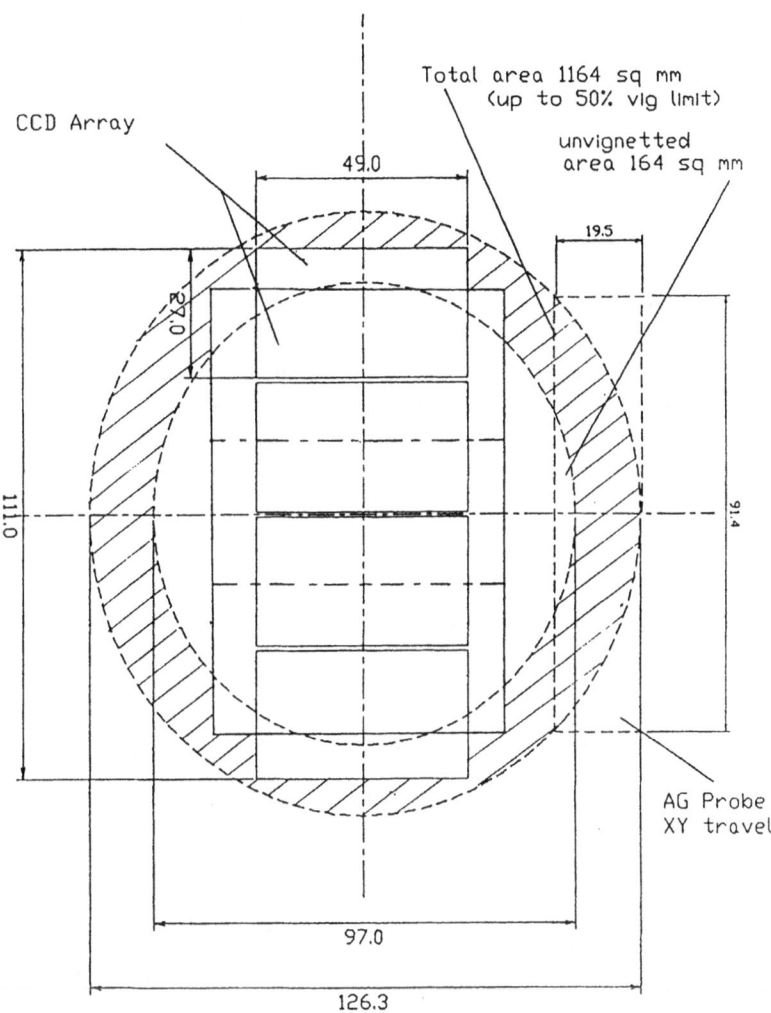

Fig. 2. An outline sketch of two possible CCD arrays mounted at the prime focus of the INT. The shaded area represents the transitional region between no vignetting and 50% vignetting.

Although the data rate per exposure would be much higher than a conventional CCD camera (up to 50 Mbytes), even with optimistic assumptions regarding exposure times and overheads a data rate per night (day) of **only** 5 Gbytes would be realised. For comparison, a UKST plate digitised at $\frac{1}{2}$" resolution corresponds to 4 Gbytes of data; whilst a measuring machine facility, such as the APM, routinely handles ~10 Gbytes of data per day. The data rates would therefore not be excessive. Consequently, in addition to merely archiving the data for later use, it would be feasible and indeed desirable to process it (e.g. to provide image lists and parameters) in pseudo–real time.

Assuming that initially only the B(dark time), R(grey time) and I(bright time) bands are used with exposure times of 900, 600 and 600 seconds respectively, an average overhead of 100 seconds per frame and an average night of 9 observing hours, the equivalent of three square degrees can be surveyed per night in all three bands. Therefore in ~50 scheduled nights >100 square degrees could be surveyed. If this seems an excessive amount of telescope time, bear in mind that current prime focus imaging takes up 30% of scheduled INT time.

4 Acknowledgements

A large number of people have been involved in a pilot study for a wide field CCD camera for the INT including: Brian Boyle, Mick Bridgeland, Pete Bunclark, Dave Carter, John Churchill, Richard McMahon and Roberto Terlevich. All of their contributions are greatly appreciated.

References

Dainty, J.C., Shaw, R., 1974. In *Image Science*, Academic Press.
Geary, J.C., Luppino, G.A., Bredthauer, R., Hlivak, R.J., Robinson, L., 1991. *Proceedings of The S.P.I.E.: CCDs and Solid State Optical Sensors II*. To be published.
Irwin, M.J., 1985. *Mon. Not. R. Astron. Soc.*, **214**, 575.
King, I., 1983. *Publ. Astron. Soc. Pacific*, **95**, 163.

Discussions

Lasker :
Your comments about the classification advantage of pixels far smaller than what one gets from Fourier arguments is based on what kind of classification?
Irwin :
The classification is based on the derived image shape parameters: moments, peak height, areal profiles. Fourier arguments are based on assuming approximately band-limited images. Whilst to first order this is correct, the plate noise has an essentially 'white noise' power spectrum and thereby causes measurable aliasing problems. Second order problems are caused by residual aliasing of the image and the rapidly varying intensity profile, since

$$\log_{10} < \text{transmission} >_{largepixel} \neq < \log_{10}\text{transmission} >_{co-addedsmallpixels} .$$

Westerhout :
In the light of this conference 'Digitised Optical Sky Surveys' it is worth noting that if you stick to the 'Sky Survey' magnitude limit of 21 mag you can do the entire POSS-II in five colours (not three) in <u>one</u> year, using existing technology and a £100 K investment!

PHOTOMETRIC CALIBRATION OF THE SOUTHERN SKY SURVEYS

S.J. Maddox and W.J. Sutherland

Department of Physics, Keble Road, Oxford OX1 3RH, U.K.

Abstract

We describe a project proposed by a large consortium of UK and Australian surveyors with the aim of obtaining accurate B and R photometric calibrations for all 270 UK Schmidt fields in the South Galactic Cap. This will have major benefits for all aspects of survey astronomy, including studies of the large-scale structure of the universe, galaxy evolution, quasar surveys, galactic structure, rare-object searches, and optical identifications of objects in surveys at other wavebands, such as the ROSAT, MIT/Parkes, and IRAS surveys.

1 Motivation

The UK Schmidt surveys scanned by APM and COSMOS are one of the most valuable existing databases for survey astronomy. Many projects have been successfully based on these surveys including searches for quasars (e.g. McMahon & Irwin 1991); searches for brown dwarfs (e.g. Hawkins 1991); optical identification programs using the MIT/Parkes survey (Savage & Wright 1991), the IRAS survey (Sutherland et al. 1991) and the ROSAT survey (Miller 1991); studies of galactic structure (e.g. Evans 1989); and large-scale galaxy surveys (e.g. Collins et al. 1989; Maddox et al. 1990a). Recently several other measuring machine groups have digitised large numbers of sky survey plates and are beginning to make use of the data as described in these proceedings.

All these surveys naturally require calibration to put the photographic magnitudes on a standard system, but to provide good calibration for each of the projects listed above would need many nights with a small telescope. Several groups have started systematic programmes to calibrate sky survey data, notably for the Space

H. T. MacGillivray and E. B. Thomson (eds.), Digitised Optical Sky Surveys 53–60.

Telescope Science Institute's HST Guide Star Catalogue. These systematic photometry programmes have concentrated on calibrating bright stars $B \lesssim 15$. Stars this
bright are heavily saturated on the deep Schmidt plates, and so these sequences
are of little use in calibrating the faint objects. So far there has been very little
systematic photometry of faint stars or galaxies. Probably the most extensive faint
photometric sequences currently available are those obtained to calibrate the large
galaxy surveys (Collins et al. 1989; Maddox et al. 1990a). Both of these surveys are
calibrated by $\lesssim 10$ objects on an average of less than 1 plate in 4. The equatorial
region wlith declination $-20° < \delta < 0°$ has virtually no faint calibrators at present.

In this project we intend to obtain a faint B and R sequence of ~ 50 galaxies
and 100 stars on every high-latitude survey field with $\delta < 0°$. This will provide
calibration accurate to ≈ 0.04 magnitudes for all of these fields. In addition to
the long-term value of providing uniform and accurate calibration for the Schmidt
data, there would be some dramatic **immediate** benefits. Two specific examples
are described in the next section.

2 Scientific Aims

From our point of view, some of the most exciting results from the Schmidt surveys have come from the analysis of large galaxy surveys. Two examples of the
great benefit that more accurate photometric calibration would provide are: (1) the
number of galaxies counted as a function of magnitude, and (2) the more precise
measurement of galaxy clustering on large scales.

2.1 Galaxy Number Counts

The number of galaxies counted in the APM Galaxy Survey as a function of magnitude suggests that there has been extremely strong evolution in the galaxy luminosity function since $z \sim 0.1$ (Maddox et al. 1990b). The slope of the number
counts between magnitudes $b_j = 16$ and $b_j = 20$ is close to Euclidean so there are
between 50% and 100% more galaxies at $b_j \sim 20$ than predicted by simple models of
galaxy evolution. The uncertainty in these galaxy number counts is dominated by
systematic errors in the photometric calibration. Our current calibration is accurate
enough to demonstrate convincingly that there is a large discrepancy between the
observations and simple galaxy evolution models, but the small uncertainties in the
calibration lead to a significant uncertainty in the observed excess over the models. The proposed extra calibrations would greatly improve the accuracy of these
measurements and so give precise measurements for modellers to aim for.

2.2 Galaxy Clustering

The angular autocorrelation function $w_{gg}(\theta)$, of galaxies in the APM Galaxy Survey
(Maddox et al. 1990a) shows more large-scale clustering than expected from the
standard Cold Dark Matter (CDM) cosmology (Davis et al. 1985). We have used
several tests to show that our $w_{gg}(\theta)$ measurements are not significantly affected by

systematic errors in the selection of galaxies for the survey. For example, in Fig. 1 we show the effect of subtracting the estimated error correlation function, $w_{err}(\theta)$, from the measurements of $w_{gg}(\theta)$ for APM galaxies in 6 independent magnitude slices. Although $w_{err}(\theta)$ is $\sim 2 \times 10^{-3}$ at $\theta = 0$, it rapidly falls to zero at larger angles, and so leaves the measured $w_{gg}(\theta)$ virtually unchanged.

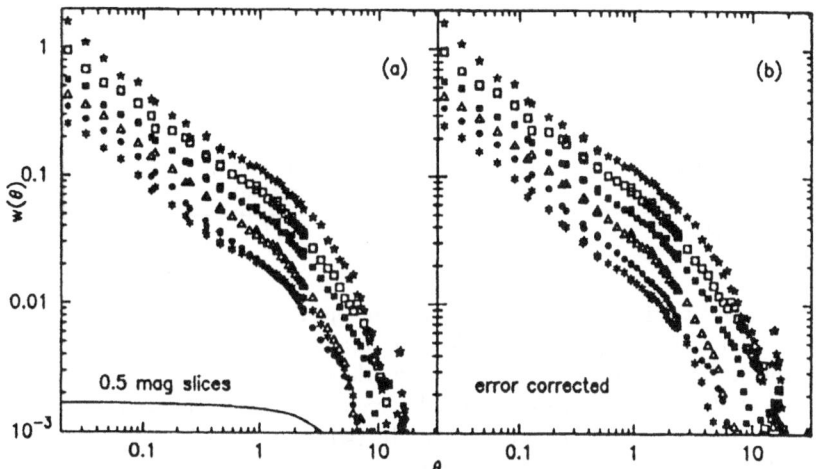

Fig. 1. Measurements of $w_{gg}(\theta)$ for galaxies in 6 independent magnitude slices from the APM survey; **(a)** before correction; **(b)** after subtracting the error correlation function. The solid line in **(a)** shows our estimated error correlation function. The differences between $w_{gg}(\theta)$ measured for galaxies in different magnitude ranges are consistent with the predictions from Limbers equation.

An even more convincing demonstration of the robust nature of our measurements is given by considering the possibility that all of the clustering signal seen for the faintest galaxies, $w_{20.5}(\theta)$, is caused by errors in the survey. The only way that errors could be larger is for them to anticorrelate with the true galaxy distribution, which is very improbable. Although using $w_{20.5}(\theta)$ represents a gross over-estimate of $w_{err}(\theta)$, it is useful to demonstrate that even the maximum possible errors do not significantly alter our conclusions. Figure 2(a) shows our six measurements after scaling to a common effective depth using the relevant factors from Limbers formula. All six slices show more clustering than the CDM prediction. In Fig. 2(b) we have subtracted $w_{20.5}(\theta)$ from the measurements for each of the 5 brighter slices, and then scaled them to the same depth. The fainter 2 of the remaining 5 measurements are much reduced by the over-correction, but the brighter 3 measurements are relatively unchanged, and still show large-scale clustering in excess of that predicted by the standard CDM model.

Using these and several other tests we are convinced that our measurements represent the best measurement of clustering at $\sim 50\,h^{-1}$Mpc scales. However, some sceptics remain unconvinced by these rather sophisticated internal error estimates

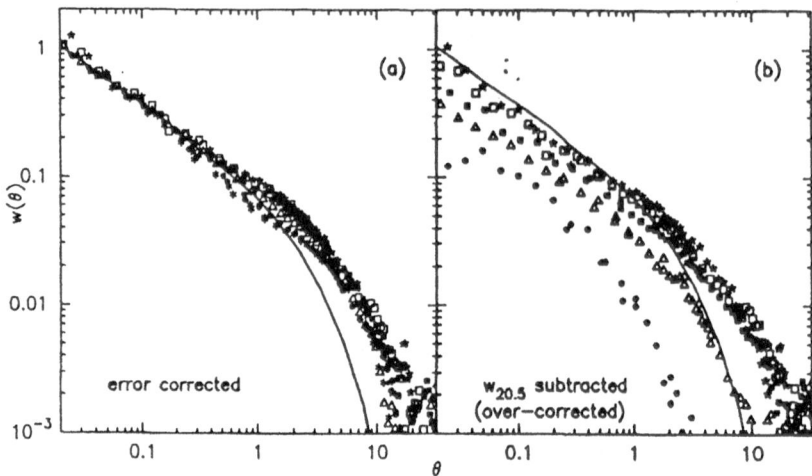

Fig. 2. Measurements of $w_{gg}(\theta)$ for galaxies in 6 independent magnitude slices from the APM survey after scaling to the depth of the Lick catalogue. **(a)** best estimate correction; **(b)** after subtracting $w_{20.5}(\theta)$. The solid line shows the prediction from the standard CDM model (Davis et al. 1985). In **(a)** there is good agreement between the six measurements, and all show strong clustering at large scales. In **(b)** the fainter 2 of the remaining 5 measurements are much reduced by the gross over-correction, but the brighter 3 measurements are relatively unchanged, and still show large-scale clustering in excess of the CDM prediction.

(e.g. Hale-Sutton et al., these proceedings). Our measurements would be given a much firmer basis by more calibrations to provide a better external check on the uniformity of our photometry. The improved large-scale calibration would not only provide confirmation of our current error estimates, but would also allow us to improve significantly over the edge-matching correction procedure currently used to produce uniform photometry. This would reduce the present matching errors and give even more precise clustering measurements.

2.3 Other Projects

The proposed observations will provide photometry and colours more accurate than possible from Schmidt plates for all images brighter than 21st magnitude in 20 square degrees of 'randomly sampled' sky. These data would be immensely useful, independent of the aim of improved calibration. A few of the interesting projects that would be possible with the new data include the study of: (a) Luminosity profiles for several hundred moderately bright galaxies, unbiased by the Local Supercluster; (b) Accurate colours for $\sim 10\,000$ galaxies to $B \sim 21$ which would provide very interesting tests of galaxy evolution; (c) Accurate colours for $\sim 20\,000$ stars well distributed at high galactic latitude, which would constrain models of our galactic disc; (d) The possible detection of rare objects via extreme colours, to a significantly fainter limit than existing surveys from Schmidt plates.

3 Proposed Observations

3.1 Requirements and Possible Strategies

The choice of B and R filters is clear: two-colour data is required to transform CCD to photographic magnitudes, and these bands are well matched to the b_J and r_F bands of the main UK Schmidt surveys. Some I frames will also be taken for selected fields, including those possessing A-grade I plates. To ensure linearity and provide a zero-point accurate to $\lesssim 0.04$ magnitudes for each field needs $\gtrsim 50$ images brighter than $B \sim R \sim 20$ with each measurement accurate to ~ 0.05 magnitudes. This needs CCD observations of $\gtrsim 100$ sq. arcmin on each of the 270 fields for the south galactic cap.

For optimal calibrations, it is clearly desirable to obtain a B and R frame on every UKST plate. However, if less time is available, a chessboard sampling will be adopted to ensure that each uncalibrated field overlaps with several calibrated neighbours. The limitation to high-latitude fields is not a major drawback since the large majority of extra-galactic survey work is carried out in this area; for galactic work at low latitudes the density of stars is high enough that small CCD frames should provide an adequate number of stars.

There are several strategies that would eventually achieve the desired photometry, and several possibilities are listed below:

1. Use a 1 meter with a standard CCD. This would cover typically just $3' \times 3'$ per CCD frame, each of which would require $\sim 3600s$ in B, and $\sim 1800s$ in R to obtain the required depth and accuracy. The total time needed is then approximately 1000 nights, which is not a feasible option.
2. Use a 1 meter with a large-format CCD. With a suitable image scale this could cover $16' \times 16'$ per CCD frame, but the integration times would still be $\sim 3600s$ in B, and $\sim 1800s$ in R. The total time needed is then approximately 50 nights. This represents a very generous allocation but may not be totally impossible. However, waiting for a large CCD on a 1-meter and then obtaining ~ 50 photometric nights, would mean that the project would not be finished until ~ 1995.
3. Use the AAT with the Thomson CCD. This would cover $16' \times 16'$ per CCD frame, with integrations $\sim 240s$ in B, and $\sim 90s$ in R. The total time needed is then only 5 nights.
4. Build a special purpose survey telescope (cf. the Princeton/Chicago proposal). Obviously this will work, but the cost is high, and the delay in waiting for the telescope would be many years.

We believe that option (3) provides the most efficient and quickest route to providing good calibration for the southern sky. Therefore we have organised a large consortium of interested UK and Australian surveyors and requested AAT time from both PATT and ATAC. We feel that it is clearly worth 5 nights of AAT time to make the best of 10 years UK Schmidt telescope time.

Parallel proposals to the SAAO 1-meter and the JKT form an essential part of the project. The aim of these is to set up a secondary sequence of intermediate brightness ($\sim 17^{th}$ mag) stars for as many Thomson frames as possible. These will serve as a check on the Thomson photometry and also provide the calibration in the case of non-photometric weather.

3.2 Technical Points

The basic aim of the project is to measure the zero point for each plate with an uncertainty < 0.04 mag, more accurate than the current APM and COSMOS overlap matching. This requires > 30 galaxies and stars per frame with $B < 21$ with random errors < 0.1 mag (i.e. substantially better than the 0.2 magnitude errors from Schmidt plates). This requires an area of > 100 sq. arcmin per plate; hence the need for the f/1 system. We anticipate 4 minute exposures in B and 1.5 minutes in R, to provide ~ 0.05 mag photometry at $B = R = 20$.

We are aware that the f/1 system suffers from chromatic aberration in the B band, but this is a minimal drawback. Spot diagrams show that this is much less serious than the effect of $2''$ seeing. Since we need only a total magnitude for each object we can use a fairly large aperture in the image analysis. This will increase the sky noise a little, but as long as our random errors per object are well below the 0.2 mag of the Schmidt plates, this is no disadvantage. For a star with $R \simeq 20$, we calculate that the S/N will be 30 in a (conservative) $5''$ diameter aperture, and for a galaxy in a $10''$ aperture the S/N will be 15.

Flat-fielding of Thomson data using the median of many frames is accurate to 0.3% according to Boyle who has commissioned the instrument. Also, as a secondary check, we can measure (CCD–Schmidt) magnitude as a function of CCD position for all 20 000 objects. The Schmidt errors will average out over the 270 plates so we will have a **very** accurate test of CCD systematics across the frame. We have also requested commissioning frames for a mosaiced area to provide tests of photometric accuracy; these will enable us to choose an optimal observing strategy in advance of the run.

The readout overhead is 2×40 seconds per field, but this is compensated by the short slews between frames and the fact that no acquisition and guiding is needed, so observing efficiency should be quite high and the project can be finished in 5 nights (divided as 3 nights PATT and 2 nights ATAC).

The recent proliferation of workstations means that data reduction will not cause serious delay in reducing the data. In fact we have two sub-groups (Sutherland and Maddox at Oxford, and Hale-Sutton and Metcalfe at Durham) who plan to reduce the data independently and perform comparisons. Objects will be detected by an automated package and overlaid on the frame on a graphics device; the only essential operator intervention is to pick out the small percentage of images affected by blending, ghosts, etc. and flag them as dubious in the final catalogue.

4 Summary and Postscript

We plan to use the AAT to obtain a $16' \times 16'$ Thomson CCD frame in B and R near the centre of every high latitude southern survey Schmidt field. Also some I

frames will be taken. This will provide photometry accurate to $\lesssim 0.05$ magnitudes for ~ 100 stars and ~ 50 galaxies to calibrate each field. For as many Thomson frames as possible we will obtain a secondary sequence of intermediate magnitude stars ($\sim 17^{th}$ mag) using the SAAO 1m and the JKT. These will serve as a check on the Thomson photometry and also provide the calibration in the case of non-photometric weather at the AAT.

Since the date of this meeting, we have heard the decisions of the time allocation committees: We were allocated 1 week of time on each of the JKT and the SAAO 1m. Due to the uncertainties with the Thomson B-band response, neither PATT nor ATAC awarded AAT time for the project, but the situation looks promising for next year when a coated blue-sensitive chip will be available.

5 Acknowledgements

This proposed project has benefitted from the support of many active sky-surveyors including: G. Efstathiou, A. Savage, R. Cannon, A. Wright, C. Collins, M. Hawkins, H. MacGillivray, L. Miller, R. Nichol, D. Evans, M. Irwin, D. Lynden-Bell, S. Raychaudhury, B. Boyle, R. McMahon, R. Ellis, C. Frenk, D. Hale-Sutton and N. Metcalfe.

References

Collins, C.A., Heydon-Dumbleton, N.H., MacGillivray, H.T., 1989. *Mon. Not. R. Astron. Soc.*, **236**, 7p.

Davis, M.D., Efstathiou, D., Frenk, C.S., White, S.D.M., 1985. *Astrophys. J.*, **292**, 371.

Evans, D.W., 1989. PhD thesis, Cambridge University.

Hawkins, M.R.S., 1991. In *Measuring Machines Newsletter*, No. 13.

McMahon, R., Irwin, M., 1991. In *Measuring Machines Newsletter*, No. 13.

Maddox, S.J., Efstathiou, G., Sutherland, W.J., Loveday, J., 1990a. *Mon. Not. R. Astron. Soc.*, **242**, 43p.

Maddox, S.J., Sutherland, W.J., Efstathiou, G., Loveday, J., Peterson, B., 1990b. *Mon. Not. R. Astron. Soc.*, **247**, 1p.

Miller, L., 1991. In *Measuring Machines Newsletter*, No. 13.

Savage, A., Wright, A., 1991. In preparation.

Sutherland, W.J., Maddox, S.J., McMahon, R.G., Saunders, W., 1991. *Mon. Not. R. Astron. Soc.*, **248**, 483.

Discussion

Schuecker :
    Do you think that one photometric CCD sequence per field is enough to reach an accuracy of $0.^m05$? I think about problems connected with field effects.

Maddox :
    Each sequence would contain ~ 100 stars and ~ 50 galaxies which should tie down the zero point for the $17' \times 17'$ area of each frame.

Obviously you need to correct for field effects if you want to use the zero point over the whole Schmidt plate. The expected errors in field correction are around the 0.05 mag level.

THE SECOND HST GUIDE STAR PHOTOMETRIC CATALOG

M. Postman [1], L. Siciliano [2], M. Shara [3], D. Rehner [3], N. Brosch [4], C. Sturch [5], B. Bucciarelli [5] and C. Lopez [6]

[1] Space Telescope Science Institute, 3700 San Martin Drive, Baltimore MD 21218, U.S.A.
[2] Cerro Tololo Inter-American Observatory, Casilla 603, La Serena, Chile.
[3] Astronomy Programs, Computer Sciences Corporation, Space Telescope Science Institute, Baltimore MD 21218, U.S.A.
[4] Wise Observatory, Tel Aviv University, Ramat Aviv, Tel Aviv 69978, Israel.
[5] Osservatorio Astronomico di Torino, Strada Osservatorio 20, I-10025 Pino Torinese, Italy.
[6] Observatorio Astronómico 'Felix Aguilar', Av. Benavídez 8175 Oeste, Chimbas, 5413 San Juan, Argentina.

1 Introduction

The second generation Hubble Space Telescope Guide Star catalog (GSC-II) is currently under construction. GSC-II will include both stars and galaxies down to the plate limits of the 1950 POSS E plates and the SERC J plates (V ≈ 21). In addition, multi-bandpass and multi-epoch data will be added as current epoch surveys in both hemispheres become available in digitized format. In order to provide accurate photometric calibration for GSC-II, the Space Telescope Science Institute is extending the BV photoelectric photometry in the Guide Star Photometric Catalog (GSPC-I) to a fainter V band limit and adding R band data. The new catalog, hereafter GSPC-II, will have 5% photometry in the V and R passbands to a limiting magnitude of V = 21. The survey is being conducted using CCD direct imaging cameras at observatories in both the northern and southern hemispheres (see Table 1).

The northern hemisphere program ($\delta \geq -6°$) consists of 651 target fields, each one centered on the faintest star in the original GSPC-I sequence. As in GSPC-I, GSPC-II will cover 0h to 24h. The fields include the 583 GSPC-I fields corresponding to the original 583 POSS fields. In addition, we are including the 72 GSPC-I fields at the equator corresponding to the 72 plates of the SERC-J Extension. The

H. T. MacGillivray and E. B. Thomson (eds.), Digitised Optical Sky Surveys 61–63.

Table 1. List of GSPC-II Survey Observatories.

Northern Hemisphere survey

	Observatory	Telescope	Coverage Region
Primary	Wise Obs.	1.0 m	+00 to +60 deg DEC
	CTIO	0.9 m	-06 to +18 deg DEC
	Megantic Obs.	2.0 m	+60 to +90 deg DEC
Secondary	McDonald Obs.	2.0 m, 0.8 m	Assorted Northern Fields
	San Diego State Univ.	1.0 m	Assorted Northern Fields
	KPNO	2.1 m, 0.9 m	Assorted Northern Fields
	Lowell Obs.	1.1 m	Assorted Northern Fields

Southern Hemisphere Survey

	Observatory	Telescope	Coverage Region
Primary	CTIO	0.9 m	-30 to -06 deg DEC
	ESO	0.9 m	-90 to -30 deg DEC

GSC-II will also be based, in part, on the new Palomar survey of the northern sky, now in production. This survey is based upon 5-degree plate centers instead of the 6-degree centers of the Quick-V Palomar survey used in the construction of the original Guide Star Catalog. Consequently, there are about 150 new POSS plate centers which will not have adequate photometric sequences. These sequences will be obtained upon completion of the survey of the primary 651 fields.

A complementary southern photometric survey ($\delta < -6°$), in collaboration with investigators from OATo (Torino), to obtain 5% BVR calibration exposures is being initiated this year at both CTIO and ESO. The 894 southern fields will be centered on the plates of the SERC Southern Sky Survey. Approximately 100 of these fields have already been observed by other programs leaving 800 fields which will require new observations.

2 Survey Strategy and Status

The availability of long term observing status on several 1 – 2m class telescopes makes this program possible. The typical observing run lasts between 7 – 14 days. This virtually assures that no run is completely lost to poor observing conditions.

For each field, a short exposure (typically 120 sec on a 1m class telescope) and a long exposure (typically 1200 sec) in each passband are acquired. The short exposures provide accurate calibration down to V = 18 mag. The long exposures get us to V = 21 mag. Long exposures are usually bracketted by the short exposures. Every star detected in the CCD frames with photometric errors less than 5% will be included in GSPC-II. About 10% of the northern fields are observed from 2 different sites to provide an estimate of any relative zero point offsets. The northern hemisphere short exposure survey is 84% complete (548 out of 651 fields have been observed). The long exposure survey is 44% complete. It is estimated that the short exposure survey will be 95% complete by the end of 1992.

The southern surveys are just getting underway. We have been allocated a total of 46 nights on the CTIO 0.9m over the next two years to cover the $-12° \geq \delta \geq -30°$ zone. We have also been awarded long term status at ESO on the Dutch 0.9m telescope to observe fields south of $\delta = -30°$. It is estimated that the southern survey will be at least 50% complete by mid 1993.

A supplemental observing program will also begin this fall to obtain I band calibration frames in the northern hemisphere.

DIGITISED OPTICAL SKY SURVEYS

Part Two:

DIGITISATION PROGRAMMES — CURRENT AND PLANNED

THE COSMOS/UKST CATALOG
OF THE SOUTHERN SKY

D.J. Yentis [1], R.G. Cruddace [1], H. Gursky [1], B.V. Stuart [1], J.F. Wallin [1], H.T. MacGillivray [2] and C.A. Collins [2]

[1] Department of the Navy, Naval Research Laboratory, Washington DC 20375, U.S.A.
[2] Royal Observatory, Blackford Hill, Edinburgh EH9 3HJ, U.K.

Abstract

We have prepared an 'object' catalog of the Southern sky south of +2.5 degrees Declination from COSMOS scans of the IIIa-J and Short Red surveys taken with the UK Schmidt Telescope. The catalog consists of upwards of 500 million objects down to the limit of the plates. A version of the catalog in compact form is available for distribution, and a database management system has been produced which allows rapid access to any part of the catalog. We also describe the creation of a catalog of clusters of galaxies derived from the galaxies appearing in the object catalog. The cluster catalog contains some 70 000 candidate clusters in the Southern hemisphere and forms the basis for further major programs of follow-up study.

1 Introduction

The collaboration between the Royal Observatory, Edinburgh (ROE) and the Naval Research Laboratory (NRL) to further develop and utilize data from COSMOS processing of the U.K. Schmidt Telescope (UKST) Southern sky survey has now completed its second year. As reported earlier by Gursky (1990), this effort has led to the creation of a complete optical object catalog of the Southern sky. The catalog is in use at the Max Planck Institute for Extraterrestrial Physics (MPE) to find optical counterparts to X-ray sources detected by the ROSAT all-sky survey satellite (see Voges, these proceedings). The catalog has also recently been installed at the Infra-red Processing and Analysis Center (IPAC) to enable optical identification of IRAS Faint Source Survey objects (see Wolstencroft et al., these proceedings). ROE and NRL are also developing a computer-selected catalog of clusters of galaxies for the entire Southern sky. In this paper we elaborate the details of the data base management system (DBMS) developed for the object catalog and the techniques being used to create the cluster catalog.

H. T. MacGillivray and E. B. Thomson (eds.), Digitised Optical Sky Surveys 67–75.

2 The COSMOS/UKST Object Catalog

The Object Catalog combines COSMOS data from scans of UKST Southern sky survey plates performed at ROE with data compression tools and a data-base management system (DBMS) developed at NRL. The UKST survey is organized into 894 6 × 6 square degree fields on 5 degree centers. The data in the catalog are obtained from glass copies of the ESO/SERC blue survey plates, J and EJ, covering the fields from declination -90 degrees to +0 degrees but excluding the region within 10 degrees of the Galactic plane and the Magellanic Clouds; the excluded regions are covered by the UKST Short Red survey (cf. UKSTU Handbook, issued 1983).

A brief description of COSMOS Image Analysis Mode (IAM) processing and the resulting IAM data file will help to make clear the design of the DBMS to access the Object Catalog. A detailed discussion of the COSMOS plate scanning machine and data processing modes can be found in MacGillivray & Stobie (1984). Each plate is aligned on the COSMOS digitizing table such that a local plate coordinate system is defined with the X-axis increasing in sky coordinates west to east and Y-axis increasing north to south. The plate is scanned in lanes by drifting the digitizing table 5.4 degrees in Y while the digitizing spot is scanned 2.3 arcmin in X. There are a total of 140 lanes. A COSMOS scan covers an area 5.4 × 5.4 square degrees of each UKST field with the scan center offset about 20 arcmin west of the field center. The scan pixel size is 16 microns (about 1 arcsec) and only those pixels with intensity above a preset local sky background threshold are recorded. Machine transmission to relative intensity conversion is carried out using spot sensitometer values present on the plates, a Look-up table being created at the start of the measurements. Image detection software joins adjacent pixels together to produce the list of objects. A crowded-field analysis (Beard et al. 1990) is applied to all the pixel information, and in regions containing overlapping images this works highly efficiently to deblend the parent pixel distribution into individual daughter components. For each object, 32 parameters are calculated including the centroid in both plate and sky coordinates, the axes and orientation of a fitted ellipse, the instrumental magnitude, and, for J survey plates, the star/galaxy image classification down to a limiting magnitude $b_j \sim 21$ (Heydon-Dumbleton et al. 1989).

Calibration of the entire data set is performed separately for galaxies and stars. For galaxies, we use the existing CCD sequences for the Edinburgh Durham Galaxy Catalogue, together with plate overlaps to extend the calibration over the remaining sky. For stars, a direct calibration is obtained from the Guide Star Photometric Catalog using extrapolation to extend the calibration to the fainter stars. Further details will be given in Wallin et al. (in preparation).

The IAM data file consists of a housekeeping header block containing plate and COSMOS information followed by the object data blocks in the order in which the plate was scanned and with the lane boundaries preserved. The total storage requirement for the UKST IAM data in the catalog is about 75 gigabytes. Prior to being archived to optical disk at ROE, the IAM data are compressed a factor of two into a new dataset referred to as the IAMC. The compression is totally without loss of information and is achieved by using scale factors, offsets, and the elimination of leading zeros. An even more compact version of the catalog, the Portable Compact Data Subset (PCDS), contains 11 image parameters derived from the IAMC by eliminating most of the COSMOS-dependent parameters (plate coordinates, etc.) and reformatting others. Celestial coordinates and magnitude information are retained to the precision of COSMOS processing. The precision of some parameters (ellipse, image area, etc.) is compromised by bit truncation while dynamic range is maintained by using logarithmic formats. Flag values are summarized where appropriate. The IAMC and PCDS files contain fixed length, unformatted records. The data structure of the IAMC is preserved in the PCDS so that both formats, as will become clear below, can be accessed by the same DBMS. The PCDS is a factor of four more compact than the IAMC and is the preferred format for most analyses and for distribution. The PCDS is the version which is currently installed at MPE and IPAC. The catalog consists of upwards of 500 million celestial objects in the entire Southern Hemisphere. Table 1 summarizes the currently available data.

Table 1. Object catalog vital statistics.

UKST Survey	Plates Processed	Size (GB) IAMC/PCDS	Class	Objects (10^6)	Per Plate	%
blue J and EJ	732(*)	24.0/6.0	galaxy	37.1	50700	11
			star	185.6	253500	52
			faint	106.6	145600	30
			'junk'	24.3	33100	7
short red	148(†)	10.4/2.6	no ID	201.1	135900	100

Notes:
* At the time of this meeting, 2 remained to be processed; the total includes 37 B-grade plates.
† 10 remain to be processed.

We chose to build a customized hierarchical DBMS rather than use a commercial relational DBMS. This decision was based on a number of considerations. First, the primary use for the DBMS would be position correlation queries; that is, extraction from the catalog of all objects in a two-dimensional region centered on a specified position on the sky. Relational DBMS generally performs poorly on this type of query. Second, the dataset is large and will no doubt become larger; it would be optimum, therefore, for the DBMS to operate on as compact a version of the dataset as possible. In addition, it is important that certain pieces of information, not necessarily explicit in the object parameters, be maintained for each object. For example, the object/plate association must be kept to allow for magnitude calibration or plate reprocessing or to identify duplicate detections of an object (which are not being eliminated from the dataset) in plate overlap regions. Also, there is sometimes the requirement to extract the parent object in a region of image overlap so that deblended image parent/daughter groups must be preserved. In developing relational DBMS tables for the catalog and in accommodating the non-standard data types in the compact formats and the other data requirements, we would have increased the total storage requirements for the catalog (about a factor of two for the PCDS) and degraded DBMS performance. Finally, a considerable amount of data analysis software had already been developed to use the compact IAMC and PCDS data formats and we did not wish to create and maintain yet another data format.

The DBMS accesses unmodified IAMC/PCDS data files and can accommodate any plate processed in COSMOS IAM mode. The DBMS uses a simple search hierarchy which takes advantage of the COSMOS lane-scan geometry. As part of the IAMC to PCDS compression step for each plate, lane and lane/block extent tables are constructed which contain, respectively, the minimum and maximum X values of each lane, and the minimum and maximum X and Y values of each data block determined from the actual object plate coordinates included in each. These tables are saved, with the COSMOS housekeeping block, in a master header file. The header file can accommodate as many plates as necessary, including multiple plates for a survey field. A header file exists for each survey. The prescription for satisfying a query of the catalog is now very simple. Given a region centered on a position (α, δ), first determine which survey fields contain any part of the region of interest. Next, for each selected field in each survey in turn, convert the position of interest to plate coordinates using the appropriate plate solution stored in the header file and determine, by examining the X,Y extent tables, firstly which lanes contain possibly relevant objects and secondly, which data blocks in selected lanes contain possibly relevant objects. Lastly, read the selected data blocks (random-access from disk or sequentially from tape), test each object in the block for actual inclusion in the region of interest and, optionally, filter for classification, magnitude, etc. Note that a successful query requires that only the IAMC and/or PCDS data files for the plates containing the region of interest be available; this allows possibly limited disk storage space to be used efficiently.

The DBMS is written in standard Fortran-77 and is installed on Digital Equipment Corporation VAX/VMS clusters (server and workstations) at NRL, ROE, and MPE. A partial port to a SUN based UNIX system is installed at the Infrared Processing and Analysis Center (IPAC). The IAMC dataset exists at ROE on optical disk and at NRL on Exabyte 8mm tape. The PCDS dataset exists at ROE, NRL, and IPAC on 8mm tape and at MPE on optical disk. UKST plates on the 8mm tape distribution are sorted in order of increasing ecliptic longitude to facilitate correlations with satellite based all-sky surveys. An interactive front-end to the DBMS has been written which produces graphic finding charts and object lists for specified positions. An example of a finding chart is shown in Fig. 1b. An automated search mode also exists which reads an extraction position and radius from a list, extracts objects from the catalog, sorts them in order of increasing distance from the extract position, and writes the results as a compressed binary data file which can be analyzed later. The automated search mode is being used at MPE in the optical identification of ROSAT survey X-ray sources, and at NRL in support of the IPAC program to identify some 400 000 southern-sky infrared sources in the IRAS Faint Source Survey list (Lonsdale 1990). The status of both these efforts is reported elsewhere in these proceedings (Voges; Wolstencroft et al.). Other sites wishing to use the catalog can make requests to the ROE or NRL.

We are currently looking at the possibility of a CD-ROM distribution of the DBMS software and a magnitude-limited subset of the PCDS format data. We are also exploring remote access to the catalog. For example, we have demonstrated the interactive finding chart generation software using the X Window System over the TCP/IP Internet between the east and west coasts of the United States and have found network performance to be quite acceptable. We are also evaluating the Astrophysics Data System (ADS), one of the more ambitious network based data systems under development by NASA. The ADS is designed to provide a 'common query' user interface to distributed astronomical data bases.

3 The ROE/NRL Cluster Catalog

The study of the properties and distribution of clusters of galaxies as a means of understanding large-scale structure in the universe requires a catalog of clusters as complete and as unbiased as possible. The best available cluster catalogs to date (for example, Abell et al. 1989) have relied on visual measurements on photographic plates. No matter how carefully prepared, these catalogs are necessarily limited by subjective biases that result in a non-uniform and incomplete sample of clusters. We have developed an automated procedure for the detection, characterization, and analysis of clusters of galaxies using data from the COSMOS/UKST Object Catalogue.

The automated search for clusters is carried out in two phases:- cluster detection and cluster screening and analysis. In the first phase, a cluster data base (CDB) of candidate cluster groups is developed; in the second phase, the CDB is reduced to a catalog of individual clusters. Two distinct computer algorithms, 'binning' and 'percolation', are applied to identify candidate clusters in each field. The binning technique detects the overdensity of galaxies above the background, while the percolation technique assembles groups of nearest neighbor galaxies. Detection thresholds, although constrained by the galaxy statistics of each field, are set deliberately low. More sophisticated techniques are used in the screening phase to select individual clusters from among the candidates and determine most of the standard cluster characterization parameters found in other catalogs.

The binning algorithm is that of Dodd & MacGillivray (1986). In this method, clusters are detected by counting the number of galaxies in fixed size bins over an entire plate and selecting those that exceed the local galaxy background density by some fixed amount. Adjacent selected bins are then joined and deblended into cluster candidate objects using the COSMOS image analysis software. For each candidate, the centroid, fitted ellipse, maximum radius, and number of galaxy members are determined and saved in the CDB. The algorithm is applied using lower and upper limit bin sizes of 2.5′ and 5.0′ respectively. A bin size of 2.5′ contains

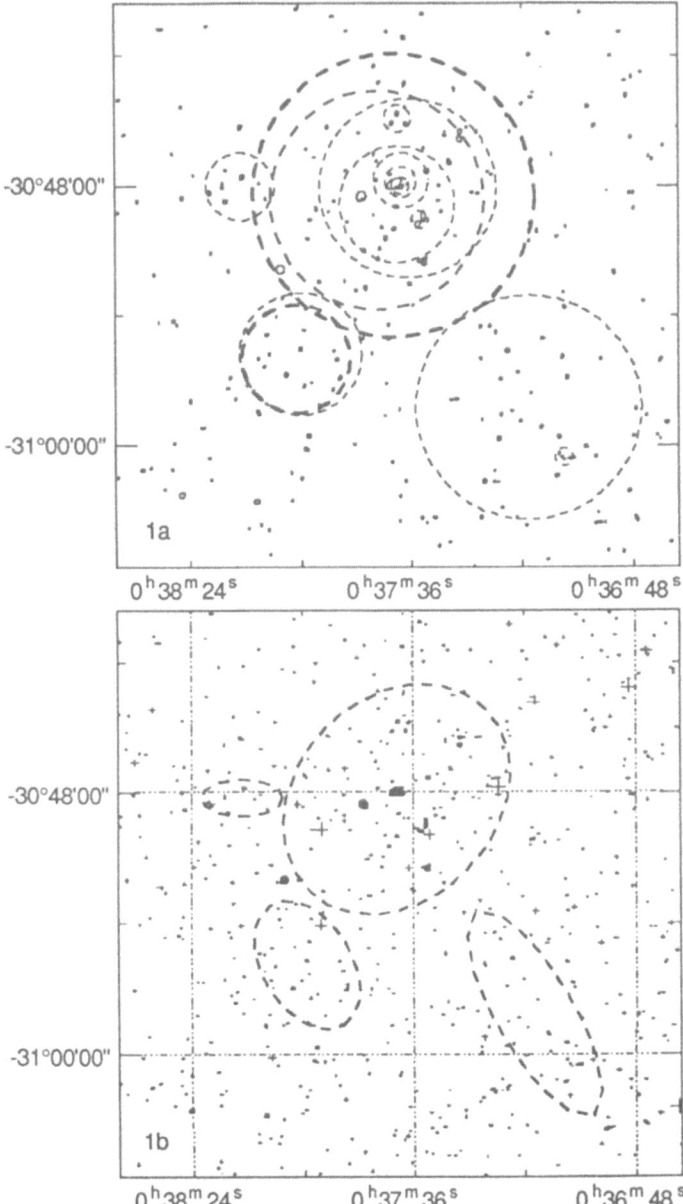

Fig. 1. a) Cluster candidate groups and galaxies in a region of field 411 near the South Galactic Pole. Cluster candidates detected by the binning and percolation techniques are shown as bold and light dashed circles respectively. **b)** A finding chart for the area in Fig. 1a generated from the Object Catalog. Stars are shown as crosses equal in size to the plate image and galaxies as ellipses. The dashed ellipses are the resolution of the cluster candidate groups in Fig. 1a into individual clusters.

about one galaxy on average. Bin sizes smaller then 2.5′ are dominated by Poisson statistics and result in too many zero bins and possible fragmenting of a cluster image. Bin sizes much larger than 5′ tend to remove detail in the galaxy distribution and result in inaccurate cluster position determination. Gaussian smoothing is applied to the binned galaxy distribution to reduce the effects of Poisson noise and allow a lower threshold for a 3 sigma detection in a bin. An estimate of the background galaxy distribution is determined by applying to the smoothed binned data a succession of three weighted median filters spanning from 50′ down to twice the bin size. This procedure effectively removes any structure (i.e. clusters) from the map smaller than 50′. The local background is subtracted from each galaxy bin and those bins above a specified threshold are saved. The background thresholds corresponding to 3 sigma detections in a single bin are 2 and 4 galaxies per 2.5′ and 5.0′ bin respectively.

The percolation algorithm begins with an arbitrary galaxy that is not a member of a candidate cluster group and selects all its neighbors within the percolation radius, R. It then proceeds recursively to include all neighbors of these neighbors and so on until no more neighbors are found. The center, radius and group size (the number of galaxies detected in the group) are determined and recorded in the CDB. Each field is processed with percolation radii between 0.4′ and 1.2′ in steps of 0.2′. The range of R must be very near the average separation of galaxies in the field; if much less, nothing will percolate, if much greater, everything will percolate. Monte Carlo simulations have been performed to establish group size thresholds for cluster candidate detection as a function of the number of galaxies N between 5000 and 100 000 objects per field and percolation radius R between 0.4′ and 1.2′. For each value of N and R, the percolation algorithm is applied to 1000 random fields and the distribution of group sizes determined. Accepting 1% false detections, that is, about one per field, we find the threshold group size varies from a few members for small values of N and R, to several hundred members for large values. For the smallest values of N and R, we have fixed a lower limit threshold at four galaxies.

Obviously, for both the binning and percolation techniques, the use of a single scale size for detection leads to the identification of clusters of a limited size. The advantage of our cluster detection approach is that with multiple bin scales and percolation radii we ensure a broad range in the size and type of cluster detected.

The CDB typically contains multiple instances of co-located and/or overlapping cluster candidates similar to those shown in Fig. 1a. By a process of elimination, the cluster screening phase resolves groups of multiple CDB entries into distinct clusters as shown in Fig. 1b. In the simplest case, where there is only one candidate in the group, the candidate becomes an entry in the catalog. Where there is more then one candidate in the group, the process compares the smallest member with the largest, then the second largest, and so on. A combination of geometric and photometric criteria are used to establish whether the co-located candidates are indeed distinct clusters or are merely multiple detections of the same cluster. If the distance between centers of the two members being compared is less than the radius of the smaller member, they are considered to be the same cluster and the smaller is rejected. This is designed specifically for instances of multiple detections of the same cluster core. If the distance is greater than the radius of the larger member, the two are considered entirely separate. For distances in between, the normalized cumulative number-magnitude distributions are compared using a Kolmogorov-Smirnov test. In this test, the maximum separation between the two distributions is used to assign a probability that the two distributions, and therefore the two clusters, are distinct. In this case, if the probability is greater than 90%, the smaller member survives this comparison. After the smallest member has been through this recursive process of elimination, the process proceeds to the next smallest member etc. until all of the members have been tested. The clusters which survive the elimination analysis become part of the catalog. Each cluster is then subjected to additional analysis steps which include, for example, the calculation of the centroid, major axis, ellipticity and orientation; the determination of photometric and morphological properties; a King-model fit using maximum likelihood analysis; an Abell type analysis to determine richness, redshift, distance; etc. Subsequent analysis further identifies duplicate entries resulting from field overlapping. A fuller description of the techniques involved in the ROE/NRL cluster catalogue will be published elsewhere (MacGillivray et al., in preparation). The catalog currently contains some 70 000 cluster identifications.

The search techniques used in our cluster analyses are still being refined and a number of problems remain to be resolved. For instance, the analysis in the vicinity of bright stellar images can produce spurious entries in the catalog. The percolation algorithm is more effective at detecting compact clusters than is the binning algorithm; however, both algorithms tend to miss large diameter, presumably nearby, low-density clusters. The binning algorithm fails in this respect because the cluster galaxies, being spread over a large area, are overwhelmed by the background galaxies in a bin; the percolation algorithm fails because the cluster galaxies percolate with the background galaxies into small groups which fall below the group size detection threshold. A solution to this problem may be to reprocess the data using only galaxies above a certain magnitude limit which would, in principle, select galaxies in more nearby clusters while reducing the number of background galaxies.

It is apparent that the techniques used in our analysis are limited by the two-dimensional view of the sky provided by photographic plates and the large variations in galaxy distribution on the sky and in clusters. This limitation can only be resolved by additional data, in particular, redshift measurements, higher resolution images, and X-ray observations. A program is underway to obtain these additional data. The Cluster Catalog is routinely being used at MPE to identify newly discovered X-ray sources in the ROSAT survey and, conversely, to confirm new clusters found in the Cluster Catalog. Additional correlation studies between the Cluster Catalog and other catalogs are in progress and are reported elsewhere in these proceedings (see the contributions by Gursky et al.). We are also currently using the catalog to locate and identify regions of sky which are overdense in clusters. We intend to use these results to provide a catalog of 'superclusters'.

4 Acknowledgements

We are grateful to the UKST Units of the ROE and AAO for allowing access to the UKST survey material.

References

Abell, G.O., Corwin, H.G., Olowin, R.P., 1989. *Astrophys. J. Suppl.*, **70**, 1.

Beard, S.M., MacGillivray, H.T., Thanisch, P.F., 1990. *Mon. Not. R. Astron. Soc.*, **247**, 311.

Dodd, R.J., MacGillivray, H.T., 1986. *Astron. J.*, **92**, 706.

Gursky, H., 1990. *Digitised Optical Sky Surveys Newsletter*, **2**, 28.

Heydon-Dumbleton, N.H., Collins, C.A., MacGillivray, H.T., 1989. *Mon. Not. R. Astron. Soc.*, **238**, 379.

Lonsdale, C.J., 1990. *Digitised Optical Sky Surveys Newsletter*, **2**, 29.

MacGillivray, H.T., Stobie, R.S., 1984. *Vistas Astr.*, **27**, 433.

THE AUTOMATED PLATE SCANNER CATALOG OF THE PALOMAR SKY SURVEY

R.L. Pennington, R.M. Humphreys, S.C. Odewahn, W.A. Zumach and E.B. Stockwell

Department of Astronomy, University of Minnesota, 116 Church Street SE., Minneapolis MN 55455, U.S.A.

1 Introduction

The Automated Plate Scanner (APS) at the University of Minnesota is a unique high speed 'flying spot' laser scanner. It uses a rotating prism to scan the laser spot and a beam splitter to produce a pair of spots which scan across a pair of plates. It scans a pair of plates in this mode in four hours with 1 micron repeatability. The APS is currently being used to scan the 936 pairs of plates comprising the Palomar Observatory-National Geographic Sky Survey. The resultant dataset will be used to produce a catalog of a billion stellar objects and several million galaxies. This catalog will be publicly available and we have undertaken the support for an on-line version which will be available to the astronomical community over the Internet. A summary of the catalog contents is given below.

- POSS-I O and E and Luyten red plates.
- Positions: (equinox 2000.0) 0.4 arcsecond accuracy for all-plate solution.
- Magnitude range: 12-22 in O and 12-20.7 in E.
- Matched images only.
- Estimated complete to 21 mag for O and 20 for E for stellar images.
- Stellar magnitudes accurate to 0.1 to 0.2 mag.
- Estimated 95% complete to 19.5 mag for galaxies.
- Integrated isophotal galaxy magnitudes accurate to 0.3 to 0.5 mag.
- Image pixels available for non-stellar objects.

H. T. MacGillivray and E. B. Thomson (eds.), Digitised Optical Sky Surveys 77–85.
© *1992 Kluwer Academic Publishers.*

2 Hardware Systems

The APS was originally designed as a specialized proper motion measuring engine for Professor W. Luyten. The machine has been redeveloped using modern electronics and computers to produce a very flexible scientific instrument. The APS has four possible scanning modes: multi-level isodensitometry, densitometry, threshold densitometry and selected area densitometry. The threshold densitometry mode is being used to scan the plates for the images in the catalog and the densitometry mode is used for the determination of the plate background. A summary of the scanning characteristics is given in Table 1.

Table 1.

Forward (Image) Scanning
> Threshold Densitometry Scanning Mode comprised of:
>> Single level Isodensitometry,
>>> Threshold level set to T=65% of local background;
>>> Fully hardware background referenced;
>>> X resolution is 5 microns;
>>> Y resolution is 3/8 micron;
>> Image densitometry,
>>> Densitometry within isodensitometry contours;
>>> No hardware backgrounding applied;
>>> 12 bit digitization of transmittance values;
>>> 5 × 12 micron sample spacing, inclusive of contour.

Reverse (Background) Scanning
> Raster densitometric mode, comprised of:
>> Full raster densitometry;
>> 10 × 48 micron sample spacing;
>> 12 bit digitization of transmittance values.

There have been significant improvements in the APS hardware in the past two years. Prior to beginning the scanning of the Palomar Sky Survey, the mechanical base of the scanner was disassembled by a field engineer from the Moore Special Tool Company, the manufacturer of the table unit. The metrological tests performed after the refurbishment showed the base to be positionally accurate to 0.3 microns through any 3cm of travel with a 1 micron total cumulative error through the full travel on either axis. This is within the original design specification of the machine. Analysis of data from before and after the rebuild shows no systematic changes in the positional characteristics of the machine. Positional repeatability, as measured by repeated scans of a POSS plate, remains unchanged after the rebuild at 1.6 microns in X and 0.9 microns in Y.

The ability to accurately measure image diameters from the POSS plates is of great significance because this measurement is used to photometrically calibrate the plate. The image diameters are measured from the single level of isodensitometry data available for every image, rather than from pixel data. This isodensitometric measure of the diameter is more accurate because of the very high resolution of the scanner in the Y direction, 3/8 micron. Repeated scans show that a 100 micron image, corresponding to m $\approx$ 18 magnitude, has an rms error of 0.8 micron or $\approx$ 0.03 magnitude.

The APS is supported by a network of computers dedicated to the task of reducing, displaying and archiving the data. The current network consists of a Sun 4 file server that interfaces to the data acquisition computer, a two processor Silicon Graphics R-4D220S for data reductions, and two Sun colour SPARC 1s, two colour X terminals and three monochrome X terminals for program development and data display and analysis. The file server has 6.4 GB of disk storage dedicated to data reduction, split into two equal sets of disks. Each set is large enough to store the raw and processed data for a pair of plates, and the sets are used alternately for data acquisition. The data reduction computer and software must be fast enough to be able to produce the background rasters from the reverse scanning for the hardware background system and doing image reconstruction for up to 2 000 000 images within the time necessary to do the scanning, about four hours. This is equivalent to maintaining a sustained processing rate of 150 KB/sec. After the completion of a scan the post processing consists of determining the astrometric solution, photometric calibration, plate-to-plate matching, star/galaxy discrimination, correction to non-stellar rasters for instrumental effects and database generation, which roughly doubles the processing requirements as this must be completed concurrently with and before the end of the next scanning session. After the reductions for a plate pair are finished, two copies of all data are written to Exabyte tapes. There is approximately 3 GB of disk storage on the Sun SPARC 1 systems for use by astronomers for small data analysis projects.

3 Software Systems

The improvements and additions to the APS software have been directed towards producing a fully automated data acquisition, reduction and archiving system and providing the tools necessary for the astronomer to access and analyze these data in a forthright manner. Figure 1 shows an outline of the APS data reduction and analysis system.

Most of the software for the initial reductions of the plates is operational on the reduction network. The images are reconstructed from the raw data stream, corrected for instrumental effects and the scanning stripes from a full plate are tied together and the plate pairs are aligned. These data are fed into the post-processing software. This software does the astrometric solution, photometric calibration and then uses a neural network to perform the star/galaxy discrimination (Odewahn et al. 1992). The fully processed data are then inserted into the StarBase database system.

StarBase is our custom database engine and is being uniquely tuned to the requirements of the APS catalog. The three major requirements that this engine had to fulfill are portability, minimized storage requirements, and the ability to

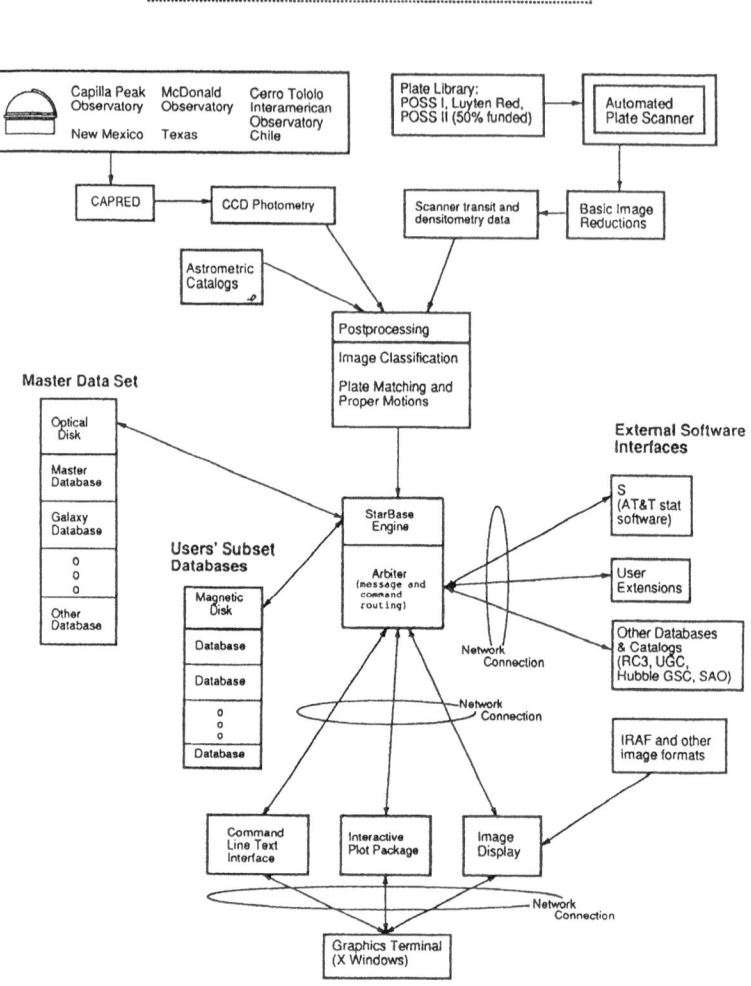

Fig. 1. The Minnesota APS data reduction and analysis system.

accommodate the full set of plates as a single entity. It is portable to any 32 bit integer machine and is written to conform to the ANSI C standard. With a database of a billion records per colour, each additional byte per record costs a gigabyte. Each record is bit-packed into 45 bytes. This contains all of the parameterized information from the APS for a stellar image, transmittance information, indices to the same image on other plates and a pointer to an extended field that may contain other information, such as pointers to other catalogs. A non-stellar image also includes all of the pixel information in the extended field. Table 2 gives the contents of each data record. The xc and yc are in the natural units of the APS, which are converted to microns by all system software (1 ere = 0.366 μm). The diameter, ellipticity and theta elements are the fitted ellipse parameters with fuz and jitter as measures of the errors in the fit. The magnitude is derived from our calibration.

Table 2. Data fields for compressed image records.

Field	Units	Bits	Notes
xc	1 ere	20	
yc	1 ere	20	
ra	$\frac{R}{n}$ sec	21	units based on plate declination
dec	$\frac{R}{n}$ arcsec	21	units based on plate scale
dia	1 ere	12	4095 $\Rightarrow$ overflow into extended field
ell	0.01	7	
theta	0.1 degree	11	
fuz	0.01 ere	9	5.11 $\Rightarrow$ overflow in extended field
jitter	0.1 ere	8	
mag	0.01 mag	11	
cen_T	0.1 trans	16	
avg_T	0.1 trans	16	
max_T	1 trans	12	
class	enum.	3	star, galaxy, unknown, special, etc.
apsname	plate, image	39	master index
id	plate, image	39	current db. index
dup	plate, image	39	image pair (current db)
starnum	NA	26	Absolute starnumber (1-indexed)
ext data fld	bytes	30	index to extended data fields

Each image on a plate that has a match to any one of the other plates is a 45 byte data record. Each colour comprises a single database of roughly a billion records. This produces a very compact database of 45 GB/colour. Searching this by brute force will require excessive amounts of time, roughly 60 hours on a Sun SPARC. This has been reduced by producing hash lists for each searchable field and two dimensional search trees for the positional information. Search times are generally seconds or minutes for reasonably sized searches.

The database is capable of operating as a stand-alone program but is designed to interface into a complete support system for the astronomer. A network based interface, the arbiter, is being developed to enable rapid sifting through huge volumes of data to search for statistically significant results or anomalous objects. The arbiter is unlimited in the sense that new client programs may be added at any time. It drives and responds to client program such as the graph package, image display package or a user extension, basically acting as an interpreter and switching center between all of the different software packages to be available. It provides a simple

C-based programming interface to allow users to produce programs that communicate with our processes and the database engine. All of the display software is being written for X windows based terminals. By writing a network based software system that displays on a standard commercially available terminal, we are producing a data analysis system that will be accessible to any astronomer with an X terminal and a connection to the Internet.

We are storing all of the data from the APS, both raw and reduced. This is currently being done by writing duplicate Exabyte tapes for every plate pair. We intend to place the dataset onto an archival media with reasonable access times. This will entail the addition of a large (1 TB) optical jukebox to the system. The quoted media life exceeds thirty years, disk change time is measured in seconds and data transfer rates approach those of magnetic disks. The size of the dataset for each colour scanned in given in Table 3.

Table 3. Size of datasets for each colour.

Type of data	Size in GB
Reduced Backgrounds	56
Reduced Images	257
Database	200

The decision to archive the reduced data as well as the catalog is to allow later reconstruction/correction of the database as improved astrometric and photometric data as well as more sophisticated algorithms for star/galaxy discrimination and image deblending become available.

4 Astrometry

The APS catalog will include the X and Y positions measured from the plate and the derived right ascension and declination for J2000.0. We are automating the astrometry by using the diffraction spikes of the reference stars for each plate to determine the center of the image. Positions on the plates are mapped using an initial rotation and translation, followed by the use of linear and radial terms in the global solution. We are currently using the Position and Proper Motion Catalog (PPM) (Roeser & Bastian 1989a, 1989b, 1990) and are testing the Astrographic Catalog of Reference Stars (Corbin & Urban 1988; Corbin 1990). The solutions based on the PPM have an internal rms error of 0.35 to 0.45 arcseconds in both dimensions for an all plate solution. We expect to further reduce this error by applying local systematic corrections either by stacking the errors from > 200 plates or by using the subplate technique for the reductions (Taff 1989; Taff et al. 1990).

The diffraction spikes are clearly visible and easily measured for the faintest astrometric reference stars ($\approx$ 11 mag). We have tried using marginal distributions and found a susceptibility to the flare that is present in the reference stars away from the center of the plate. We are testing the technique of using the diffraction spikes of the bright stellar images to determine the image center. The two diffraction spikes are treated independently and their intersection defines the center of the image. The line that is fit to a spike is derived by fitting parabolas to all of the available cross-sections of the spike. The ridgeline of the spike is then defined as the best fit line to the location of minima (transmittance data) of all of the cross-sections.

The plate constants are determined in a two step process. The first step locates all of the brighter ($<$ 9.5 mag) astrometric reference stars using a standard mapping to the plate and searching a 1 cm box centered on the expected position. This yields 30-40 images that are then used to compute a new plate center and initial rotation. This fit is then used to locate all of the remaining reference stars on the plate with a 250 micron search radius. The centers of these images are then found using the diffraction spike method and the final plate constants are computed iteratively. The three worst matches are rejected and the constants are re-computed until the final termination criteria are met. The program then becomes interactive and the astronomer inspects the quality of the solution.

While the diffraction spike approach works for stars brighter than $\approx$ 12 mag, it cannot be used for the fainter stars. For these stars we are using the geometric x and y centroid determined from the isodensitometric data. We need to determine if any difference in image centers exists for these two methods. To do this, we are using the Carlsberg Meridian Circle Catalog (Carlsberg Consortium, see Morrison et al. 1988) which contains fundamental positions for stars down to 14 mag. This will provide a consistent dataset for images with and without diffraction spikes. In this way, we can compare the astrometric solutions and positions of the stars from the two methods.

5 Photometry

One of the biggest uncertainties in the Palomar Sky Survey has been the accuracy and uniformity of magnitudes estimated from the plates and prints based on published sequences for a few scattered fields. This problem was partially alleviated with the publication of the Guide Star Photometric Catalog (GSPC) (Lasker et al. 1988) which has photometric standards for all of the POSS I fields but only to a limiting magnitude of 14.5 mag in B and V. Because the POSS I plates reach six magnitudes fainter than this catalog and include a red emulsion, we are obtaining direct photometric BVR calibration of about one-third of the Sky Survey fields. We have been obtaining CCD frames with the Capilla Peak Observatory 24'' telescope (University of New Mexico), the Cerro Tololo Interamerican Observatory 0.9m and the McDonald Observatory 36'' (University of Texas). The CCD integrations have all been selected to give photometry accurate to $\pm$0.1 mag for images 20-22 mag from the Cerro Tololo and McDonald Observatories and $\pm$0.1 mag for 19-20 mag images from the Capilla Peak Observatory. We have completed the observations and reductions for 230 POSS fields and expect to complete 300 fields as part of this project.

The large number of CCD observations required a more automated reduction process than is currently available with most astronomical photometry packages. We have written a semi-interactive program called CAPRED that assembles the necessary bias, dark, and flat field frames, processes each program frame and uses a signal thresholding algorithm to search each frame for potential stellar candidates. The stellar photometry code uses an optimized circular aperture and a median filtered background annulus to produce instrumental stellar magnitudes. Although this technique lacks the sophistication of a two dimensional point spread function fitting routine, such as DAOPHOT as discussed by Stetson (1987), it is quite adequate for producing better than 3% photometry in uncrowded regions and is able to run with absolutely no user interaction. The final phase of the CAPRED program involves automatic calculation of extinction and transformation coefficients. At this point in the reduction, we rely upon the interactive approval of the astronomer to verify the validity of the solutions and to maintain quality control. A table of V, B-V and V-R photometry and their estimated errors are then produced for each program field.

The results of Humphreys et al. (1991) suggest that a common function will describe the magnitude-diameter relation for the entire Palomar Sky Survey when corrected for zero-point variations. This study is the first photometric intercomparison of many Sky Survey fields. BVR photoelectric sequences for 49 Sky Survey fields were obtained for an investigation of the degree of linearity and uniformity of the magnitude-diameter relation for the Palomar Sky Survey. The diameters of the sequence stars for 28 of these fields was determined from scans with the APS. These diameters are measured from the best fit ellipse to the isodensitometry data for the stars. The magnitude-diameter relations for both the O and the E plates for all 28 fields have essentially the same form with small zero-point variations. Although they can best be fit by a smooth curve, the magnitude-diameter relations are very nearly linear over several magnitudes with very similar slopes, from 13.5 to $\approx$ 19 mag for the E plates and 14 to $\approx$ 19 mag for the O plates. A mean relation has been determined from the results of these 28 plates.

When corrected for the zero-point variations from field to field, these O and E mean magnitude-diameter relations can be readily used to estimate the magnitudes from any of the Palomar plates. We can use the stars from 12-15 mag in the GSPC to determine the zero-point shift for any field. When we do this for several sample fields and apply the mean relation, the magnitudes for fainter stars (14-19 mag) in our sequences have an rms deviation from their expected values of 0.15 mag.

6 Summary

We are currently scanning the glass copies of the Palomar Sky Survey with the APS at the University of Minnesota. We expect to finish the scanning of the high latitude plates in early 1992 and most of the low latitude plates by mid 1992. We will also be scanning the Luyten red plates and expect to finish those by late 1992. All of the detected images will be cataloged with positions, shapes, magnitudes, colours and classified as to star, galaxy, nebula or other. The astrometric positions will be accurate to ≈ 0.4 arcsecond. The photometric calibration will have stellar magnitudes accurate to 0.1 to 0.2 magnitudes. There will be more than a billion stellar objects and a separate galaxies catalog of several million pixel images. We are producing a complete software system that will allow access to an on-line version of the catalog across the Internet using X Windows terminals. The raw data and the catalog data will both be publicly available to the astronomical community using this software system.

References

Corbin, T.E., 1990. *'Initial Coordinate Systems on the Sky'*, *Proceedings of IAU Symposium* No. 141, p. 465, eds. Lieske and Abalaken.

Corbin, T.E., Urban, S.E., 1988. *'Mapping the Sky'*, *Proceedings of IAU Symposium* No. 133, p. 287, eds. Debarbat et al.

Humphreys, R.M., Landau, R., Ghigo, F.D., Zumach, W., 1991. *Astron. J.*, **102**, 395.

Lasker, B.M. et al., 1988. *Astrophys. J. Supp.*, **68**, 1.

Morrison, L.V., Helmer, L., Quinano, L., 1988. *'Mapping the Sky'*, *Proceedings of IAU Symposium* No. 133, p. 369, eds. Debarbat et al.

Odewahn, S.C., Stockwell, E.B., Pennington, R.L., Humphreys, R.M., Zumach, W.A., 1992. *Astron. J.* In press.

Roeser, S., Bastian, U., 1989a. In *Highlights of Astronomy, 20th General Assembly of the IAU*.

Roeser, S., Bastian, U., 1989b. *Introduction to the PPM*.

Roeser, S., Bastian, U., 1990. *Introduction to the PPM – South*.

Stetson, P.B., 1987. *Publ. Astron. Soc. Pacific*, **99**, 191.

Taff, L.G., 1989. *Astron. J.*, **98**, 1912.

Taff, L.G., Lattanzi, M.G. Bucciarelli, B., 1990. *Astrophys. J.*, **358**, 359.

Discussion

Lasker :

Would you care to comment on the spatial uniformity of the POSS-I plates (i.e. with regard to vignetting, non-uniformities, etc.)?

Pennington :

The POSS-I copies seem to show roughly the expected vignetting, being constant to 2.7° and increasing parabolically to about 0.3 mag at the corner. This is far from perfect for any given plate and we will be stacking many plates later to look for a generalized plate background. There are many small irregularities on different scales present on these plates, particularly near the edges.

Taff :

Do you have a physical explanation for the zero-point offset in the photometry? Is it exposure time, emulsion variability, or seeing conditions?

Pennington :

We know that some of the offset is due to the copying process. Five different copies of one plate all have different zero points. There may well be differences in the originals but we do not know that yet.

DIGITISED SKY SURVEYS AT THE SPACE TELESCOPE SCIENCE INSTITUTE

B.M. Lasker

Space Telescope Science Institute, 3700 San Martin Drive, Baltimore MD 21218, U.S.A.

Abstract

Space Telescope Science Institute (ST ScI) activity in the general area of 'Digitised Optical Sky Surveys' (DOSS) is directed, on the operational side, to supporting the use of Hubble Space Telescope (HST) by providing guide stars for pointing and calibrated images for observation-planning, and, on the scientific side, to applying the same resources to selected programs in community service and astrophysics.

1 The GSC I Program

The first Guide Star Catalog, now called GSC 1.0 to distinguish it from planned updates, was reported upon at the previous DOSS meeting (Lasker et al. 1989), and has since been published (Lasker et al. 1990; Russell et al. 1990; Jenkner et al. 1990). A fourth paper in this series (McLean 1992) is in preparation. While a complete review of the applications of and critiques to GSC 1.0 has not been complied, attention is called to two analyses thereof: Ratnatunga (1990), which addresses photometric performance in three southern selected areas, and Taff et al. (1990), which addresses plate-based systematics in the astrometry and the potential for redressing them.

A GSC 1.1, scheduled for 1991 release, includes the cross matches against several catalogs of bright objects; these matches were a collaborative effort of E. Egret and B. McLean, based on compilations in the Hipparcos input catalog.

Additionally GSC 1.1 addresses three classes of errors: a number of errata, mostly gross image processing errors and plate defects, have been discovered in the routine use of the GSC both by us and by users in the community; they are being corrected, generally by removal. (We repeat our thanks to those who have reported

H. T. MacGillivray and E. B. Thomson (eds.), Digitised Optical Sky Surveys 87–94.

GSC errors and remind others intending to do so that the preferred mechanism is E-mail to user GSCDIST at the ST ScI.) A second class of errors consists of false entries around bright stars, specifically those brighter than 9 mag from the southern J plates and 3 mag from both the northern and the southern V plates; and the third class involves multiple entries with different names for the same object, either because of noise problems in the blend resolver or because of object matching errors in the plate-overlap areas. Statistical tools are being used to identify and remove these.

A GSC 1.2, scheduled for 1992 release, will recalibrate the astrometry according to the precepts reported by Taff elsewhere in this volume, with the expectation of positions better than 0.25″ (per coordinate) and a minimization of the plate-based systematics cited above. As the GSC 1.2 calibrations are computationally quite different from those in GSC 1.0, appropriate changes in the astrometric support for image display and manipulation packages [GASP (Guide Star Astrometric Support Package) and ST SDAS (Space Telescope Science Data Analysis System)] will be issued concurrently.

Planning for GSC 1 maintenance beyond revision 2 remains informal and will probably be limited to issues of completeness and errata.

2 Publication of the GSC-I Scans

The GSC-I scans are archived as 400 Mbyte files on optical disks at the ST ScI (see Lasker et al. 1990 for details). From the earliest days of the GSC project it was clear that the community regards these scans as a resource of importance sometimes approaching the catalog itself. The wider applications underlying this interest come from programs of object identification (especially against non-optical catalogs), planning for ground-based programs, operation of multi-aperture spectrographs, and investigations of peculiar objects, etc.

For many years, requests to access the scans have been made at a level far in excess of our capabilities to support them; and in response to the as yet unsatisfied demand for access to these scans, serious planning to distribute them in a lightly compressed form on CD ROMs is in progress. The compression approach is based on imperfect encoding of parts of the data that are most likely to be noise. The algorithm, together with some tests of its astronomical properties (specifically with respect to photometry, astrometry, and object classification) are described in detail by White et al. elsewhere in this volume. In brief it appears that most of the demands from the professional astronomy community can be satisfied with a compression factor averaging about ten.

In the configuration being proposed by White et al., an all-sky set will occupy about 100 volumes and, assuming a subscription for 100 sets, will be sold on a non-profit basis for $6000, or less if the interest is greater (but do note that this estimate does not yet constitute an offer). A community survey letter describing the project and asking for serious but still tentative expressions of interest has been distributed. On the basis of the responses to this letter, a final plan and a firm offer will be made, hopefully within the year.

One may also note that even more highly compressed images (e.g. 50X; see White et al., these proceedings) are useful for some purposes (quick look telescope operations, browsing, as well as the amateur and education communities). Such a set would occupy about 20 volumes, and will be produced as part of the larger program if the interest is sufficient.

3 Overview of the GSC II Program

Beyond the maintenance program leading to the GSC 1.n series of catalogs and the publication of the GSC-I scans, current ST ScI activities are organized about a set of activities generically called the GSC-II program. The motivations for these are to provide guide star positions for HST pointings in the presence of accumulating proper motions and to provide all-sky coverage for planning purposes to the limit imposed by modern Schmidt art (and technology), as well as to support all the calibration functions underlying these facilities.

The most serious proper motion requirements arise in the south, where the initial resource was the circa 1975 epoch SERC J survey. Taking a 0.25″ (per coordinate) precision at 1975, simulations by White indicate an acquisition failure rate of about 10% in 2005 for the FOS (acquisition aperture 4.3″, even larger for instruments with smaller apertures). The data for satisfying this will be obtained from the SES survey (see Morgan et al., elsewhere in this volume).

The other GSC-II requirement, deep all-sky coverage, leads to the need to supplement the northern Palomar 'Quick V' survey (limit V=19) with the deeper data obtainable from the second Palomar Sky Survey. If all available plates from both the POSS-II and the SES are scanned, and allowing some margin for safety, the scanning element of the GSC-II program requires scanning about 3500 plates over an appropriate interval, tentatively taken as 5 years.

While the 25 μm sample interval used in GSC-I was a marginal choice for the older Palomar material and a poor choice for the modern materials from both Schmidts, it was required to meet the original GSC schedule imposed by HST mission constraints. As the new scanning will be done under less constraining schedules, we revisited the question of the sampling interval, the goal being to adopt a sample interval which is small enough to collect most of the information from the Schmidt plates and to perform well with conventional image processing algorithms, while still large enough that the scanning throughput is reasonable and that the (uncompressed) scans can be stored on existing 1 Gbyte optical disks. This study, which led to the adoption of a 15 μm sampling interval, is described in the paper by Laidler et al. elsewhere in this volume; one also notes that Stobie et al. (1984) reached a similar conclusion from somewhat different arguments.

This decision led to the adoption of a preliminary PDS configuration for second generation scanning, based on an initially lowered throughput goal and another modification of the Wisconsin amplifier (Anderson et al. 1983). The corresponding throughput is about 1 plate in 48 hours, and our goal is to decrease this interval to 12 hours with additional modifications to the PDS.

4 Photometry

Our photometric calibrations, initially based on the first Guide Star Photometric
Catalog (GSPC-I; Lasker et al. 1990), is being supplemented with a second catalog,
the GSPC-II. The GSPC-II plan is to cover the sky in V and R with CCD images
located near the faintest star in each GSPC-I sequence, to a limit of about 21
mag. Details of this program, which is based on observations taken at a network of
collaborating observatories, is described by Postman et al. elsewhere in this volume.

The GSPC-II program, while large, is significantly incomplete in that it defines
the (V, R) photometric calibration at each sequence but only as far from the sequence as one trusts the uniformity of the photographic material. Extensions of the
photometric calibrations to more positions (near the plate edges and corners) and
to other passbands are an obvious topic for coordination among the various groups
at this conference.

In the GSC-I, photometric calibrations were based on a non-linear calibration
of the GSPC magnitudes against integrated photographic densities. This technique
works well within the magnitude range defined by the sequence, but extrapolates
poorly. Therefore, our plan is to base further work on the Bunclark-Irwin (1983)
calibration algorithm, which is based on the fact that all stars have the same PSF.
Both the version kindly furnished by P. Bunclark circa 1988 and a locally coded,
more modular IDL version of the algorithm are in use.

In applying this procedure to the GSC-I scans (25 μm pixel size), a small non-
linearity in the stellar calibration has been noted. The Fourier analyses by Laidler
et al. elsewhere in this volume attribute this to data-dependent fluctuations in the
photometric collection efficiency for larger pixels; and the effect is expected to be
much smaller for the GSC-II scans (15 μm pixels).

5 PDS Hardware Modifications

Currently, the most serious limitation to PDS performance is the logarithmic amplifier in the photometry system. Electronics has changed greatly since the implementation of that design, and mixed analog-digital approaches now give much better
performance. An implementation following this concept and presently nearing completion at the ST ScI is based on the maximal use of commercial components. The
design consists of (1) a front end made with a rather conventional low noise ten-
times linear amplifier, specifically the same one that is used as the first stage in the
present 'modified Wisconsin' system; (2) band-limiting with a digitally controlled
Bessel filter (Frequency Devices 848P8L-4); (3) a floating point analog-to-digital
converter (FP ADC; Micro Networks MN5420), and (4) a look up table (in read-
only memory) to convert the floating point numbers to photographic density.

The FP ADC converts a 0-5 volt signal into a 4 bit exponent and a 12 bit man-
tissa. With full scale corresponding to density one, a one bit increment corresponds
to 0.009 density units at density 5 and to 0.087 density units at density 6. This
system has a maximum conversion rate of 3 microseconds, but we shall probably
use it at a 50 to 80 microsecond rate. Note however that higher rates could be used

if one wanted to oversample and use a digital filter in the output, e.g. to control the beam profile.

It is expected that making copies of this unit will be straightforward. After testing is completed, the design and artwork for this subsystem will be available on request.

A second class of limitations on PDS performance includes the ability to autolock (maintain the slow scan position while running the fast scan axis) and to perform the turn-arounds at the ends of the scan lines efficiently. These are associated with the servo system, which is hard to tune optimally and which fails to exploit fully the capabilities of the motors. In analogy to the situation in the photometry subsystem, servo technology has changed dramatically since the design of the PDS; and a number of commercial units with smart damping functions (programmable or with learning modes) are available. We expect to select and install one such unit later this year.

With the amplifier and servo changes in place, the performance limitation at rates of about 1 plate per day will be primarily imposed by photon noise. Plans for increasing the light level, tentatively by the use of laser illumination with optical concepts borrowed from scanning systems existing elsewhere, and for adding a second (possibly many) channel are under development; when implemented, the throughput goal of 1 plate per 12 hours will be achieved.

6 Software

The second generation software efforts have been directed to understanding the properties of the image processing routines, to perfecting a production pipeline, and to developing database concepts. A few details on selected aspects of this work follow:

- PDS shearing, caused by fixed delays in the photometric system (as opposed to data dependent delays, which are beyond linear redress) have different effects for lines of different parities in a boustrophedontic scan. This is normally compensated by tuning the line origins or a delay element in the control system. After suffering considerable efforts in doing this tuning by visual inspection of scanned images, we developed a cross-correlation approach (left-going against right-going lines) which measures offsets as low as 0.05 pixel robustly. This is now used both for system tuning and for scan control (as well as to set up the occasional repair of a sheared scan).
- A deblender coded on the COSMOS design (Beard et al. 1990) has been adopted. We find best results when this is run on all objects [as opposed to only those that have been identified by a classifier as blends (because such classifiers are not as robust as the deblender)]
- We are now routinely making small area cutouts around each object inventoried, and storing them in a random access file, i.e. to support convenient reprocessing by supplemental algorithms, such as tests for diffraction spikes, the deblender, the correlation classifier, alternate astrometric centroiders, and galaxy typing and photometry routines.

- As always, scatter plots of image features remain a vital tool for understanding (and even classifying) inventories. A new tool, built on the cutout and deblender capabilities, provides a capability to 'click' a cursor on a point in a scatter-plot and immediately see the image; we find that this greatly facilitates both production and diagnostic work.
- A new centroider, based on correlations with reference images, is under development in collaboration with Luciano Lanteri and Luca Pividori (Osservatorio Astronomico di Torino). The goal is to find procedures that work stably and close to optimally for images which are not well suited to function fitting, e.g. because of their complexity or asymmetry.
- Much of our recent work on catalog development has led to large data sets stored on 8-mm tape. This situation is not conducive to either long-term archiving or to convenient access; and (most probably like everyone else at this meeting) we are also devoting increasing attention to the problems of creating, maintaining, and distributing a database of about a billion objects. We continue to explore the concept presented by Jenkner et al. (1988) of storing the measurements, the calibrations, and the addenda separately, then using a 'smart' server to combine the materials when requested. Additionally, the possibility of using compression techniques on the measurements is under study.

7 Science Programs

In addition to the primary program, plate scanning in preparation for a second GSC containing colors and proper motions, a number of parallel science programs are being conducted.

L. Drissen and M. Shara are using the overlap areas of the southern plates to search for large amplitude variables. Beginning with a 100 sq. degree search in the 4-fold overlap areas, they have selected about 100 objects, mostly cataclysmic variables and optically violently variable quasars, for which spectroscopic follow up is now in progress.

M. Postman, V. Laidler and the present author are using scans from the POSS-I E plates to produce a catalog of northern galaxies of intermediate brightness. The initial version of the catalog, to be completed this year, will cover a 24-plate area near the north celestial pole. It will reach to about $R \approx 17$ and will contain about 24000 galaxies. Photometry, based on densitometry from the Bunclark-Irwin algorithm, calibrated against the GSPC-I and GSPC-II reference stars, is expected to initially be at the 0.2 mag level. Future efforts will be directed to working fainter, over a larger area, to greater photometric precision, and in additional passbands.

A stellar statistics program by M. Lattanzi and the author, which is still in the definition phase, will address colors and proper motion studies, both as probes of galactic structure and as tools to identify rare objects (Population II in particular).

B. McLean and J. MacKenty are completing a statistical study of all Markarian galaxies (and a control sample) using GSC-I scans of the Palomar 'Quick V' plates. The goals are to discover morphology-environment correlations, including Seyfert activity and IRAS properties. A by-product of this work will be an image atlas of the Markarian objects, together with revised astrometry and photometry.

Additionally, a program of ROSAT identifications is addressed by McLean et al. elsewhere in this volume.

8 Acknowledgements

The author is pleased to acknowledge the individuals who contributed to the work reported in this review, grouped (roughly) by technical area, as follows:

- a core GSC group consisting of J. Doggett, H. Jenkner, V. Laidler, B. McLean, J. Phillips, M. Postman, M. Shara, and C. Sturch;
- image processing contributors I. Szapudi and R. White;
- the astrometry group, B. Bucciarelli, M. Lattanzi, and L. Taff;
- the photometry group, N. Brosch, A. Ferrari, J. Koornneef, C. Lopez, G. Massone, D. Rehner, and L. Siciliano;
- the engineering team, K. Ray and A. Evzerov;
- and advisors P. Bunclark, R. Burg, M. Damashek, L. Drissen, R. Settergren, and D. Silverberg.
- Special thanks are also due to Riccardo Giacconi for his vision in supporting the integration of HST operational work with parallel astrophysical programs.

The Space Telescope Science Institute is operated by the Association of Universities for Research in Astronomy, Inc., under contract to the National Aeronautics and Space Administration. This paper is based in part on plates taken with the Oschin Schmidt, operated by the California Institute of Technology, and with the UK Schmidt, operated by the Royal Observatory Edinburgh until June 1988, thereafter by the Anglo Australian Observatory, as well as on photometry obtained with the facilities of CTIO, ESO, KPNO, Lowell Observatory, Megantic Observatory, San Diego State University, and Wise Observatory.

References

Anderson, C.M., Slovak, M.H., Michalski, D.E., 1983. In *Astronomical Microdensitometry, NASA Conf. Proc.* No 2317, p. 163, ed. D.E. Klinglesmith.

Beard, S.M., MacGillivray, H.T., Thanisch, P.F., 1990. *Mon. Not. R. Astron. Soc.*, **247**, 311.

Bunclark, P.S., Irwin, M.J., 1983. In *Proc. Statistical Methods in Astronomy, ESA SP-201*, p. 195, ed. E. Rolfe.

Jenkner, H. et al., 1988. In *I.A.U. Symp.*, No. 133, p. 23, eds. S. Debarbat, J.A. Eddy, H.K. Eichhorn and A.R. Upgren.

Jenkner, H. et al., 1990. *Astron. J.*, **99**, 2019.

Lasker, B.M. et al., 1989. In *Proc. of the First Conference on 'Digitised Optical Sky Surveys', Bull. d'Information du CDS*, No. 37, p. 15, ed. C. Jaschek.

Lasker, B.M. et al., 1990. *Astron. J.*, **99**, 2019.

Ratnatunga, K., 1990. *Astron. J.*, **100**, 280.

Russell, J.L. et al., 1990. *Astron. J.*, **99**, 2059.

Stobie, R.S., Okamura, S., Davenhall, A.C., MacGillivray, H.T., 1984. *Occ. Rept. R.O.E.*, No. 14, p. 219.

Taff, L. et al., 1990. *Astrophys. J. Letters*, **353**, L45.

Discussion

Odewahn :

Capaccioli and collaborators have shown the great power of the HAAR transform in faint galaxy surface photometry. They apply this technique to an intensity matrix. Do you encounter problems in applying this to density arrays?

Lasker :

Our work is being done in density space. As much of the compression occurs near the sky, where the photographic response of well-exposed plates is not too non-linear, I would not expect a big gain from working in intensity space.

Humphreys :

You are going to scan the POSS-II with 15μm resolution. What is the resolution for your POSS-I red plate scans?

Lasker :

We will do all the second generation scanning at 15μm. All our first generation work, including the 1950 POSS-E, is at 25μm.

THE SELECTION OF A SAMPLING INTERVAL FOR DIGITISATION: HOW FINE IS FINE ENOUGH?

V.G. Laidler [1], *B.M. Lasker* [2] *and M. Postman* [2]

[1] Computer Sciences Corporation at Space Telescope Science Institute.
[2] Space Telescope Science Institute *, 3700 San Martin Drive, Baltimore MD 21218, U.S.A.

Abstract

A critical property of digitized sky surveys prepared from photographic materials is the sample interval, δ, used by the microdensitometer. At issue is a tradeoff between coarse sample intervals that, while economical, do not allow the faithful reproduction of the information in photographic material, and fine intervals, which can record all of the information, but at a prohibitive cost. We investigate these issues by conducting Fourier analyses of images characteristic of the imaging process in the absence of noise and by examining the properties of object inventories for a test field digitized at a range of sampling intervals. Both analyses indicate that $15\mu m$ is an adequate sampling interval for modern Schmidt plates on Type III emulsions.

1 Fourier Analysis

The two plates used for the Fourier analysis were a glass copy of the southern field 885 $[20^h 40^m + 00°$, UK Schmidt Telescope number J6136, J passband (filter GG395), 65 min] and an original plate of the same region taken in the north (Oschin Schmidt Telescope number SJ01919). A selected area of each plate was scanned with the STScI PDS microdensitometers (Lasker et al. 1990) with a sample interval of $10.1\mu m$ using a $30\mu m$ soft-edged aperture.

---

* The Space Telescope Science Institute is operated by the Association of Universities for Research in Astronomy, Inc., for the National Aeronautics and Space Administration.

H. T. MacGillivray and E. B. Thomson (eds.), Digitised Optical Sky Surveys 95–101.
© 1992 *Kluwer Academic Publishers.*

Standard inventory software was used to catalog objects from each scan. Objects visually confirmed to be clean stars were grouped by image radius and position-shifted (by phase adjustment in the Fourier domain) such that the discrete sample space had a common relation to the centroids (i.e. the centroids were all moved to the center of the central pixel). Then averages were made within each group. As the average images are smooth and nearly noise-free, they may be regarded as describing the telescope-atmosphere-plate combination, but with the effects of plate noise removed.

The group averaged images, $I(x, y)$, are Fourier transformed to obtain $F(f_x, f_y)$. Then we compute a Fourier amplitude, $(FF^*)^{1/2}$, and impose cylindrical symmetry to obtain a Fourier amplitude as a function of frequency only, $A(f_r)$. (Note that because of the cylindrical symmetry, the amplitude in an annulus about f_r is $f_r A(f_r)$.)

The total image amplitude spectrum between zero and f_r is simply the cumulative,

$$C(f_r) = \int_0^{f_r} A(f_r) f_r df_r \; . \tag{1}$$

Figure 1 shows a typical radially symmetric Fourier amplitude and corresponding cumulative for an averaged image.

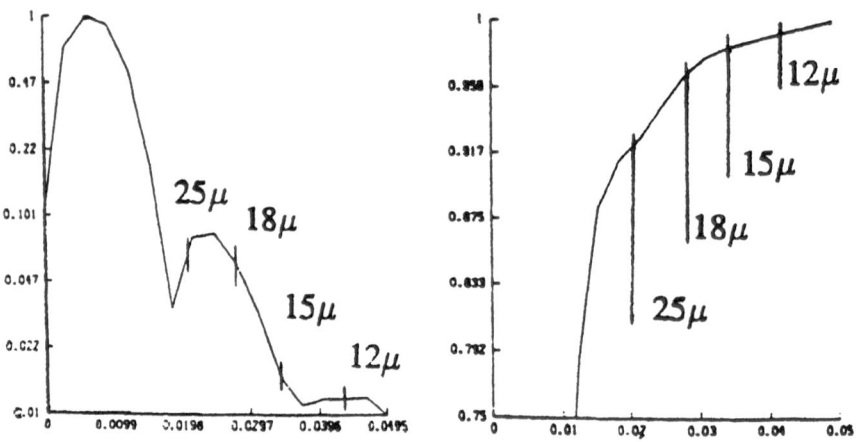

Fig. 1. A typical radially symmetric Fourier amplitude (a) and cumulative (b), plotted against spatial frequency. This example is taken from a mid-sized POSS-II averaged image, diameter approximately 150 microns. Cutoff frequencies for the sampling intervals of interest are marked.

In analogy to the Stobie et al. (1984) studies, one may adopt f at the frequency where most of the information is included to obtain a sampling interval. However, additional insight is provided by an examination of the fraction of the cumulative as a function of object size, parameterized by sample interval (Fig. 2). For the GSC–I sampling interval, 25μm, the function is seen to rise steeply and non-linearly as one looks from the threshold objects to the larger well-exposed stars.

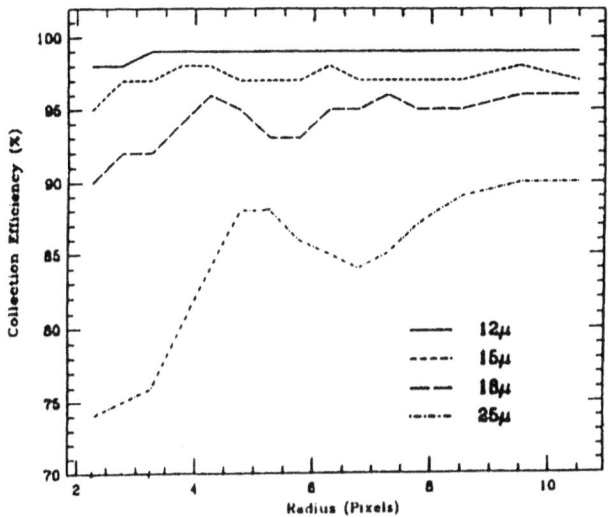

Fig. 2. Collection efficiency (cumulatives at selected intervals), as a function of object size, parameterized by sampling interval, for the SERC images.

The complexities of the cumulative at this sample interval are the likely explanation of the difficulties experienced by our group (at the 0.1 mag level) in using the method based on the point-spread function (Bunclark & Irwin 1984) to photometrically linearize stellar data scanned with 25μm samples.

2 Spatial Analysis

We perform the spatial analysis on an original J-band plate of field 330 from the second Palomar Sky Survey (POSS–II, Oschin Schmidt Telescope number SJ01840, $16^h02^m +40°$). A region 0.85 square degrees in area near the center of the plate was scanned with PDS stepsizes of 12μm, 15μm, 18μm, and 25μm using the appropriate PDS 'cat-eye' apertures.

Each PDS scan was processed using the latest version of the Faint Object Classification and Analysis System (FOCAS; Valdes 1989). Merged objects were deblended and object parameters were computed. The density-to-intensity relation for each digitized image was determined from the calibration spots on the plate. The FOCAS object classification algorithm was then applied to each catalog.

Results from the Palomar faint cluster survey (Gunn et al. 1990) were used to define the 'truth' for object detection, photometry, and classification. These data, taken with the four-shooter CCD camera on the Hale 5m telescope, are complete to V=24, and the four-shooter scale is 0.34″/pixel, corresponding to 5μm pixels on a Schmidt plate. The final PDS object catalogs were then matched with the CCD survey catalog which was also created with the FOCAS software.

The matched catalog is first used to determine the dependence of detection completeness on scan stepsize. It is clear from Fig. 3 that all four PDS object catalogs are complete to the same limiting V magnitude, $V_{lim} \approx 21.2$ ($J_{lim} \approx 21.7$). The flat tails in the magnitude histograms at V magnitudes greater than 23 are due to spurious object detections being matched with faint CCD objects. The frequency of spurious detections is not strongly correlated with scan stepsize.

We cannot, at this time, reliably measure the dependence of absolute photometric accuracy on stepsize because the CCD survey and the POSS plate are in different passbands and color information is not readily available. However, we show in Table 1 the magnitude residuals of the objects in the coarser scans with respect to those in the 12μm scan.

Table 1. Internal Magnitude Residuals.

Scans	Residuals ($14 \leq M_{12\mu m} \leq 16$)	Residuals ($17 \leq M_{12\mu m} \leq 19$)
12μm : 15μm	0.04	0.13
12μm : 18μm	0.08	0.18
12μm : 25μm	0.12	0.22

We assume that the object classifications in the CCD survey are 100% accurate. We show in Fig. 4 the fraction of objects correctly classified as stars and as galaxies, respectively, for the four PDS scans as a function of CCD V magnitude.

3 Conclusion

The procedures we have used lend themselves to a general answer to the question 'How fine is fine enough?' From the Fourier analysis, the sampling interval is fine enough when 'most' of the information is collected, and the fraction of the information collected is well-behaved as a function of image size. From the spatial analysis, it is fine enough when object detection and classification are 'mostly' correct. Our results indicate that a 15μm sampling interval meets these criteria for modern Schmidt plates on Type III emulsions.

4 Acknowledgements

The authors thank the Palomar Observatory for making Oschin Schmidt plates SJ01919 and SJ01840, which are part of the POSS–II program, available in advance

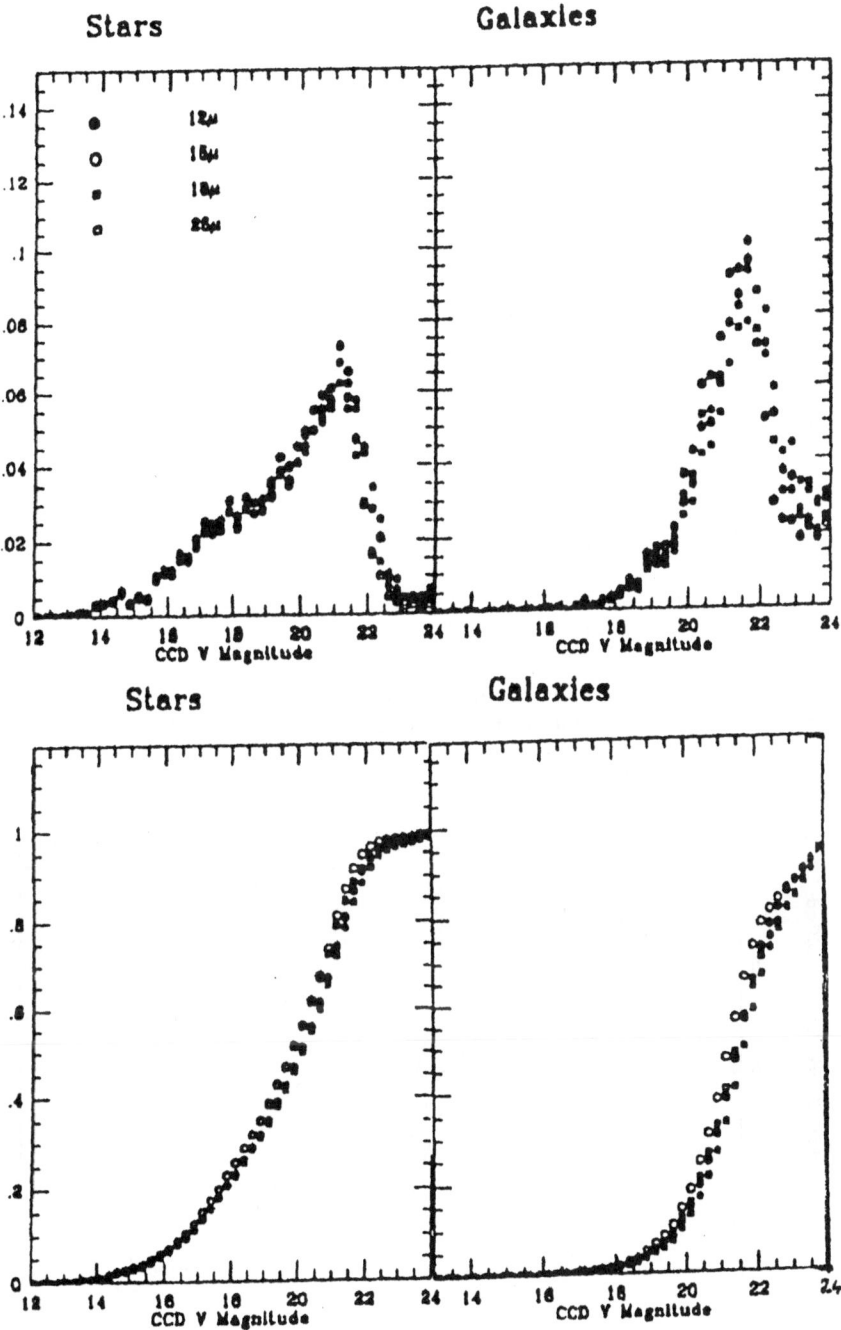

Fig. 3. (a) the differential sample fraction of stars and galaxies detected in the PDS scans as a function of CCD V magnitude; **(b)** The cumulative sample fraction of stars and galaxies detected in the PDS scans as a function of CCD V magnitude.

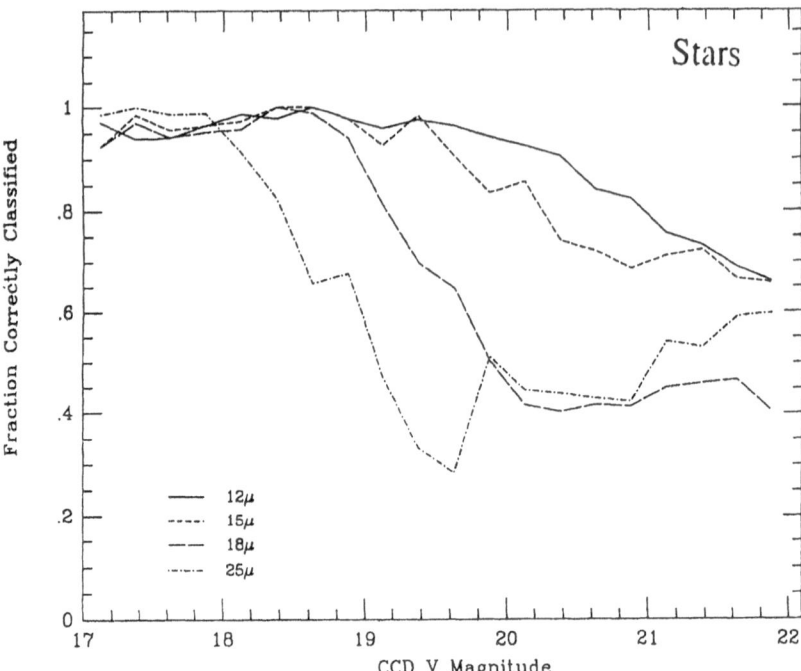

Fig. 4. (a) Ratio of the fraction of objects correctly classified as stars in the 15μm, 18μm and 25μm scans to the fraction of stars correctly classified in the 12μm scan.

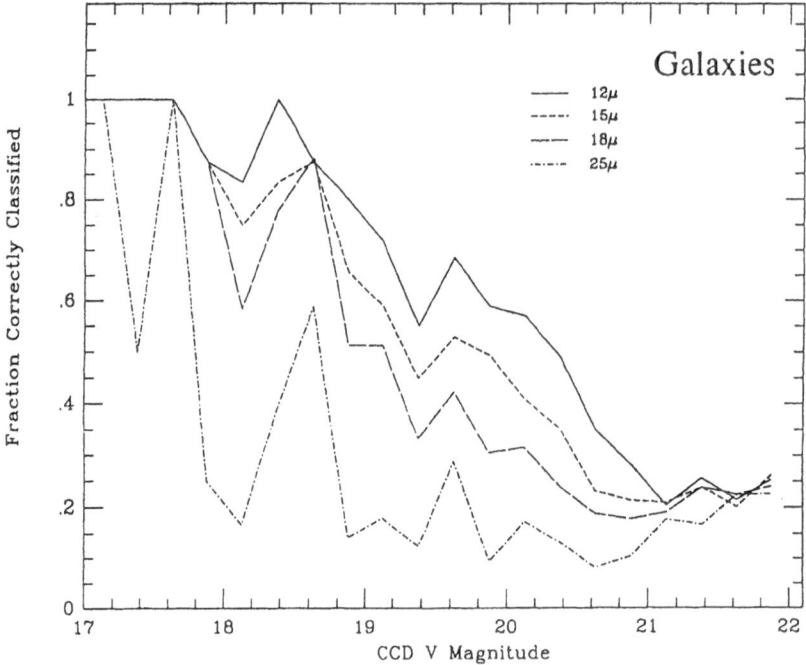

Fig. 4. (b) Ratio of the fraction of objects correctly classified as galaxies in the 15μm, 18μm and 25μm to the fraction of galaxies correctly classified in the 12μm scan.

of the completion of the survey. We also thank the Royal Observatory Edinburgh for making the EJ survey available in a manner consistent with the GSC–I development.

References

Bunclark, P.S., Irwin, M.J., 1984. *'Astronomy with Schmidt-type Telescopes'*, *IAU Colloquium* No. 78, p. 147, ed. M. Capaccioli, D. Reidel, Dordrecht.

Gunn, J.E., Postman, M., Oke, J.B., Hoessel, J., Schneider, D.P., 1990. Work in progress.

Lasker, B.M., et al., 1989. *'Digitised Optical Sky Surveys'*, *CDS Bull.*, No. 37, p. 5, ed. C. Jaschek.

Lasker, B.M., Sturch, C.R., McLean, B.J., Russell, J.L., Jenkner, H., Shara, M.M., 1990. *Astron. J.*, **99**, 2019.

Stobie, R.S., Okamura, S., Davenhall, A.C., MacGillivray, H.T., 1984. *Occ. Rept. R.O.E.*, **14**, 219.

Valdes, F., 1989. Private communication, NOAO Computer Support Division.

THE MAMA FACILITY:
A SURVEY OF SCIENTIFIC PROGRAMMES

J. Guibert

Institut National des Science de L'Univers, Centre National de la Recherche Scientifique, 77 Avenue Denfert-Rochereau, 75014 Paris, France.

1 Introduction

MAMA *(Machine Automatique à Mésurer pour l'Astronomie)* is a fast and accurate multichannel microdensitometer developed and operated by INSU and located at the Observatoire de Paris. MAMA processes in a few hours photographic plates up to 14″ × 14″ with a positional accuracy of 1μm (repeatability: 0.2μm) and a photometric accuracy of 2% over a dynamical range of 3 densities. The detector is a RETICON CCPD array with 1024 photodiodes. X and Y coordinates are measured with Heidenhain encoders. Autofocus is achieved through a maximisation of the plate grain noise, and is accurate to better than 4 microns. The plate can be digitised either in a systematic way by lanes 10.24 mm wide, or in a random access mode from a catalogue of preliminary positions. The basic pixel size (and sampling step) is 10μm. Oversampling down to 2μm can be used to digitise spectra; for some applications, pixels of 20, 30...80 μm can be synthesized in real time.

The digitised images can be processed according to three main modes. On-line processing leads, through a multi-level thresholding technique, to a catalogue of positions, areas, fluxes and second-order moments. Off-line processing is possible on-site using DEC-3100 or SUN SPARC 2 workstations, and a VAX 8250 computer; the available software includes MIDAS and a number of tools specially designed to extract the best from the astrometric and photometric capabilities of the machine. Finally, the user can of course take the pixels with him to process using his own facilities.

Information about MAMA and reduction techniques can be found in the paper by Berger et al. (1991).

2 Scientific Projects

A wide variety of scientific projects are currently being carried out using MAMA (see Tables 1 to 4). Several long term programmes dealing with solar physics are

H. T. MacGillivray and E. B. Thomson (eds.), Digitised Optical Sky Surveys 103–108.
© 1992 *Kluwer Academic Publishers.*

based on spectral images from Pic du Midi, Teide Observatory (Tenerife), Sacramento Peak and Meudon where spectroheliograms have been accumulated since the beginning of the century. Work concerning the solar system, stellar populations and galactic structure as well as extragalactic astronomy are mainly based on Schmidt plates from Palomar, Siding Spring, ESO, Calar Alto, Tautenburg and CERGA.

Table 1. Extragalactic research.

Programme	Laboratories	Instruments	Technique
Search for quasars candidates	IAC/Paris/ Meudon/ Montpellier	Palomar/CERGA Schm. Telescopes	multicolour photometry; variability
Survey of galaxies with bursts of star formation	Marseille	ESO Schm. telescope; direct and O.P. plates	multicolour photometry; radial velocities
Clusters of galaxies	Meudon/ Montpellier	SERC and CERGA Schm. telescopes	astrometry; photometry
SO galaxies	Nice	K.P.N.O. 4m; ESO/CERGA/SERC Schm. telescopes	photometry
Identification photometry of IRAS sources	Institut d'Astrophysique /IAC	ESO/SERC/ Palomar Schm. telescopes	astrometry; photometry
Counterparts of radio sources in galaxies	IRAM/Meudon	Calar Alto/ Palomar Schm. telescopes	astrometry; photometry
Stars in LMC/SMC	Marseille	ESO 152 +Boller- Chivens spectrograph	radial velocities

Galactic structure surveys conducted with MAMA take advantage of the astrometric accuracy of the machine. Using plates taken over 40 years, *absolute* proper motions are obtained by C. Soubiran (1991) for high numbers of stars with an accuracy of 1.5 milli-arcsec/year down to magnitude B=18 (see Fig. 1), which compares favourably with the accuracy of HIPPARCOS. The technique is described by O.

Table 2. Galactic structure.

Programme	Laboratories	Instruments	Technique
Luminosity function of low mass stars	IAC/Meudon	CERGA/Palomar/ Tautenburg Schm. telescopes	proper motions; photometry
Galactic stellar populations	Besançon	ESO/CERGA Schm. telescopes	proper motions photometry
Meridian section of the galaxy	Potsdam	Tautenburg Schm. telescopes	proper motions
Study of nearby stars	Bordeaux	CERGA Schm. telescope	astrometry
Search for high velocity stars	Strasbourg	Carte du Ciel Astrograph	proper motions
Open and globular clusters	Bonn Potsdam IAC	Bonn/Paris astrographs; CERGA/Tautenburg Schm. telescopes	proper motions
Atlas of galactic open clusters	Lausanne/ Meudon	ESO/CERGA Schm. telescopes; astrographs	astrometry; photometry
Extinction in molecular clouds	Meudon/ Sao Paulo	ESO/Palomar Schm. telescopes	Star counts; photometry
Variable stars in the galactic centre	Lyon	ESO Schm. telescope	photometry
Search for macroscopic dark matter through microlensing	CEA/IN2P3/IAP/IAC Marseille	ESO Schm. telescope	variability

Table 3. Solar physics.

Programmes	Laboratories	Instruments	Techniques
Small scale solar magnetic field	Athens	Scramento Peak solar tower	photometry
Solar cycle and solar dynamo	Meudon	Meudon spectroheliog.	rotation of sunspots
Fine structure and kinematics of Chromosphere/ photosphere	Meudon/ Toulouse	Pic du Midi /Teide D.P.S.M. spectrographs	photometry; radial velocity
Solar protuberances	Verrieres le Buisson Ondrejov/ Wroclaw	Sacramento Peak solar tower	photometry
Polarization of Ha line during solar flares	Meudon	Polarization analyzer	photometry
Solar eclipse of 10 July 1991	IAP/ Verrieres	C.F.H.T. prime focus	photometry

Table 4. Planetology.

Programme	Laboratories	Instruments	Technique
Weak satellites of Jupiter	O.C.A. (Grasse)	CERGA Schm. telescope	astrometry
Occultation by Neptune Arc Rings	Meudon	ESO Schm. telescope	astrometry
Search for asteroids	Bruxelles	ESO Schm. telescope	astrometry; photometry

Bienaymé (these proceedings), and can be also used for astrometric work in the Solar System domain (see, on Table 4, 'Occultations by Neptune Arc Rings').

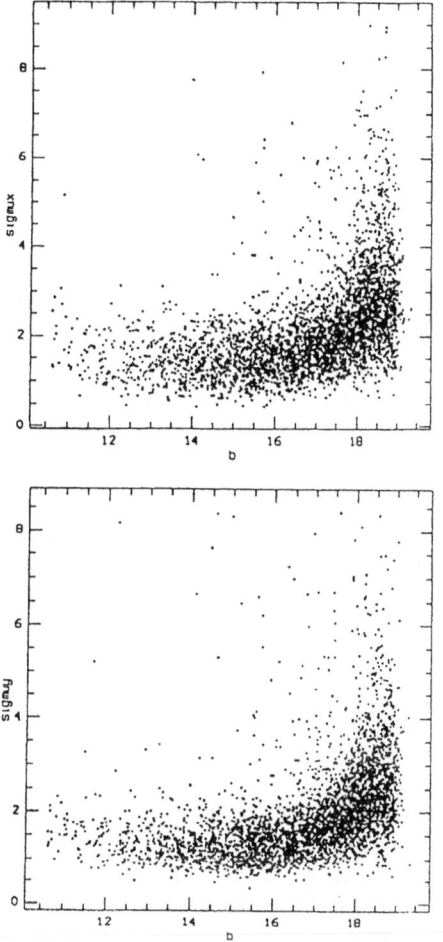

Fig. 1. Accuracy of absolute proper motions obtained by C. Soubiran towards the North Galactic Pole. Material: 11 Schmidt plates from Palomar, Tautenburg and CERGA telescopes; time scale: 40 years. Abscissae are B magnitudes; ordinates: r.m.s. errors on proper motions along both axes (milli arc. sec. per year).

This geometric accuracy of MAMA is also quite appreciable when reducing objective-prism images, since the quality of radial velocity determination strongly depends on the geometry of the measuring machine. An accuracy of 100 km/s is for instance obtained in a study of galaxies with bursts of star formation (G. Comte et al. 1991), using direct and objective prism plates from the La Silla Schmidt telescope.

The photometric accuracy allows stellar magnitudes to be determined to within

0.05 mag., provided good sequences are available. This feature is of course interesting for the study of stellar populations as well as for extragalactic programmes. Among the latter is an extensive search for quasar candidates mainly based on multicolour photometry in the North Galactic Pole region; Schmidt plates taken at various epochs will also be used to investigate the variability of the detected galactic and extragalactic objects.

Among the programmes currently underway, we quote an extensive project based on plates taken with the ESO Schmidt telescope, and aimed at the search for baryonic dark matter in the Galactic halo. The technique consists of monitoring the magnitude of a large number of stars of the LMC, the light of which could be amplified by microlensing when passing close to small and massive halo objects, such as jupiters, brown dwarfs or small black holes (see contribution by Moniez in these proceedings). The ESO Schmidt telescope is used to search for deflectors with masses in the range $10^{-1} - 10^{-4}$ solar masses; a companion programme, using an assembly of CCD detectors, is aimed at the detection of masses in the $10^{-4} - 10^{-6}$ solar masses range. In the galactic domain, we can also mention several studies of stellar populations, and interstellar extinction via star counts.

MAMA is widely used for the identification, astrometry and photometry of optical counterparts of sources detected at other wavelengths: radio, X, infrared. The machine provides equatorial coordinates presently accurate to 0.25 arc. sec. (the accuracy of the PPM catalogue). We are confident of reaching 0.1 arc. sec. (together with a better uniformity in the photometric calibration) as soon as the TYCHO catalogue will become available, giving, for each Schmidt field, 700 stars known to a few hundredths of a second of arc and to a few hundredths of a magnitude.

References

Berger, J., Cordoni, J.-P., Fringant, A.-M., Guibert, J., Moreau, O., Reboul, H., Vander-riest, C., 1991. *Astron. Astrophys. Supp. Series*, **87**, 389.

Comte, G., Surace, C., Schuster, H.-E., Guibert, J., 1991. '*A New Survey of Starburst Galaxies using the ESO Schmidt Telescope*', *The Messenger*. In preparation.

Soubiran, C., 1991. *Astron. Astrophys.* In preparation.

ASTROSCAN II:
THE NEXT LEIDEN MICRODENSITOMETER

E.R. Deul

Sterrewacht Leiden, P.O. Box 9513, 2300 RA Leiden, Netherlands.

1 Introduction

In Leiden we have had the ASTROSCAN I microdensitometer operating for about fifteen years now. This machine has been described in detail by Swaans (1981, PhD thesis). As the name of the measuring machine suggests, a major part of its functionality is to accurately measure photographic densities. Although the old epoch Palomar Observatory Sky Survey (POSS) plates cannot be used for accurate photometry, many other plates, including the new epoch POSS plates, are photometrically correct. To accurately obtain photometric measures, one also needs to couple into that the positional tolerances of the measuring machine, thus obtaining high positional precision.

The ASTROSCAN I machine can only measure $\frac{1}{3}$ of the total area of a POSS plate, so that one needs to rotate the plate three times in the holder to scan the full-sized plate. Furthermore, the detector array is a rather small, 128-element photodiode array which makes scanning of a complete plate a time-consuming undertaking. Also, the positional accuracy and scanning speed of the ASTROSCAN I machine is limited by its, by current standards, less refined mechanics.

The above considerations prompted us to design a more accurate and powerful measuring instrument to overcome the problems encountered with the older instrument. The requirements for the new machine directly originate from the deficiencies of the old machine and the expected astronomical desires for the coming years. The development phase for the new machine, the ASTROSCAN II, has two distinct parts. The mechanical structure was designed and later built in an early stage of the project. This could be done because the development of new techniques in the world of mechanics has not evolved as fast as that for the other part of the development phase, the electronics. In fact it was only during the later stages of the development phase, when the mechanical structure was being assembled, that we began designing the electronic hardware and the underlining software. That way we took advantage of the most recent technology.

H. T. MacGillivray and E. B. Thomson (eds.), Digitised Optical Sky Surveys 109–114.
© 1992 *Kluwer Academic Publishers.*

The ASTROSCAN II project is now part of a large national project. The Expertise Center for Astronomical Image processing (Expertise Centrum Astronomische Beeldverwerking – ECAB) is currently being established as an organization that embraces all Dutch Astronomical observatories and institutes. This national facility has as its main goals:

1. the standardization of image processing environments in the Netherlands;
2. the installation, documentation, and maintenance of standard national and international image processing packages (e.g. AIPS, IRAF/STSDAS, MIDAS, GIPSY) at all collaborating observatories and institutes; and
3. provision of standardized access to national and international databases (e.g. Westerbork Radio Telescope Archives, La Palma Archives).

It is in particular in the context of the last point that the ASTROSCAN II measuring machine fits into the Expertise Center. As part of the national facility, the Dutch observatories and institutes have acquired a complete copy of the new epoch POSS and ESO plates. The copy is to reside at Leiden Observatory, where it can be accessed and digitized using the new ASTROSCAN II measuring machine. The POSS copy ordered consists of a full set of glass J plates, and the remaining survey on film. It has been shown that film copies are photometrically useful, but astrometrically less reliable. To bootstrap the astrometry for the films, the glass plates can be used. This economic choice preserves the full astronomical information content of an all-glass copy.

2 Requirements

The ASTROSCAN II measuring machine should provide the Dutch (or international) astronomer with a digitized copy of a photographic plate to such an accuracy that there is no loss of information during the digitization process. The no-loss-of-information clause in the framework of astronomy means that both the photometric and positional astronomical information that reside on the plate must be digitized without loss of that astronomical information. This means that all information is retained, including any plate errors, deficiencies, etc. The end-user generally has his own opinion about which algorithms should be used to extract the astronomical information from the digitized images. Therefore, the principal output (product) coming from the ASTROSCAN II measuring machine will be pixellated data, not a parameterized source list.

From the viewpoint of the ECAB, the ASTROSCAN II measuring machine will provide access to digitized copies of the POSS or other plates. Furthermore, it should provide a standard reduction facility which e.g. results in a parameterized source list of objects found on the digitized plate. The ASTROSCAN II reduction software should also provide an environment in which to implement externally developed astronomical algorithms which can then be run to process the pixellated data while it is being obtained.

To meet the digitization criteria mentioned above, the ASTROSCAN II measuring machine should perform within strict specifications. For the positional accuracy, the rms/pixel value should be less than one micron, which translates to well below a tenth of a micron for an ensemble of pixels (generally used in the determination of object positions). The photometric accuracy we wish to obtain with the ASTROSCAN II microdensitometer should be well below a thousandth of a unit, while in extreme cases we hope to gain another factor of a hundred in accuracy. In connection with this, the size of the detector elements should be less than twice the photographic material's intrinsic point spread function, thus of order ten microns either side (see Lasker, these proceedings). Depending on the true noise level of the photographic material and the internal noise level of the detector elements we should at least measure to densities of 2.5.

Furthermore, it is required that a complete POSS plate can be scanned in a single pass, without the cumbersome plate rotation scheme as is employed for the ASTROSCAN I machine. Finally, one would always want a measuring machine that does plate scanning in an absolute minimum of elapse time, provided the above criteria are first met.

3 Hardware Specifications

The hardware specifications follow directly from the measuring machine requirements.

The positional accuracy requirements impose high precision, high dynamic-range distance determination and table-positioning mechanics. With current techniques this means using laser interferometry to measure absolute distances. Unfortunately, the travel distance of the table on which one mounts the photographic plate is of such magnitude that only special lasers with extremely long coherence length can be used. Laser interferometry provides only a high precision, relative distance measure; to obtain an accurate and reproduceable zero-point we use the moiré pattern produced by a diffraction grating and the detector elements to lock phase with the interferometer. Because the expansion coefficient differs among materials used in the construction and because the wavelength of the laser light depends on temperature, air pressure, and air humidity, we have built a temperature controlled room with a maximum temperature variation of 0.2°C. Temperature, along with the other quantities, is measured and incorporated in the position determination.

Another point to make in connection with the positional accuracy is that we have chosen to measure photographic densities from a plate that is not in motion. Several artifacts are produced during the measuring of photographic densities from a plate that is in motion. Depending on the hardware implementation of the detector arrays, a cross-coupling between position and signal is introduced through the non-parallel manner in which the individual elements of the array are read out/initialized. This causes variations of positions across the array; e.g. the detector measures a line that is not perpendicular to the scanning direction. To make things worse, the deviation from perpendicularity of the measuring line is dependent on the direction of scanning; forward and backward scanning will produce different results. Furthermore, the intrinsic point-spread function of the detector elements is smeared out in the direction of scanning, producing an elongated point-spread function with a shape that is highly dependent on the scanning speed. These considerations prompted us to choose a move-stop-measure strategy, contrary to most other measuring machines.

A final point to make with regard to position-signal cross coupling is that we have chosen to perform the density-to-intensity conversion (logarithmic transformation) in software and not in hardware as is done in all PDS and some other machines. Only through ingenious phase correction filters is it possible to minimize the phase transformation induced by the logarithmic analog-to-digital (AD) conversion hardware. Therefore, we use a linear AD convertor which does not show any phase transformations and a software logarithmic transformation. This technique has produced highly accurate results for the ASTROSCAN I machine.

The photometric accuracy we wish to obtain principally puts constraints on light path, type, and the detector array type. To minimize the correlation length of the illuminating light, thus to minimize the diffraction pattern strength, we use a very wide band illuminating light source (wavelength range: 550 to 1100 nm). To also reduce the stray radiation, we have reduced the optical lensing system to a bare minimum of four glass-to-air surfaces; the illuminating light is brought into the optical beam using fiberglass tubes. To obtain the highest possible signal-to-noise ratio, detector stability, and linearity, we prefer a Reticon diode array rather than a CCD array. Each diode in a reticon (either 1-D or 2-D) is read out individually; there is only a two-point correlation between adjacent diodes which is a simple deconvolution problem; for CCD arrays the principle of bucket memories produces complex convolution functions that are expensive to correct. Currently the ASTROSCAN II machine is being built with a one dimensional detector, but provisions for a two dimensional detector array have been made so that at a later stage this step can be easily made.

The self-imposed constraint on the measuring strategy (move-stop-measure) and the relatively high measuring speed required implies that we have to build light-weight tables that have almost no friction. These are obtained using an aluminium frame x- and y-table, both mounted on air-pressure bearings and driven by linear motors. The current y-table has a weight of about 50 kg; its air-pressure bearings are hollow yellow-copper cylinders around steel rods. Through the cylinders, air is blown which spreads out evenly between the cylinder and the rod through specially designed air vents and canals. This way, a 7 micron gap is created so that both the x- and y-tables fly freely, without friction. Furthermore, linear motors are used to move the x- and y-tables. This type of motor is brushless so that the armature does not touch the mounting, causing the motors to move without friction; and there are no other mechanical components, like spindles, that can cause play in the construction. The tables are designed to have a lowest resonance frequency of order 64 Hz, the electronics are designed with a lowest resonance frequency near 75 Hz. It is thus possible to operate the move-stop-measure strategy at a rate of 25 Hz.

4 Current Status

At the time of writing this contribution, the following progress report applies. All mechanical hardware has been built and shipped to the Leiden observatory. There a special room has been constructed to provide an air-controlled environment. The installed cooling equipment can provide a stable temperature within a range of 0.2°C. A rather tight temperature control is required because different parts of the mechanical structure have been constructed from different materials (steel, aluminium, yellow-copper).

The ASTROSCAN II mounting is now being erected. The base has been set up and the steel rods are currently aligned so that at a later stage the x-table can be mounted. This is a rather tedious process because of the small tolerances on the position of the one meter long rods (less than a few microns).

Most of the electronic hardware has been designed. The electronics consist of three major parts: the open-loop linear motor control, the reticon diode readout electronics, and the domestic equipment control (focus motors, filter wheel, slit positioning, etc). Those parts that after prototyping have proved to work satisfactorily are built for permanent use. Construction is on its way to house the different electronic components.

The electronic hardware control was designed to be performed by a chain of 803(2)86 cpus. They are linked together with a special purpose parallel communications channel to allow distributed processing of measuring commands and a local ethernet to allow data transport. A 80286 computer controls the table positioning and all incidental motor steering (the domestic equipment). Above it in the hierarchy is a 80386 computer that reads out the reticon diode array, does the density-to-intensity conversion and transports the information up the tree to a higher level. There, another 80386 computer accumulates and reshuffles a complete square matrix of measured points, which on completion is transported to the top level machine. This is a workstation that does the astronomical interpretation and provides the user interface to the ASTROSCAN II measuring machine.

All interfaces between the individual computers (cpus), along with the accompanying software, have been prototyped and found appropriate. The interfaces between the cpus and the electronic hardware are currently being prototyped.

The software design can be separated into two parts: the part that concentrates on the steering and communication with the ASTROSCAN II hardware, and the part that does the astronomical interpretation of the measurements. The latter is the simpler one from the software engineering point of view because it consists of a user interface (such as those already developed for standard reduction packages) and a well-defined set of astronomical interpretation algorithms. The software design is such that it should be rather easy to hook a new algorithm into the existing package.

The software responsible for controlling the hardware is more complicated and touches the realm of real-time process control. It has been designed using the Yourdon-de Marco technique, since the main purpose of the ASTROSCAN II machine is the production and transportation of data, for which the above software engineering tool is particularly useful. Software engineering has gone down to the lowest level of the Data Flow Diagrams and is currently being evaluated. The next step involves actual coding.

Discussion

Gursky :

We have heard of three groups scanning the POSS plates, if I counted correctly. Did you consider purchasing a data base from one of these groups rather than committing to scanning the plates yourselves?

Deul :

The aim of the Astroscan II machine is to digitise a POSS plate without the loss of astronomical/photographic information. Full understanding of the machine characteristics is required to extract every bit (per pixel) of information in the digitisation.

MacGillivray :

Do you intend to systematically scan the POSS II plates or will you purely be providing scans according to demand?

Deul :

The latter; our machine's principal goal is to provide a digitised copy of the POSS without loss of (astronomical/photometric) information. This can only be done with our mechanical structure and start-stop technique in 1.5 days, which is too long to save/measure all plates.

Cormack :

Do you feel that the 12 bit A/D is adequate to cover your 2.5D dynamic range?

Deul :

Within the framework of future developments, a higher dynamic range A/D conversion is planned. Might I ask, however, if you can give me a reference where astronomy is done using > 3D dynamic range!

Parker :

Can you comment on the time-scale for Astroscan II actually coming on-line?

Deul :

The 'first light' is planned to happen at the end of this year (1991). The full user interface should be available at the end of the Expertise Center term, thus beginning in 1993.

A NEW ASTROMETRIC MEASURING MACHINE: DESIGN AND ASTRONOMICAL PROGRAMMES

Chr. de Vegt, L. Winter and N. Zacharias

Hamburg Observatory, Gojeubergsweg 112, 2050 Hamburg 80, Germany.

Abstract

The design of a new type of astrometric measuring machine, recently installed at Hamburg Observatory is described. The measuring system consists of a very compact high-precision, air-bearing, granite x-y measuring table; the photoplate is digitised in a frame-by-frame mode using the VIDEK-KODAK MEGAPLUS CCD-camera which provides a frame size of 1035×1320 square pixels of 6.8×6.8 micrometers at unit magnification. The machine design is based on a fully modular concept; machine operation and camera data are handled by independent computer systems. The maximum measuring area is about 270×270 mm^2, typical astrograph plates (measuring area $\approx 220 \times 220$ mm^2) can be digitised in about 30 minutes. First measuring programs will concentrate on the Hipparcos groundbased extragalactic reference link and various catalogues, in particular remeasurement of AC zones and the AGK2 plates.

1 Introduction

Photographic astrometry is still the most adequate and efficient technique to provide precise positions and proper motions of millions of stars on a global basis. This technique is capable of extending the present reference frame of bright stars, without any problems, by at least an order of magnitude, which opens the exciting possibility of providing for the first time a dense system of reference stars for a rigorous and uncompromised astrometric calibration of the large Schmidt-telescope based sky surveys, and hence the determination of precise proper motions of millions of very faint stars. To achieve this goal, the precise measurement of large batches of plates from a great variety of astrometric and non-astrometric telescopes has to

115

H. T. MacGillivray and E. B. Thomson (eds.), Digitised Optical Sky Surveys 115–121.

be accomplished in the most efficient way. In the last few years, the availability of new technological developments has provided all means for the construction of new types of astrometric measurement machines by taking advantage of highly efficient but moderately priced mechanical and electronic components.

2 Design of the Measuring Machine

To follow-up the rapidly changing technological developments, our measuring machine has been designed on a strictly modular concept. Furthermore, the design priorities are directed to highest metric stability and optimal determination of astrometric plate parameters, whereas photometry is a secondary concern. As a consequence, a very compact mechanical layout with minimum thermal distortions was a major design driver. Table 1 summarizes the main machine components.

Table 1. Main components of the Measuring Machine.

Granite air bearing table	size $1100 \times 900 \times 200$ mm^3
Two axes servo drives	free programmable modes (scanning, start-stop mode)
Heidenhain linear scales	LID 350C, 0.1μm resolution
Measuring speed	< 67 mm/sec at 0.1μm resolution < 134 mm/sec at 0.2μm resolution
Illumination system	Fibre optics-cable, diffuse illumination
Optical system (mapping)	variable magnification
Solid state detector system	CCD-Camera: VIDEK-KODAK-Megaplus 1035×1320 pixels, 6.8 square $<=>$ 63mm^2 at V=1. Output 8 bit parallel, 256 grey levels + analog video sign
Digitising speed	10-14 frames/sec, < 14 Mb/sec
Measuring area	about 260×260mm^2
Measuring accuracy	about 0.3 μm absolute calibration

All major components of the machine (Fig. 1) have been designed on a modular basis. For example the present Fiber Optics illumination system could be exchanged easily by a microdensitometer type system. The camera is mounted onto a granite bridge which carries a heavy duty motorized focusing unit, which in turn is attached to the camera mount and is adjustable in 3 axes. The camera therefore can be replaced also at any time by a more advanced system (e.g. Kodak's new $2 \times 2K^2$ model) or a microdensitometer system. A plateholder has been constructed which can handle very different plate sizes, the plate itself being mounted in a dedicated frame; when attached to the plateholder, the emulsion surface is already adjusted perpendicular to the camera axis and close to the final camera focus position. To maintain maximum stability, the plateholder is kept in a fixed position on the plate carriage; change of plate orientation by 180 deg. is achieved by re-inserting the frame in the opposite orientation.

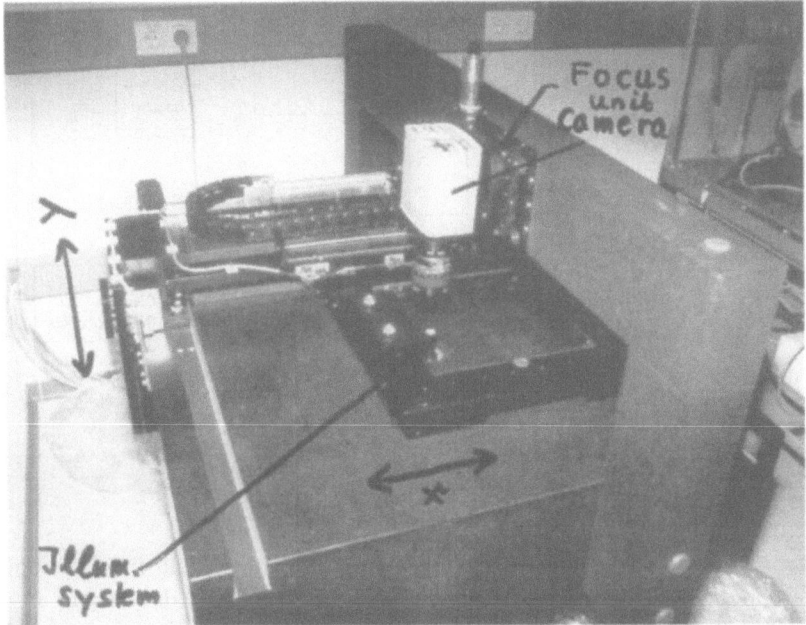

Fig. 1.

Concerning machine operation and detector system, a modular concept has been utilized too: the complete machine and camera operation is handled by fully independent computer systems. Whereas the machine operation does not give rise to any data transfer and computing speed problems and therefore can be handled easily by a specialized microprocessor, the high data rates of the CCD-camera and following image analysis pose a special problem. The camera computer is based on a VME-bus System, using several Motorola 68030/40 processors under the OS9 realtime operating system; details are presented in Fig. 2. (see also Winter et al., these proceedings). Two consecutive camera frames can be processed independently in a quasi parallel mode. Each of the two identical branches has its own 68030 processor connected in a memory-mapped mode to a CSPI QUICK CARD vector processor for model fitting by non linear Least Squares algorithms (Levenberg-Marquardt), whereas the star-subframe selection and frame data compression is accomplished by the **AFG2** frame grabber subsystem 68020 processor (for further details see Winter et al., these proceedings).

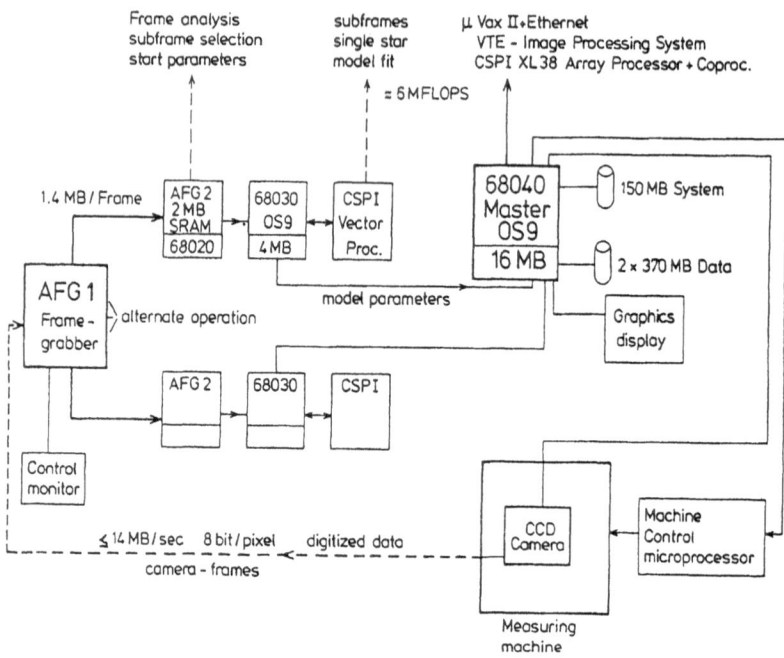

Fig. 2. Measuring machine computer system.

At present we have finished about 70% of the camera and machine operation software and the camera is operated currently on real plate data using an experimental measuring table setup. The machine will be used mainly in two operation modes: a star-list mode, moving from star to star by following a precomputed measuring list and taking a single frame at the star position, or a frame-mode, where the whole plate is digitised by taking successive frames with or without overlap. In both modes, all camera frames are taken while the machine is at rest.

Using the frame-mode, a whole plate (typical measuring area $\approx 220 \times 220 \text{mm}^2$) can be digitised in about 0.5 hours. If only a few hundred stars have to be measured, the star-list mode is preferable. Most plates can be digitised probably at V=1 magnification with a projected plate pixel size of 6.8×6.8 micrometers and resulting in a single frame size of more than 60mm^2; however the present optical system can be used at other fixed magnifications V=1.5 and V=0.5. The determination of optimal magnification to obtain the maximum astrometric accuracy and to extract all useful plate information is a matter of further detailed investigation at present.

The new measuring machine is operated in a temperature and humidity controlled room of our laboratory building and is mounted on a separate pier for decoupling from environmental vibrations. The machine itself is shielded directly by a ventilated housing. In addition, several high resolution temperature probes are attached at various parts of the machine and read out into the computer at short time intervals. Plate measuring is performed strictly at 20°C and about 50% rel. humidity; all unavoidable electronic heat sources in the room are controlled by additional cooling, if necessary. No operator is present during measuring and all plates are stored in the room for temperature adjustment before measurement. As the measuring machine is situated close to the large plate archive (which is likewise temperature and humidity controlled), plates are mostly already adjusted to measuring temperature. Under these very favourable astrometric measuring conditions, sub-micrometer measuring accuracy is possible on a routine basis. As soon as the whole measuring system has reached its full performance, we expect to achieve a digitisation rate of about 8 plates/day, all plates being measured in two orientations.

3 Astronomical Research Programmes

One of the most urgent problems of photographic astrometry is the extension of high precision position and proper motion work to much fainter magnitudes on a global scale to meet the requirements of many astrophysical and space research projects in the coming years. In particular, for the determination of proper motions, a multitude of old epoch plate material is available which could be remeasured to extract fully its great astrometric potential; important examples are selected zones of the Astrographic Catalogue (AC), the AGK2 plates and zones of the Yale catalogs. Furthermore, high precision new epoch catalogue projects, covering the whole sky to at least 14th magnitude, are urgently needed to densify the primary reference frame of bright stars as provided by the FK5 fundamental catalogue. A detailed description is given in Eichhorn (1974) and de Vegt (1988, 1989a, 1989b, 1991a).

These catalogues will provide, for the first time, an adequate dense net (typically 150-200 stars/sq.deg., $m_v = 12-14$) for a rigorous and uncompromised astrometric calibration of the large Schmidt Telescope sky surveys.

With the new measuring machine, large plate quantities (some 10^3) which are typical for those projects can now be measured and processed within a few months. Some examples of our future measuring programmes may be addressed briefly:

1. Extragalactic optical-radio reference frame. Precise optical positions of about 450 extragalactic radio sources, displaying optical counterparts, will be determined in the FK5 system and future Hipparcos instrumental system to provide a link to the inertial VLBI based radio reference frame and in particular to provide a final absolute orientation of the Hipparcos stellar net. About 1600 wide field astrograph plates, taken with two similar astrographs on both hemispheres will be measured and a system of 100-150 secondary reference stars ($m_v = 12-14$, ±0.05 arcsec) for each extragalactic field derived for the reduction of deep plates, obtained from large telescopes, which contain the faint optical counterparts (quasars, BL-Lacs, galaxies) of the radio sources. For further details see Ma et al. (1990), de Vegt et al. (1991b) and Russell et al. (1991).

2. Proper motions from AC-catalogue plates. A pilot program, to remeasure completely selected AC-fields and zones will be started to study the ultimate astrometric potentialities of this important global old epoch plate material to determine precise absolute proper motions to 2-3 mas/yr accuracy for millions of faint stars to 12-13th magnitude. Corresponding new epoch positions will be obtained with high quality astrographs. For first test fields, plate material from the extragalactic programme will be measured completely to the plate limit.

3. Remeasurement of the AGK2 plates. The favourable limiting magnitude of the 2000 AGK2 plates, covering the whole northern hemisphere with a two-fold overlap at epoch 1930, provides a great potential for the determination of proper motions of all stars down to m_b=11-12. The plates, which have two exposures of 3 and 10 minutes respectively, have been measured originally only for the much brighter AGK1 stars forming the AGK2/AGK3 catalogue (de Vegt 1988). The complete remeasurement of the AGK2 plates, which are stored under excellent conditions in our plate archive, will provide a major contribution to the construction of a homogeneous system of absolute proper motions of fainter stars for galactic research and catalogue updating.

4. The USNO-Hamburg global astrometric faint star catalogue project. Plans have been worked out to photograph the whole sky with a four-fold plate overlap, using two high quality astrographs placed on optimal sites in the northern and southern hemispheres respectively. Both instruments will observe in parallel with a common zone of additional overlap at the celestial equator. A limiting magnitude of at least m_v=13.5 is envisaged. The whole project, including plate measuring and catalogue construction, could be accomplished in five years. The resulting catalogue, containing about 7-10 million stars with an individual positional accuracy of at least 0.05 arcsec and a system accuracy of 0.01 arcsec will be based on the future Hipparcos global inertial reference frame. Further details can be found in de Vegt (1988, 1989a).

Because of current financial problems at both observatories, the project unfortunately is pending.

4 Acknowledgements

Chr. de Vegt gratefully acknowledges financial support from the Bundeministerium für Forschung und Technologie (BFMT) under grant 0100013-8 (Hipparcos).

References

de Vegt, Chr., 1988. 'Status of Photographic Catalogs. Available Material and future Developments'. *'Mapping the Sky'*, *Proc. IAU Symp.*, No. 133, p. 211.

de Vegt, Chr., 1989a. 'Global High Precision Position and Proper Motion Catalog Work'. *Conference on Digitised Sky Surveys*, *Bull. Inf. CDS*, Genf 1989, No. 37, p. 21.

de Vegt, Chr., 1989b. 'A New Type of Astrometric Telescope for the Construction of a Global High Precision Star Catalog'. *Contr. Van Vleck Obs.*, **8**, 51.

de Vegt, Chr., 1991a. 'Modern Photographic Catalog Work. Astrometric Techniques and Reduction Methods'. *Astrophys. Space Science*, **177**, 3. (= Proc. IAU Coll. 100, Belgrade 1987)

de Vegt, Chr., Zacharias, N., Johnston, K.J., 1991b. 'Groundbased Optical and Radio Astrometry of Hipparcos Extragalactic Link Objects'. *Adv. Space Res.*, **11(2)**, 133.

Eichhorn, H., 1974. *Astronomy of Star Positions*. F. Ungar Publ.C. NY.

Ma, C., Shaffer, D.B., de Vegt, Chr., Johnston, K.J., Russell, J., 1990. 'A Radio Optical Reference Frame I. Precise Radio Source Positions Determined by Mark III VLBI. Observations from 1979 to 1988 and a Tie to the FK5'. *Astron. J.*, **99**, 1284.

Russell, J.L., Johnston, K.J., Ma, C., Shaffer, D., de Vegt, Chr., 1991. 'A Radio-Optical Reference Frame II. Additional Radio and Optical Source Positions in the Northern Hemisphere'. *Astron. J.*, **101**, 2266.

HARDWARE AND SOFTWARE ASPECTS OF CCD CAMERA-BASED ASTROMETRIC PLATE MEASUREMENTS

L. Winter, Chr. de Vegt, M. Steinbach and N. Zacharias

Hamburg Observatory, Gojeubergsweg 112, 1050 Hamburg 80, Germany.

Abstract

This paper discusses the application of CCD cameras for digitisation of astrometric plate material. Details of the astrometric plate measuring systems and their design principles at Hamburg Observatory are discussed.
First results concerning performance and obtainable measurement accuracies are presented.

1 Introduction

CCD cameras are widely used in astronomy nowadays, because of their high dynamic range and quantum efficiency. To obtain this high dynamic range it is necessary to keep them at low temperatures.

In our application, CCD cameras are used at room temperature (20°C), which restricts their dynamic range but maintains their stable and homogeneous metric properties. To make full use of these advantages, two measuring systems (the first one being the prototype) were designed, always keeping in mind the special needs of astrometry. In particular the computer systems and software have been optimized to meet this goal, thereby keeping speed and measuring accuracy a prime concern. Table 1 gives an overview of both systems.

2 Hardware

Just by looking at one of our astrograph plates, it is obvious that most of the plate is empty — the tiny star images cover less than one percent of the whole plate area. In astrometry our main goal is to determine precise star positions, whereas

H. T. MacGillivray and E. B. Thomson (eds.), Digitised Optical Sky Surveys 123–131.
© 1992 Kluwer Academic Publishers.

magnitudes are needed only for identification purposes and to correct for possible magnitude-dependent position errors.

Taking this into account in the construction of our measuring systems, we decided to get rid of most of the image data as soon as possible and try to optimise the systems for on-line reduction. Nevertheless it is possible to store all image data, if a particular problem is not solvable by on-line reduction methods. Figure 1 gives an overview of our system. The plate is illuminated by a diffuse light source, consisting of a fibre optics waveguide for cold illumination and a diffusing screen as near to the emulsion surface as possible. The emulsion is imaged onto the CCD chip by a microscope lens (system 1) or a Nikkor macro lens (system 2, see also de Vegt et al., these proceedings). Digitisation is achieved in approximately 0.5 seconds and the frame-grabber is used simultaneously for preprocessing the image-data (system 2 only).

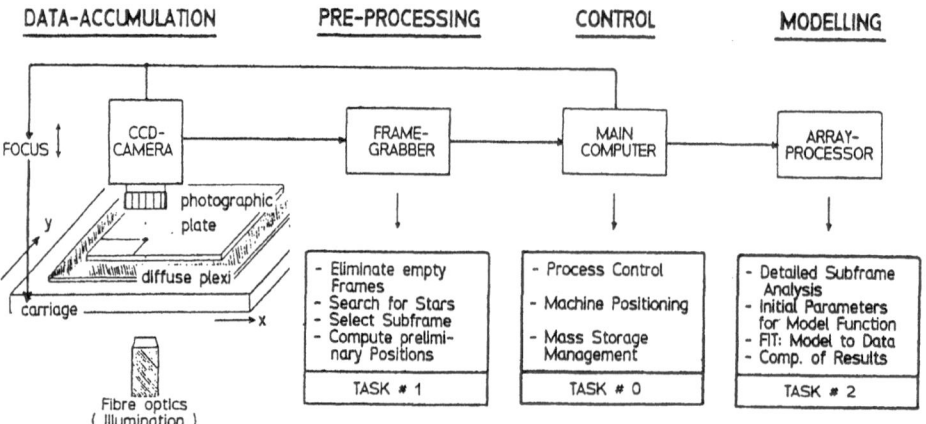

Fig. 1. Hardware overview.

As soon as the preprocessed images have been compressed, i.e. only the pixels belonging to the star images and their surrounding area have been saved, the plate carriage is moved to the next position, while the star image modelling is done in parallel by an array processor.

Designing the system in this way, the basic time frame consists of successive carriage movement and digitisation steps, which take less than two seconds. Due to parallel processing, all further tasks have to be finished in the same time interval.

2.1 Measuring Table

Positioning of the plates is one part of the measuring process, the other part is image centering, which will be discussed below. Therefore one has to monitor carefully the geometric properties of the plate carriage and machine base, because of its substantial influence on the measuring accuracy.

Table 1. Measuring System Characteristics.

	System 1	System 2
Measuring table		
Type (date)	MANN 422F (1969)	Anaspec Transglide (1989)
material	metal	granite, air bearings
metrology	Heidenhain LID 310/10	Heidenhain LID 350 C
	(0.1μm resolution)	(0.1μm resolution)
calibration	< 0.4μm	(to be completed)
reproducibility	< 0.2μm (marks)	(to be completed)
	0.5μm (stars)	(to be completed)
CCD camera		
Type	Hamamatsu C1700	Videk-Kodak Megaplus
used pixel	256 × 256 pixels (27μm)	1024 × 1280 pixels (6.8μm)
area on plate	1 mm^2 (scale 4.25μm/pix)	66mm^2 (scale 7.1μm/pix)
optics	Zeiss microscope lens	Nicon Nikkor macro lens
dynamic range	47dB (128 grey steps)	63 dB (256 grey steps)
	1.5 diffuse density	> 1.6 diffuse density
digitisation	8 bit, linear	8 bit, linear } built into
synchronisation	by C1000 camera controller	internal } camera
centering accur.	0.01 pixels (cal. marks)	0.01 pixels (cal. marks)
Computer system		
Type	Digital microVAX II	Redstone PME 68-31, 32,40
operating system	VMS 4.4, Digital	OS9 2.4, Microware
array processor	CSPI miniMAP XL38	CSPI Quickcard (two PCBs)
frame grabber	Hamamatsu C1000	EyeTec AFG1, AFG2 (two PCBs)
Measuring speed	(including plate handling)	
star-list mode	1 plate / hour	2 plates / hour
mapping mode	—	1 plate / hour
	(typical plate field size > 220 × 220 mm^2)	

For precise measurement, it is necessary to keep the photographic plate and all parts of the measuring machine at constant temperature — for example our measuring room has a temperature control to 20°C ± 0.1°C and relative humidity is kept stable at about 55%.

Plate positioning can be done in steps of 2.5 (1.0 for system 2) micrometers with an accuracy of 0.2 micrometers. The plate carriage is at rest (start-stop frame mode), while taking frames with the CCD camera.

2.2 Plate Illumination

Obviously centering of the star image, which means determining the position of the projected image on the CCD-chip, depends on the accuracy with which the profile of the stellar image can be measured (Figs. 2 to 5). The measurement itself consists of a geometric and a photometric component. Therefore it is necessary to calibrate the CCD camera photometrically as well as geometrically.

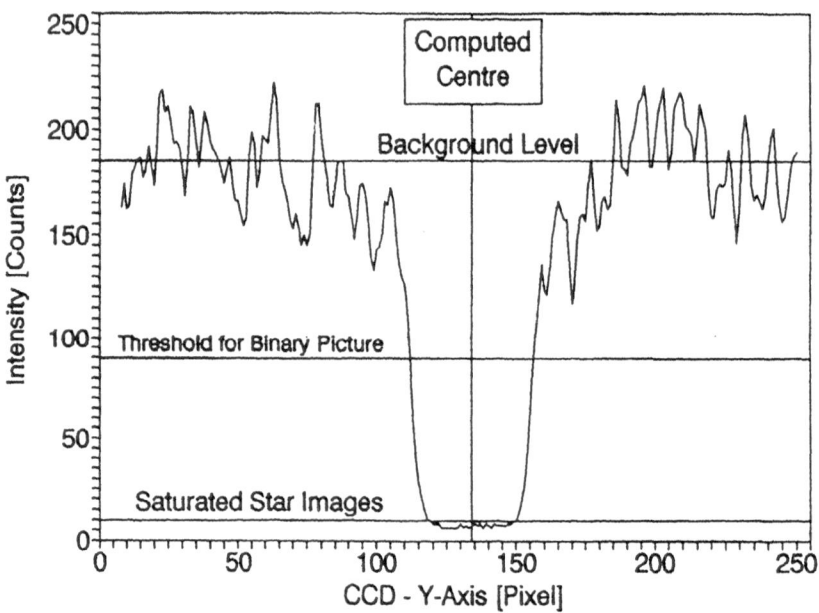

Fig. 2. Cross section of raw data.

Figure 4 shows the photometric calibration, obtained using a KODAK step wedge. To convert these diffuse densities to specular densities, they have to be multiplied by 1.32 (Lee & van Altena 1983). Compared to a microdensitometer the dynamic range in density is smaller, but by using total diffuse illumination and taking into account the influence of the graininess of the emulsion (i.e. Kodak 103aG), which has a high Callier coefficient (Callier 1909), this will be partly compensated. A further improvement of dynamic range is possible, if required, by increasing the magnification and thereby oversampling the star images even more.

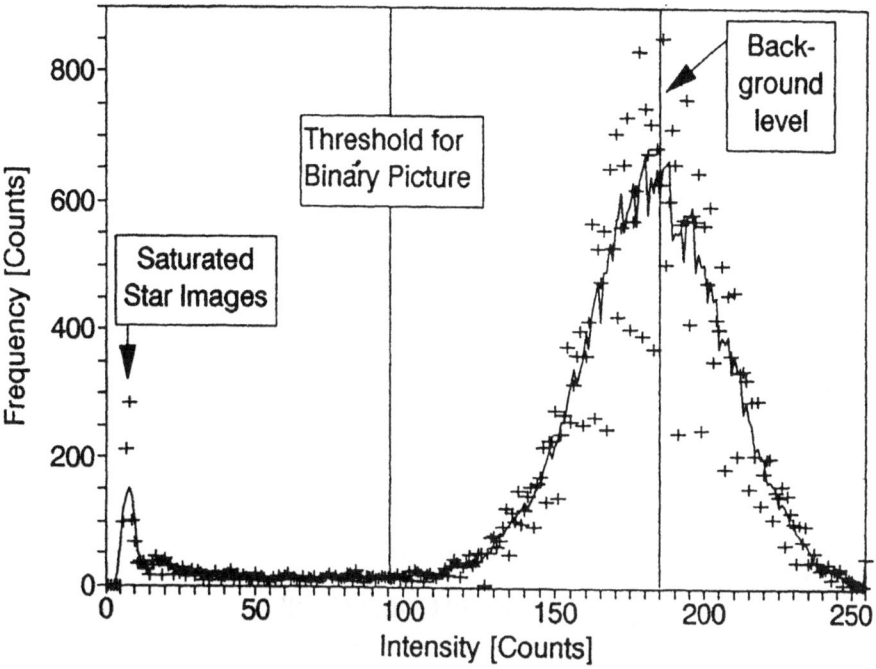

Fig. 3. Histogram of raw data.

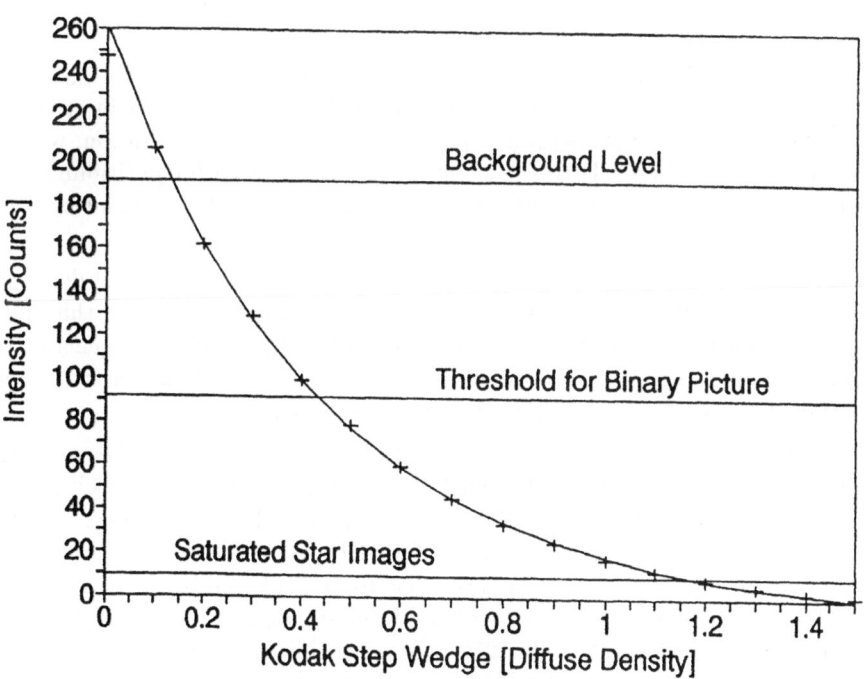

Fig. 4. Density calibration.

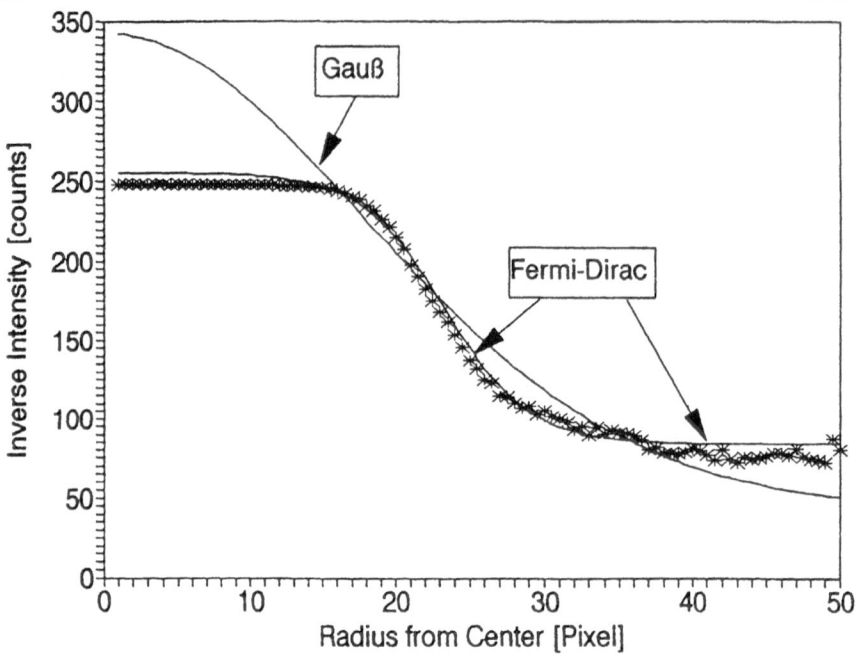

Fig. 5. Centering (both models).

2.3 CCD camera and frame-grabber

The geometric calibration of the CCD camera can be done either by measuring metric standard scales, or by moving a well-defined spot across the field of view of the CCD camera. The first method will result in an absolute calibration, whereas the second one ties the CCD to the same metric that is used for positioning. Our systems were calibrated using both methods with an internal agreement of about 0.05 micrometers.

Depending on the adopted CCD chip, the homogeneity of the pixel definition is better than 1/100 of a pixel. After correction for optical distortions this limits the ultimate measuring accuracy. Within the limits of resolution in general, the measuring accuracy can be improved by changing the magnification in a wide range, restricting the field of view of the CCD camera of course.

By looking at the electronics in more detail, it is obvious that the geometric stability of the camera is limited by synchronizing errors between CCD pixel clock and analog-to-digital converter (ADC) timing. For example system 1 uses a 6 MHz pixel clock, resulting in 0.17 microseconds for each pixel. To reach an accuracy of 1/100 of a pixel, the synchronization error has to be kept well below two nanoseconds. This is difficult to achieve with a band-limited transmission line of 10 m and electronic noise anywhere (Winter 1989). To solve this problem, the ADC has to be close to the CCD chip and both have to be timed by the same oscillator. An example is the KODAK-Megaplus camera, where CCD chip and ADC are mounted on the same printed circuit board (Chang et al. 1986).

3 Software

As was mentioned above, one of our goals in designing the measuring systems was to make all calculations as fast as possible to achieve a nearly on-line performance. The objects we are interested in mainly are well exposed or nearly saturated star images, recorded on different emulsions and plates taken with different telescopes, preferably astrographs or prime focus modes of larger telescopes with well-defined point-like images (de Vegt 1988, 1991).

Furthermore the selection criteria of the software have to be very strict: only undisturbed single star images will result in accurate positions, because photographic neighbouring effects like the Eberhard effect tend to change the position of the star in a somehow unpredictable way if there are close companions in the frame or unresolved double stars. These objects have to be singled out or at least marked for a separate, more sophisticated, astrometric reduction later on.

Another point to be considered is the continuous monitoring of all important hardware functions, which may degrade the measuring accuracy, for example changes in temperature or illumination strength.

Our solution is to generate quality parameters describing shape, illumination etc. for each star, besides the measured coordinates and diameter, which represent the astronomical information. The software will be expanded to cope with galaxy images as well.

3.1 Search Algorithms

The first step to be accomplished by the software (see Fig. 6) is to decide whether the camera frame is empty or not. The easiest way to decide this is to use a histogram (Fig. 3) displaying the frequency of each intensity level in the whole frame. Furthermore the histogram is used to compute threshold levels for the search routines and to monitor the homogeneity of the plate illumination. After analysis of the histogram, the software searches for all objects exceeding the threshold level. These objects are classified as mentioned above and initial parameters for the modelling are calculated. The modelling itself is carried out only for those objects we are actually interested in.

3.2 Modelling and Centering

Figure 2 shows a cut through the raw data of a star image. The plate noise is very dominant, which is even more obvious in the histogram (Fig. 3). We made some tests with binominal and median filtering. Both techniques reduced the noise and did not change the final star position within the error limits, as we expected. There was, however, no obvious effect in reducing the scatter in the astrometric positions, which indicates a dominance of the intrinsic geometric distortions on the whole plate surface. We therefore decided to use no filtering at present to reduce computation time.

Instead, the modelling is carried out with the preprocessed raw data by applying a non-linear least squares algorithm (Levenberg-Marquardt) and using a weighted

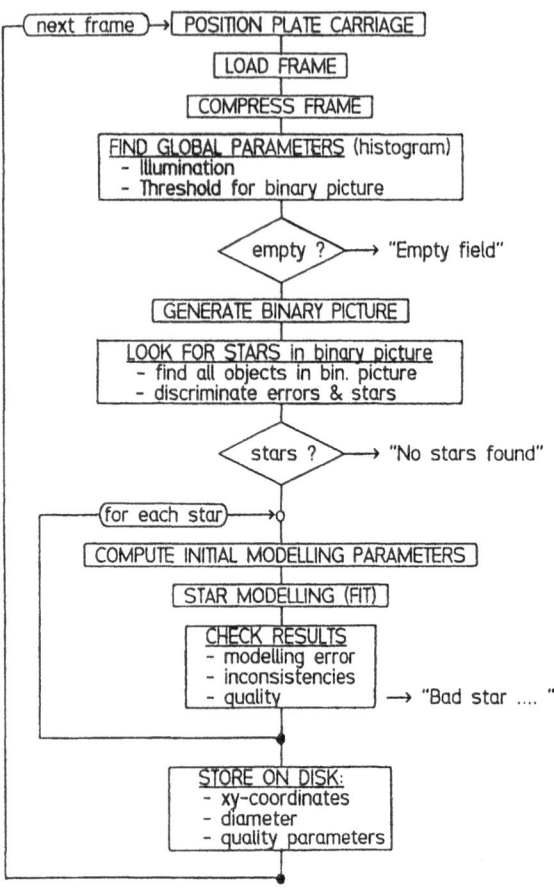

Fig. 6. Software.

two dimensional symmetric gaussian, which turned out to give the best astrometric results, or a weighted Fermi-Dirac-model (giving the best photometric results). Figure 5 illustrates the behaviour of both models, displaying data (∗) and fitted model function along a radius. The astrometric position parameters obtained from both models are in excellent agreement.

The centering accuracy is better than 0.4 micrometers for our astrograph plates, in agreement with the limits expected from the grain noise (Farrell 1967). For fine grain emulsions or marks on our calibration plates the centering error drops below 0.05 micrometers, which is also the intrinsic limit of the CCD geometry.

4 Concluding Remarks

System 1 is measuring plates on a production level now, whereas system 2 still needs to be calibrated, to obtain the full accuracy.

The stability of our software turns out to be very satisfactory. All images of astrometric interest could be measured, but the classification of images, which could not be measured, has to be improved further.

Finally we are planning to upgrade our old system (1) as soon as the new system is in full production rate, which is estimated for early 1992.

5 Acknowledgements

Chr. de Vegt acknowledges financial support from the Bundesminister für Forschung und Technologie (BMFT) under grant No. 100013-8 (Hipparcos).

References

Callier, A., 1909. *Zeitschr. f. wissensch. Photogr., Photophys. u. Photoch.* **7**, p. 257, eds. H. Kayser and K. Schaum.

Chang, W.C., Tredwell, T.J., Stevens, E.G., Nichols, D.N., 1986. 'High-density solid-state Image Sensors'. *SMPTE Technical Conference*, New York 1986.

de Vegt. Chr., 1988. In *'Mapping the Sky'*, *Proc. IAU Symp.*, No. 133, p. 211, eds. S. Debarbat, J.A. Eddy, H.K. Eichhorn and A.R. Upgren.

de Vegt, Chr., 1991. *Astrophys. Space Science*, **177** Proc. IAU Coll. 100, p.3.

Farrell, E.J., 1966. *Journal Opt. Soc. Am.*, **56**, 578.

Lee, J.-F., van Altena, W., 1983. *Astron. J.*, **88**, 1683.

Winter, L., 1989. Diplomarbeit, University of Hamburg.

SUPERCOSMOS

L. Miller, W. Cormack, M. Paterson, S. Beard and L. Lawrence

Royal Observatory, Blackford Hill, Edinburgh EH9 3HJ, U.K.

1 Introduction — Why Fast Measuring Machines?

We describe the design of and progress on a new, highly accurate machine for digitising photographic plates which is under construction at Edinburgh.

Before starting to build such a machine, we should ask whether this is the most effective way of achieving the scientific aims, or whether it would be better to pursue the alternative approach of surveying large areas of sky directly with large format CCD cameras. The use of CCDs in this way has a number of advantages and a number of disadvantages, which we discuss below. It is worth noting that the largest problem associated with surveying large sky areas in a finite time, the rate at which data needs to be processed, arises regardless of whether this is done via photographic plates and a measuring machine or via a direct CCD camera.

To make a comparison of the surveying methods, we must first define the problem. Let us discuss what is needed to survey the entire sky of 40 000deg^2 at a spatial resolution of $\frac{2}{3}$ arcsec, requiring the processing of about 3×10^{11} pixels. A wealth of multi-colour, multi-epoch information is stored photographically over that area. The sky is covered by 1716 UKST/POSS II photographic fields: in the Northern hemisphere each field is covered by 5 surveys in 3 wavebands; in the Southern each is covered by 4 surveys in 2 wavebands. In addition, there are tens of thousands of additional plates which have been taken for specific surveys in particular fields. The power of this dataset is illustrated by the observation that 30 percent of U.K. optical observational programmes on large telescopes are based on data from the U.K. measuring machines, even without the vast input of information from the new second-epoch surveys in the North and South. The photograph stores that information with high astrometric precision and with high dynamic range.

A new CCD survey would have the advantage that the photometric information would be on a linear magnitude scale, although zero-point calibration would still be needed as it is for photographic surveys. The dynamic ranges of present CCDs are not as great as those of photographic emulsions, but are still comparable to those achievable by fast measuring machines. Producing accurate astrometry from a mosaic of CCD exposures would not be trivial, but is likely to be of comparable

H. T. MacGillivray and E. B. Thomson (eds.), Digitised Optical Sky Surveys 133–139.
© *1992 Kluwer Academic Publishers.*

difficulty to deriving accurate astrometry from Schmidt plates. To make such a survey in a finite time would require a dedicated telescope for a few years, with the time required varying inversely with the area covered by the CCD camera per exposure. One interesting timescale is the time taken to read out the CCD: large 2-D CCDs currently need readout times about 20 μs per pixel to allow acceptable charge transfer. So if n CCDs were used simultaneously to survey the sky, the readout time alone would amount to an overhead of $200/n$ clear nights per waveband. It seems clear that to make optimum use of facilities it would be best to continue to survey the entire sky photographically at bright magnitudes, with the aid of an accurate measuring machine, and to utilise also the wealth of archive survey photographic material. Direct CCD data have insufficient advantages to make a bright multicolour CCD all-sky survey worthwhile, but large format CCDs could be put to very effective use making longer exposure, faint multi-colour surveys covering smaller sky areas.

2 The Performance of the Photographic Emulsion

The specification to which the machine should be built depends on the performance of the emulsion and the plates which are being measured.

2.1 The Astrometric Accuracy of Emulsions

A fundamental limit on the astrometric accuracy attainable is set by emulsion granularity. Lee & van Altena (1983) have calculated the astrometric accuracy expected taking into account the variation in emulsion noise with density. To first order, the accuracy in μm is independent of the image size (and hence the plate scale) provided the image is adequately oversampled, and depends on the emulsion type and the density of the image being measured. For stars with $B \lesssim 19$ on deep UKST plates with IIIa-J and IIIa-F emulsions the limit is $\sim 0.4\mu$m, which converts to 0.03 arcsec. The limit for the IIa-O emulsion used for POSS I is only 20 percent worse.

One criterion which must be satisfied is that the data are adequately oversampled. Lee & van Altena (1983) tested their analytical results by PDS measurements of Schmidt plates. The theoretical limit was attained with an oversampling of 3.

Attaining that accuracy over the full field of a UKST plate is difficult because of distortions introduced by bending the plates in the focal plane of the telescope, because of expansion of the plate under changing temperature conditions, and, most importantly, because of the effect of changing differential atmospheric refraction during a tracked exposure. The last effect can amount to several arcsecs at large airmasses in the U and B bands. A critical factor in being able to correct for these systematic errors is the coherence length over which they occur. It is worth noting that the full granularity accuracy has been achieved from astrometry of 4-m plates over sizes of order 15 cm (Chiu 1977; van Altena et al. 1988).

The manner in which these effects are corrected depends on the application. Obtaining accurate absolute positions can only be obtained using an accurate grid of reference stars with known positions, with a high enough sky density to oversample the coherence scale length of the plate errors. But for many purposes such an approach is unnecessary. To measure proper motions, either absolute or relative, we only need a background distribution of objects in common between plates making up the proper motion survey of sufficiently high sky density to oversample the scale length. Using background galaxies, which are known to have zero proper motion and which have a high sky density, would enable absolute proper motions to be determined. Stars could be used as a reference frame for measuring relative proper motions.

2.2 The Dynamic Range of Emulsions

Although photographic emulsions have non-linear response, they are capable of high dynamic range: IIIa emulsions contain information up to about 5 densities.

3 The Performance of Measuring Machines

To maximise the usefulness of the photographic data we would like a machine which adds no significant noise to either the astrometric or photometric performance of the emulsion, and which retains the full dynamic range of the emulsion. With a new generation machine such as SuperCOSMOS, it is now possible to achieve the first aim and to come close to achieving the second. Before describing how SuperCOSMOS achieves that, we need to define what we mean by astrometric accuracy.

We define two terms:

- **absolute accuracy** is the accuracy to which a position can be measured with respect to a standard length. Good absolute accuracy is needed when measuring plates with different field centres or when tying in to astrometric reference positions.
- **repeatability** is the accuracy to which a measured position can be repeated on successive measurements, which may be separated by long or short times. Good repeatability is acceptable when measuring relative positions on plates with the same field centres.

At accuracies down to $\sim 1\mu$m the table is usually the most important source of error. At greater accuracies the imaging system and temperature fluctuations also become important. The flying spot system of COSMOS and possible temperature fluctuations of $\sim 1°$C limit the repeatability to $\sim 1\mu$m. Temperature fluctuations of $1°$C cause expansions of $\sim 3\mu$m over 0.5m, and small temperature fluctuations can have an unpredictable effect on measurement accuracy and repeatability.

4 The Design of SuperCOSMOS

SuperCOSMOS has been designed taking into account the above considerations, and has the following design goals:

- 0.3μm metrology on plates up to 0.5 m square;
- < 2 hours to scan and parameterise UKST or Palomar Schmidt plates;
- good dynamic range – 2.3 diffuse densities (3.6 specular) above sky background;
- modular construction with low maintenance.

The absolute accuracy has been achieved by purchasing a Tesa-Leitz air-bearing granite table and installing it in a class 100 clean room which is temperature stable to ±0.05°C. The accuracy of the table has been assessed by using the Leitz mechanical probes to measure gauge blocks whose lengths have been accurately determined by the National Physical Laboratory. Measurements along the x–axis are shown in Fig. 1. The top figure shows the mean deviation of the measured lengths from the standard lengths over a range of 500 mm. The error bars are the standard error of a set of ten measurements. The overall rms error is $\sim 0.2\mu$m on both axes; the short-timescale repeatability (lower figure), as measured by the standard error, is $0.05 - 0.1\mu$m.

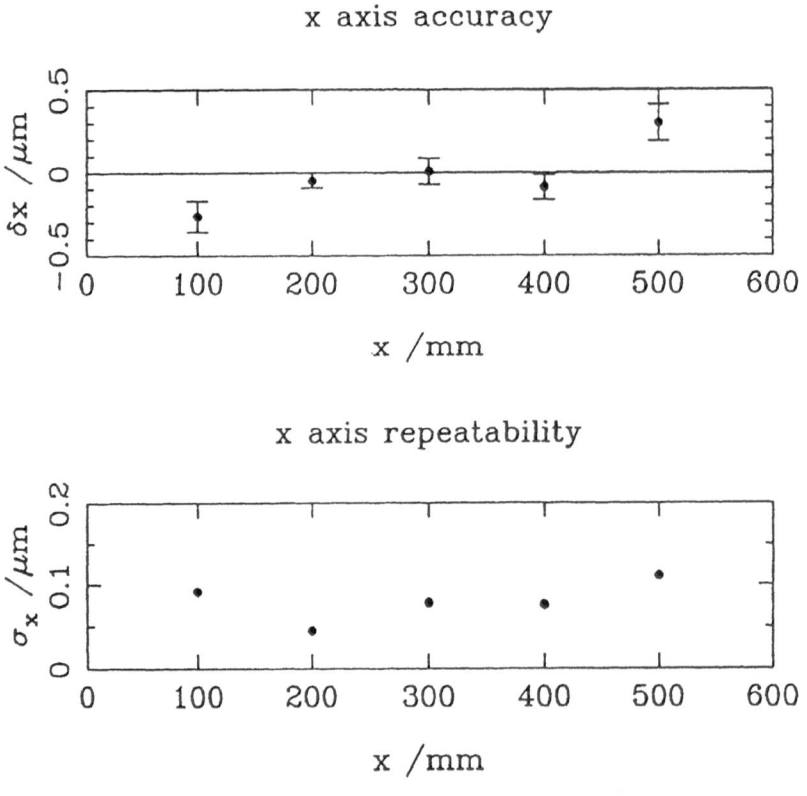

Fig. 1. Accuracy tests of the SuperCOSMOS table, showing absolute accuracy (top) and repeatability (bottom).

Having ensured that the fundamental part of the machine meets the accuracy requirements, we must not degrade that performance by using a flying-spot optical system. The solution is to use a fixed geometry imaging system which is mounted on the table by a rigid, lightweight (< 12 kg) bridge structure. The plate is illuminated by a variable intensity, external incoherent light source shining through a slit which is imaged onto the emulsion. The slit cuts out unnecessary scattered light. The illuminated portion of the emulsion is imaged onto a CCD camera by a 1:1 telecentric lens. The telecentric lens provides a good depth of field and, since it only accepts on-axis rays, cuts out scattered light extremely effectively, although at the expense of resulting in a highly specular measurement system. The CCD array itself is a linear device. The length used is 2cm, and 2048 pixels are read out every 8 ms. With such short integration times, the CCD can be operated at room temperature. The pixels are 10μm square, which corresponds to $\frac{2}{3}$ arcsec on UKST plates. Hence typical image sizes $\gtrsim$ 30μm will be oversampled by factors greater than three. The imaging system introduces only minor degradation of the resolution. The linear CCD conveys a number of advantages over a 2-D device: it has high dynamic range (peak/digitisation noise of 4.5 densities); it is cosmetically perfect; and it has a short readout time of 2μs — the data from the illuminated part of the array are dumped into shift registers and then subsequently read out while the array continues to integrate.

In operation, the CCD dark current and incident light level are continually monitored to an accuracy of 0.1 percent. This is essential for the accurate detection of images which are faint compared with the sky background level.

In order to achieve a plate measurement time of two hours, we need to process the data at a rate of 0.5 Mbyte sec^{-1}, from collecting the CCD data through to the end result of parameterised images. One practical limit on how fast data can be processed is the speed of analogue to digital conversion. SuperCOSMOS uses a 16-bit A/D converter with a convert time of 2μs pixel^{-1}. One consequence of this limit is that increasing the number of pixels in the CCD array would not enable the machine to scan faster: hence it does not matter that the array only uses a small number of pixels compared with 2-D CCD arrays. The machine is controlled and data processed by a low cost transputer front-end (see the article by Paterson, in these proceedings) and image parameterisation is carried out on a VAXcluster with accelerator.

5 Progress on SuperCOSMOS

The environmental chamber has been working to specification since Autumn 1990. The Tesa-Leitz table has been commissioned and the optics components are complete. The mechanical system being constructed at ROE to mount the optics on the table is nearing completion and the Mark I CCD camera has been completed and tested. A more advanced CCD camera is presently under test. The transputer system is working, and software implementation is ongoing. Software which presently parameterises COSMOS data on the VAXcluster is being modified for use with SuperCOSMOS data. It is planned to replace COSMOS by SuperCOSMOS on 1st April 1993, when the new machine will be operating at full speed but with only about 1μm absolute accuracy. Further work on pixel error-mapping will be undertaken to achieve the full astrometric accuracy of the machine in the raw pixel data.

References

Chiu, L-T.G., 1977. *Astron. J.*, **82**, 842.

Lee, J-F., van Altena, W., 1983. *Astron. J.*, **88**, 1683.

van Altena, W., Lee, J.T., Tsay, W-S., Lopez, C.E., 1988. *Astr. Soc. Pacific Conference Series*, **1**, 346.

Discussion

Taff :

You spoke of high measurement accuracy leading to extremely good proper motions, 10 mas/year. How do you plan to really achieve this in the presence of $0.°5 - 1°$ scale-length correlations on Schmidt plates?

Miller :

To achieve accurate absolute positions, we would need a large number of stars with accurately known positions per scale-size of any astrometric variation. To achieve accurate proper motions we can readily calibrate out those variations by measuring the positions of faint stars (for relative motions) or faint galaxies (for absolute motions).

Russell :

It seems to me that the new measuring machines are now capable of greater astrometric precision than is generally present on 14 inch Schmidt plates. On the UKST, we cannot control our environment so precisely and also certain techniques used in processing probably introduce small random emulsion shifts. Would you like to comment?

Miller :

The presence of useful information at the 0.4μm level has been empirically demonstrated, at least over scales of 10-20 cm. It should be quite feasible to extract proper motions to this accuracy over the entire plate even in the presence of correlated errors on 10cm scales. Note that temperature stability at the telescope is far less crucial than temperature stability during the plate measurement.

Cannon :

SuperCOSMOS has a pixel size of 10μm. We heard yesterday about a new film which may well give us images with sizes of less than 20μm in good conditions, in which case 10μm scanning pixels will be barely adequate for extracting all the information. Will it be feasible to modify SuperCOSMOS to work with smaller pixels if this becomes desirable?

Miller :

UKST images very rarely have FWHM smaller than 25μm at present, owing to seeing, telescope and tracking effects. I am sceptical that using a finer grain emulsion would lead to systematically smaller images. But if that did happen we could replace the present camera, although that would place an even greater load on the data storage requirements.

Baruch :

Concerning new fine grain films. Is it not true that to produce a photographic grain, two photons are required? In that case the theoretical resolution of fine grain film for faint objects is better than a CCD and can exceed the seeing limit because there is a much higher probability of getting two photons on a grain in the centre of the image than in the wings?

Miller :

If such an effect did operate, the profiles of images of a given peak exposed density would be independent of the grain size. But in practice there are a number of reasons why such 'super-resolution' is not observed — and even if profiles could be sharpened in this way the information content would have to be less than that of an unaffected image.

Russell :

Regarding the difficulty of calibrating photographic plates compared to CCDs. I would like to comment that 4415 film seems to be better in this respect than the IIIa emulsions with a longer linear section and less pronounced toe on its characteristic curve.

Westerhout :

I cannot let the comment pass by that CCDs will never be able to do full sky surveys. I predict that within 5-7 years we will be more efficient with CCDs than with photography. Compare the price of plates, measuring machines, manpower etc. with the price of arrays of CCDs: now and five years hence. Several CCD systems <u>now</u> under development will be able to do the equivalent of POSS-II in real time in one year. NOW, not 5 years from now. If the funds now spent on buying plates and building new measuring machines were spent on all-sky CCD development, we would be ready to start such a new POSS-II next year.

REAL-TIME DATA PROCESSING FOR SUPERCOSMOS

M. Paterson

Royal Observatory, Blackford Hill, Edinburgh EH9 3HJ, U.K.

Abstract

The high data rates achieved on SuperCOSMOS and the large data volume on a Schmidt plate place significant demands on traditional computer systems. A multi-domain Transputer system which can process large volumes of data at the required rate has been designed for SuperCOSMOS and is currently under development. The processes which must be implemented on such a system are discussed, and the methodology which supports the design is outlined.

1 Introduction

The COSMOS Group at the Royal Observatory Edinburgh is responsible for the design and development of the SuperCOSMOS advanced plate-measuring machine. The machine uses a linear CCD in the imaging system and, to minimise mass on the Z-axis, the CCD is run uncooled. As a result, in order to preserve the signal-to-noise performance of the machine, the CCD must be read out every 8ms. With 2048 pixels on the detector, this means that for 16-bit data the camera is generating data at 500 Kbytes per second.

2 The Data Required

COSMOS is an acronym for **C**o-**O**rdinates, **S**izes, **M**agnitudes, **O**rientations and **S**hapes. The off-line Image Analysis software which calculates these parameters for astronomical objects on the photographic plate requires information on the intensity values of image pixels. An additional constraint on the data produced is the use of the most significant bit to denote special values, such as end-of-lane. This means that the off-line software copes best with 15-bit data. However, 15 bits are sufficient to extract all the useful data using the current CCD camera; so, in the present configuration, no information is lost due to this constraint.

H. T. MacGillivray and E. B. Thomson (eds.), Digitised Optical Sky Surveys 141–145.
© 1992 *Kluwer Academic Publishers.*

3 Real-Time Processes

3.1 Data Acquisition

The main purpose of the system is to acquire data from the CCD camera electronics. The A/D converter produces data in the form of inverse 16-bit 2's complement words; the rest of the processing chain requires data in the range 0 to 32767, so some data conversion is required.

3.2 System Control

The Transputer system is responsible for machine control in SuperCOSMOS. The 3-axis co-ordinate table has its own microprocessor-based control system capable of executing primitive commands, but the higher-level table control algorithms are under the control of software running in the Transputer system. The major control tasks are plate positioning, x-axis drift control, y-axis dynamic correction (for lane straightness) and z-axis operation for dynamic focus control. In addition, the optical path-length must be set according to the thickness of the photographic plate under measurement, and this is done under Transputer control.

3.3 Signal Processing

There are three types of digital signal processing to be performed on the data acquired from the camera electronics. To obtain the highest positional accuracy from SuperCOSMOS, CCD triggering is performed by the table control electronics. This ensures that the pixel x-dimension is constant (normally 10 microns) but, because of minor variations in x-axis velocity, there are variations in integration time between CCD scans. The first process applied to the data normalises the d for this variation, as well as for fluctuations in the intensity of the light sou. used to illuminate the emulsion. The second is scan correction of the data (flat-fielding) to take account of effects such as pixel-to-pixel non-uniformity in quantum efficiency, and vignetting within the SuperCOSMOS optics. This is achieved by applying values derived from the scan signature obtained when scanning on clear air. The third is subtraction of the counts contributed by the dark current within the CCD detector. Data from the CCD include dark pixels, and these are used to calculate the dark current.

3.4 Background Determination

It is standard practice in machines of this type to determine the sky background on photographic plates. On SuperCOSMOS, square skyblocks of 64 × 64 pixels are used, and the first estimate of sky background is derived using the median of the transmission values within the block. Once the complete background map has been established, a weighted median filter is used to remove erroneous values caused by bright and extended objects on the plate.

3.5 T/I and T/D Conversion

The CCD camera is used to measure the amount of light transmitted through the photographic emulsion. However, of greater interest to the user of SuperCOSMOS data are the intensity values of the pixels. Lookup tables are generated from the spot sensitometer on the plate, and these tables can be used for Transmission to Intensity conversion or for Transmission to Density conversion if required.

3.6 Disk Interface

The background map, housekeeping data and corrected, normalised and converted pixel data are then written via a SCSI interface to disk.

3.7 Interpolation of Background Map

The sky background determination process gives a fairly coarse-grained estimate of the skymap for the plate, with discontinuities at the boundaries between skysquares. To smooth out these effects, background values are calculated for individual pixels by interpolating from surrounding skysquare values.

3.8 Pixel Thresholding and Run Encoding

On a typical UK Schmidt plate, between 75 and 95 percent of the pixels measured represent sky data of little or no direct interest to the user, as well as representing a significant overhead in storage and data transfer costs. To eliminate these sky pixels, each measured pixel is compared with the interpolated sky background value for the pixel. If the values differ by more than some predetermined level (a certain fraction above the sky intensity level), then the pixel is regarded as a true image pixel; otherwise the pixel is assigned a null value and can be ignored. Sequences (or runs) of adjacent image pixels are gathered and given a header to define the position of the run, and these sequences are then passed on for storage or further processing if required.

3.9 Positional Error Correction

This process is necessary to extract the full positional accuracy of the SuperCOS-MOS table. Mechanical deficiencies in the CCD detector, and residual systematic errors in the table mean that the pixel centres do not lie on a regular rectilinear 10 micron grid. Initial work (Miller 1990) shows that the full SuperCOSMOS accuracy can be obtained by correcting the data using error maps of the CCD and table. This is, however, a highly CPU-intensive task, requiring a throughput of over 30 million pixels per second, and will be implemented using the next generation of Inmos Transputers, the T9000 series.

3.10 VAX Interface

The VAX interface process provides a means of communicating between the Transputer system and the VAXcluster used as a host. All commands, non-plate data and status information are routed via this process.

3.11 Disk Farm Control

The disk system, for maximum bandwidth, is split into several logical subsystems, and these are connected to the individual domains within the Transputer system via multiple SCSI interface modules. Control of these connections and the issuing of commands is carried out by the disk farm control process.

3.12 Data Transfer to Image Analysis Cluster

All image analysis and off-line image parameterisation is carried out in an accelerated VAXcluster, which also hosts and controls the Transputer network. Transfer of run-encoded data and of commands and status information is via this process, using an interface between Transputer links and the Ethernet. However, should there be a need to transfer Mapping Mode (i.e. complete pixel information) data to the VAXcluster, it is unlikely that the Ethernet connection would be able to sustain the 500 Kbytes/s SuperCOSMOS data transfer rate in addition to the normal Ethernet traffic involved in clustering and networking. To this end, the option is available to make a direct connection between the SCSI interface to the disk farm, and the SCSI bus on a VAXstation within the cluster.

4 Design Methodology

The methodology adopted by the project team has been described elsewhere (Cormack et al. 1988; Paterson et al. 1989), but is based on the Yourdon/DeMarco method of analysis, with realtime enhancements. Analysis and documentation are carried out using a commercial package, 'Software through Pictures', running on a VAXstation 3100 under DECwindows.

5 Implementation

The hardware used to implement the design consists of 40 Inmos Transputers. The mixture of T414 integer devices and T800 floating point Transputers is divided into 3 domains. This allows full exploitation of algorithmic and operational modes of parallelism in the processing of SuperCOSMOS data.

Available to the network is a disk farm of 8 Gbytes capacity with a sustainable data transfer bandwidth of over 4 Mbytes/s. The entire system is under the control of tasks running within a VAXcluster, and communication between the cluster and Transputer system is primarily over an Ethernet connection.

6 Performance

The system hardware is capable of very high computing performance. It can execute some 400 million instructions per second concurrently with 40 million floating point operations per second. In addition, without impinging significantly on the processor performance, the system has a bisection communications bandwidth of over 350 Mbytes per second.

All processes have been tested at the required data rate of 500 Kbytes per second, and for evaluation purposes, some have been run with data rates of 1.7 Mbytes per second which is the maximum unidirectional data rate along any single Transputer communications link.

Future enhancements to the system will include the incorporation of Inmos T9000 Transputers to handle error correction for stage 2 of the project. It is expected that this will give a five-fold increase in computing power along with simplification of the network topology.

7 Conclusions

The system currently under integration on SuperCOSMOS is the most powerful computer system in the world dedicated to the acquisition of wide-field astronomical data. The use of programmable components and the adoption of structured techniques have given the project team rapid access to a highly flexible system during the analysis, design, development and test phases of the SuperCOSMOS project. The system has also proved to be scaleable up to the limitations imposed by current Transputer link technology. As this handles data significantly faster than current 16 bit A/D converters can operate and will, in any case, soon be superseded by faster Transputer links, it is expected that the system will be capable of handling the requirements of SuperCOSMOS and its future enhancements for the foreseeable future.

8 Acknowledgements

It is a pleasure at this point to record my thanks to the rest of the SuperCOSMOS project team, in particular Bill Cormack as Project Engineer, for their assistance in defining the requirements of the system and for advice and encouragement during the development stages.

References

Cormack, W.A., Herd, J.T., Beard, S.M., Paterson, M.J., 1988. *'Real-Time Processing of Large Volume Data from Photographic Plate Measurements'*, *Proceedings of the 8th Technical Meeting of the Occam User Group*, p. 125.

Miller, L., 1990. *'Error Mapping'*, ROE internal paper.

Paterson, M.J., Cormack, W.A., Herd, J.T., Beard, S.M., 1989. *'An Astronomical Imaging Application Using Transputers'*, *Proceedings of the 1989 IEEE International Conference on Acoustics, Speech and Signal Processing*, p. 1528.

A COLOUR/PROPER MOTION STUDY OF THE SOUTHERN SKY – A PLANNED DIGITISATION PROGRAMME ON SUPERCOSMOS

M.R.S. Hawkins

Royal Observatory, Blackford Hill, Edinburgh EH9 3HJ, U.K.

1 Introduction — the COSMOS Survey of the Southern Sky

The combination of all (southern) sky surveys in various passbands and at different epochs from the UK Schmidt telescope in Australia, and the measuring power of the COSMOS machine at Edinburgh have for some time held out the prospect of a comprehensive, digitised, multi-colour proper motion Southern sky survey. Although much of the photographic work is still to be completed, there already exists complete coverage of the Southern Sky ($\delta < 0°$) in the IIIa-J passband for high galactic latitude and in IIIa-F near the galactic plane. These plates have been measured by COSMOS and are available both in the form of pixel data at a sampling interval of 16μm and as parameterised image data. In the latter case images are detected above a pre-computed threshold related to the sky background and some 32 parameters output, including position, integrated density and parameters describing the shape and compactness of the image (see Yentis et al., these proceedings). The data are generally available to users and are stored in various ways, including optical disc, and in compressed form on Exabyte tape. So far, there is no external calibration for the entire survey, but a provisional calibration is available based on brighter standards for stars, and for galaxies measures of the density step wedges together with plate overlaps and CCD sequences. This calibration is in the form of coefficients for each plate which may be applied to the COSMOS measures of the IIIa-J survey.

To date, the most widespread use of these data have been in the determination of large-scale galaxy and cluster catalogues (see papers by Gursky et al. in these proceedings) and in the programmes of optical identification of the IRAS Faint Source Survey and ROSAT X-ray sources (see Wolstencroft et al. and Voges, both papers in these proceedings).

H. T. MacGillivray and E. B. Thomson (eds.), Digitised Optical Sky Surveys 147–150.
© 1992 *Kluwer Academic Publishers.*

2 The Future Planned Survey of the Southern Sky on SuperCOSMOS

There are now comprehensive plans to extend the survey into other photometric wavebands, and to incorporate proper motions. For the most part, the galactic plane region will not be covered by the new surveys. The survey of this region will comprise the already existing I band plates together with matching short exposure R (IIIa-F/RG630) plates. The combination of passband and exposure have been chosen so that in all but the most crowded regions the majority of stars are not contaminated by their neighbours. The high galactic latitude region of the Southern sky has been divided into two sectors, with $-20° < \delta < 0°$ and $\delta < -20°$. The equatorial belt will be covered by the UKST equatorial survey in the IIIa-J/GG390 and IIIa-F/RG590 passbands. These are sky-limited plates and are virtually complete in IIIa-J, with the IIIa-F expected to be complete within a couple of years. It is also hoped to complement these with a set of I plates, and a number have already been taken around the equator itself.

By far the most comprehensive coverage will be available for $\delta < -20°$. The Southern Sky Survey in the IIIa-J/GG395 passband has already been completed and issued as a photographic atlas. The UKST is currently engaged in a '2nd epoch' survey in the IIIa-F/RG590 passband, and this in conjunction with the existing ESO IIIa-F/RG630 survey will provide two epochs for the measurement of proper motions. In addition, a high galactic latitude I survey is being undertaken albeit rather slowly to date. Finally, plates from a survey for UVX objects will be available to provide U magnitudes. The IIa-O/UG1 passband plates have a 60 min exposure and will not go to the full depth of the other passbands. The expected survey limits in the various photometric bands are expected to be as follows:

$$
\begin{aligned}
U &= 18.5 \\
B &= 21.5 \\
R &= 20.0 \\
I &= 18.5
\end{aligned}
$$

The plates will be measured by SuperCOSMOS, the successor to the COSMOS measuring machine at ROE. The very high speed of the new machine will allow the entire measuring programme to be completed in a relatively short time, around 18 months depending on the machine's other commitments. The data will be parameterised with a similar set of parameters to those produced by COSMOS, and should allow much improved image description and classification. The sampling interval will be 10 microns and the expected positional accuracy for well exposed images should be approximately 0.3 microns.

3 Calibration

It is planned to calibrate the survey photometrically using CCD sequences in each survey field. These should go to the plate limit, and will complement the brighter magnitude already published as part of HST guide star catalogue. There will thus be a catalogue of most of the southern sky in UBRI colours. Proper motions will be obtained from the two sets of R plates with a baseline of about 10 years. This will give an accuracy of a few milliarcseconds for well exposed images. The astrometry will be tied to the extragalactic reference frame using measures of galaxies and possibly quasars. This will simplify large scale kinematic studies of the Galaxy as well as enabling searches for relatively high proper motion objects.

4 Applications

The digitised survey should lead to a wide range of applications in many areas of astronomy. Among the more obvious programmes are large scale searches for brown dwarfs, which will utilise both colour and proper motion data. Similarly, reduced proper motion diagrams will make possible comprehensive searches for cool white dwarfs and halo stars for which only inadequate surveys exist at present. Large amplitude variables should also be systematically discovered to faint magnitudes using the two R epochs. A by-product of the survey will be a complete catalogue of galaxies in UBRI to extend and complement the existing data for the Southern sky.

Discussion

Pennington :
> You have described a very ambitious program. Can you tell us how many plates will be scanned for this survey, when it will start, how long the scanning will take and when the data might be available?

Hawkins :
> We will start scanning when SuperCOSMOS comes on line, between 1 and 2 years from now. The scanning and processing time for the survey will be of the order of 2 years, and the data should be made available soon after, or possibly in smaller intermediate blocks.

Lasker :
> As many people in this room have at least complementary astrophysical goals and there may be opportunities for coordination (albeit better discussed at the next meeting), it seems appropriate to ask how you intend to begin.

Hawkins :
> The fields will be measured with a priority depending upon the availability of complete sets of plates. We don't expect astrophysical considerations to play an important part.

Clowes :
> What happened to the Kron and Chiu quasars in which they claimed to detect proper motions (!) — including one at 7σ?

Hawkins :
> I believe it must have been an astrometric error.

Monet :
> Yes. This result is a known astrometric peculiarity which (I believe) arises

from the interaction of the QSO spectrum with an astrometric chromatic term calibrated by stars. There is little reason to take this detection as evidence for a detection of a significant QSO proper motion.

DIGITISED OPTICAL SKY SURVEYS

Part Three:

DEVELOPMENTS IN TECHNIQUES — COMPRESSION, ASTROMETRY, PHOTOMETRY AND OBJECT CLASSIFICATION

DEVELOPMENTS IN COMPRESSION TECHNIQUES FOR COSMOS/SUPERCOSMOS DATA

J.E.F. Baruch [1], *A.J. Crolla* [1] *and R.D. Boyle* [2]

[1] Department of Electrical Engineering, University of Bradford, Bradford BD7 1DP, U.K.
[2] Department of Computer Studies, University of Leeds, Leeds LS2 9JT, U.K.

1 Introduction

The atlases of sky photographs produced by the Schmidt telescopes (e.g. the UK Schmidt Telescope — UKST) have served as the fundamental reference atlases for astronomical research. A typical hemisphere atlas is 900 fields covering approximately 6 degrees square and including all objects brighter than about 20th magnitude per square arc second.

One major area of use for the atlases is as finder charts. The authors are particularly interested in producing finder charts for robotic telescopes, including all objects accessible to the telescope and brighter than about 18th magnitude per square arc second. The objective is to make such a finder chart available on a single CD ROM.

The development of high-quality, digital image-acquisition systems such as COSMOS and SuperCOSMOS (MacGillivray & Stobie 1984; MacGillivray 1989) have enabled astronomers to extract valuable science from the Schmidt astronomical photographs. In addition, they have generated enormous amounts of image data. A single Schmidt plate produces about 2 Gbytes of digital information at 10μm resolution. The storage of this digitised information places heavy demands on existing hardware. The problem of image archiving is partially soluble with costly optical WORM drives, and at less expense with very high density tape drives or video cassette systems. These technologies do not aid the rapid transmission of data to remote users, or its distribution via wide and local area networks. In order to supply the astronomers with this valuable source of information, a strategy to lessen the great data-load is required. Image compression is the basis of this strategy (Baruch 1989). Image compression is also the only way of providing an atlas of the whole sky down to 18th magnitude on a single CD ROM.

H. T. MacGillivray and E. B. Thomson (eds.), Digitised Optical Sky Surveys 153–165.
© 1992 *Kluwer Academic Publishers.*

2   A CD ROM Atlas of the Sky

The generation of such a CD ROM will satisfy other needs in astronomy. Current catalogues go down to about 9th magnitude. The Hubble Space Telescope Guide Star Catalogue lists objects down to about 16th magnitude (Lasker et al. 1989). Some of the uses of a CD ROM atlas can be listed as follows:

1. Finder charts generated on a scale and orientation to match any telescope and produced on floppy disc.
2. The use at the telescope for on-line field confirmation.
3. Comparison of multiband overlays for position checks.
4. Use at the telescope for guide-star selection.
5. Mapping star surface density and magnitude distribution for galactic structure or extinction studies.
6. To identify changes between epochs, or nights: minor planets, large proper motion stars, supernovae, comets, large amplitude variables.
7. To estimate low precision magnitude limits for exposure calculations.
8. For general recreational astronomy: to zoom and pan around the sky.

The immediate objective of providing a star chart for robotic telescopes is to provide target acquisition assurance. Robotic telescopes working with CCD photometers could return images to the observer with the target object marked. This is inefficient and hardly practical if 1000 observations are required. The objective here is to use pattern recognition programmes to define the field around the object in both the telescope CCD field and the CD ROM field of the same coordinates. The pattern recognition parameters associated with the target object can then be returned with the observations. This will reduce the size of the data file associated with the target assurance from possibly 200 kbytes to about 20 bytes.

As an illustration, Fig. 1 shows a relatively crowded field. Assume the astronomer wishes to obtain photometric information on the varying object A and be assured that the brighter object B has not been observed nor the similar objects C, D or E. Pattern parameters generated by the astronomer using the CD ROM atlas can be matched to pattern parameters generated by the CD ROM atlas at the telescope and the pattern parameters generated from the observed field. The astronomer can obtain a high degree of assurance of the validity of the object acquisition.

2.1   COSMOS Data

COSMOS typically reproduces the photographic plate with 16 micron square pixels producing 14 bits per pixel. This produces about 600 Mbytes per plate. The background is averaged using a 32 micron binned histogram of pixel values to produce the threshold mapping mode (TM) data at about 60 Mbytes per plate. The TM data is currently stored on optical disc with a capacity of about 1 Gbyte per side, storing about 30 plates on each disc. Alternatively the data can be supplied on magnetic tape.

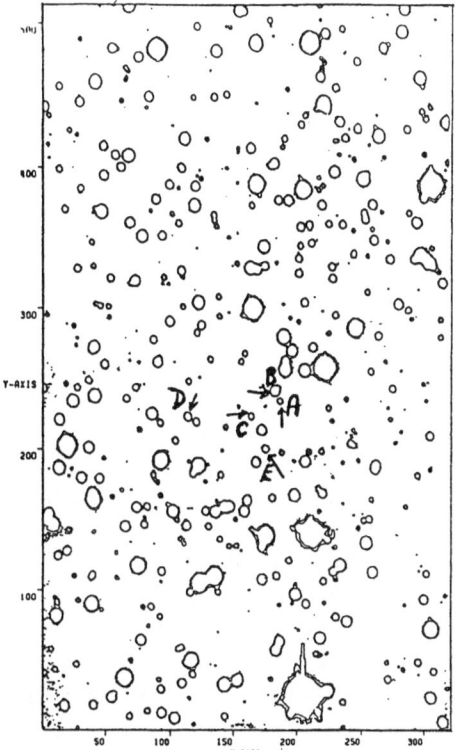

Fig. 1. Thresholded image of a crowded field.

The astronomical images under evaluation in this paper are Palomar-Schmidt photographic plates digitised by COSMOS. The image data is derived from the mapping mode (MM – i.e. all pixels recorded) measurement technique, scaled to produce 256 shades of grey. This effectively generates nearly a factor 2 increase in the compression ratios achieved below, which are all calculated starting with the scaled data.

3 Data Compression Algorithms for Astronomy

Data compression has been progressively improving over the past several decades. Techniques divide into two distinct sets. The first set provides lossless data compression, that is to say that the original data can be reproduced precisely when decompressed.

The second set of techniques makes the assumption that there is redundancy of data. The entropy of the image data source is not known and therefore we are no longer constrained by classical information and coding theory. These techniques lead to lossy data compression, the original image data is not reproduced exactly. This method of compression is clearly only valid when some degree of image degradation can be tolerated. The toleration of data loss is clearly very specific to the application for which the data is required. Every use may therefore require a different compression algorithm.

One route to compression is to extract the objects of interest from the image and produce a catalogue list. Typically, each entry in the list provides RA, DEC and a description of the object which requires a minimum of about 20 bytes. A highly compressed image requires about 2 bytes per star. In this way compressed images can increase the storage on a 600 Mbyte CD ROM from about 19 million catalogue objects on the Hubble Guide Star Catalogue to an atlas of about 300 million objects. This can only be achieved by deciding which losses are acceptable.

Classical information theory constrains lossless compression to a saturation level of around 8:1 (Jain 1981) depending on the source of the data. Such techniques have been implemented using Huffman and arithmetic coding techniques which are suitable for images which contain a degree of repetition e.g. stars (Kunt 1985). When it is possible to lose some of the information i.e. the grain information of the film then the compression ratio can be increased considerably. Fractal compression techniques (Barnsley 1989) have indicated that very large compression ratios may be possible. These have only been demonstrated in images which can be parameterised as edges and texture, and they are not considered further here.

The image compression objective, therefore, is to minimize the number of bits necessary to represent and reconstruct a faithful duplicate of the original image. The nature of this reproduction is governed solely by the application for which it is intended. Much previous valid research work has been undertaken in the wide field of image data compression; this however does not imply that it has immediate relevance to astronomical image data. Due to the lossy nature of the proposed compression scheme, it is imperative to select a technique which removes data considered to be unimportant, but retains valid information content e.g. stars, background details and galaxies. We must therefore evaluate the unique information content and attempt to tailor the compression scheme to the problems only associated with astronomical images. Regrettably, little valuable work has been completed in this niche.

4 Applications of COSMOS Data

The COSMOS machine is a project-driven tool, providing information for many areas of the astronomical community. The requirements of different users places different demands on an image compression scheme. It is necessary to consider the degradation of the image and the compute time for both compression and decompression.

The archiving of the data and the user requiring total fidelity in all aspects of the data are of no interest here. Our evaluation of the detailed data stream indicates that lossless compression of the COSMOS data can only yield compression ratios of up to 3:1. At this level, it is debatable whether compression is useful, especially considering the extra computational demands of coding and decoding. There are other interests in compressed images for distribution over the computer network; e.g. for e-mail or for eavesdropping on remote observing. The techniques developed here may be useful for such distribution but are mainly directed towards astronomers requiring basic detailed sky information for automated telescope finder-charts.

In an attempt to quantify the requirements, consider a Schmidt plate digitised by COSMOS yielding around 60MB of data. One hemisphere is covered by approximately 1000 such plates producing 60 Gbytes of uncompressed information. A compression ratio of 100:1 would enable complete storage on a CD ROM. This would provide an extremely cost effective way of supplying digitised sky surveys, and a far greater penetration of this information into the astronomical community than conventional methods would allow.

A compression ratio of 100:1 is a high objective to achieve whilst retaining an acceptable reproduced image clarity. The techniques described in this paper attempt to attain this level of compromise within the context of the application for which it is intended.

5 Image Compression Techniques

The solution to the image compression objective discussed in the previous sections is to develop an image representation that removes a significant amount of the inherent redundancy in the image data. The canonical form of the image is an extremely inefficient representation of the information it contains. In general terms, we require a transformation of the image data such that the transformed image consists ideally of uncorrelated data. This new representation can be subsequently stored more efficiently, yielding the desired compression ratio. The distortion artifacts introduced by this transformation must be minimised to lessen the resultant degradation. The Karhunen-Loeve transform (Kunt 1985) is the classical linear transform to produce uncorrelated coefficients. It is not normally used because of its large computational requirements. Alternatives are the Discrete Cosine Transform (DCT) and the Haar Transform. Neither of these are dependent on the statistics of the input image and they both work with manageable blocks of data 128^2 or 8^2, other solutions requiring much larger blocks. Small blocks facilitate rapid rebuild of the image and the implementation of functions such as zoom and pan.

It is considered that currently the most effective method of astronomical image data compression is to use a hybrid technique shown in Fig. 2 (Baruch & Crolla 1990). A transformation provides uncorrelated image data which may be thresholded and encoded very efficiently using a lossless technique. The transformation stage of the process, to produce an intermediate transform, can be successfully implemented using one of a number of common techniques. These include the Fast Fourier Transform (FFT), the Discrete Cosine Transform and the Haar Transform. The FFT is more compute intensive and has none of the advantages of the DCT or the Haar for astronomical images. It is not considered further. Table 1 shows a comparison of the relevant features of DCT and Haar transforms. After suitable scaling and thresholding this intermediate transform is encoded using an arithmetic or Huffman-based lossless technique to compress the coefficients that have not been discarded.

Figures 3, 4 and 5 show a section of original COSMOS data and compression ratios of 116:1 and 348:1 respectively. The transformation of the original image matrix is achieved using the Haar transform (Longair et al. 1986). The result of the Haar is to produce another matrix containing a handful of values which are

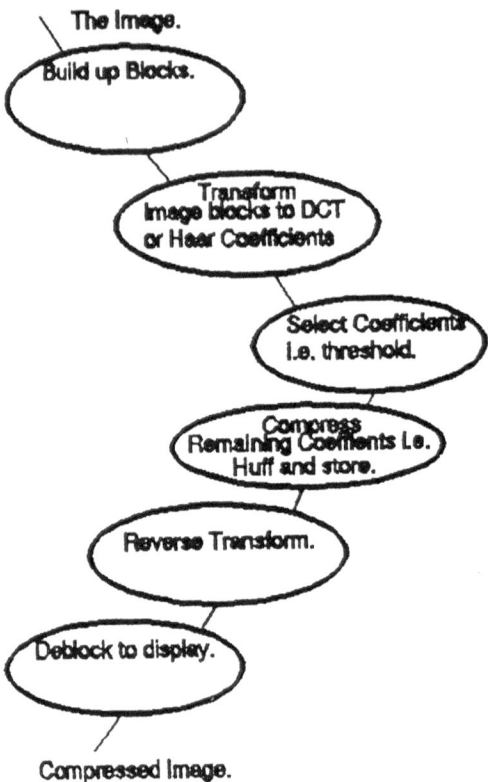

Fig. 2. The image compression process.

significantly larger than the others. The smaller values can be thresholded to some arbitrary value, often zero. The more values thresholded, the higher the compression ratio at the expense of final picture quality. The intermediate transformed image is in itself no smaller than the original image; however it contains less data which has the advantage of also being uncorrelated. A Huffman coding scheme has been used to encode this remaining data very efficiently producing compression ratios as high as 348:1. It can be seen that even at these very high compression ratios most of the significant detail necessary for finder charts is still present.

6 The Quality of the Compressed Image

The key parameters which define the quality of the compressed image in this case can be described as the problems of Confusion, Background and Artifacts. Confusion addresses the problem of how close two stars can be to maintain their separate identity with increasing compression ratios. Confusion is also a function of the rel-

Table 1. A comparison between Haar and DCT compression algorithms.

Haar	DCT	Comments
Integer Arithmetic	Floating Point Arithmetic	Haar Easier and Faster
Blocks sized in powers of 2. i.e. $(2^n)^2$ pixels. COSMOS 128^2	Blocks in 8×8 pixels	Haar less blocky and well suited to COSMOS scan width.
Very poor for TV images.	Good for TV images.	DCT implemented on a JPEG standard chip. Available and very fast.
Decompression time with available technology — 10's seconds.	Decompression time with available technology — milliseconds.	DCT will facilitate zoom and pans.

ative brightness of the two objects. Figure 6 shows a 8x zoom on part of the image. From the key, object B which is the brightest one is relatively unaffected by the compression. Object A which is faint but well separated from the others against a relatively homogeneous background is clearly retained. Objects C and D are tending to merge as the compression ratio is increased but still at 98:1 they are individual objects. As the compression ratio is increased the background grain disappears whilst the brightness and signal-to-noise ratio remain approximately constant.

The problem of background is to retain that part of the background that is essential for recognition, especially extended galaxies and nebulae. Background levels also affect the way in which faint stars are merged into the background as the compression ratio increases. Figure 7 again shows an 8x zoom on very faint objects. The grain fades but the objects and the signal to noise ratio is retained as the compression ratio increases.

Artifacts are essentially the blocking generated by the compression algorithms but there are other interactions with the image. Figure 8 shows an 8x zoom picture around a bright object. At even low compression ratios the grain disappears, the halo is lost but the 'cross' artifact remains even in the highest compression ratio pictures. It absorbs faint nearby objects e.g. star D and emphasises the blocking effect. The fainter star B just outside the halo is able to retain its identity even at the highest compression ratios. The very faintest isolated star C, fades out as the compression grows above 116:1.

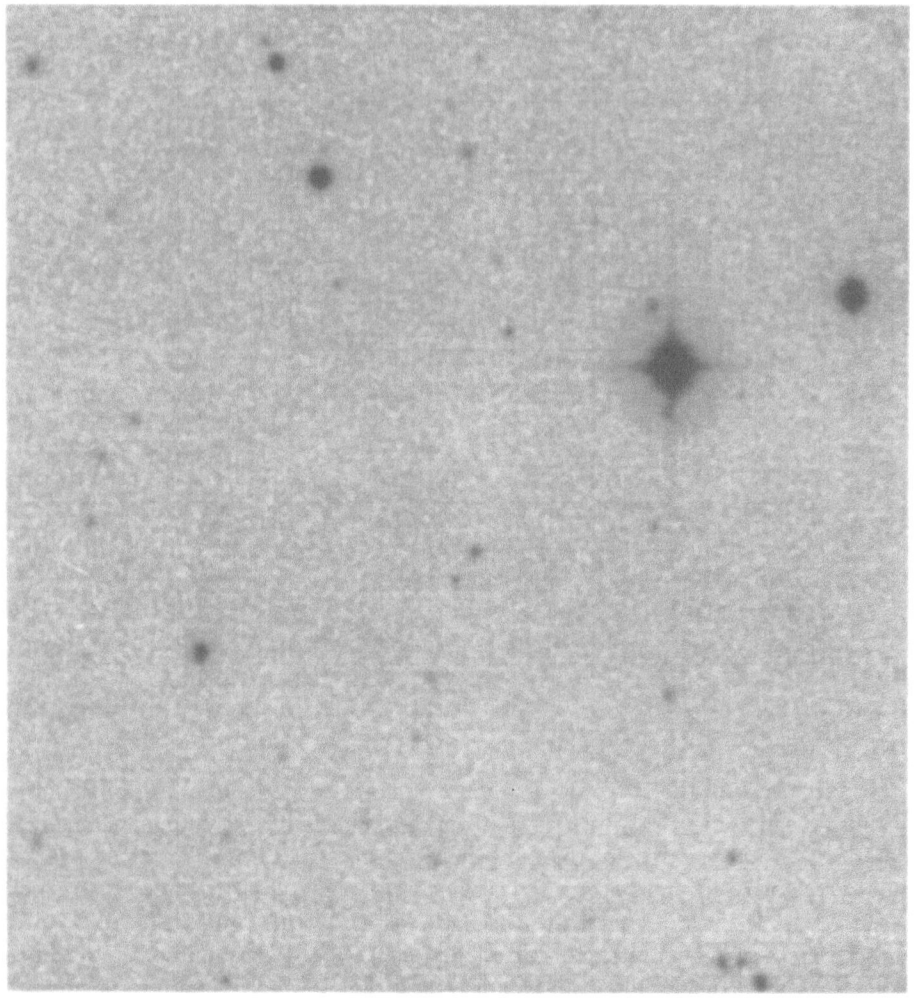

Fig. 3. 512 × 512 original COSMOS MM image. 256 grey scale.

One aspect of the artifact problem and blocking is the potential for smoothing. For images that are going to be used by robotic telescopes it is questionable whether the blocking is significant. For images for the human eye then some form of smoothing would appear to be desirable. Experiments with simple smoothing algorithms make very little difference, since they are tuned to particular spatial frequencies. It is thought that the most effective method would be to tune the smoothing to the particular Haar coefficients or use a fractal method. This problem is currently the subject of further work.

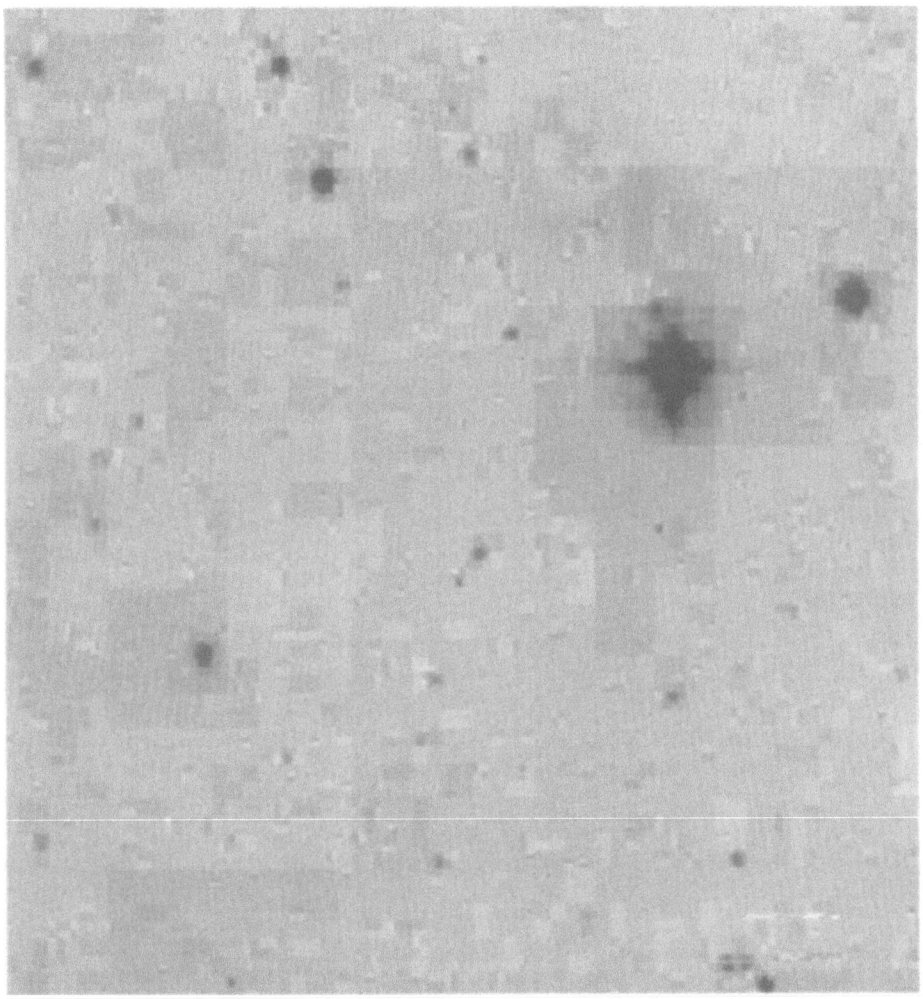

Fig. 4. As Fig. 3, 116:1 compression ratio.

7 Discussion

It can be seen above that compression ratios above 100:1 are viable. For complex fields containing extended galaxies the compression ratio will have to be reduced to ensure that sufficient data for recognition is retained (Baruch & Crolla 1990). For fields illustrated here it is clear that compression ratios well above 100:1 are quite acceptable. The images used here are from a Palomar Schmidt Telescope plate with estimated limiting magnitudes fainter than 20.

Hidden advantages of this compression technique also exist. The Haar transform can operate efficiently over any size of image block providing that it is a power of

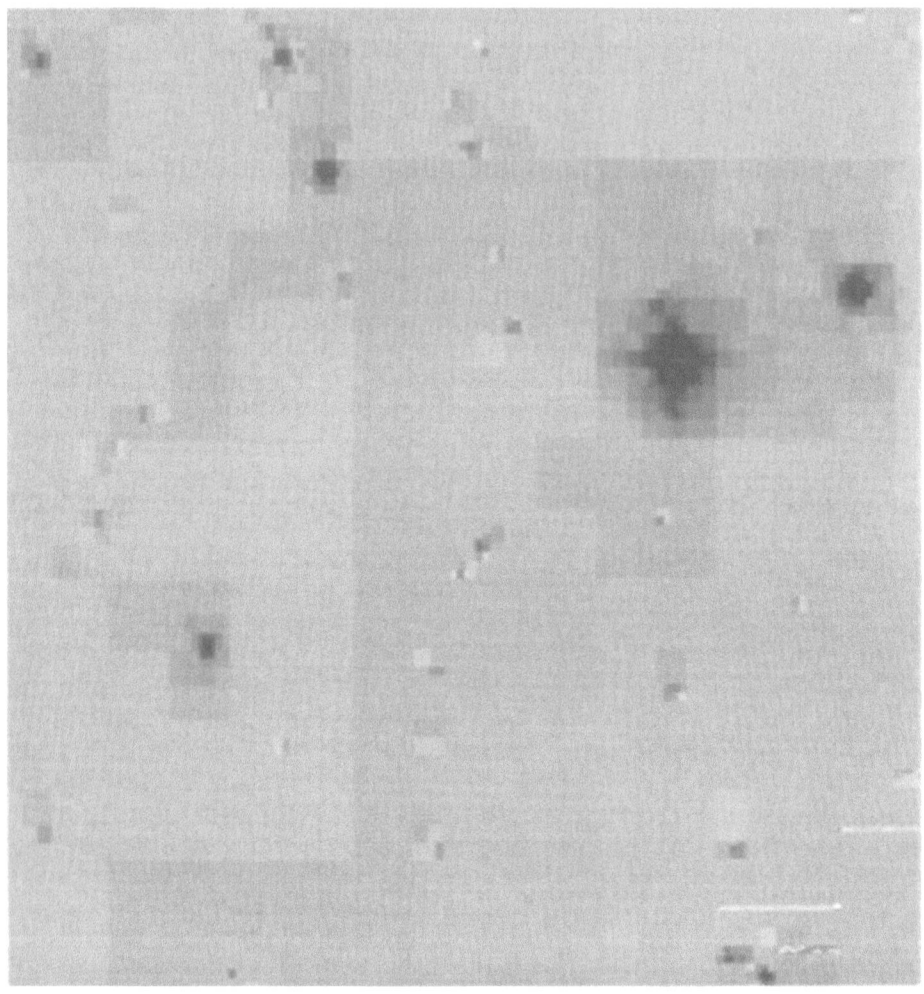

Fig. 5. As Fig. 3, 348:1 compression ratio.

2. The COSMOS machines digitises plates in lanes of typically 128 pixels. This means that the data can be compressed as it is produced by the machine. Other techniques will only work efficiently over large image matrices. This requires large areas of the plate to be built up before compression can take place. In addition, as the data is available in small, manageable chunks it is very easy for applications using the stored compressed data to locate a particular area quickly without the necessity of decompressing a large image matrix. This has enormous advantages in developing, say, an application to pan around the sky. The Haar transform, unlike other similar transforms, can be implemented using integer arithmetic only. This not

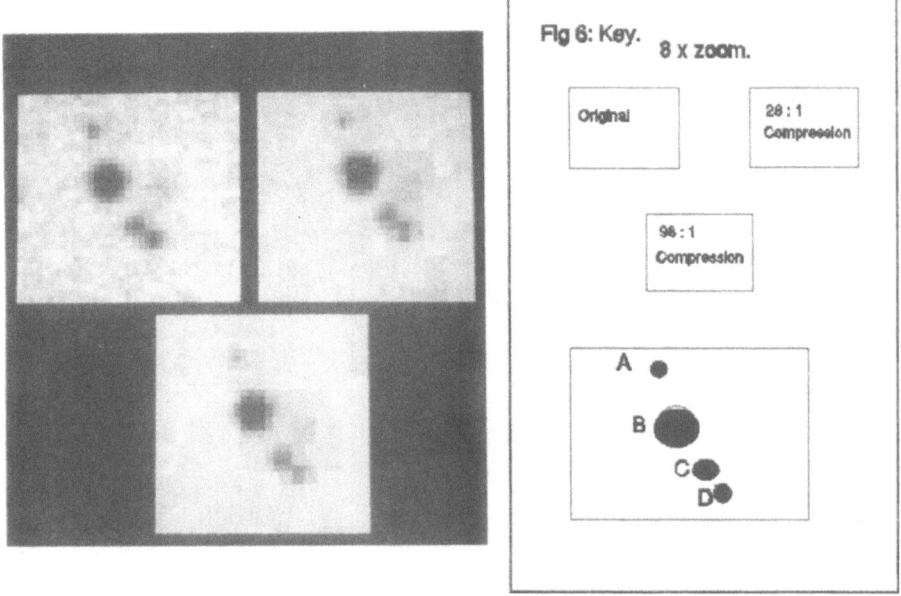

Fig. 6 Merge resistance with increasing compression ratios.

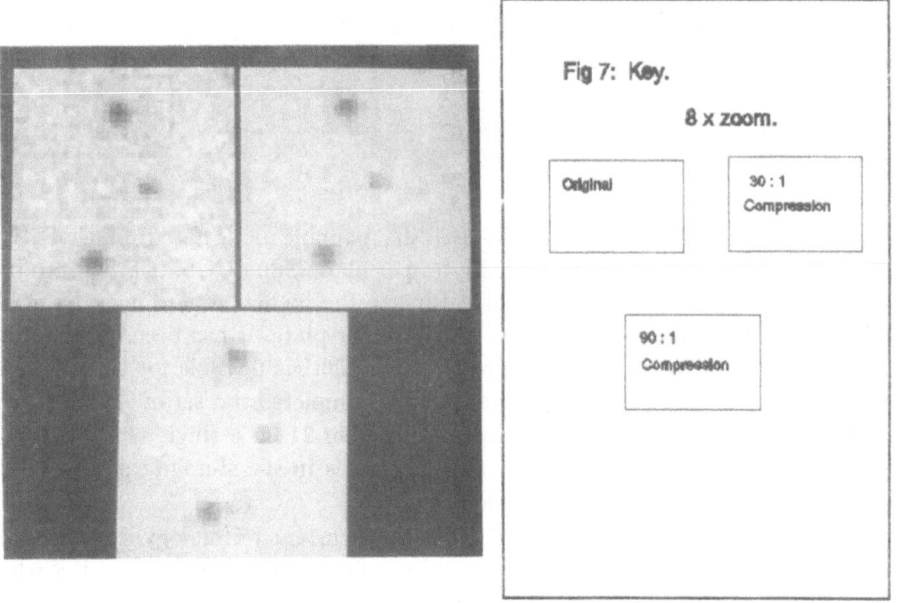

Fig. 7. Faint object retention with increasing compression ratio.

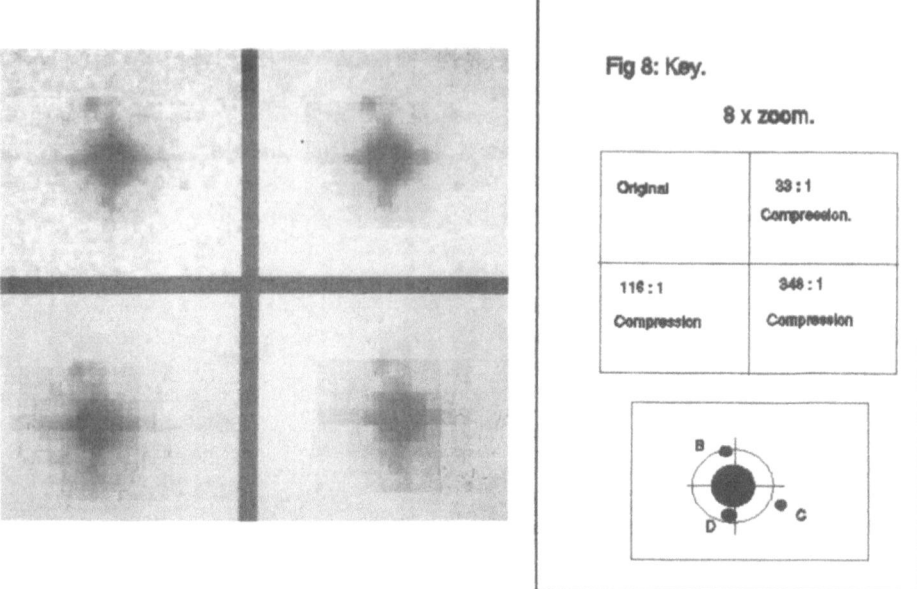

Fig. 8. Object loss resistance under complex adverse conditions.

only means that it can transform data faster than its floating point counterparts, but an implementation as a custom chip would be simplified.

8 Conclusions

The experimental work described here used the mapping mode data from COSMOS. This data form occupies about 600 Mbytes per plate. SuperCOSMOS will produce data in a similar format but more of it (due to the higher resolution to be used). In its basic form there will be about 2 Gbytes per plate. It has been shown above that an average compression ratio greater than 100:1 is possible for these images. It is clear that such methods could not put the complete atlas set of 900 plates per hemisphere including all objects to a magnitude of 21 on a single 600 Mbyte CD ROM. It is unlikely that there will be no increases in the storage capacity of the CD ROM over the next couple of years.

To achieve the aim of a single CD ROM with current technology it is necessary to increase the above compression ratios by a factor of about 20 to put a whole hemisphere on a single CD ROM. This would be achieved in two ways. Firstly by a reduction in the number of bits stored per pixel and secondly by an on-line software generated increase in the pixel size. The 10 micron sampling frequency of SuperCOSMOS would be increased to about 30 microns. This would increase the brightness of the faintest images stored in the atlas to about eighteenth magnitude.

References

Bailey, D.J., 1989. *'Image Compression using a Discrete Cosine Transform Processor'*, *Electronic Engineering*.

Barnsley, M., 1989. *'Fractals Everywhere'*, Pubs. Iterated Systems Inc., 5550-A, Peachtree Parkway, Suite 545, Norcross, GA 30092, U.S.A.

Baruch, J.E.F., 1989. *'Compressed Schmidt Images for Astronomy'*, *Digitised Optical Sky Surveys*, Newsletter No. **1**, 23.

Baruch, J.E.F., Crolla, A.J., 1990. *'Opportunities for Image Compression in Astronomy'*, *Digitised Optical Sky Surveys*, Newsletter No. **2**, 44.

Jain, K., 1981. *'Image Data Compression: A Review'*, *Proc. IEEE*, **69**, 349.

Kunt, M., 1985. *'Second Generation Image Coding Techniques'*, *Proc. IEEE*, **73**, No. 4.

Lasker, B.M., Sturch, C.R., McLean, B.J., Russell, J.L., Jenker, H., Shara, M.M., 1989. 'The Guide Star Catalogue 1', ST ScI preprint.

Longair, M.S., Stewart, J.M., Willliams, P.M., 1986. *'The UK Remote and Service Observing Programme'*, *Q. Bull. of the R.A.S.*, **27**, 153.

MacGillivray, H.T., 1989. *'Surveys with COSMOS'*, *Digitised Optical Sky Surveys*, Newsletter No. **1**, 9.

MacGillivray, H.T., Stobie, R.S., 1984. *Vistas in Astronomy*, **27**, 433.

Martin, R., Hartley, K., 1985. *'Remote Operation of Telescopes from the UK'*, *Vistas in Astronomy*, **28**, 555.

COMPRESSION OF THE GUIDE STAR DIGITISED SCHMIDT PLATES

R.L. White [1], *M. Postman* [1] *and M.G. Lattanzi* [1,2]

[1] Space Telescope Science Institute, 3700 San Martin Drive, Baltimore, MD 21218, U.S.A.
[2] European Space Agency, Villafrance, Madrid, Spain.

Abstract

An effective technique for image compression may be based on the H-transform (Fritze et al. 1977). The method that we have developed can be used for either lossless or lossy compression. The Guide Star digitised sky survey images can be compressed by at least a factor of 10 with no major losses in the astrometric and photometric properties of the compressed images. The method has been designed to be computationally efficient: compression or decompression of a 512 × 512 image requires only 4 seconds on a Sun SPARCstation 1.

1 Introduction

Astronomical images consist largely of empty sky. Compression of such images can reduce the volume of data that it is necessary to store (an important consideration for large scale digital sky surveys) and can shorten the time required to transmit images (useful for remote observing or remote access to data archives).

Data compression methods can be classified as either 'lossless' (meaning that the original data can be reconstructed exactly from the compressed data) or 'lossy' (meaning that the uncompressed image is not exactly the same as the original). Astronomers often insist that they can accept only lossless compression, in part because of conservatism, and in part because the familiar lossy compression methods sacrifice some information which is needed for accurate analysis of image data. However, since all astronomical images contain noise, which is inherently incompressible, lossy compression methods produce much better compression results.

H. T. MacGillivray and E. B. Thomson (eds.), Digitised Optical Sky Surveys 167–175.

A simple example may make this clear. One of the simplest data compression techniques is run-length coding, in which runs of consecutive pixels having the same value are compressed by storing the pixel value and the repetition factor. This method is used in the standard compression scheme for facsimile transmissions. Unfortunately, it is quite ineffective for lossless compression of astronomical images because even though the sky is *nearly* constant, the noise in the sky ensures that only very short runs of equal pixels occur. The obvious way to make run-length coding more effective is to force the sky to be exactly constant by setting all pixels below a threshold (chosen to be just above the sky) to the mean sky value. However, then one has lost any information about objects close to the detection limit. One has also lost information about local variations in the sky brightness, which severely limits the accuracy of photometry and astrometry on faint objects. Worse, there may be extended, low surface brightness objects that are not detectable in a single pixel but that are easily detected when the image is smoothed over a number of pixels; such faint structures are irretrievably lost when the image is thresholded to improve compression.

2 The H-Transform

Fritze et al. (1977, see also Richter 1978 and Capaccioli et al. 1988) have developed a much better compression method for astronomical images based on what they call the *H-transform* of the image. The H-transform is a two-dimensional generalization of the Haar transform (Haar 1910). The H-transform is calculated for an image of size $2^N \times 2^N$ as follows:

- Divide the image up into blocks of 2×2 pixels. Call the 4 pixels in a block a_{00}, a_{10}, a_{01}, and a_{11}.
- for each block compute 4 coefficients:

$$h_0 = (a_{11} + a_{10} + a_{01} + a_{00})/2$$
$$h_x = (a_{11} + a_{10} - a_{01} - a_{00})/2$$
$$h_y = (a_{11} - a_{10} + a_{01} - a_{00})/2$$
$$h_c = (a_{11} - a_{10} - a_{01} + a_{00})/2$$

- Construct a $2^{N-1} \times 2^{N-1}$ image from the h_0 values for each 2×2 block. Divide that image up into 2×2 blocks and repeat the above calculation. Repeat this process N times, reducing the image in size by a factor of 2 at each step, until only one h_0 value remains.

This calculation can be easily inverted to recover the original image from its transform. The transform is exactly reversible using integer arithmetic if one does not divide by 2 for the first set of coefficients. It is straightforward to extend the definition of the transform so that it can be computed for non-square images that do not have sides which are powers of 2. The H-transform can be performed in place in memory and is very fast to compute, requiring about $16M^2/3$ (integer) additions for a $M \times M$ image.

3 Compression using the H-Transform

If the image is nearly noiseless, the H-transform is somewhat easier to compress than the original image because the differences of adjacent pixels (as computed in the H-transform) tend to be smaller than the original pixel values for smooth images. Consequently fewer bits are required to store the values of the H-transform coefficients than are required for the original image. For very smooth images the pixel values may be constant over large regions, leading to transform coefficients which are zero over large areas.

Noisy images still do not compress well when transformed, though. Suppose there is noise σ in each pixel of the original image. Then from simple propagation of errors, the noise in each of the H-transform coefficients is also σ. To compress noisy images, divide each coefficient by $S\sigma$, where $S \sim 1$ is chosen according to how much loss is acceptable. This reduces the noise in the transform to $0.5/S$, so that large portions of the transform are zero (or nearly zero) and the transform is highly compressible.

Why is this better than simply thresholding the original image? As discussed above, if we simply divide the image by σ then we lose all information on objects that are within 1σ of sky in a *single* pixel, but that are detectable by averaging a *block* of pixels. On the other hand, in dividing the H-transform by σ, we preserve the information on any object that is detectable by summing a block of pixels! The quantized H-transform preserves the mean of the image for every block of pixels having a mean significantly different from that of neighbouring blocks of pixels.

As an example, Fig. 1 shows a 128×128 section (3.6×3.6 arcmin) from the E-plate of the Palomar Observatory–National Geographic Society Sky Survey (hereafter POSS) containing the Coma cluster. Figures 2, 3 and 4 show the resulting image for $S \simeq 0.5$, 1 and 2. These images are compressed by factors of 10, 20 and 50 using the coding scheme described below. The image compressed by a factor of 10 is hardly distinguishable from the original. In quantizing the H-transform we have adaptively filtered the original image by discarding information on some scales and keeping information on other scales. This adaptive filtering is most apparent for high compression factors (Fig. 4), where the sky has been smoothed over large areas while the images of stars have hardly been affected.

The adaptive filtering is, in itself, of considerable interest as an analytical tool for images (Capaccioli et al. 1988). For example, one can use the adaptive smoothing of the H-transform to smooth the sky without affecting objects detected above the (locally determined) sky; then an accurate sky value can be determined by reference to any nearby pixel.

4 Efficient Coding

The quantized H-transform has a rather peculiar structure. Not only are large areas of the transform image zero, but the non-zero values are strongly concentrated in the lower-order coefficients. The best approach we have found to code the coefficient values efficiently is quadtree coding of each bitplane of the transform array. Quadtree coding has been used for many purposes (see Samet 1984 for a review);

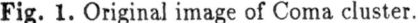

Fig. 1. Original image of Coma cluster.

Fig. 2. Result of compression by factor of 10.

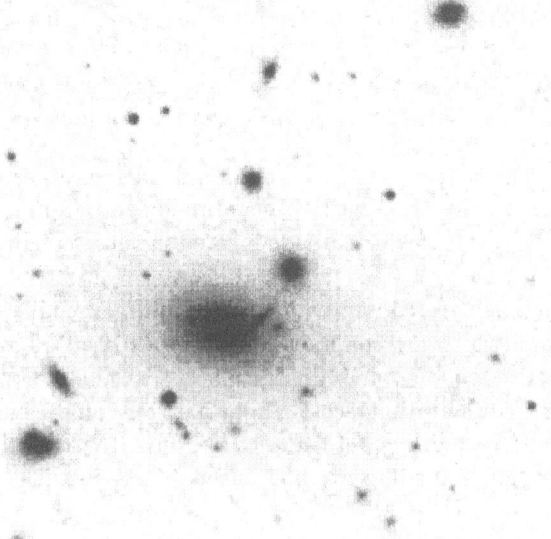

Fig. 3. Result of compression by factor of 20.

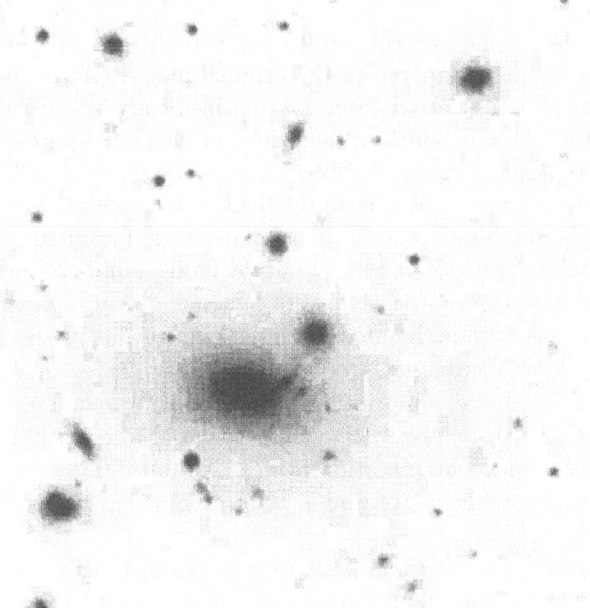

Fig. 4. Result of compression by factor of 50.

the particular form we are using was suggested by Huang & Bijaoui (1991) for image compression.

- Divide the bitplane up into 4 quadrants. For each quadrant code a '1' if there are any 1-bits in the quadrant, else code a '0'.
- Subdivide each quadrant which is not all zero into 4 more pieces and code them similarly. Continue until one is down to the level of individual pixels.

This coding, (which Huang and Bijaoui call 'hierarchic 4-bit one' coding) is obviously very well suited to the H-transform image because successively lower orders of the H-transform coefficients are located in successively divided quadrants of the image.

We follow the quadtree coding with a fixed Huffman coding which uses 3 bits for quadtree values which are common (e.g. 0001, 0010, 0100 and 1000) and uses 4 or 5 bits for less common values. This reduces the final compressed file size by about 10% at little computational cost. Slightly better compression can be achieved by following quadtree coding with arithmetic coding (Witten et al. 1987), but the CPU costs of arithmetic coding are not, in our view, justified for a 3-4% better compression.

For completely random bitplanes, quadtree coding can actually use more storage than simply writing the bitplane directly; in that case we just dump the bitplane with no coding.

5 Astrometric Properties of Compressed Images

We have performed some experiments to study the degradation of astrometry on the compressed images compared to the original images. A test field of 512×512 pixels (25μm each) was selected from the digitized copy of the Palomar 'Quick-V' plate N002 some 8 arcmin south of the center of NGC 188 (as defined in Sandage 1962). The plate limit is about $V = 19$. Two sets of 10 stars each grouped within about 10 arcminutes were chosen from the field, being careful to avoid overcrowded spots. The bright sample covers the magnitude range (as directly estimated from Sandage photometric data) $V = 14.2 - 15.3$; the faint sample covers the interval $V = 16.2 - 18.1$. The twenty stars selected were centroided using a circularly symmetric 2-dimensional Gaussian function on the original image and two compressed images: C_1, compressed by a factor ~ 12 (equivalent to Fig. 3), and C_2, compressed by a factor ~ 64 (even more compression than that shown in Fig. 4). The average magnitudes of the two samples were chosen to avoid images that were too bright (strong saturation and contamination from spike formation and filter ghosts) or too faint (too close to the plate limit), thus minimizing the contribution of the centroiding errors.

The astrometric test consisted of computing the distances among the ten stars of both the bright and the faint samples on images C_1 and C_2. Those distances were then compared to the corresponding quantities derived from the uncompressed image. The mean and the RMS about the mean of the distance residuals are given in Table 1.

Table 1. Astrometric test results.

Sample	Mean	RMS
	(pixels)	
C_1(12×, bright)	0.002	0.014
C_1(12×, faint)	0.037	0.273
C_2(64×, bright)	-0.004	0.064
C_2(64×, faint)	0.176	0.419

Individual distance residuals were inspected for possible outliers. Given the nature of this set (each star enters 9 times in the calculation of the set statistics), poor results during the centroiding stage would sensibly bias the values reported in Table 1. Only one residual in the C_1(12×, bright), one in the C_1(12×, faint), and one in the C_2(64×, faint) sets appear close to the 3σ limit. Also, the residuals were inspected for possible correlations with the distance and the magnitude difference between the pairs. None of the plots showed hints of such correlations.

That the quality of the image centering is a function of the compression factor is evident. Also not surprising is that the quality of the astrometry changes noticeably with magnitude. Much of the error for both the bright and faint stars on image C_1 may be attributed to centering error. If we assume that the additional errors in the compressed images add in quadrature to the centering errors (which is not necessarily true), then the astrometry errors in the C_2 image are at least 0.062 and 0.32 pixels (0.1 and 0.5 arcsec) for the bright and faint samples.

From these results we can conclude that compression factors as small as 12 (image C_1) give astrometric results on relatively bright (down to $V = 15 - 15.5$) stellar images hardly distinguishable from those obtained for the same objects on the uncompressed originals. On the other hand, high precision astrometry on fainter objects will indeed be *different* if done on more highly compressed copies. More experimentation based on image simulations is in progress to determine whether the astrometry is systematically worse and how fast it deteriorates.

6 Photometric Properties of Compressed Images

We have also performed some experiments on photometry on the compressed images. Sections 512 pixels square from digitized POSS E-plates covering Abell clusters A1016 (a high latitude cluster, $|b| \sim 52°$) and A3627 ($|b| \sim 7°$) were analysed by finding, classifying, and calculating isophotal photometry (in density units) for all objects on the image. Surface photometry was also performed on larger galaxies. Similar analyses were carried out using the original images and images compressed by factors (for A3627) of 6, 10 (cf. Fig. 2), 18 (cf. Fig. 3), 32 (cf. Fig. 4) and 67.

For the sparser A1016 image, the compression factors were larger: 9, 17, 78, 170 and 300.

A very interesting result is that the number of detected objects in the two fields actually increases slightly for modest compression factors! This is a result of the adaptive smoothing of the image, which flattens the sky and smoothes faint extended objects, making the detection of objects near the plate limit easier. More analysis of these results is needed to determine how the detection parameters should be modified to optimize discrimination against noise in the compressed images. For compression factors less than that shown in Fig. 3, essentially all of the objects in the original image are detected. For the higher compression factors some faint objects are lost.

Even for compression factors similar to that in **Fig. 4**, magnitude residuals between the compressed images and the original image are only about the same size as the intrinsic internal uncertainty in the photometry on the original image ($\sim 10\%$). Surface photometry on extended galaxies is unaffected by the compression until one goes beyond a compression factor equivalent to that in Fig. 4.

We conclude that, as one would expect from the design of the algorithm, it is possible to do accurate, unbiased photometry on even very highly compressed images.

7 Conclusions

In order to construct the Guide Star Catalog for use in pointing the Hubble Space Telescope, the ST ScI scanned and digitized Schmidt plates covering the entire sky. The digitized plates are of great utility, but to date it has been impossible to distribute the scans because of the massive volume of data involved (a total of about 600 Gbytes). Using the compression techniques described in this paper, we hope to distribute our digital sky survey on CD ROMs; about 100 CD ROMs will be required if the survey is compressed by a factor of 10.

The algorithm described in this paper has been shown to be capable of producing highly compressed images that are very faithful to the original. Algorithms designed to work on the original images can give comparable results on object detection, astrometry, and photometry when applied to the images compressed by a factor of 10 or possibly more. Further experiments will determine more precisely just what errors are introduced in the compressed data; it is possible that certain kinds of analysis will give more accurate results on the compressed data than on the original because of the adaptive filtering of the H-transform (Capaccioli et al. 1988).

This compression algorithm can be applied to any image, not just to digitized Schmidt plates. Experiments on CCD images indicate that lossless compression factors of 3-30 can be achieved depending on the CCD characteristics (e.g. the readout noise). A slightly modified algorithm customized to the noise characteristics of the CCD will do better. This application will be explored in detail in the near future.

8 Acknowledgements

We acknowledge a grant from NASA headquarters supporting this work. ST ScI is operated by AURA with funding from NASA and ESA.

References

Capaccioli, M., Held, E.V., Lorenz, H., Richter, G.M., Ziener, R., 1988. *Astron. Nachr.*, **309**, 69.
Fritze, K., Lange, M., Möstl, G., Oleak, H., Richter, G.M., 1977. *Astron. Nachr.*, **298**, 189.
Haar, A., 1910. *Math. Ann.*, **69**, 331.
Huang, L., Bijaoui, A., 1991. *Experimental Astron.*, **1**, 311.
Richter, G.M., 1978. *Astron. Nachr.*, **299**, 283.
Samet, H., 1984. *ACM Computing Surveys*, **16**, 187.
Sandage, A., 1962. *Astron. J.*, **135**, 333.
Witten., I.H., Radford, M.N., Cleary, J.G., 1987. *Commun. ACM*, **30**, 520.

PROPER MOTIONS OF M.A.S./YEAR ACCURACY WITH PHOTOGRAPHIC SCHMIDT PLATES

O. Bienaymé [1] *and C. Soubiran* [2]

[1] Observatoire de Besançon, 41 bis Av. de l'Observatoire, F-25010 Besançon, France.
[2] Observatoire de Paris (DASGAL), 61 Av. de l'Observatoire, F-75014 Paris, France.

1 Introduction

Currently accepted ideas concerning the structure and evolution of stellar populations in our Galaxy are mainly based on detailed observations of stars in the close solar neighbourhood. Faint star counts have long been advocated only to get an overview of the large scale features (Van Rhijn 1965). Apart from the pioneering work of the Basel Halo-Disc program (Becker 1965), only recently have the modern digitization and processing facilities opened the way to getting medium accuracy photometry for complete faint star probes in large selected fields (Chiu 1980; Sandage 1987; Gilmore & Wyse 1985; Mohan & Crézé 1987), thus giving access to the properties of star samples out of the solar neighbourhood. In some cases (Chiu 1980; Spaenhauer 1989; Reid 1990) the measurements include both photometry and proper motions.

We present two new photometric and astrometric (relative proper motion) surveys. The first survey was obtained in a 1.78 square degree field centered on l=3°, b=47°, near the globular cluster M5. The proper motion accuracy is below 3 mas/year (Bienaymé et al. 1991). The other survey (Soubiran 1991), near the NGP, was obtained in a 9 square degree field, the relative proper motion accuracy ranging from 1.4 to 2 mas/year.

2 Plate Material and Measurements

The photographic material used for these astrometric measurements are CERGA, Tautenburg plates, and first epoch POSS glass copies. All plate measurements have

H. T. MacGillivray and E. B. Thomson (eds.), Digitised Optical Sky Surveys 177–183.

been performed on the digitizing machine at the Observatoire de Paris (MAMA) which is optimized for astrometric performance. The resulting accuracy at least for strictly differential proper motions supports the conclusion that plates from different telescopes and even glass copies can be successfully used without major degradation of the astrometric results.

Plates are digitized with a 10 micron pixel size. 11 plates have been used for the near NGP field (largest timebase being 40 years), while 3 plates have been used for the near M5 field (largest timebase 31 years). Gravity centers of each stellar image are computed on-line by MAMA. Digitized images are stored on magnetic tape and reprocessed off-line for star detection and star centering (correlation centers — Bienaymé et al. 1988). In the latter field, the centering accuracy turns out to be about 0.5 micron on either coordinate. No evidence is found that using a smaller pixel size would improve this result.

Our direct tests of the astrometric capability of this program do confirm that this half micron limit is presently the state of the art of the MAMA machine. The repeatability over 24 hours is of the order of 0.5 micron. The coordinates given by the machine have been controlled using a Nanomask plate covered by large dots. Scans with different plate orientations over a 12×12 cm^2 area provide external calibration of the MAMA reference system: dot positions are again obtained with an accuracy of 0.5 micron. More extensive tests performed by the MAMA team show that this accuracy is achieved over the whole scanning table.

3 Centering Accuracy

The centering accuracy strongly depends on the plate quality. We have determined magnitude biases by comparison of plates taken at the same epoch. Biases are found to be negligible (smaller than 0.4 micron from V=12 to V=17) except for the brightest stars. Differential atmospheric refraction creates a colour term in proper motion determined from plates taken at different zenith angles. Our astrometric plates have been taken at small zenith angles (below 40°) and the effect is found to be at a 0.2 micron level between the bluest and the reddest stars.

4 Proper Motion Solution

A mathematical transform is used to model the transformation between plate coordinates from two epochs. For this purpose we assume the mean proper motion over the considered field to be null or constant. This assumption is valid so long as there is no star streaming in any part of the field. The mathematical functions used are Legendre polynomial expansions.

$$X(x,y) = \sum a_{i,j} L_i(x) L_j(y) \tag{1}$$

$$Y(x,y) = \sum b_{i,j} L_i(x) L_j(y) \tag{2}$$

where L_i is a Legendre polynomial of order i, x and y are the plate coordinates of about 2000 stars over 50 cm^2 (for the M5 survey), the $a_{i,j}$ and $b_{i,j}$ are coefficients to be determined, and X and Y are the transformed coordinates in the reference system of the other plate. Legendre polynomials do provide an orthogonal basis for the representation of functions on a rectangular window. It is not strictly true for the subspace represented by stars, although they do provide a reasonable approximation to orthogonality in so far as stars involved in the transform are densely and uniformly distributed throughout the rectangular window. Proper motions are given by:

$$\mu x = x_{Epoch2} - X(x_{Epoch1}, y_{Epoch1}) \qquad (3)$$

$$\mu y = y_{Epoch2} - Y(x_{Epoch1}, y_{Epoch1}) \qquad (4)$$

Coefficients of X and Y transforms are estimated to minimize the rms proper motions. The selected stars (used to compute the mathematical transform) have been chosen after rejection of stars with large proper motion (15 microns), of stars that have been improperly centered at least one time, and of the faintest stars.

The large number of stars allows the computation of transforms with high order polynomials and the correction of plate distortions at any scale length. A polynomial of order 3 or 4 for the L_i (respectively 16 or 25 $a_{i,j}$ or $b_{i,j}$ coefficients) gives the best fit with a proper motion rms of 5 microns.

Additional solutions based on subsets including only one half of the reference stars randomly chosen have been tried to check for the significance of the transform. Differences between two such solutions are plotted in Fig. 1; a reasonable agreement is obtained except near the border. This comparison indicates that the fourth order transforms are significant at the 0.5 micron accuracy level.

Higher orders (up to 10 corresponding to 121 coefficients) have been tried but the fit is not significantly improved, and the computed proper motions are only slightly modified. The average difference between orders 4 and 7 in proper motions is 0.5 micron in the Y transform but 1.0 micron in the X transform (largest changes remain near the borders where the transforms are loosely constrained).

Subsamples with different magnitude ranges or different B-V indices allow verification that there is no differential effect over the field. Transforms from each subsample do not deviate from the reference one by more than a constant and a 0.5 micron rms spread. The constant is related to the kinematic differences of the considered sample of stars.

In conclusion, plate distortions can be modelled with 4th order X,Y transforms with an average accuracy of 0.5 micron without adding sizeable systematic errors. Higher order transforms will not improve the transform (i.e. distortions with scale lengths smaller than about 1 cm on the plate cannot be modelled). However transforms are poorly defined at the edges of the field.

5 Overall Error Budget

This purely differential approach brings the following advantages already noticed by Chiu (1980):

1. We avoid determining and modelling the plate constants.

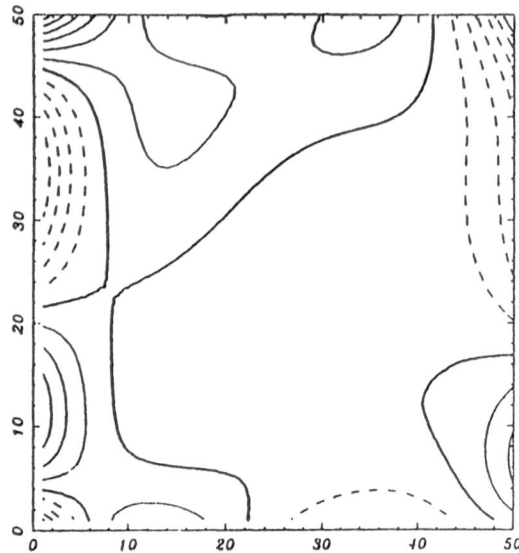

Fig. 1. Isocontour plot showing the differences between two 4th order X transforms. The two transforms are determined with two independent subsets of stars, and they define identical transforms within better than ±1 micron over most of the field. Differences give the amount of error in the transforms. The distance between isocontours is 1 micron. The thick line gives the zero level.

2. The solution does not depend on the accuracy of celestial coordinates of any intermediate catalogue.
3. The exact accuracy of the reference system depends only on the number of stars and on the rms proper motions.
4. It is possible to measure proper motions using plates from different instruments, even glass copies and different plate centers.

The proper motion accuracy is comparable with the accuracy expected for the HIPPARCOS mission, although only relative. The key conditions for such a performance are:

1. An astrometric machine like MAMA giving access to submicron accuracies.
2. The choice of centering algorithms: gravity center methods (modified or not) are not the best choice and would decrease the accuracy for faint stars.
3. Plate to plate transforms using polynomial expansions allow modelling of small scale distortions on plates; this method is restricted to fields with a sufficient star density and would fail to detect local streaming motions.

The M5 survey based on 3 plates over 31 years has an accuracy of 3 mas/year on proper motions for brightest stars (V<16). For the near NGP survey, based on 11 plates taken over 40 years, the accuracy ranges from 1.4 to 2 mas/year (with V from 12 to 17.5).

6 Discussion

Since there is no substantial overlap of our catalogues with any data of comparable accuracy, the only way to get an assessment of the error budget is to give some comparisons of the distribution of observable quantities with model prediction. The Besançon galactic model (Bienaymé et al. 1988; Robin & Oblak 1987) provides such a suitable framework.

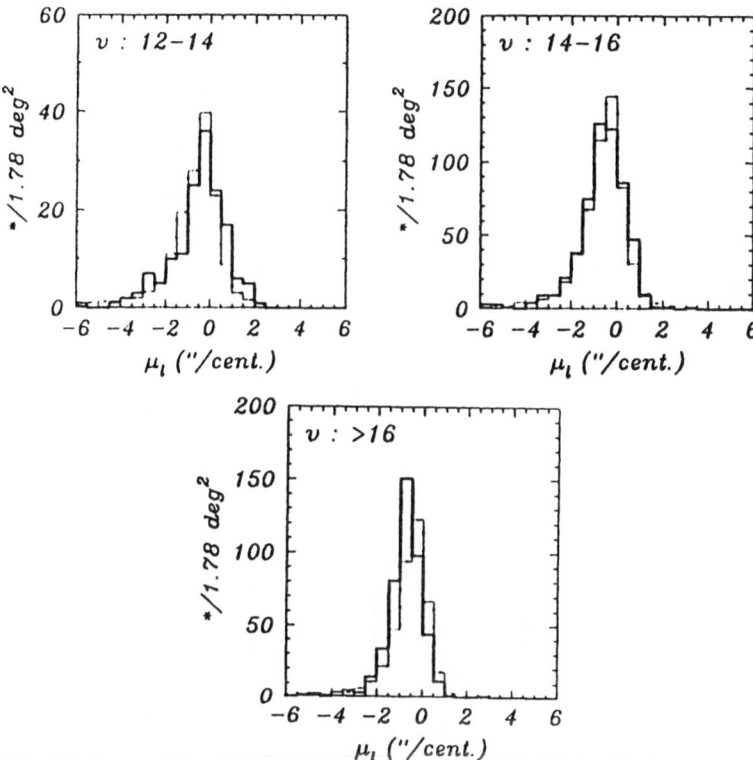

Fig. 2. μ_l observed proper motion histograms for three v instrumental magnitude intervals: data (solid lines) and model prediction (thin dotted lines).

From the first field (l=3°, b=47°), μ_l histograms of observed and model predicted proper motions in three magnitude ranges are plotted in Fig. 2. The asymmetric shape of the distribution of longitude proper motions is clearly visible in both data and model. According to the model, this asymmetry results from the asymmetric drift. Most stars contributing to the histograms are found a few hundred parsecs away from the sun; at this distance the drift reflects in proper motions by roughly 1 arcsec per century. The standard error of proper motions has been set to 0.3 arcsec per century in the model, as a result of the above analysis of the astrometric accuracy. If the error analysis were grossly wrong, the asymmetric drift effect would disappear as can be seen in Figs. 3a, b. In the same way, in Fig. 3c, model velocity dispersions have been arbitrarily increased by a factor of 1.5. Larger velocity dispersions, although resulting in an increase of the dispersion of proper motions, leaves the signature of the asymmetric drift unaltered. So the quality of the astrometric data alone allows one to measure with some accuracy important kinematic features.

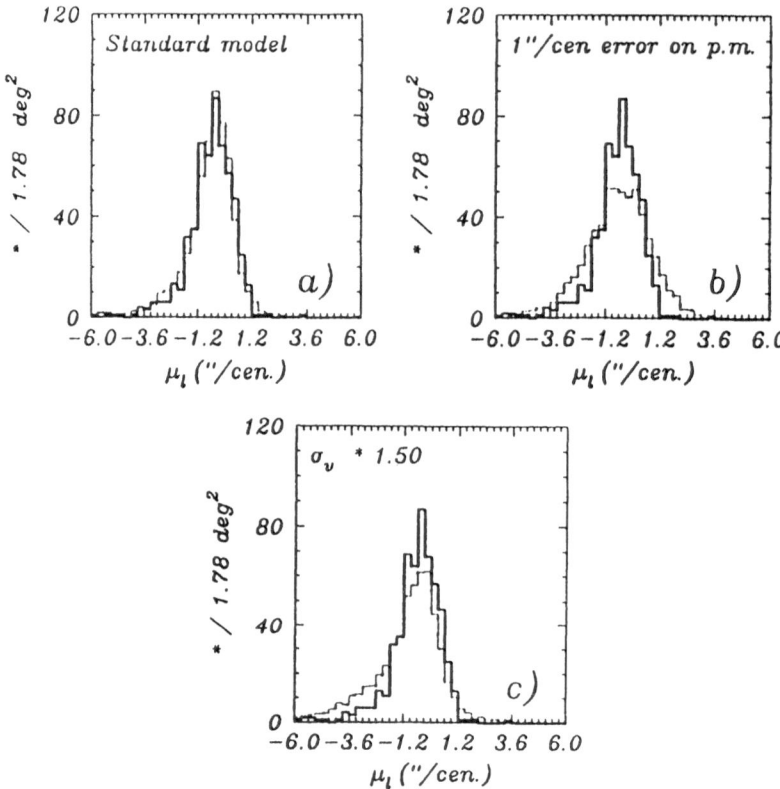

Fig. 3. μ_l proper motion histograms for stars with $14 < v < 16$. Data (solid lines) can be compared with model predictions (thin dotted lines): **a)** standard model with an assumed 0.3″/cent accuracy in proper motions; **b)** with an assumed 1.0″/cent accuracy in proper motions; **c)** as (**a**) but velocity dispersion of stars is increased by 50%.

References

Becker, W., 1965. *Z. Astrophys.*, **62**, 54.

Bienaymé, O., Robin, A.C., Crézé, M., 1987. *Astron. & Astrophys.*, **157**, 71.

Bienaymé, O., Motch, C., Crézé, M., Considère, S., 1988. *'Mapping the sky'*, *Symp IAU*, **133**, p. 389, Eds S. Debarbat et al., Kluwer Acad. Press.

Bienaymé, O., Mohan, V., Créze, M., Considère, S., Robin, A.C., 1991. *Astron. & Astrophys.* Submitted.
Chiu L-T., G., 1980. *Astrophys. J. Suppl.*, **44**, 31.
Gilmore, G., Wyse, R.F.G., 1985. *Astron. J.*, **90**, 2015.
Mohan, V., Créze, M., 1987. *Astron. Astrophys. Suppl. Ser.*, **68**, 579.
Reid, I.N., 1990. *Mon. Not. R. Astron. Soc.*, **247**, 70.
Robin, A.C., Oblak E., 1987. *Publ. Astron. Inst. Czech. Acad. Sci.*, **69**, 323.
Sandage, A.R., 1987. *Astron. J.*, **93**, 610.
Soubiran, C., 1991. In preparation.
Spaenhauer, A., 1989. *'The gravitational force perpendicular to the galactic plane'*, *Proc. of a Conference held at Danbury*, p. 45, eds A.G.D. Philip, P.K. Lu , Davis Press.
van Rhijn, P. J., 1965. *'Stars and Stellar Systems'*, *Vol V Galactic Structure*, p. 27, eds A. Blaauw, M. Schmidt, The University of Chicago Press.

Discussion

Taff :

Are these 6.5° × 6.5° plates? If so, then why do you say there are only ∼ 50 reference stars per plate? With the PPM or ACRS one can get ∼ 350 stars per plate. How can you validate your proper motion errors? Your formal estimates for the propagation of error are worthless in the presence of large systematic errors.

Bienaymé :

The field near M5 is 2 square degrees. The field near M3 is limited by the Tautenburg plates: 3° × 3°.

All known sources of systematic errors have been examined and measured. Moreover the combination of 11 plates from 3 different telescopes has not shown any systematic effects greater than 1μm.

Irwin :

The large number of stars and galaxies on a Schmidt plate can be used to map the differential systematic errors as a function of plate position when pairing up plates to look for proper motion. The only assumption necessary is that the proper motion is itself independent of plate position. The galaxies can then be used to put the proper motion on an absolute scale.

UPDATING THE AGK3 REFERENCE STAR CATALOG WITH ASTROMETRICALLY REDUCED SCHMIDT PLATES

L.G. Taff [1], M.G. Lattanzi [1,2,3], B. Bucciarelli [1,2] and D. Daou [1]

[1] Space Telescope Science Institute, 3700 San Martin Drive,Baltimore, MD 21218, U.S.A.
[2] On leave of absence from Torino Observatory.
[3] Affiliated with the Astrophysics Department, Space Sciences Division, ESA.

1 Introduction

Figure 1 shows a typical pattern of residuals (Taff et al. 1990c), over an entire plate, from a cubic plate model normally used for the astrometric reduction of wide field-of-view Schmidt plates (e.g. 6.5×6.5 deg^2). Their presence explains the generally poor results obtained by previous workers when they attempted to astrometrically reduce large portions of Schmidt plates. This systematic form of the residuals also shows that precision has been meaningless since the problem is dominated by issues of accuracy. Any quadratic or cubic plate model will successfully randomize some small area, nominally a square degree, of any Schmidt plate. Without the appropriate version of Fig. 1, one can not know where this good section is; it does move around the plate as the terms included in the plate model are varied.

We have been able to discover these patterns in the residuals from quadratic, cubic, and so on polynomial plate models, and to devise techniques which successfully ameliorate them (Taff 1989; Taff et al. 1990a; Taff et al. 1990b; Bucciarelli et al. 1991; Lattanzi & Bucciarelli 1991). Thus, for the first time, the appellation 'astrometric' is deserved when speaking about positions obtained from the global reduction of large field-of-view Schmidt plates.

One consequence of this work is that the Hubble Space Telescope Guide Star Catalog (Lasker et al. 1990; Russell et al. 1990; Jenkner et al. 1990) will be completely re-reduced by these methods (Version 1.2). The anticipated precision is $< 0.\!''25$ per equatorial coordinate with an accuracy at the $0.\!''01$ level (whereas

185

H.T. MacGillivray and E. B. Thomson (eds.), Digitised Optical Sky Surveys 185–191.
© 1992 Kluwer Academic Publishers.

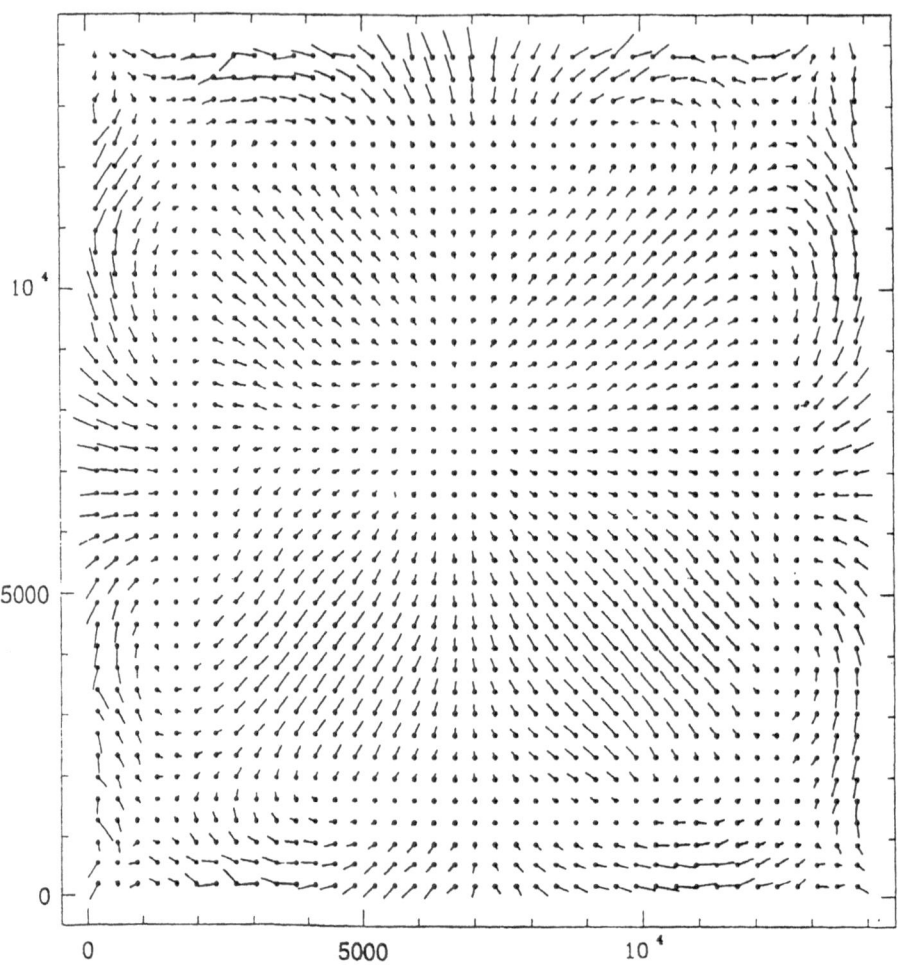

Fig. 1. Vector residual map of CAMC stars from the AGK3R cubic pre-corrected quadratic polynomial model. The longest line represents about $1''$. A 2×3 moving average was used to smooth the raw version.

heretofore, for Schmidt plates in general, the accuracy was typically $0.''5$). A second consequence of this advance is that we have been able to construct an updated version of the AGK3 by combing the AGK2 and AGK3 material with the re-reduced northern hemisphere Guide Star Catalog 'Quick V' plate material. This uniquely successful integration of Schmidt plate material with astrographic plate material has resulted in the best northern hemisphere photographic catalog ever created. A third consequence of this work is that it is appropriate to consider a Schmidt Plate Compilation Catalog — a deep, all sky, multi-color star catalog based solely on Schmidt plate materials, with each plate reduced by a common method, using the same high quality reference star catalog (and, therefore, naturally in a homogeneous reference frame), and spanning 40 years of time so that the proper motions are reasonably good too. This catalog could be constructed at one observatory and then distributed to its community of contributors as well as to the astrophysical community in general. We briefly review these three consequences with emphasis on the latter two as published references already amply document the former.

2 Astrometry

The pattern in Fig. 1 was constructed by summing the vector residuals [ie. $(x, y) = (\Delta\alpha\cos\delta, \Delta\delta)$] from 764 northern hemisphere HST GSC plates — reduced with a cubic pre-corrected quadratic polynomial plate model (without color index or apparent magnitude terms) — in a plate-based coordinate system. On the average, 42 AGK3R stars were used to determine the six plate model parameters. The residuals are *not* from the plate model but, rather and much more meaningfully, represent the differences between the plate model predicted positions and the actual positions of 40 419 CAMC stars (for a mean of 27 stars per vector residual in Fig. 1). Thus, except for the perimeter of the plate, there is almost no 'small number statistics noise' in the figure. The relative quiet in Fig. 1, that is the lack of statistical vector residual-to-vector residual fluctuations (on a 10' scale), proves that this pattern is relatively independent of the hour angle of an individual exposure, of the zenith distance of an individual exposure, of the meteorological conditions peculiar to an individual exposure, and so on.

The combination of our plate model, the reference catalog which was used to construct it, and Fig. 1 is an invariant. Whence, while your version of the figure might be different, the sum total of your contributing components will also represent the same, intrinsic, physical deformation of the Schmidt plate. The task of astrometrically reducing Schmidt plates is completely dominated with breaking the embrace of the systematics. Because of the aforementioned invariance, the simplest way to do so is to combine the residual map, which we colloquially refer to as the 'mask', with the plate model that produced it.

Our original idea (Taff 1989) works much better in practice, especially in the outer parts of large field-of-view Schmidt plates where the gradients in the pattern of systematics is largest in amplitude. The key concept is to subdivide the physical Schmidt plate into imaginary pieces we call 'subplates'. Each square, equally-sized, subplate should be small, preferably $\frac{2}{3}$ deg or less. This size is below the scalelength of the bulk of the systematics (Lattanzi & Bucciarelli 1991). Therefore, the subplate plate model need only be linear. Each of the aforementioned 764 plates has been re-reduced by both techniques and in 80.8% of the cases the subplate method was superior (as judged by the CAMC stars; formal, internal estimates were never used).

In more detail, Figs. 2 show the positional error histograms for the subplate method re-reduced GSC material. Figure 2a contains the results for the stars found in the core of the plates, Fig. 2b is for the stars near the edges of the plates, and Fig. 2c contains the statistics for the entire plate. The 'core' and 'edge' regions were defined by dividing the Schmidt plate into two equal area sections; a central, centered, circular area (of radius 2.6 deg) and the remainder. Since the CAMC stars are uniformly distributed, this provided us with two equal number samples (on the average) to use to judge the efficacy of the subplate and mask methods. The choice between the two was influenced by superiority in the edge region. The mask version of Figs. 2 looks similar to the subplate version; the principal difference is in the edge subset which has a higher population beyond 0.″5.

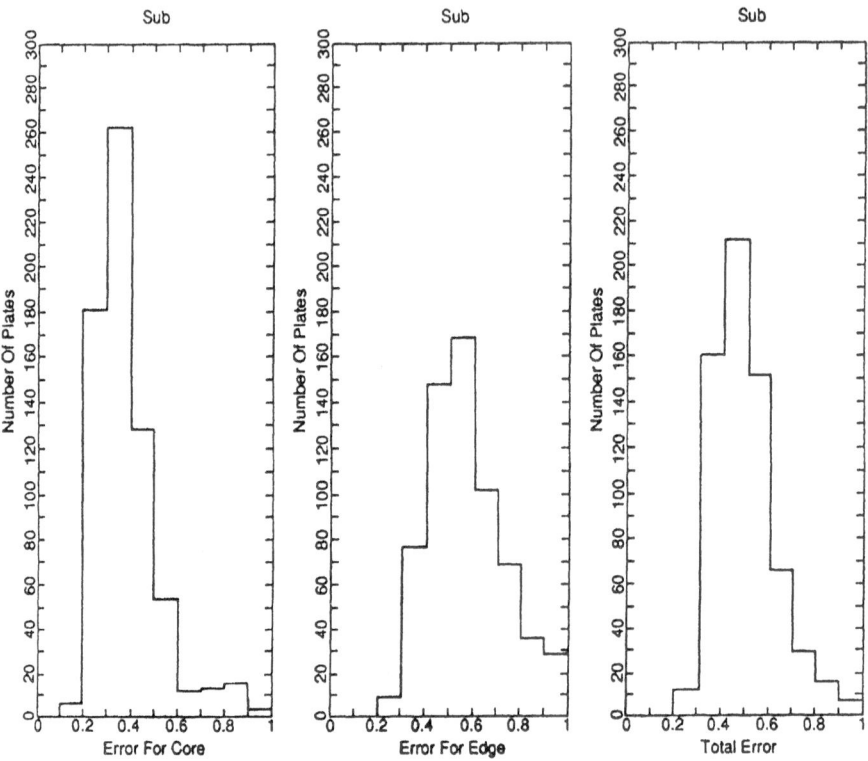

Fig. 2. Histograms of CAMC star positional errors from the subplate method for the **(a)** central core of the plate, **(b)** the surrounding edge area, and **(c)** the entire plate. Units are arc seconds.

To support the subplate reduction, we used the Positions and Proper Motions catalog of Roeser & Bastian (1989). Still using the CAMC stars to objectively assess the quality of the reduction, the average coordinate mean error (i.e. standard deviation about the mean) — for each equatorial coordinate — from our re-reduction of the GSC 'Quick V' plate material is 0.″33. The accuracy is at the 0.″01 level. Unfortunately, the PPM proper motions are 2.7 times worse than is claimed in the PPM documentation (see Bucciarelli et al. 1991 for the details). Hence, the real imprecision in a typical 1983.3 (the mean epoch of the GSC material) PPM position is 0.″87 rather than 0.″34. Therefore, with a much better reference catalog, we believe that our re-reduction would have been much better too (until we reach the errors contributed by the centering algorithms; Lasker et al. 1990).

3 The Updated AGK3: the AGK3U

Nonetheless, we have successfully integrated our re-reduction of the GSC material with the AGK2 and AGK3 material. Our realistic, external errors can be compared to the errors quoted in the Introduction to the AGK3 since theirs comes from a plate-to-plate comparison. When we combine the AGK data with ours in the usual weighted least squares way for proper motion determination, the mean epoch of the updated AGK3 becomes 1950.62. At that instant the AGK3U positions typically have a mean error of $0.''17$ and their angular speeds have mean errors of $0.''82/\text{cy}$ (i.e. these standard deviations are two-dimensional).

Figure 3 shows the individual equatorial error distributions for each re-reduced GSC plate. These values include the contribution of the CAMC star errors (about $0.''12$ per coordinate). They also include multiple entries for about 30% of the stars because of plate overlapping. Our final version of the northern hemisphere GSC re-reduction contains normal point values for these stars. The $0.''33$ per coordinate mean error mentioned above reflects the gains made by averaging to the form the normal points as well as the removal of the CAMC star contribution. The AGK3U is available through both the NSSDC at GSFC and the Strasbourg Data Center.

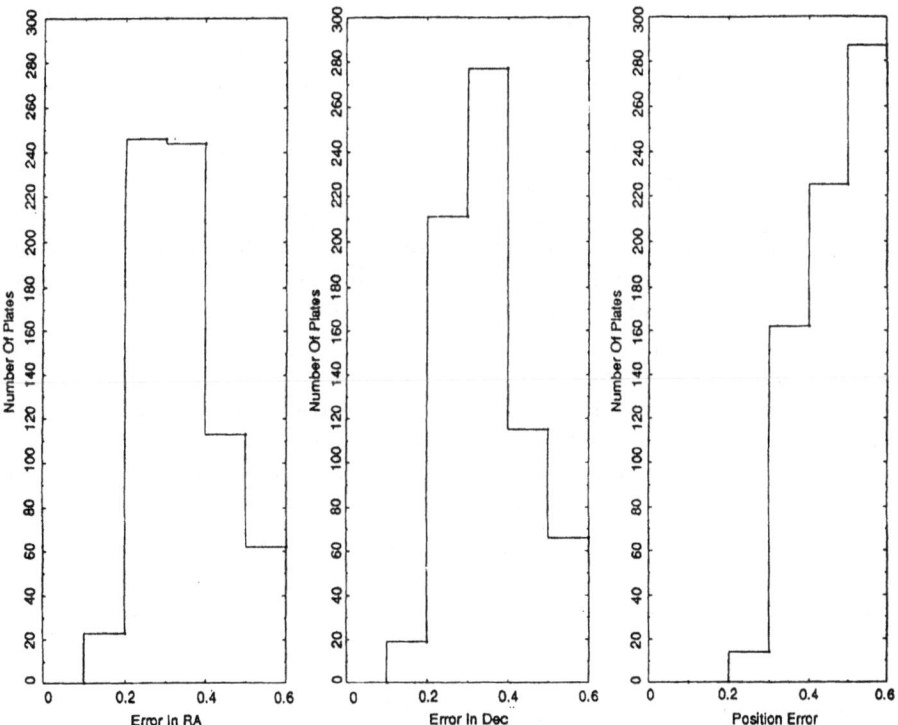

Fig. 3. Histograms of re-reduced HST GSC Quick V Schmidt plate equatorial co-ordinate errors for CAMC stars. **(a)** Right ascension, **(b)** declination, and **(c)** position. Units are arc seconds. Also includes multiple entries and the contribution from the CAMC errors.

The ACRS (Corbin & Urban 1988) has just become available. It appears to be almost as good as Corbin has claimed (i.e. mean errors of 0.$''$09 per equatorial coordinate at 1949.9 and proper motion mean errors of 0.$''$5/cy per coordinate). Thus, at the mean epoch of our plates, ACRS positions should be good to 0.$''$3. We anticipate a factor of 2 improvement (directly from the PPM/ACRS ratio) when the GSC material is again re-reduced — with the same x, y positions (Lasker et al. 1990), with our techniques, and the ACRS. Thus, we are contemplating Schmidt plate *positional* precisions below 0.$''$25 (with 0.$''$01 accuracy).

4 A Schmidt Plate Compilation Catalog

The ability to achieve this level of precision with one, homogeneous, all-sky reference catalog implies that the results from any Schmidt plate could also be on the same, homogeneous, reference frame. Thus, one can, for the first time, envision a uniform Schmidt plate reduction scheme. Our software and subplate technique, coupled with the ACRS as a reference catalog, is utilized in conjunction with the CAMC stars for quality assurance/error analysis. In parallel, our software and mask technique, coupled with the AGK3R and SRS as reference catalogs, is utilized in conjunction with the CAMC stars for mask construction and quality assurance/error analysis. One would continue to choose the best reduction on a plate-by-plate basis. One can also conceive of integrating the reductions of all Schmidt plates together to form a Schmidt Plate Compilation Catalog. Given the diversity of Schmidt plate collections, especially with regard to the time span they cover and the passbands of their emulsions, a Schmidt Plate Compilation Catalog could be a deep ($B = 16$ or 17 or 18 mag), all-sky, complete, multi-color, 0.$''$2 mean error star catalog with $< 0.$$''$5/cy proper motion mean errors.

Such a catalog would be updated whenever new plate measurements were forwarded to the observatory producing it. Moreover, this observatory, with a minimum of organization, could also produce an observing list of those portions of the sky which needed additional coverage in one wavelength region or those portions of the celestial sphere for which additional coverage would greatly improve the proper motion mean errors. Observatories which have large repositories of measured Schmidt plates would have all their astrometric data reduction performed including being supplied with the information necessary to retrieve positions for all the other objects of interest to them. With a fixed time interval to protect their science, and first distribution rights on the SPCC, the benefits to the contributors should be sufficient to induce their cooperation.

Something like the SPCC would be a great benefit to both satellite mission planning and galactic kinematics. Imagine having a *complete* catalog, with good positions and proper motions, all on the same homogeneous reference frame, to the sixteenth magnitude (plus color indices!). On the other hand, imagine trying to do mission planning for a new astronomical satellite with the SPCC — especially if it were cross-referenced to all other wavelength catalogs of interesting objects. Target selection, new object identification, and so on would be very easy. Finally, once an observatory had been chosen to create the SPCC, there could be natural growth into actually measuring plates, developing centroiding algorithms, advancing the retrieval of photometric and astrometric information, and so on. Something like the SPCC is a fitting embodiment of all the work done to create the Schmidt plate repositories we now have.

5 Acknowledgements

This work was sponsored in part by NASA under grant NAGW–2597.

References

Bucciarelli, B., Daou, D., Lattanzi, M. G., Taff, L. G., 1991. *Astrophys. J.* Submitted.
Corbin, T., Urban, S., 1988. *'Mapping the Sky', IAU Symp.*. No. 133, p. 75, ed. H. Eichhorn, Reidel, Dordrecht.
Jenkner, H., Lasker, B.M., Sturch, C.R., McLean, B.J., Shara, M.M., Russell, J.L., 1990. *Astron. J.*, **99**, 2082.
Lasker, B.M., Sturch, C.R., McLean, B.J., Russell, J.L., Jenkner, J., Shara, M.M., 1990. *Astron. J.*, **99**, 2019.
Lattanzi, M.G., Bucciarelli, B., 1991. *Astr. Astrophys.* In press.
Roeser, S., Bastian, V., 1989. PPM. *Positions and Proper Motions of 181731 Stars North of −2.5 Degree Declination for Equinox and Epoch J2000.0*, Astron. Rechen-Institut, Heidelberg.
Russell, J.L., Lasker, B.M., McLean, B.J., Sturch, C.R., Jenkner, H., 1990. *Astron. J.*, **99**, 2059.
Taff, L.G., 1989. *Astron. J.*, **98**, 1912.
Taff, L.G., Bucciarelli, B., Lattanzi, M.G., 1990a. In *Errors, Bias, and Uncertainties in Astronomy*, p. 35, eds C. Jaschek and F. Murtagh, Cambridge Univ. Press, Cambridge.
Taff, L.G., Lattanzi, M.G., Bucciarelli, B., 1990b. *Astrophys. J.*, **358**, 359.
Taff, L.G., Lattanzi, M.G., Bucciarelli, B., Gilmozzi, R., McLean, B.J., Jenkner, H., Laidler, V.G., Lasker, B.M., Shara, M.M., Sturch, C.R., 1990c. *Astrophys. J.*, **353**, L45.

Discussion

Bienaymé :

The α, δ determination, absolute p.m. and relative p.m. measurements are not the same problems. Relative p.m. measurements are not affected by plate distortions as long as the scale length of these distortions is big enough. For this reason relative p.m. can be measured more accurately than α, δ coordinates.

Cannon :

Given that most of the swirling distortion you find is probably caused in the plate holders, I suspect you might find quite significant differences between plates taken with different UK Schmidt Telescope plateholders. I suspect that there will also be some large scale distortions due to non-parallelism of the filters.

CALIBRATION OF LARGE BATCHES OF PHOTOGRAPHIC IMAGES *

Z. Liu [1], *C. Sterken* [2], *H. Hensberge* [3] *and J.-P. De Cuyper* [3]

[1] Beijing Observatory, China.
[2] University of Brussels VUB, Belgium.
[3] Royal Observatory, Uccle, Belgium.

1 Introduction

During the 1985-1986 appearance of Comet Halley, hundreds of photographic images were taken by the authors. These images were obtained with several telescopes, using different photographic emulsions and filters. All photographic images have subsequently been digitised with a PDS Microdensitometer. This paper describes the calibration procedure developed.

2 The Observations

2.1 Xinglong Station

From September 20, 1985 to January 11, 1986, Comet Halley was observed with the 600/900-mm Schmidt telescope at Xinglong Station of Beijing Astronomical Observatory. Plate sizes of 16×16 cm^2 were used, each covering 5 degrees on a side, at a plate scale of 115 arcsec per mm. The plate-filter combinations were Kodak IIa-O, IIa-O+UG2, IIa-O+GG13, IIa-D+GG14, and 103a-F+RG1. From September 20, 1985 to November 12, 1985 the plates were hypersensitised with hydrogen. Calibration images were made with a spot sensitometer with 9 steps.

---

* Based on observations obtained at Beijing Observatory, China, Mt John Observatory, New Zealand, and the European Southern Observatory, Chile.

H. T. MacGillivray and E. B. Thomson (eds.), Digitised Optical Sky Surveys 193–198.
© 1992 *Kluwer Academic Publishers.*

2.2 Mt John Observatory

From March 7 to May 2, 1986, the comet was photographed using a 300 mm Schmidt camera (MPT300 Lichtenknecker Optics, f=576 mm) mounted on the tube of the astrograph telescope at Mt John Observatory (University of Canterbury, Christchurch, New Zealand). All photographs were taken with a 24mm-reflexcamera on Kodak TP2415 film. Hypersensitisation was done with nitrogen-baking. Images were taken without filter, as well as with red and blue filters (resp. Kodak Wratten 47 and 23A).

2.3 European Southern Observatory

From April 30 to July 13, 1986, the comet was observed with the 400/4000 mm GPO-astrograph at ESO La Silla. Unsensitized 160 mm x 160 mm IIa-O and O 98 plates (with RG630 filter) were used. Some calibration plates were taken with an external 21-spot sensitometer.

3 The Characteristic Curves

The characteristic curve is sensitive to differences in sensitization of the emulsion, temperature and humidity of storage before and during exposure, uniformity of the plate, elapsed time between exposure and development, and of course to the whole developing process. Because of the very heterogeneous character of our basic dataset due to the differences in emulsions and to the pre- and post-observational handling, large variations are to be expected in the forms of the characteristic curves.

A widely used analytical form of the characteristic curve is the Honeycutt-Chaldu function (Tsubaki & Engvold 1975)

$$\log(I/I_0) = Q_1 + Q_2 D + Q_3 \ln(\exp(Q_5 D^{Q_6}) - 1) + Q_4 \exp(Q_5 D^{Q_7}) \qquad (1)$$

where I are the known intensities of the calibration spots, I_0 is an (unknown) scale factor in the intensities, D are the measured calibration densities, and Q_i are the parameters to be determined by a least-squares fit to the calibration data.

This function represents well all parts of the characteristic curve: the second term corrects the slope of the linear part predetermined by the sum of the third and fourth term, which respectively determine the shape of the toe and of the shoulder. Q_1 shifts the curve along the axis of $\log I$. Values of Q_6 and Q_7 are respectively (1,0), (1,1) and (1,3) according to whether there is no saturation, saturation of emulsion, or saturation of densitometer (densities exceeding the measurable density). An iterative procedure for determining Q_5 gives rapid convergence.

Kormendy (1973) proposed the construction of an average characteristic curve from several plates for which a same emulsion, same filter and identical treatment (hypersensitization and dark room processing) have been used. This can be done when the differences between the characteristic curves of one group are not larger than the effects of processing errors, and when the differences in the overall intensity level of each set of spots can be eliminated by shifting in log I the whole sequence of spots.

Our scanned data have been reduced with MIDAS software at ESO, Garching, and at the Royal Observatory of Belgium. For each plate, a density-log intensity table was made and the resulting relation inspected graphically. Calibration curves from identical emulsions and with similar characteristics (irrespective of filter) were merged together, and this yielded 8 different subgroups. Groups 1 to 6 contain data from glass plates; groups 7 and 8 are from film emulsions; groups 9 and 10 are ESO blue and red plates. For each of these groups, a fit with the Honeycutt-Chaldu relation was made, and the differences between observations and predictions were calculated. Figure 1 shows all data for groups 1 to 4, together with the fitted curves. As one can see, the fits are acceptably close to the experimental data.

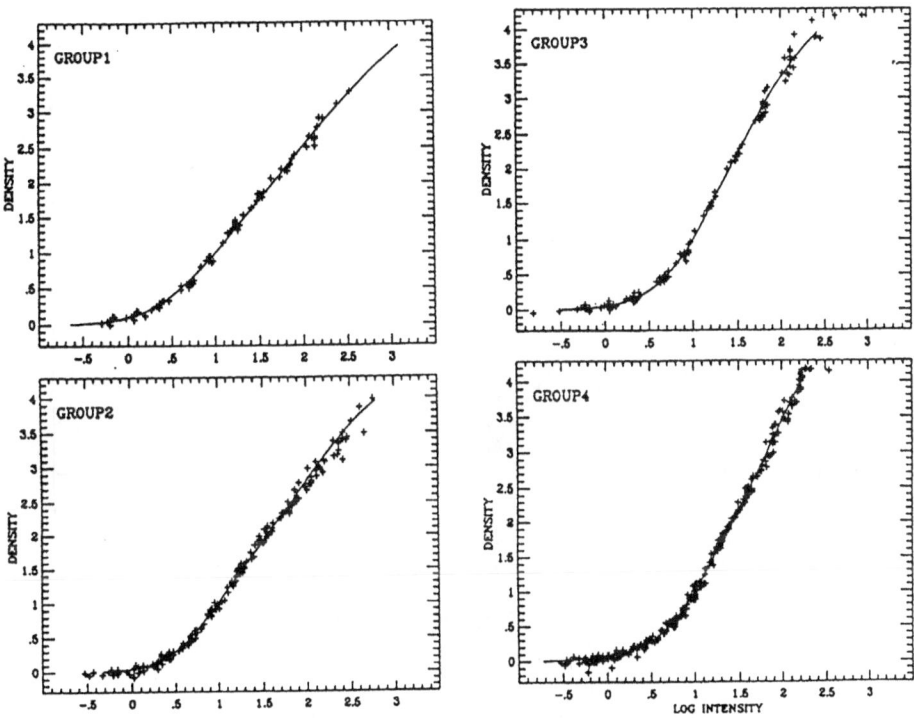

Fig. 1. Calibration data for groups 1 to 4, with the fitted equation (1).

4 Discussion

Figure 2 gives the standard deviations for the differences between observed and calculated densities for all data belonging to one group, as well as for the toe, linear and shoulder regions. The standard deviations of the linear and toe part are quite consistent for groups 1 to 4, but strongly increase for the last four groups. The shoulder part of the curves seems to be less well defined. It is not surprising that the first groups give better results than the last ones, since the selection in natural groups was performed by visual inspection, and the later groups clearly contain data of plates which yielded results of lower standard; these groups were also the smallest. The film material (groups 7 and 8) yields a higher degree of heterogeneity than is present in the data from the plate material.

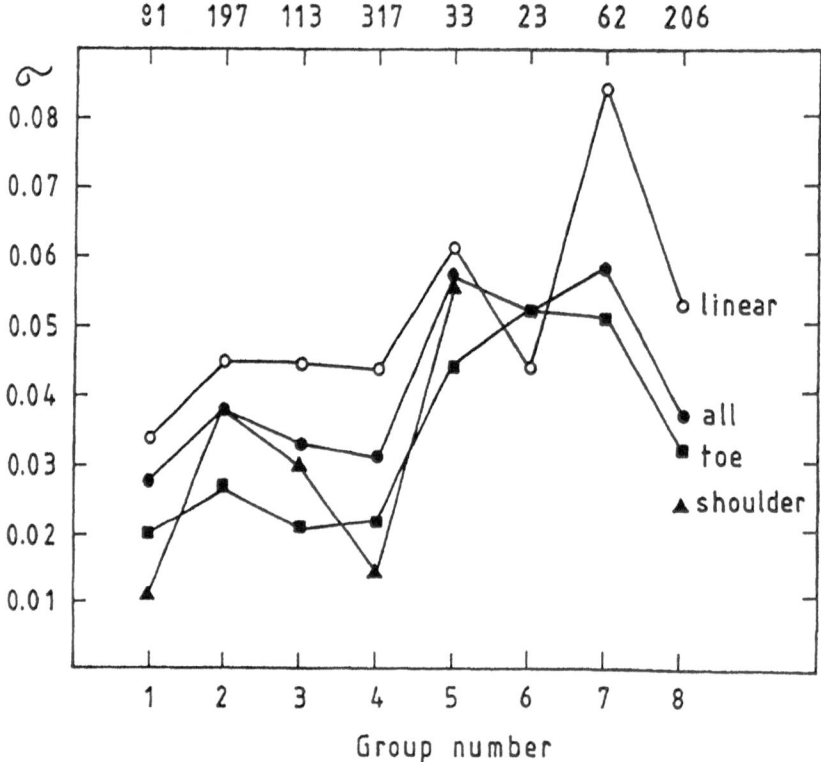

Fig. 2. Standard deviation from the fit for all data, as well as for toe, shoulder and linear part of the characteristic curve. The numbers along the top horizontal line represent the numbers of available calibration images.

In the right part of Fig. 3 we display the derivative $\Delta D/\Delta\log I$ (i.e. the steepness of the linear part of the curve) as a function of a 'toe parameter' which describes how fast the linear relation is lost at low densities (and which is defined by the difference between log I values for which the slope of the curve repectively equals $\frac{2}{3}$ and $\frac{1}{3}$ of the slope of the linear part of the curve). The left part gives the correlation between dynamic range and the steepness gradient. It is clear that the film has a much smaller dynamic range than do the glass plates (almost 5 magnitudes in group 1 but only 2 magnitudes in group 8). However, different glass plates cover also a wide variety in dynamical range, and moreover, plates with a similar dynamical range (groups 3, 4 and 5) may have a differently shaped toe. In other words: the two parameters displayed in the left part of Fig. 3 seem to be quite independent.

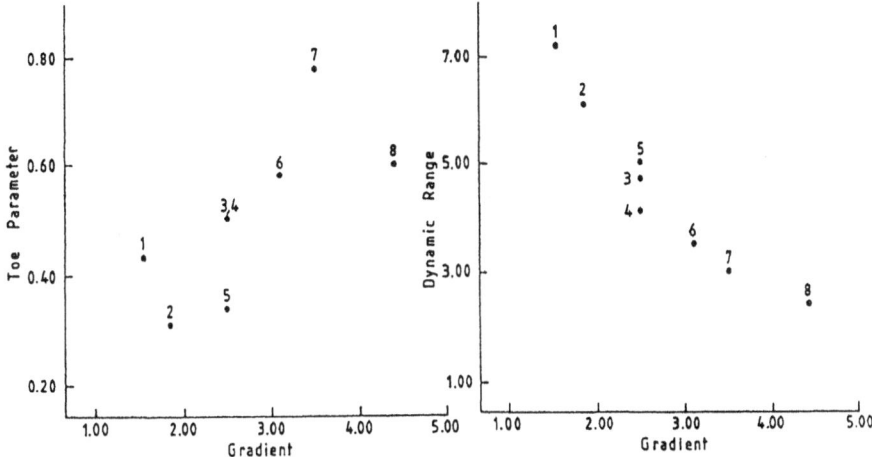

Fig. 3. Left: steepness gradient versus toe parameter (see text). Right: Dynamic range versus gradient. Open circles are for films, filled circles are for glass plates.

The top part of Fig. 2 indicates a larger degree of noise in the linear part for the film data, if expressed in density units, but the smaller dynamical range means that the corresponding noise in the images transformed to a magnitude scale is similar for film and glass plates. The small groups 5 and 6, consisting of those few plates that did not fit into any of the well-defined groups, might for that reason be somewhat less homogeneous, but such a conclusion is weak because of low-number statistics.

It is interesting to examine the quality of the Honeycutt-Chaldu fit by comparing the measured and calculated values of D. Plots of residuals in log D with respect to the Honeycutt-Chaldu fit (Fig. 4) show a small but definite systematic effect with log I. While the effect would not be obvious from a small number of plates, the gain made by averaging the residuals over many plates allows detection. The 'wavy' pattern strongly suggests that the function fitted is at this level not a perfect model for the characteristic curve. Even when combining all plate groups and all film groups, the pattern does not disappear. In principle, one could remove these small systematic effects by defining a global density-dependent correction term to (1). Its importance should not be overestimated, as its amplitude is not larger than the uncertainty in individual densities.

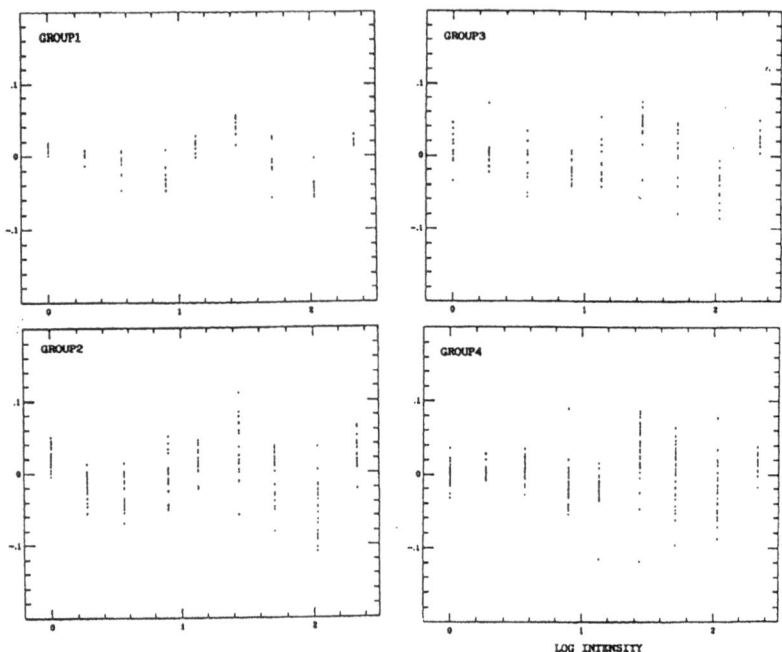

Fig. 4. Deviations from the fit as a function of log *I* for some groups of glass plates.

5 Conclusion

The main conclusion is that, whereas our Honeycutt-Chaldu representation of the
characteristic curve gives realistic results in the case of single images, large batches
of photographic plates can not only help to avoid errors in the determination of
the parameters, but might also motivate the introduction of a density-dependent
correction term that would gain importance in applications wherein many images
are averaged before the final analysis is carried out.

References

Kormendy, J., 1973. *Astron. J.*, **78**, 255.
Tsubaki, T., Engvold, O., 1975. *AAS Photo-Bulletin*, **9**, 17.

AN AUTOMATED IMAGE MEASURING SYSTEM

M. Doi [1], N. Kashikawa [1], S. Okamura [1], K. Tarusawa [2], M. Fukugita [3], M. Sekiguchi [4] and H. Iwashita [4]

[1] Department of Astronomy, University of Tokyo, Bunkyo-ku, Tokyo 113, Japan.
[2] Kiso Observatory, Institute of Astronomy, University of Tokyo, Mitake-Mura, Kiso-Gun, Nagano 397-01, Japan.
[3] Yukuwa Institute for Theoretical Physics, Kyoto University, Kyoto 606, Japan.
[4] National Astronomical Observatory of Japan, Mitaka, Tokyo 181, Japan.

Abstract

An outline and the status of development are presented for a new Automated Image Measuring System (AIMS), which has been designed for data reduction of Schmidt plates and for a large format CCD camera. The image analysis programme developed for AIMS is currently being used for an extensive photometric study of galaxies in the Coma cluster region. First results for the central 5° × 5° region are presented.

1 Introduction

It is now widely recognised that wide-field, digitised, sky survey material forms one of the most important bases for a variety of astronomical research work. For the moment, photographic plates taken with Schmidt telescopes are the major source of digitised sky survey data.

Recently we have been developing a data reduction system for Schmidt plates which consists of a PDS-2020GMS microdensitometer and image analysis software (Automated Image Measuring System, AIMS for short). The software is designed so that it can be applied straightforwardly to a mosaic CCD detector which is currently under development.

In this paper, we report primarily on the software and its first application to a study of bright galaxies in the Coma cluster region based on a plate taken with the 105-cm Kiso Schmidt telescope. While most of the software is now fairly standard,

H. T. MacGillivray and E. B. Thomson (eds.), Digitised Optical Sky Surveys 199–208.

a new method has been developed for deblending overlapped images, and most doubly-blended images are separated successfully. For the science output, an emphasis is given to the automated morphological classification of galaxies into early (E–S0/a) and late (Sa–Irr) types. The status of our development of the 8 × 8 mosaic CCD camera will also be described briefly.

2 Software for AIMS

Image analysis in AIMS proceeds as follows. We first (1) digitise the plate with a PDS-2020GMS microdensitometer, and (2) convert photographic density to intensity using the characteristic curve. In the case of CCD data, flat-fielding replaces these procedures. Next we (3) determine the background level, (4) subtract the background from the image, and (5) normalise the subtracted intensity by the local sky value. We then (6) smooth the data with a variable-width Gaussian filter, and (7) identify objects using a connected pixel method to obtain the peak position, the peak intensity, and the moment parameters up to the second order. The procedures (3) to (7) are fairly standard as described in the literature (e.g. MacGillivray & Stobie 1984). Unlike many of the fast plate-scanning systems, we retain all the scanned data, even after the image detection, for further measurements. We (8) search for blended objects by multi-level detection (Irwin 1985; Beard et al. 1990), and (9) deblend them. Finally, we (10) measure various surface photometric parameters. In the present paper, we describe the deblending procedure and surface photometry measurements.

Double images are deblended automatically, but triple and more complicated images are deblended manually in an interactive processing. For a double image, it is shown that the image can be deblended uniquely if (1) the centres of both components are known, (2) both components are invariant under a 180° rotation, and (3) the flux decreases to zero in the outer region (Doi 1991). Our algorithm is a straightforward application of this finding. We determine the flux of each pixel inward from outside like 'onion peeling'. A similar technique also applies to more complicated images (e.g. triple images), and an automated deblending algorithm is now under development. After deblending, surface photometry measurements are carried out. We measure D_n, r_e, and μ_e using the growth curve, a, b/a, and the position angle by sinusoidal ellipse fitting (Kent 1983). The concentration index C_{in} is also measured using the equivalent (areal) profile. In our work we define the index by

$$C_{in} = \frac{\text{flux within } 0.3 \times r_{25.5}}{\text{flux within } r_{25.5}} ,$$

where $r_{25.5}$ is the equivalent radius at $\mu_B = 25.5$ mag arcsec^{-2}. Other parameters, if necessary, can easily be measured by simply adding appropriate subroutines.

3 Photometric Study of the Coma Cluster

We have been carrying out a photometric study of galaxies located in a 10° × 10° region centred on the Coma cluster using the image analysis programme and Kiso

Schmidt plates in the B band (IIa-O emulsion + GG385 filter). We show in this section preliminary results for the central $5° \times 5°$ region. The field centre is $\alpha_{1950} = 12^h 57^m 30^s$, and $\delta_{1950} = 28°15'$.

The plate (K6147) was scanned with the PDS-2020GMS at the Kiso Observatory. An aperture of 17 μm square was used with a sampling pitch of 16 μm (1.0 arcsec). A total of 5106 objects were identified above our limiting magnitude set at $B_{25.5} = 16.5$ mag, where $B_{25.5}$ is an isophotal magnitude at $\mu_B = 25.5$ mag arcsec^{-2}. The number of blended objects among these is 482, of which 393 were deblended by the automatic mode and 89 by the manual mode.

Star/galaxy separation may be made using a variety of methods (Sebok 1979; Jarvis & Tyson 1980; Yamagata 1985; Heydon-Dumbleton et al. 1989). In this study we used a surface brightness ($SB_{25.5}$) versus concentration index (C_{in}) diagram, which will be described below, and obtained 4690 stars and 416 galaxies. Results of star/galaxy separation were checked by examining visually the images of objects in the central $1° \times 1°$ region on a print of a large-scale (11 arcsec mm^{-1}) plate taken with the du Pont 2.5-m telescope at the Las Campanas Observatory. This examination yielded a misidentification rate of 0.8%. All Zwicky galaxies (218 galaxies) in the $5° \times 5°$ region were detected except for one, which was fainter than our magnitude limit after deblending.

The zero point of the magnitudes, i.e. the sky brightness, was obtained by comparing photoelectric magnitudes with the corresponding photographic magnitudes. We used 245 photoelectric aperture measurements ($\geq 20''$) of 51 galaxies taken from Longo & de Vaucouleurs (1983) and photoelectric magnitudes of 5 stars taken from Lasker et al. (1988). The mean sky brightness was found to be 22.24 ± 0.09 (1σ) mag arcsec^{-2}. Figure 1 shows the regression between these two magnitudes. We find an excellent linearity from 13 to 16.5 mag with $\sigma = 0.09$ mag. If we allow for different sky brightnesses for different objects, the dispersion decreases to $\sigma = 0.05$ mag, which may be a typical accuracy of present photoelectric aperture photometry of galaxies (cf. Kodaira et al. 1990). We did not detect any systematic variation in the sky brightness for individual objects with respect to the position on the plate, however.

A comparison has been made between the present magnitudes and those obtained by Godwin et al. (1983) for the sample of 112 non-blended galaxies. The sample galaxies were taken from the catalogue of Mazure et al. (1988). The comparison is shown in Fig. 2. We obtained a good linear correlation between the two magnitudes with a dispersion as small as 0.10 mag in the range of B $\sim 14 - 18$ mag. The agreement is rather striking, as they are obtained independently from different plate material and by different data reduction systems. (As for the zero point our $B_{25.5}$ is ~0.02 mag brighter than $b_{j26.5}$ by Godwin et al. (1983), while we expect $B_{25.5}$ to be fainter than $b_{j26.5}$ by $0.04 - 0.06$ mag. This offset is mostly explained by adoption of different zero points for calibration.)

A key feature of our image analysis programme is the capability for morphological classification of galaxies. We used the $SB_{25.5} - C_{in}$ diagram for morphological type classification as well as for the star-galaxy separation, where $SB_{25.5}$ is the mean surface brightness within the isophote of 25.5 mag arcsec^{-2}, the lowest detection level of this study. The effectiveness of the diagram of the concentration index versus

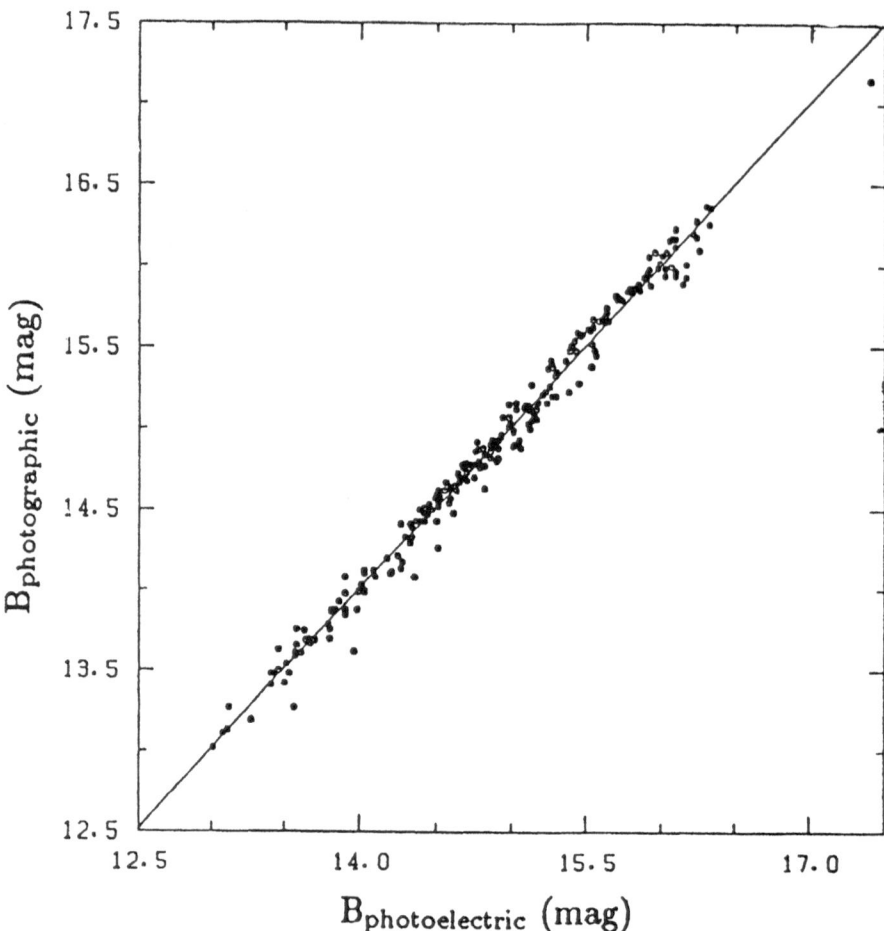

Fig. 1. Comparison between photoelectric magnitudes (Longo & de Vaucouleurs 1983; Lasker et al. 1988) with photographic magnitudes integrated within corresponding apertures. The number of data points is 245 and the rms scatter about the regression line is 0.09 mag.

surface brightness for the morphological classification has already been advocated by Okamura et al. (1984). We use here, however, a simpler version of such a concentration index. Figure 3 shows the 58 UGC galaxies with given morphologies in the $SB_{25.5} - C_{in}$ diagram. The diagram shows that a morphological classification such as early-type (E−S0/a) versus late-type (Sa−Irr) is quite feasible and promising. In our work, we try to separate the galaxies into early and late types by drawing a line as in Fig. 3. Table 1 summarises the correlation between the type classification thus obtained and the morphological type in the UGC. The inconsistency is 9 − 12%. We present in Fig. 4 the spatial distribution of early-type(160) and late-type (256) galaxies in the sky.

For our example, a full cycle of analysis required ∼200 hours in plate digitization

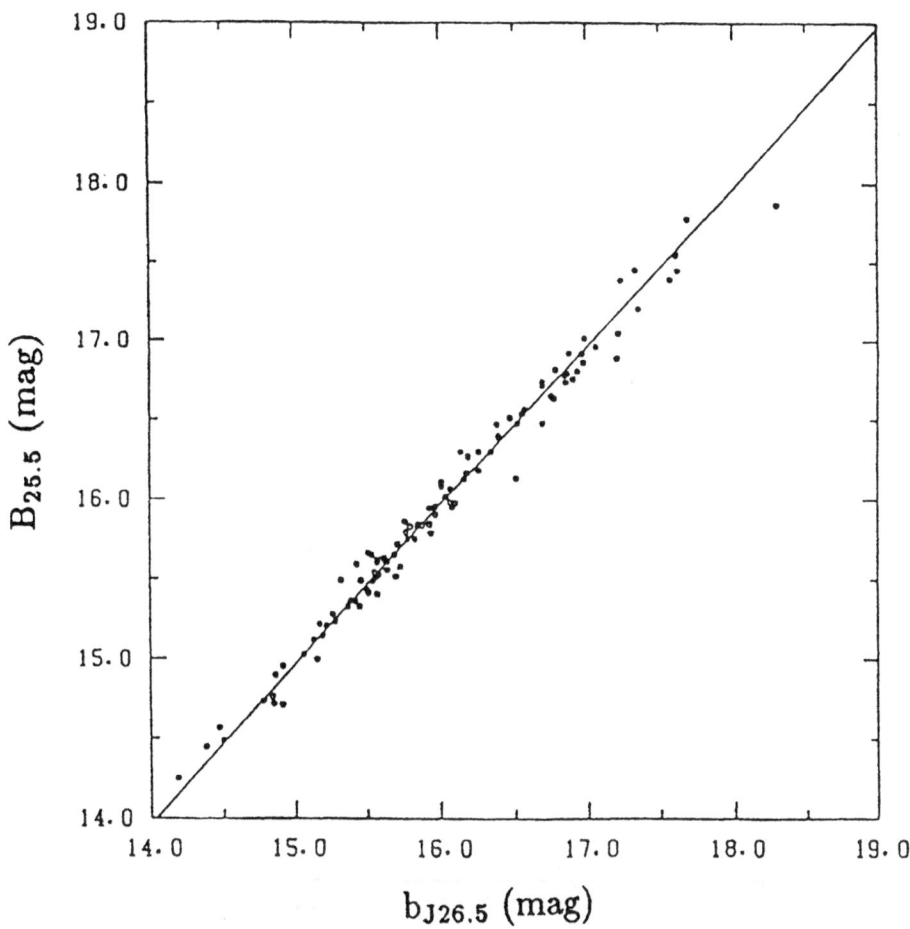

Fig. 2. Comparison between $b_{j26.5}$ taken from Godwin et al. (1983) and our $B_{25.5}$ for 112 galaxies. The regression line is $B_{25.5} = b_{j26.5} - 0.03$ with rms scatter of 0.10 mag.

with the PDS and 30 hours in computation with a FACOM M-780 computer ($\sim$80% of the time being used for smoothing). The required main memory was 32MB for $\sim$700 MB PDS data.

4 Mosaic CCD Camera for the Kiso Schmidt Telescope

We now plan to carry out a wide-field galaxy survey with a mosaic CCD camera designed for the Kiso 105-cm Schmidt telescope combined with our AIMS software for the data reduction.

The camera is designed to use 64 TI-TC215 CCD chips arranged in 8 columns with 8 chips in each column (Fig. 5). Each chip consists of 1000 × 1018 pixels of 12 × 12 μm^2 in size. The field of view of a chip is 13′ × 13′. The chip spacing is slightly less than the chip size ($\sim$ 30 % of the total area is covered.) The chips will

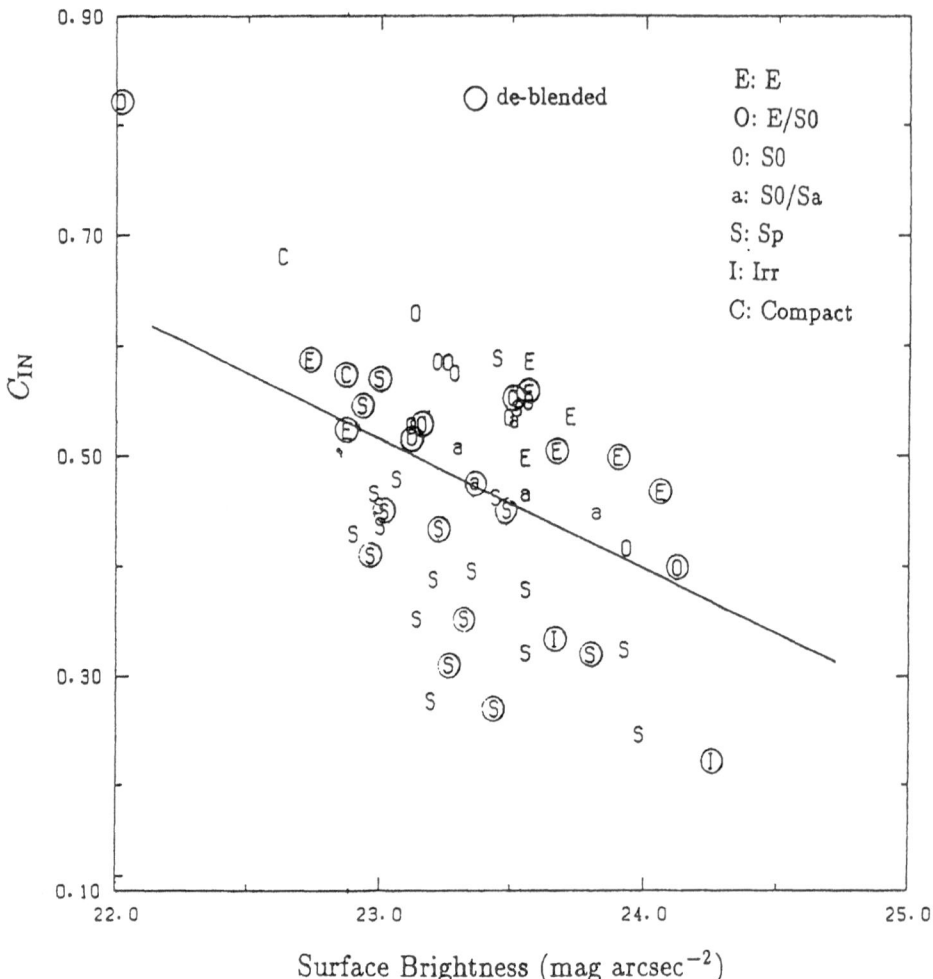

Fig. 3. Plot of mean surface brightness $SB_{25.5}$ versus concentration index C_{in} for 58 UGC galaxies. UGC morphological types are indicated by the symbols as given in the figure. Circles represent the galaxies that were subjected to deblending. The straight line is our criterion to separate early-type (E–S0/a) galaxies from late-type (Sa–Irr) galaxies.

be mounted on a face–plate carefully to follow the curved focal plane of the Schmidt telescope. A single complete picture will be obtained by composing four shots with the field center shifted to four directions in a regular grid pattern. The effective field of view covered by 4 shots is $3.°3 \times 3.°3$, which is almost half of that for a Schmidt plate. The control electronics is based on a general-purpose CCD controller, which has highly modular architecture (Sekiguchi 1991). The chips in each column are read sequentially, but read–out for the 8 columns proceeds in parallel.

We have already taken 'first light' on June 6, 1991, with two chips in a column (Fig. 6). We will complete 2 columns and start observations in autumn 1991. We

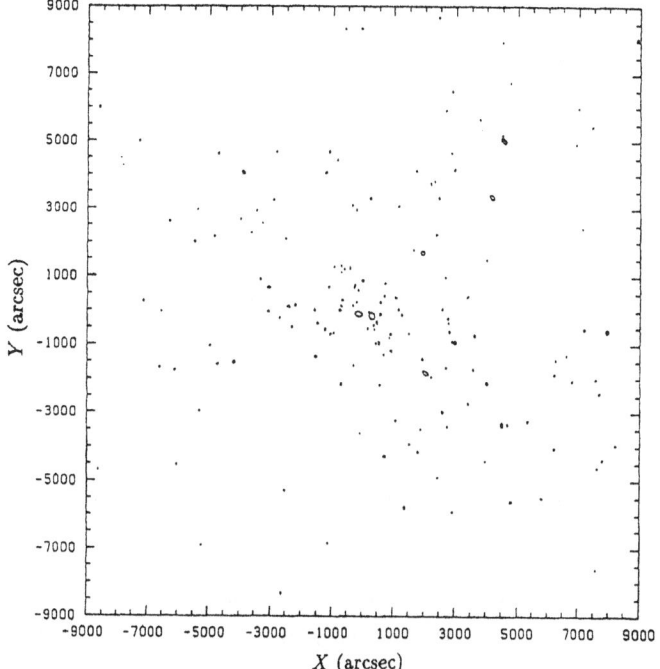

Fig. 4. a) Spatial distribution of early-type galaxies with $B_{25.5} < 16.5$ in the $5° \times 5°$ region centred on the Coma cluster. North is up and East is to the left. The two cDs are indicated by ellipses.

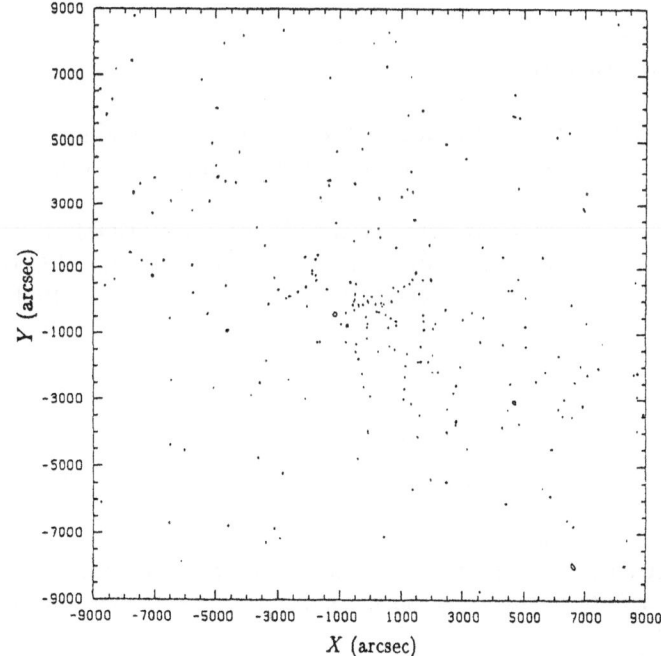

Fig. 4. b) As for Fig. 4a but late type galaxies.

Table 1. Comparison of the morphological type in the UGC catalogue with that determined by AIMS. The numbers in parenthesis show the uncertainty arising from galaxies close to the separation line in Fig. 3.

		A I M S	
		E - S0/a	Sa - Irr
U G C	E - S0/a	29 (-1)	1 (0)
	Sa - Irr	4 (+3)	24 (-2)

expect to have the full 8 × 8 array ready around 1993.

The camera is also a prototype of the wide-field CCD camera used at the prime focus (F/2.3; FOV $0.5°\phi$) of the 8-m Japanese National Large Telescope (JNLT), which will be completed in 1998 on the summit of Mauna Kea, Hawaii. AIMS, together with a wide-field CCD camera and dedicated computer system for quick data analysis, will be an integral part of the system which will enable us to conduct deep survey observations with the JNLT.

5 Acknowledgements

We wish to thank the staff of the Kiso Observatory for their kind collaboration with the AIMS project and Dr. K. Wakamatsu for making the Las Campanas plate available to us.

The analyses and calculations were performed with the FACOM M-780/10S at the Astronomical Data Analysis Center of the National Astronomical Observatory of Japan, and in part with the FACOM VP-200E at the Nobeyama Radio Observatory.

References

Beard, S.M., MacGillivray, H.T., Thanisch, P.F., 1990. *Mon. Not. R. Astron. Soc.*, **247**, 311

Doi, M., 1991. In preparation.

Godwin, J.G., Metcalfe, N., Peach, J.V., 1983. *Mon. Not. R. Astron. Soc.*, **202**, 113.

Heydon-Dumbleton, N.H., Collins, C.A., MacGillivray, H.T., 1989. *Mon. Not. R. Astron. Soc.*, **238**, 379.

Irwin, M.J., 1985. *Mon. Not. R. Astron. Soc.*, **214**, 575.

Jarvis, J.F., Tyson, J.A., 1980. SPIE, **264**, 222.

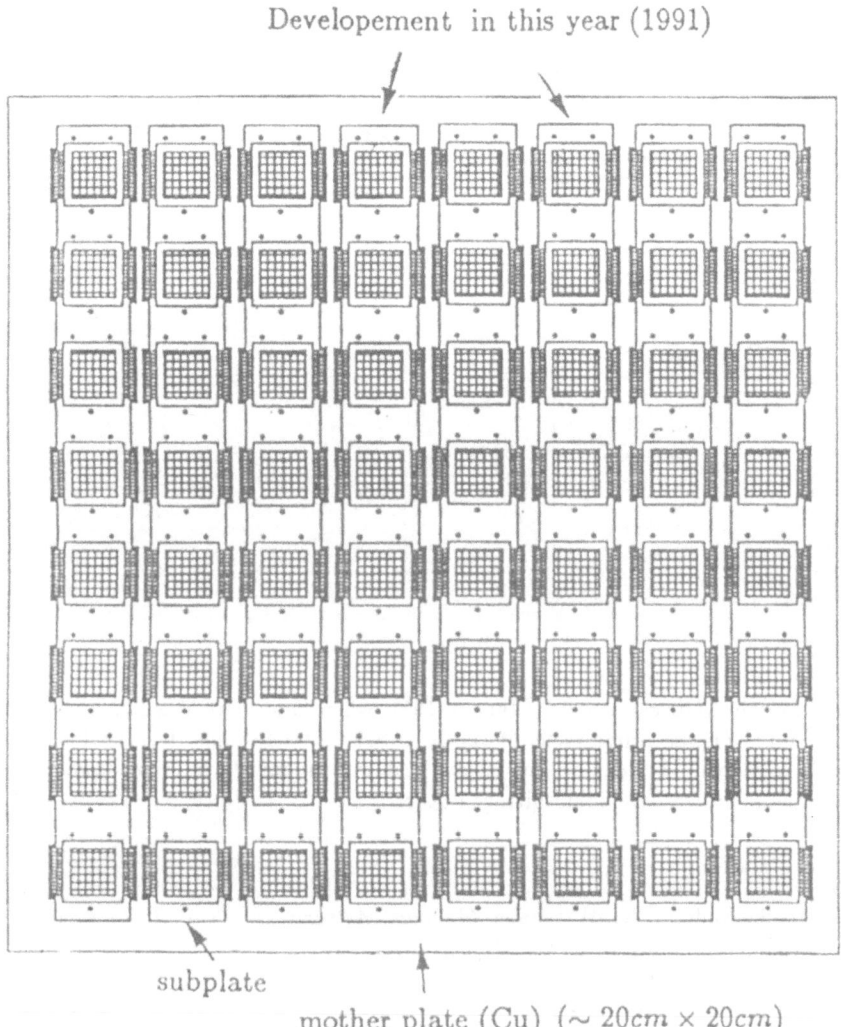

Fig. 5. Schematic view of mosaic CCD chips mounted on the face plate. Eight CCD chips are mounted on a long subplate. Two subplates will be completed by the autumn 1991.

Kent, S.M., 1983. *Astrophys. J.*, **266**, 562.

Kodaira, K., Okamura, S., Ichikawa, S., Hamabe, M., Watanabe, M., 1990. *Photometric Atlas of Northern Bright Galaxies* (Tokyo: Univ. Tokyo press).

Longo, G., de Vaucouleurs, A., 1983. The University of Texas Monographs in Astronomy, No.3 (Austin: Univ. Texas).

Lasker, B.M. et al., 1988. *Astrophys. J. Suppl.*, **68**, 1.

MacGillivray, H.T., Stobie, R.S., 1984. *Vistas in Astronomy*, **27**, 433.

Mazure, A., Proust, D., Mathez, G., Mellier, Y., 1988. *Astron. & Astrophys. Supp.*, **76**,

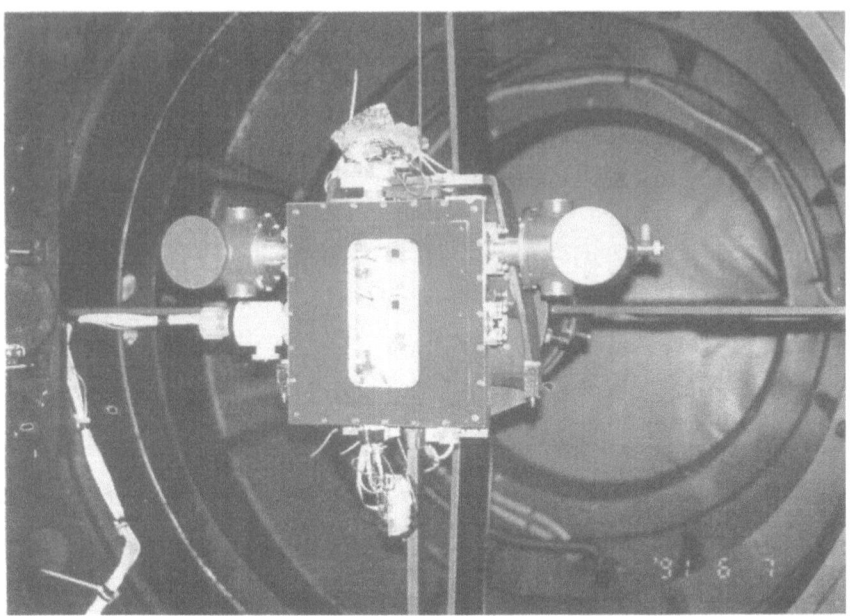

Fig. 6. Mosaic CCD camera as of June 6, 1991, when 'first light' was taken. The camera was mounted on the prime focus of the Kiso Schmidt telescope. Two CCD chips were actually equipped. They can be seen inside the dewar window. Cylindrical dewars attached on both sides are used as liquid nitrogen tanks.

339.
Okamura, S., Kodaira, K., Watanabe, M., 1984. *Astrophys. J.*, **280**, 7.
Sebok, W.L., 1979. *Astron. J.*, **84**, 1526.
Sekiguchi, M., 1991. In preparation.
Yamagata, T., 1985. *Ann. Tokyo Astron. Obs.*, **21**, 31.

Discussion

Odewahn :

Your galaxy type classification is very impressive. How do you define your mean surface brightness parameter?

Doi :

We use just the one at our detection (lowest) level (25.5 mag in the B–band)

MacGillivray :

How faint can you reliably star/galaxy classify with your classifier on KISO Schmidt plates?

Doi :

We have measured objects for $m_B \leq 16.5$, and have not tested fainter than this limiting magnitude. We think we should, of course.

AUTOMATED MORPHOLOGICAL CLASSIFICATION OF FAINT GALAXIES

G. Spiekermann

Astronomisches Institut der Universität Münster, Wilhelm-Klemm-Straße 10, D-W 4400, Münster, Germany.

1 Introduction

The Muenster Redshift Project (MRSP) analyses large scale structures of galaxies by studying film copies of direct and objective prism Schmidt plates (Seitter et al. 1989). For the digitization of these plates, two Microdensitometers PDS2020GM[plus] are available. All steps of data reduction on both direct and objective prism plates are performed automatically. From the objective prism plates the galaxy redshifts are determined (Schuecker 1988). The direct plates are used for image segmentation, star-galaxy-separation (Horstmann 1988) and the derivation of various properties characterizing the image. The description of a further step, the automated morphological classification of galaxies, is given here (see Fig. 1).

The two main procedures in automated morphological classification of galaxies are the analysis of the parameter structure of prototype galaxies and the classification itself.

The basic principle for a classification program which works quite similarly to a human classifier, is to classify galaxies by comparing them to prototypes (template matching). Five morphological classes, equivalent to the Hubble-types, are used : E, S0, Sa, Sb, Sc/Ir. Prototype galaxies from digitized and segmented IIIa-J Schmidt plates are selected interactively using a colour workstation. Five intervals of apparent magnitudes and two groups of different inclinations are distinguished.

2 Parameter Extraction

For the description of galaxy structure, ten different criteria and twelve intensity-levels are used. The parameters represent symmetry, intensity profiles and shapes of the isophotes. All parameters are normalised so that the absolute sizes of objects do not affect the values of the parameters directly. Only the decrease of the

H. T. MacGillivray and E. B. Thomson (eds.), Digitised Optical Sky Surveys 209–213.
© 1992 *Kluwer Academic Publishers.*

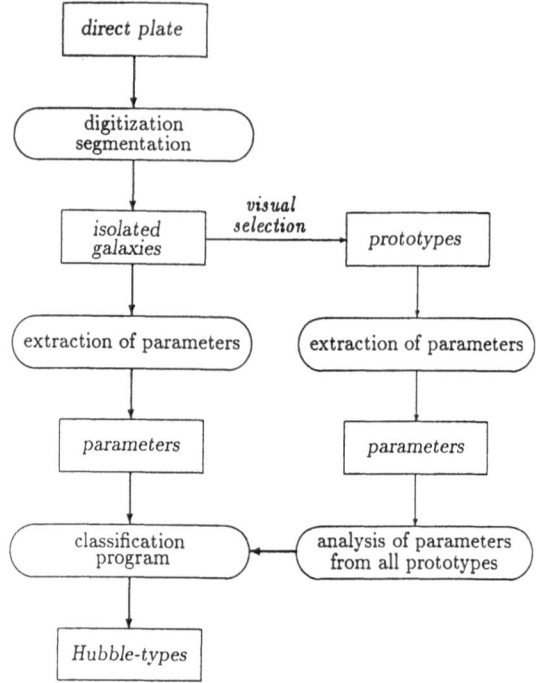

Fig. 1. Schematic illustration of automated morphological classification of faint galaxies (14 < m < 19). The classification program is trained by visually selected prototype galaxies.

signal-to-noise ratio with fainter magnitudes causes changes in parameter structure. Altogether 98 parameters are extracted for each galaxy:

1. M3, normalised moment of 3rd order (for each isophote);
2. distance between the centre of an isophote and the centre of the whole object (for each isophote);
3. changes of position angle of the major axis at different intensity levels;
4. parameters of fit to the intensity distribution;
5. $M4_a$, normalised moment of 4th order relative to the major axis (for each isophote);
6. $M4_b$, normalised moment of 4th order relative to the minor axis (for each isophote);
7. ratio of length of isophote and fitted ellipse (for each isophote);
8. area ratio of isophote and fitted ellipse (for each isophote);
9. ratio of length to area of an isophote (for each isophote);
10. symmetry of the edge structure of each isophote.

There exist more parameters useful for morphological classification but considering the small size of most of the objects, the most stable and safest criteria are selected. Because of the low signal-to-noise ratio, complex analyses such as Fourier-transformations are not used.

3 Fuzzy Classifier

In the classification program, a classifier based on fuzzy-set-theory is used. The basic idea of fuzzy-set-theory is to avoid *yes* or *no* decisions and to use the terms *more* or *less* instead. Especially for galaxy classification, this strategy seems to be very effective. The membership function $\mu_A(x)$ indicates the membership of point x to the set A. Between 0 for 'no member' and 1 for 'perfect member', all values can be adopted.

For every criterion and threshold, S-functions (Zadeh 1975) determine the membership of a galaxy to the prototype groups early-type (E) and late-type (Sc) galaxies. These functions are defined by the shape of the distribution for each threshold, criterion and prototype group. The functions do not contain information about the statistics of the whole sample, in contrast to 'normal' distribution functions.

Two values are attributed to each object, μ_{P_E} and μ_{P_S}, representing the membership to early- and late-type galaxies based on a single criterion. The quality of each criterion is determined by the size of the overlap regions of the prototype distributions. Together with the quality measures, two basic parameters are defined. These values build a feature space in which the position of a galaxy is used to determine its morphological type.

4 Tests

Using the results of visual morphological classification of galaxies in the range $14 < m < 17$ on 16 ESO-SERC fields listed in the ESO-Uppsala Catalogue (Lauberts & Valentijn 1989), 258 galaxy standards are obtained (the completeness limit of the catalogue is given as $14.5m$). The misclassification rate of this sample is determined to 11.2% assuming an error tolerance of one Hubble-type. For 101 galaxies with $14 < m < 15$ the error is 10.9%, for 130 galaxies with $15 < m < 16$ the error is 10.8%. 27 galaxies with $16 < m < 17$ show an error of 15%, in agreement with our own results obtained by visual inspection of several magnitude limited test samples ($m < 18$) containing more than 2 000 galaxies, also from ESO-SERC fields.

5 Application

Classification in the range $14 < m < 18$ is based on visually selected prototypes. The normalised structure of criteria and feature space motivated the extrapolation to $18 < m < 19$. From all examined galaxies between $14m$ and $19m$ the morphological mixture is determined as E:S0:S/Ir = 15/21/64(%) showing no significant variations with magnitude. In clustered areas the number of E galaxies is higher than in cluster-poor fields. For the ESO-SERC field 411, with prominent clusters, we obtain E:S0:S/Ir = 20/23/57(%), in the cluster-poor field 294 late type galaxies dominate, E:S0:S/Ir = 11/19/70(%) (see Table 1.).

Table 1. Morphological mixture of galaxies with $14 < m < 19$ near the South Galactic Pole.

Type	E	S0	Sa	Sb	Sc/Ir	total
16 fields	14.5	20.8	12.1	29.8	22.8	103 394
field 294	11.3	18.9	11.8	32.3	25.7	6 773
field 411	19.6	23.2	11.6	27.3	18.3	8 087

The two-point correlation function for $0.02 \, \text{deg} < \Theta < 10 \, \text{deg}$, confirms these results. From early to late type galaxies, the values of the two-point correlation function decrease systematically, indicating the more clustered distribution of early-type galaxies (see Fig. 2).

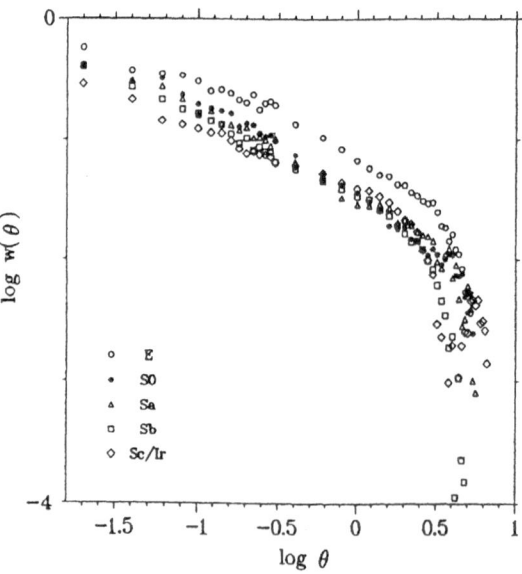

Fig. 2. The two-point correlation function of galaxies in five morphological classes determined in the range $14 < m < 19$ on 16 ESO-SERC fields.

References

Horstmann, H., 1988. In *'Large-Scale Structures in the Universe – Observational and Analytical Methods', Lecture Notes in Physics*. No. 310, p. 111, eds. Seitter, W.C., Duerbeck, H.W., Tacke, M. Springer-Verlag, Berlin.

Lauberts, A., Valentijn, E.A., 1989. *The Surface Photometry Catalogue of the ESO-Uppsala Galaxies*. ESO Garching.

Schuecker, P., 1988. In *'Large-Scale Structures in the Universe – Observational and Analytical Methods', Lecture Notes in Physics*. No. 310, p. 142, eds. Seitter, W.C., Duerbeck, H.W., Tacke, M. Springer-Verlag, Berlin.

Seitter, W.C., Ott, H.-A., Duemmler, R., Schuecker, P., Horstmann, H., 1989. In *'Morphological Cosmology', Lecture Notes in Physics*. No. 332, p. 3, eds. Flin, P., Duerbeck, H.W. Springer-Verlag, Berlin.

Spiekermann, G., 1991. To be published.

Zadeh, L.A., 1975. In *Fuzzy Sets and Their Applications to Cognitive and Decision Processes*, eds. L.A. Zadeh, K.S. Fu, K. Tanaka & M. Shimura. Academic Press, London.

AUTOMATED STAR/GALAXY DISCRIMINATION WITH NEURAL NETWORKS

S.C. Odewahn, E.B. Stockwell, R.L. Pennington, R.M. Humphreys and W.A. Zumach

Department of Astronomy, University of Minnesota, 116 Church Street SE., Minneapolis MN 55455, U.S.A.

1 Introduction

Many of today's most relevant astrophysical problems concerning galactic structure and dynamics, environmental effects on galaxy formation and maintenance, and the large-scale distribution of matter in the Universe are approached in a statistical fashion using deep surveys of stars and galaxies over large areas of the sky. Some of the deepest and most complete galaxy catalogs such as those by Shane & Wirtanen (1967), Zwicky et al. (1961-1968), and Nilson (1973) have been compiled through visual inspection of photographic surveys. More recent efforts by Dickey et al. (1987), Heydon-Dumbleton et al. (1989), Slezak et al. (1988) and Rhee (1990) have used fast scanning machines and automated image detection and classification techniques to compile galaxy catalogs in specific areas of the sky.

A major contribution to both stellar and galaxy surveys will be the catalog of images contained on the 936 plate pairs of the first epoch Palomar Sky Survey currently being generated with the University of Minnesota Automated Plate Scanner (APS). Detailed discussions of this project are given by Humphreys & Pennington (1989). Machine classification is the only viable route in handling the massive data sets encountered in Schmidt plate surveys; but perhaps more importantly, the use of automated classification techniques will produce a more homogeneous sample selection. In this work we shall discuss the development of a novel approach to the problem of star/galaxy separation for the APS automated survey.

2 APS Image Data

We have experimented with a variety of methods for effectively parameterizing the images detected by the APS in threshold densitometry mode. Much of this work fol-

H. T. MacGillivray and E. B. Thomson (eds.), Digitised Optical Sky Surveys 215–224.

lows the methodology of Dickey et al. (1987), Rhee (1990), and Heydon-Dumbleton et al. (1989), but unlike the latter two works we are not using a density to intensity transformation in the image classification stage in order to decrease machine processing time. The image parameters adopted for use in this work are summarized in Table 1. The plate transmission, T, is used directly for the calculation of simple image moments and gradients. Each threshold contour is fit by an ellipse during the scanning process. In a post-processing phase, all pixels with transmission values above the scanning threshold (65% of the sky background) are used to reconstruct the image and compute additional parameters.

Table 1. Image parameters.

Diameter	dia
ellipticity	1 - b/a
average transmission	Tav
central transmission	Tc
(ellipse area) / (area from pixel count)	c1
log (area from pixel count)	c2
$(\Sigma T/r)/(\Sigma T)$	moment 1
rms error of ellipse fit to transit endpoints	fuzz
Y centroid error	jitter
$(T_4 - T_1)/(r_1 - r_4)$	gradient 14
$(T_3 - T_1)/(r_1 - r_3)$	gradient 13
$(T_2 - T_1)/(r_1 - r_2)$	gradient 12
$(T_3 - T_2)/(r_2 - r_3)$	gradient 23
$(T_4 - T_3)/(r_3 - r_4)$	gradient 34

The peak transmission, Tc, and average transmission, Tav, have proven to be very effective discriminators when combined with some measure of image size such as the diameter (see Fig. 1). Other useful parameters for image discrimination are:

1. $c1 = 2\pi a^2/A$ where a is the ellipse semimajor axis length, and A is the image area derived by summing the number of image pixels.
2. $c2 = \log A$.
3. $mom1 = (\sum(T(x,y)/r))/(\sum T(x,y))$ where r is the pixel radius measured from the image center (X_c, Y_c).
4. $Gij = (T_j - T_i)/(r_i - r_j)$, a simple image gradient in which T_i is the median transmission value in an elliptical annulus (having a shape which matches the threshold isophote) and semimajor axis length r_i.

Five distinct image gradients are formed using 4 image radii, to give a total of 14 classification parameters. These quantities comprise the input vector for the neural network classifiers to be discussed in section 3.

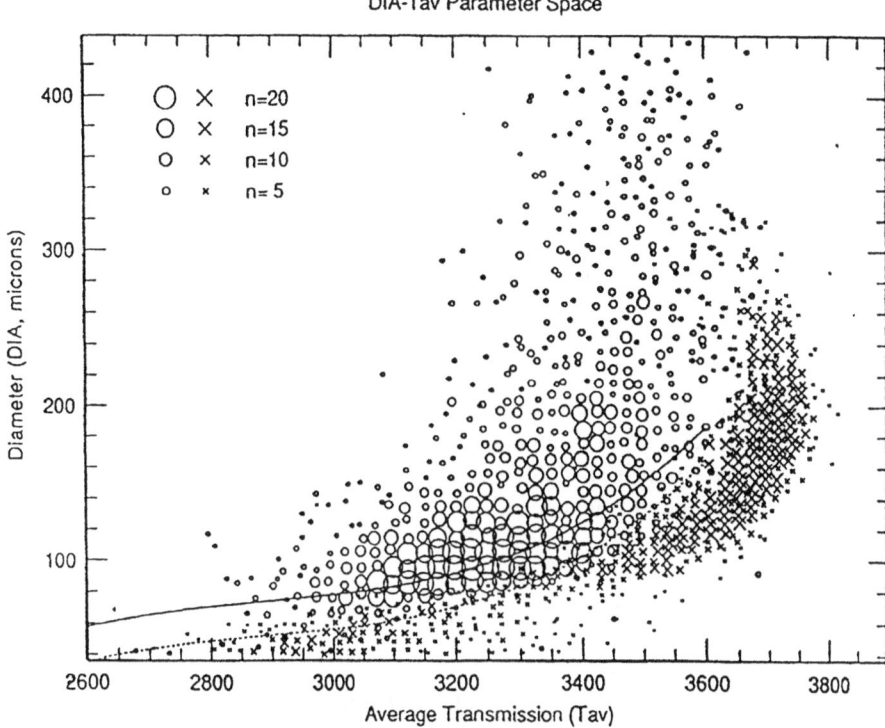

Fig. 1. The distribution of galaxies (open circles) and stars (crosses) in the diameter (DIA) vs. average transmission (Tav) parameter plane. The point size scales linearly with the number of objects occupying that position in parameter space. Notice that galaxies generally possess a larger diameter than stars for a given average transmission value. The solid curve may be used to select galaxies contaminating a star sample. All objects lying above this line have a very high probability of being a nonstellar source. The dashed line is used in a similar fashion to select contaminating stellar objects in a nonstellar sample.

Initial samples of galaxy and star images were collected from two regions of equal area on P323 (the POSS plate containing the Coma cluster of galaxies). As a final check, a smaller area, which extends from the center of the plate to the edge, was also surveyed and used purely as a source of test data for the image classification software. Region 1 was centered on the core of the Coma cluster. Galaxies were collected in this portion by cross-matching APS image coordinates with positions of Coma galaxies in the deep catalog of Godwin et al. (1983), hereafter referred to as GMP83. In this initial study we have restricted our sample to a minimum APS image diameter of $D = 40\mu$m (corresponding to B = 21 for a stellar image) and galaxies brighter than $B_{26.5} \leq 20.0$, where $B_{26.5}$ is the isophotal B band magnitude from GMP83 which measures the flux integrated within the $\mu_B = 26.5$ B-mag arcsec^{-2} isophote. A sample of stellar objects was selected in this region by locating objects not matched to the GMP83 catalog and obeying the diameter restriction. This provided an initial list of approximately 2400 galaxies (B $\leq$ 20) and 750 stars

$(11.75 \leq B \leq 21)$.

An area having the same size as Region 1, but located in the opposite plate quadrant, provided the **Region 2** sample. Incomplete samples of objects in progressively smaller image diameter intervals (down to $D = 40\mu m$) were classified as galaxies or stars by visual inspection. This produced a list of 290 galaxies ($B \leq 20$) and 1350 stars ($11 \leq B \leq 21$). The last area surveyed, which is used as test data only, is referred to as **Region 3**. In Region 3, we visually classified all images in a $94' \times 205'$ rectangular box extending roughly from the plate center, with diameters greater than $73\mu m$ (corresponding to approximate B band magnitudes of 19.5 for stars and 20.1 for galaxies). In this case, a total of 4135 detected objects were classified into 5 major categories: galaxies, stars, merged stellar images, plate defects and uncertain types. For this last survey, all classifications were done on the glass copy plate using a $8\times$ loop magnifier for images with diameters larger than $130\mu m$, and a variable magnification binocular microscope for all smaller images. The resultant sample was comprised of 2380 stars and 936 galaxies to be used exclusively for testing the automated image classifiers developed in this work.

3 A Neural Network Approach

Neural networks are a family of artificial intelligence techniques that are capable of performing difficult pattern classification tasks. Their design and development have been inspired by biological neural networks, but the algorithms we have used do not accurately model real biological systems. Many introductory references on neural networks are available: Lippmann 1989 and Rumelhart & McClelland 1988.

We have use the perceptron and backpropagation neural network algorithms to create accurate classifiers for separating star and galaxy images and to inspect their parameter spaces. Backpropagation and perceptron networks are supervised learning techniques. The networks start from a random initial state. A set of training patterns is used to 'teach' the network to perform the desired classification function. The training set must contain a representative sample of patterns for each class. Starting from a random initial configuration, the network is used to classify each pattern in the training set. Each time a pattern in the training set is misclassified, an error term is computed and used to adjust the network's configuration. The patterns in the training set are repeatedly presented to the classifier in this manner until the entire training set is correctly classified or until the network is unable to learn any more patterns.

A perceptron is a simple classifier that forms a hyperplane in parameter space to separate the two classes. It is trained using a simple gradient descent procedure to minimize an error criteria function. The *Perceptron Convergence Theorem* (Duda & Hart 1973) guarantees that training will converge to a solution vector if the classes are linearly separable.

Backpropagation networks are capable of learning much more complex functions and are not restricted to linearly separable classification problems. A backpropagation network consists of one or more layers of nodes ('neurons'). Each node computes the following function *:

$$output = \frac{1}{1 + \exp -(\vec{i} \cdot \vec{w} + b)} \qquad (1)$$

where $\vec{i}$ is the node's input. This is either the network's input vector or the output from the previous layer. The vector $\vec{w}$ and b are the weight vector and bias term and are unique for each node. Initially they contain random values. The weight and bias values are modified by the training process according to the generalized delta rule (Rumelhart & McClelland 1988).

Nodes are arranged in layers. The number of nodes and layers in a network determines the complexity of the function that it computes, as well as the amount of information that it contains. We have experimented with networks consisting of two and three layers of nodes. Figure 2 illustrates a network with an input vector of length five: four nodes in the first layer, two nodes in the second layer, and three in the output layer. As a shorthand, such a network can be written as $\{5 : 4, 2, 3\}$. To use a backpropagation network as a classifier, each class is assigned to an output node. The node with the greatest output value determines the classification of the input pattern. While a simple perceptron classifier is restricted to two classes, a backpropagation network can be used for an arbitrary number of classes.

Several different networks were generated. We have trained a perceptron for a small diameter regime (73μm $\leq$ D $\leq 137\mu$m), referred to as SP; and one for a large diameter regime (146μm $\leq$ D $\leq 330\mu$m), referred to as LP. Using the same diameter regimes, we have trained $\{14 : 14, 13, 2\}$ backpropagation networks which are referred to as S1 for the small regime, and L1 for the large regime. These are the primary networks discussed in this paper. Additional networks were also trained to investigate specific features. Networks SR (small reduced) and LR (large reduced) were trained with smaller training sets to explore the number of prototypes necessary to produce an accurate classifier. A backpropagation network which was allowed fewer training passes and was used to experiment with using the network's output values as a measure of the confidence of image classification is referred to as SC. The training sets and network configuration for the classifiers are summarized in Table 2.

Considering the network trained with the full set of training data (SP, LP, S1, L1), the networks had greater difficulty learning to separate the star and galaxy classes for small diameter images than for large diameter images. This is indicated by the higher rate of misclassified training patterns for both the small diameter backpropagation and perceptron networks. Also, S1 required many more training passes than L1 before it reached its final state.

Our results suggest that the parameter space for diameters greater than 137μm separates well and is nearly linearly separable. Smaller images have a more complex parameter space and a linear classifier is not adequate. This also demonstrates the superior learning capacity of the backpropagation network in comparison to the perceptron algorithm. One possible drawback is that if errors exist in the training

---

* This function is most commonly used. Other functions with a similar shape such as $x/(c + |x|)$ may also be used (Stockwell 1991).

Input   First   Second   Third (output)
Vector   Layer   Layer   Layer

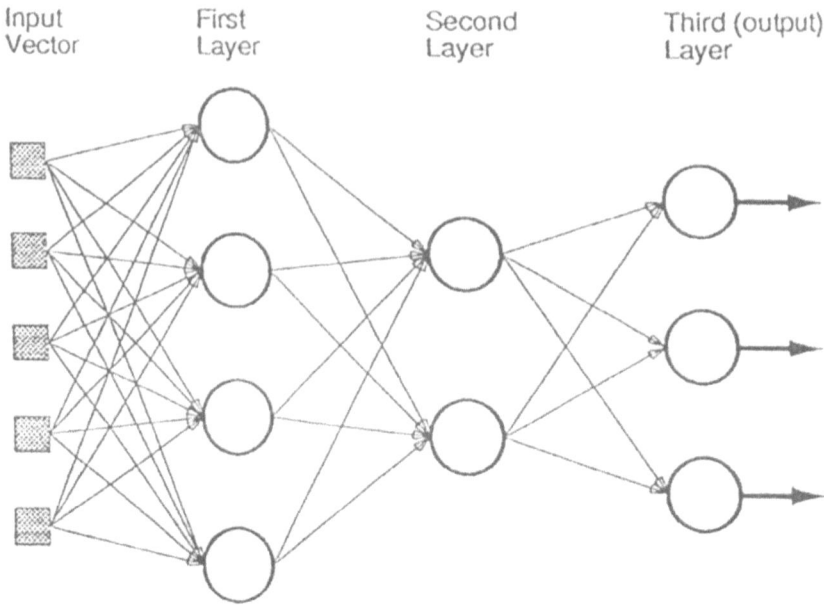

Fig. 2. A schematic illustration of a network with an input vector of length five: four nodes in the first layer, two nodes in the second layer, and three in the output layer. As a shorthand, such a network can be written as: {5:4,2,3}.

Table 2. Neural network training parameters.

| Network | Diameter | Patterns in Training Set | | | Training |
	Range (microns)	Galaxies	Stars	Total	Passes
L1	$d > 137.3$	1116	1050	2166	925
LP	$d > 137.3$	1116	1050	2166	50000
LR	$d > 137.3$	300	300	600	834
S1	$146.5 > d > 73.2$	1584	719	2303	6540
SP	$146.5 > d > 73.2$	1584	719	2303	50000
SR	$146.5 > d > 73.2$	300	300	600	3873
SC	$146.5 > d > 73.2$	1584	719	2303	218

set, the backpropagation network is more capable of learning these as well. The perceptron, by virtue of its limitation to simple division of parameter space, is required to learn a more general function.

For the classification results listed so far, the image classification for the backpropagation networks is based on the node with the highest output value. Since the network is trained with idealized target outputs of zero and one, a pattern that produces an output between 0.1 and 0.9 must deviate from the patterns that the classifier learned when it was trained. Although many images are correctly classified with output values that fall in this intermediate range, the frequency of misclassification is higher. Although it cannot be used directly as a probability of class membership, it is clear that the network's output value can provide some measure of classification confidence and a means of producing data sets that are relatively free from contamination.

4 Analysis of the Classifier Success Rate

In order to test the quality of the final perceptron and backpropagation networks, which were trained using galaxy and star images collected in P323 Regions 1 and 2, a third independent set of test images was established. As described previously, over 4500 images with diameters larger than 73μm were classified in order to provide an adequately large test sample. The image parameter sets for these images were measured and normalised in the same manner as the training data. Depending on the diameter of a test image an appropriate network was then used to classify the test image. We judge the success of the automated network classifiers by comparing their results to those classifications established by visual inspection. It should be stated explicitly that all visual classifications were made prior to image analysis by the network.

In Fig. 3 we plot the success rate, S=(number of successes/total number) × 100, as a function of the O plate B band magnitude. The curves were established by binning the data in equal number diameter bins containing roughly 30 objects each in the case of galaxies, and 80 objects each in the case of stars. The point symbols represent mean points established in binning intervals of 0.5 magnitudes. The success rate of the backpropagation network classifier remains above 90% until B ≈ 20.0 for galaxies. The success rate for stars fluctuates in the range between 95% and 100% out to B ≈ 19.5. Very similar results were obtained with the perceptron classifiers.

It should be noted that while these results are very satisfactory, implying that our POSS galaxy catalog will be as deep as the often used Lick survey, it is unclear whether the fall-off in galaxy classification success rate at the faint end is due to lack of information in the small image parameter sets, inadequacies of the neural network method when applied to noisy data, or lack of precision by the human classifier in the case of small, faint images. Efforts are underway, using deeper, higher resolution imagery to differentiate these effects. Even if the success rate function implied in Fig. 3 is dictated by the deepness and resolution of the POSS copy plate material, the fundamental advantage of the automated classification approach is evident when we consider that the human classifications were made over several weeks of tedious work, while the plate digitization, image parameterization and network classification were carried out in a matter of hours.

One disadvantage of neural network classifiers is that they give no direct information regarding why an object is assigned a particular classification. In an attempt to determine the significance of the various parameters in our 14 element input vector, we have used an empirical approach in which each component of the input vector is distorted in such a way that its information input is nullified. In

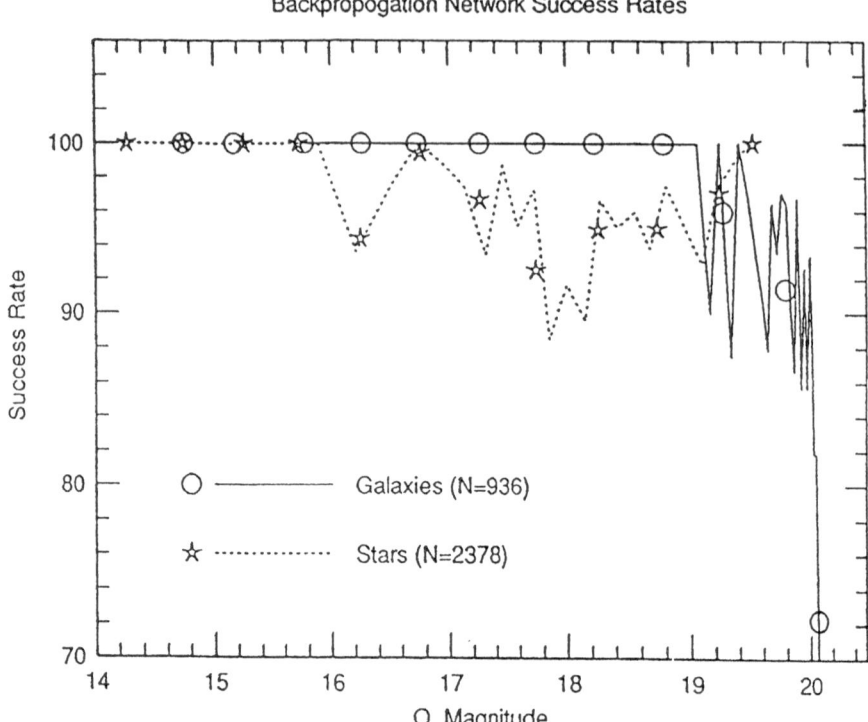

Fig. 3. The classification success rates of the backpropagation network developed in this work. The curves are derived by binning equal numbers of objects into diameter bins. The symbols denote mean success rates in binning intervals of 0.5 magnitudes. Note that both stars and galaxies are classified with success rates above 90% down to the relatively faint O magnitude of 19.5.

our case, the network is run on 14 separate test data sets, each set having a percentage level of random noise added to one of its components. A parameter, X, is adjusted using the relation $X_a = X \pm \beta X$, alternating the sign of the incremental adjustment for every other image. If a parameter carries high weight in the network, then the rate of classification success should be decreased when that information is effectively removed from the calculation. This method is simplistic in its approach in that the network might rely upon interdependencies between several sets of parameters, and removing just one component at a time may not severely cripple the network's ability to make accurate classifications. This is one of the positive features of a neural network which produces a very robust classifier in the presence of observational scatter. This simple empirical method should allow a rough determination of which input parameters are most important to our present networks.

We have run eight sets of calculations, varying the input β value, for each diameter regime. So as not to bias the results of this experiment, we have used approximately equal numbers of stars and galaxies in each diameter regime. Classification success rates decrease with increasing β value. The four most significant image parameters for each diameter regimes (those parameters which produce the fastest success rate degradation) were found to be:

1. Tav, g34, g14 and g13 for the large diameter set,
2. Tav, Tc, c2 and g14 for the small diameter set.

The results of this simple test are encouraging in that they confirmed our intuitive feeling that image contrast parameters such as the gradient and moment terms would be most useful in the large images. For small images, the calculation of these parameters is complicated by poor resolution, and image classification must be based on simpler global parameters. This exercise also serves to demonstrate the robust nature of the neural network classifiers. A large level of noise ($\beta = 0.5$ and greater) was required to significantly degrade the classifier success rate.

5 Conclusion

A key step in the compilation of an automated survey such as that being conducted with the APS is the discrimination between stellar and nonstellar images. The development of a machine automated technique for performing this task ensures that the tremendous flow of data generated by such a project may be processed in a reasonable period of time. Perhaps more importantly, such a technique produces a very homogeneous catalog whose constituents are selected in an objective and consistent fashion. In this work we have presented a rather novel automated image classification technique employing a neural network which shows great promise. Classifications into stellar and nonstellar categories are based upon a 14-element image parameter set. We have shown that a number of parameter spaces formed with these vector elements are effective in separating a sample of images into the two basic populations of stellar and nonstellar objects. The application of a neural network to this problem allows a large number of image parameters to be used simultaneously in distinguishing a classification.

We have experimented with a simple linear neural network classifier known as a perceptron, as well as with a more sophisticated backpropagation neural network with the result that we are able to attain classification success rates of 99% for galaxy images with B $\leq$ 18.5 and above 95% for the magnitude range $18.5 \leq B \leq 19.5$. Based on an analysis of 3601 galaxy images and 4460 stellar images on the POSS field containing the Coma cluster, we have determined the success rate of these classifiers as a function of image diameter and integrated magnitude. Simple numerical experiments illustrate the robust nature of this method and identify the most significant image parameters used by the networks in distinguishing image class. The results from these experiments are extremely promising and indicate that the APS survey of the POSS copy plates will reach a limiting magnitude fainter than that of the Lick integrated counts (Shane & Wirtanen 1967).

References

Dickey, J.M., Keller, D.T., Pennington, R., Salpeter, E.E., 1987. *Astron. J.*, **93**, 788.

Duda, R.O., Hart, P.E., 1973. *Pattern Classification and Scene Analysis*, John Wiley & Sons, New York.

Godwin, J.G., Metcalfe, N., Peach, J.V., 1983. *Mon. Not. R. Astron. Soc.*, **202**, 1123.

Heydon-Dumbleton, N.H., Collins, C.A., MacGillivray, H.T., 1989. *Mon. Not. R. Astron. Soc.*, **283**, 379.

Humphreys, R.M., Pennington, R.L., 1989. *Workshop on Digitized Optical Sky Surveys*, p. 1, eds. C. Jaschek, H.T. MacGillivray.

Lippman, R.P., 1989. *IEEE Communications Magazine*, **27**, 11.

Nilson, P., 1973. *Uppsala General Catalog of Galaxies*, Uppsala Offset Center.

Rhee, G., 1990. *The Structure of Rich Clusters of Galaxies: Clues to Formation and Origin*, PhD thesis, University of Leiden.

Rumelhart, D.E., McClelland, J.L., 1988. *Parallel Distributed Processing*, Vol 1, MIT Press, Cambridge, Mass.

Shane, C.D., Wirtanen, C.A., 1967. *Publications of Lick Observatory*, Technical Report 22, Lick Observatory.

Slezak, E., Bijaoui, S., Mars, G., 1988. *Astron. Astrophys.*, **201**, 9.

Stockwell, E.B., 1991. *IEEE Transactions on Neural Networks*. Submitted.

Zwicky, F., Herzog, E., Kowal, C.T., Wild, P., Karpowicz, M., 1961-1968. *Catalogue of Galaxies and Clusters of Galaxies*, Pasadena: California Institute of Technology.

Discussion

Deul :

How much time does the training of the network take, and is your final network fully connected (no bypasses, no feedbacks)?

Odewahn :

For training our {14:14,13,2} backpropagation networks on a sample of approximately 5,000 objects has required three to four (human) hours on an unloaded Sun4. We have not yet experimented with disconnecting various nodes in the network.

Parker :

Are you going to have to retrain the neural network for each new plate that you wish to scan, as of course plates come in a variety of different qualities, etc.?

Odewahn :

We are investigating the plate to plate variations present in POSS copies in general. We believe that a set of systematic parameter space corrections will be applied before treatment by the neural network classifier. Hence, only one extensively trained network (in theory!) will be required. Certainly this is an area which will require much more work.

A COMPARISON OF STAR/GALAXY CLASSIFICATION APPROACHES ON DIGITISED POSS-II PLATES

N. Weir and A. Picard

Division of Physics, Mathematics and Astronomy, California Institute of
Technology, Pasadena, CA 91125, U.S.A.

Abstract

We have compared the use of so-called 'parametric' and 'resolution'-based
object classifiers on digitized images from the Second Palomar Observatory Sky Survey. We find that the latter approach, involving the fitting of
two-dimensional templates to each detected object, can provide reasonable
discrimination at levels approximately a magnitude fainter than the former.
We will employ this technique to significantly increase the depth of a galaxy
survey using Palomar Schmidt plates.

1 Introduction

The accuracy of star/galaxy separation generally determines the limiting magnitude of galaxy surveys. We have therefore taken great interest in evaluating the
effectiveness of various object classification algorithms, in hopes of squeezing every
last bit of information from imaging data. Our intent with this work was to determine the relative advantage of applying classification methods which rely upon
full access to pixel data to those which solely use pre-measured object parameters.
Only relatively recently has computing and storage technology reached a level that
we could consider using the pixel data from Schmidt plate surveys for extensive
off-line analysis. The question remained, however, exactly how much was gained by
having it available?

We are primarily interested in applying such methods to the construction of
object catalogs from digitizations of the Second Palomar Observatory Sky Survey
(POSS-II: Reid et al. 1991). A galaxy catalog covering approximately 386 square
degrees, and complete to a magnitude of 19 in Gunn r, has recently been completed

H. T. MacGillivray and E. B. Thomson (eds.), Digitised Optical Sky Surveys 225–230.
© 1992 *Kluwer Academic Publishers.*

by Picard (1991 a, b, c) using object catalogs from scans of IIIa-F POSS-II plates provided by COSMOS. Soon we will have access to the scanned pixel data for an even larger region, including both IIIa-F and IIIa-J plates. This work is a part of our investigation of reduction techniques for the analysis of this data set, and eventually the digitized POSS-II in its entirety.

2 Data

The photographic material used in this project consists of a IIIa-F plate obtained at the Oschin 48″ Schmidt as a part of POSS-II. The plate is of designated field 681, which contains two Abell clusters of richness class one: Abell 68 and 73. It is one of nine plates in the Southern field analyzed by Picard as a part of his PhD. thesis. In terms of image quality, this plate may be characterized as average to good, with a limiting magnitude of 21.0 in Gunn r.

The plate was scanned using the COSMOS machine at the Royal Observatory, Edinburgh, which is described extensively in MacGillivray & Stobie (1984). The deblending software described by Beard et al. (1990) was employed in constructing the final COSMOS catalog. This catalog consists of a set of 32 parameters for each object detected on the plate. The isophotal threshold employed for detection and photometry was 7% of the local sky.

A portion of the plate enclosing both of the Abell clusters was also scanned at JPL using a PDS microdensitometer. The step size employed was 15 μm ($\approx 1.1''$), compared with 16 for COSMOS, and a spot size of 16 μm, compared with 32. A 5000×5000 pixel image was obtained, in addition to scans of the calibration spots. A density to intensity transformation was performed on the PDS image using the spots.

CCD calibration frames were obtained covering both Abell 68 and 73 in Gunn g and r as a part of the galaxy survey conducted by Picard (1991c). The data were acquired at the Palomar 60″ telescope using a TI 800×800 CCD array with a re-imaging lens in order to obtain a suitable field of view. This system yields a pixel size of 0.615″ and a total field of view of 8′. The exposures were for 1200 seconds in r and 1800 seconds in g, yielding an estimated completeness limit of 22.0 in r and 23.0 in g.

3 Star/Galaxy Classification

Two different classification methods were applied to the data obtained from the COSMOS and PDS scans. The two approaches are often characterized as the 'parametric' and 'resolution'-based classification methods. The former may be applied to a catalog of objects for which a set of attributes have been pre-measured, while the latter requires access to the pixel data to form an empirical estimate of the local point spread function (PSF), which is then fit to all local objects. Both are actually parametric in the sense that they eventually involve establishing divisions in some parameter space distinguishing stars, galaxies, blends, etc. Resolution classifiers, however, make fuller use of the pixel data to provide an extremely sensitive and discriminating set of additional parameters.

3.1 The Parametric Classifier

The star-galaxy separation method employed on the COSMOS catalogs is described in detail by Picard (1991c), and is similar to one described by Collins et al. (1989). It is based on the fact that stars, being determined uniquely by the instrumental point spread function, occupy a well determined region in the measured parameter space. Galaxies, however, being resolved objects, have their own intrinsic profiles and thus tend to deviate from the region occupied by stars. The idea is therefore to separate stars from galaxies by establishing boundaries, or 'cuts', dividing the two within parameter space. All objects quickly merge into one ill-defined ridge at faint magnitudes, regardless of which parameters we choose. Incompleteness and contamination will therefore always be a part of faint object catalogs. Accordingly, we must take care that the method for cutting up the parameter space remains uniform and objective throughout the survey. Picard accomplishes this by establishing a composite classification parameter ϕ which is approximately uniform in his survey. The classification of objects in the COSMOS catalog were derived using a cut in this parameter space which was found by Picard (1991c) to provide the most reliable classifications.

3.2 The Resolution Classifier

The star-galaxy separation on the PDS pixel data was performed using the classifier incorporated in FOCAS (Faint Object Classification and Analysis System: Valdez 1982; Jarvis & Tyson 1979). This algorithm is fully described in Valdes (1982). In a nutshell, the program employs a maximum likelihood technique to fit the light distribution of each image to templates representing stellar, resolved, and noise images, and then based upon the best-fitting template parameters, establishes a classification for each object. Given an estimate of the PSF, the templates are parameterized by only two values, representing the fraction and scale of a non-point source like component needed to obtain the best-fit to a given image.

An essential task in employing this technique is, therefore, to establish an accurate estimate of the PSF. Only after this is obtained can the classification parameters be measured. With the PDS scans, we first selected a group of candidate stellar objects using a simplified parametric approach similar to that employed above. We then checked each candidate by eye and compiled a final list of objects which we were fairly certain were isolated and unresolved. These stars, approximately 10 in each field, were averaged to form the PSF template. We are currently investigating robust techniques for automatically generating lists of PSF stars from POSS-II fields, to eliminate the need for human interaction in this process.

4 Results

To assess the accuracy of the above classifications, we processed the deeper CCD images using FOCAS and followed-up with visual checks. We then only judged objects in the plates overlapping the approximately $8' \times 8'$ CCD fields. We generally found FOCAS to reliably classify objects in the CCD frame to within a magnitude of the plate detection limit. For Abell 68, visual inspection verified that the r band image was virtually 100% accurate to a limiting magnitude of 20 in r, two magnitudes above the CCD detection limit. Because the total number of objects in each field was so low (88 in Abell 68 and 93 in Abell 73 to a magnitude of 20 in r), we were able to check nearly every classification in all four CCD fields, paying special attention to those objects where the g and r classifications disagreed. There were some notable discrepancies in the Abell 73 frames where the re-imaging optics failed to maintain a constant PSF across the chip. Because FOCAS assumes a space-invariant PSF, it was not possible to accept every FOCAS CCD classification at face value, and several reclassifications had to be performed by inspecting the two CCD images simultaneously by eye. Because these images were of sufficiently high signal/noise relative to the plate data, however, we could conservatively estimate the (occasionally modified) FOCAS CCD classifications to be more than 95% accurate up to the faintest magnitude cut we applied to the plates. We therefore accepted them as 'truth' in assessing the quality of the plate classifications.

Our essential results are presented in Figs. 1 and 2. They give the estimated completeness (fraction of true galaxies classified as such) and contamination (fraction of objects mistakenly classified as galaxies) as a function of limiting r magnitude for the two plate classification methods employed. We note that these estimates suffer from small number statistics, and that, as demonstrated by Picard (1991 b, c), the actual average completeness is approximately 94% at $r = 19$ mag for the parametric approach using COSMOS data. This was the magnitude to which he limited his galaxy survey. However, the FOCAS resolution classifier provides consistently high completeness and low contamination all the way to an r magnitude of 20. The results are consistent for both of the independently analyzed fields. From these results, it appears we will be able to extend the limiting magnitude of forthcoming catalogs derived from digitized scans of POSS-II a full magnitude beyond previous limits, with a similar degree of completeness and contamination.

We have yet to fully establish to what extent this improvement is due to the classification method employed, or possible differences in the quality of the scans themselves. Given the similarity of the scanning parameters used (with the possibly important exception of the spot size), and the relatively faint magnitude range of the galaxy sample analyzed (i.e. saturation effects are negligible), we do not anticipate that pixel data differences dominate. Nonetheless, an investigation of this factor is in progress.

5 Conclusion

We have compared the results of applying two common object classification methods to digitized scans of POSS-II plates. We find that the resolution classifier approach, such as that implemented in FOCAS, may achieve star/galaxy separation, to a given degree of completeness and contamination, a magnitude fainter than is possible with the sort of parametric approaches employed in most Schmidt surveys to date. The former method is more computationally expensive and requires greater access to the scanned pixel data, but it will be readily possible with the forthcoming digitized POSS-II.

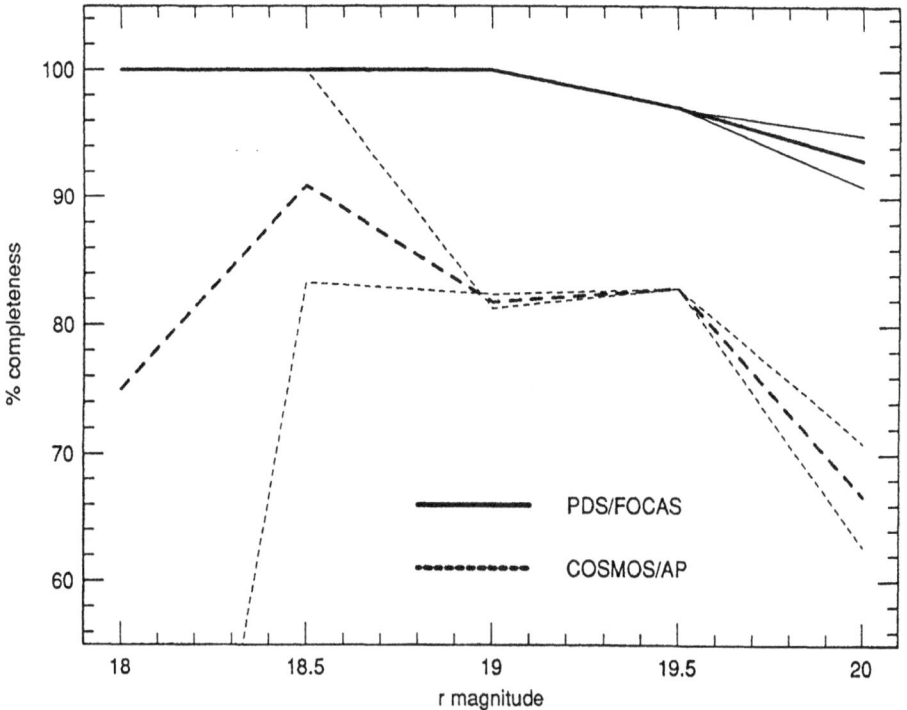

Fig. 1. The completeness of the galaxy catalogs obtained using the methods described in the text as a function of limiting magnitude. The bold lines are for the complete data set, while the fainter lines are for each Abell cluster field processed separately. Discrepancies at the bright end are not significant, as the sample is extremely small.

6 Acknowledgements

We would like to thank the COSMOS group at the ROE, in particular H. T. MacGillivray, and J. Fulton at JPL for providing the scan data for this work. Also, many thanks to the Sky Survey team for their expertise and effort in acquiring the plate material. This work was supported in part by an NSF graduate fellowship (N.W.).

References

Beard, S.M., MacGillivray. H.T., Thanisch, P.F., 1990. *Mon. Not. R. Astron. Soc.*, **247**, 311.

Collins, C.A., Heydon-Dumbleton, N.H., MacGillivray, H.T., 1989. *Mon. Not. R. Astron. Soc.*, **236**, 7.

Jarvis, J.F., Tyson, A.J., 1979. *SPIE Proc. Instrumentation in Astronomy*, **3**.

MacGillivray, H.T., Stobie, R.S., 1984. *Vistas in Astronomy*, **27**, 433.

Picard, A., 1991a. *Astrophys. J. Lett.*, **368**, L11.

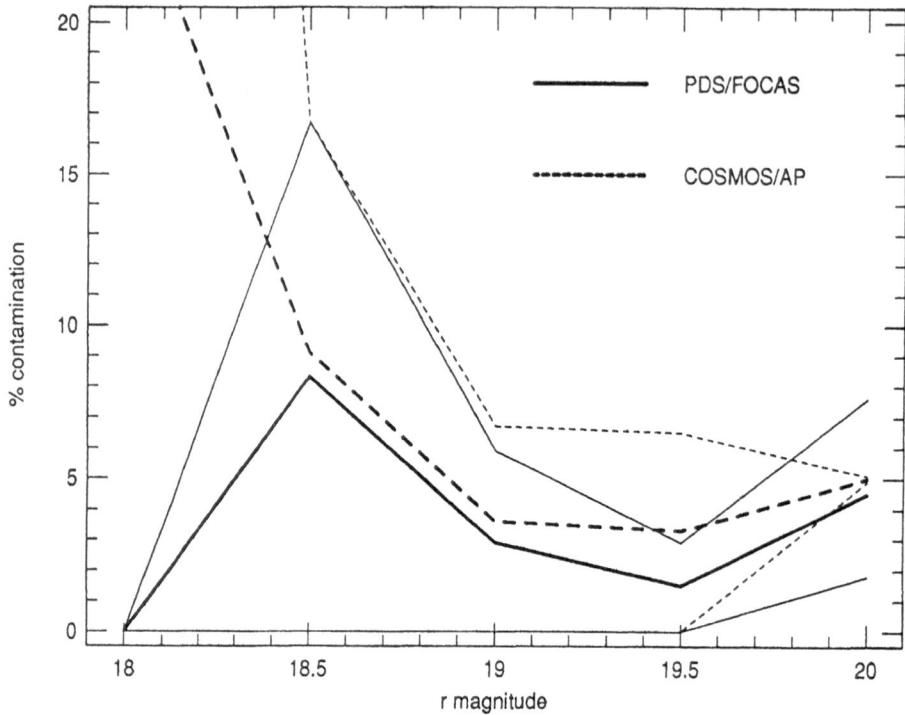

Fig. 2. The contamination of the galaxy catalogs obtained using the methods described in the text as a function of limiting magnitude. The line-types are the same as in Fig. 1.

Picard, A., 1991b. To appear in *Astron. J.*

Picard, A., 1991c. PhD thesis, Caltech.

Reid, I.N., Brewer, C., Brucato, R.J., McKinley, W.R., Maury, A., Mendenhall, D., Mould, J.R., Mueller, J., et al. 1991. *Publ. Astron. Soc. Pacific*, **103**, 661.

Valdes, 1982. *Instrumentation in Astronomy* IV, SPIE **331**, 465.

AUTOMATED STELLAR SPECTRAL CLASSIFICATION VIA STATISTICAL MOMENTS

R.J. Dodd and S. Legget

Carter Observatory, P.O. Box 2909, Wellington, New Zealand.

Abstract

A method for classifying stellar spectra using statistical image moments is described.

1 Introduction

A typical objective prism plate obtained using the 120cm UKST may contain over 10^5 images of spectra (Savage et al. 1985). Classifying such spectra by eye or manually using microdensitometer scans is very time consuming. To overcome this problem, various methods of classifying spectra automatically have been tried (cf. Schmidt-Kaler 1982). LaSala (1988) has categorised the different automated techniques into two classes which he calls criterion-evaluation and pattern-recognition respectively. The method described in this paper falls into the latter category.

2 Statistical Moments

Statistical moments have been used to provide objective data on direct Schmidt plates measured by the COSMOS machine. Stobie (1980) showed how they may be used to distinguish and parametrize different image types and Dodd & MacGillivray (1986) extended the technique to search for clusters of galaxies.

To see if a moments analysis method would be of use in classifying stellar spectra, tests were carried out using the published spectral data of Straizys & Sviderskiene (1972). These tables, listing $F(\lambda)$ in uniform intervals of λ, were converted to a nonlinear form to match the $800\text{Å}/mm$ at $H\gamma$ dispersion of one of the UKST prisms. This produced a table of $I(x)$ against x where x is the measured distance from a datum point within the spectrum and $I(x)$ is the measured intensity at that point.

The following intensity weighted image moments were computed:

H.T. MacGillivray and E.B. Thomson (eds.), Digitised Optical Sky Surveys 231–233.

1. M_1 - mean $(\sum I(x)x)/(\sum I(x))$
2. M_2 - variance $(\sum I(x)x^2)/(\sum I(x)) - M_1^2$
3. B_1 - skewness
4. B_2 - kurtosis

Approximately 100 sample points uniformly spaced in x were used.

3 Results

The most promising relationship between spectral type and statistical moment was found for luminosity class V using M_1 versus Spectral Type (see Fig. 1).

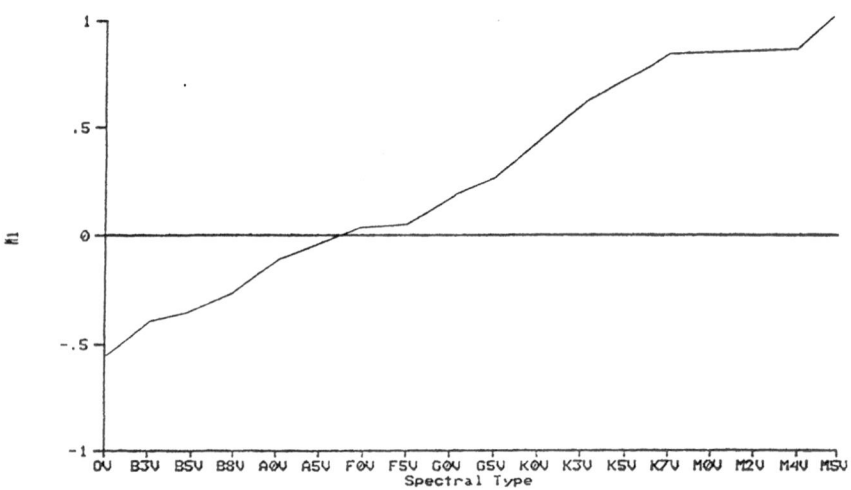

Fig. 1. Plot of M_1 (mean) versus Spectral Type for luminosity class V stars.

Plots were also constructed of values of different statistical moments plotted against each other for luminosity classes I, III and V. The most useful relationships for all luminosity classes was found between M_1 and M_2 (see Fig. 2).

4 Tests on Schmidt Plates

So far, no tests have been carried out on high speed measuring machine output of a Schmidt plate, though a few trials using a film copy have been carried out at the Carter Observatory using a PC controlled J-L Chromoscan III microdensitometer. As with most other methods these tests suggest that the statistical moments may well prove of use for ideally exposed spectra but that problems will occur for under and over exposed images. Problems may also occur with reddened spectra, and at low dispersions luminosity class discrimination is not possible.

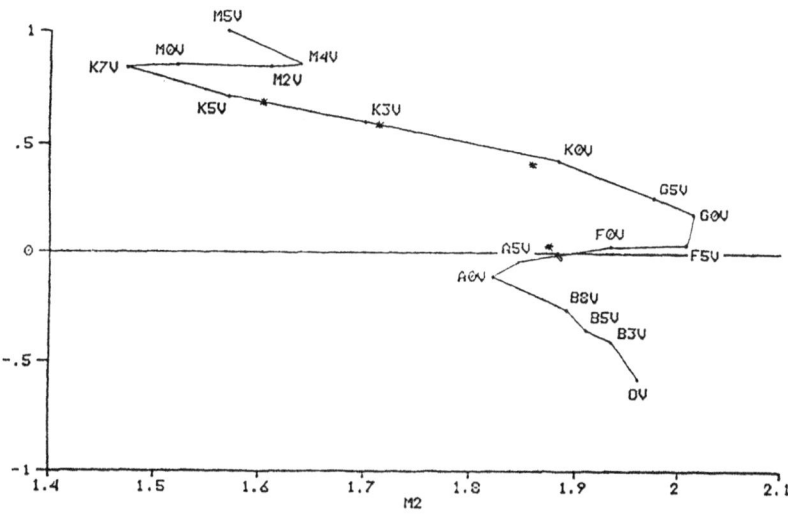

Fig. 2. Plot of M_1 (mean) versus M_2 (variance) for luminosity class V stars. (* actual stars measured).

5 Acknowledgements

We wish to thank Messrs Andrews, Louden and Priestly for assistance rendered and the New Zealand Lottery Board for a grant to purchase the microdensitometer.

References

Dodd, R.J., MacGillivray, H.T., 1986. *Astron. J.*, **92**, 706.
LaSala, J., 1988. *'Astronomy from large databases'*, *ESO Conference Workshop Proc.*. No. 28, p. 127, eds. F. Murtagh and A. Heck, ESO, Garching.
Savage, A. et al., 1985. *'The UK Schmidt Telescope Objective Prisms II and III'*, Royal Observatory Edinburgh.
Schmidt-Kaler, Th., 1982. *Bull. Inf. CDS* No. **23**, 2.
Stobie, R.S., 1980. *J. Brit. Interplanetary Soc.*, **33**, 323.
Straizys, V, Sviderskiene, Z., 1972. *Vilnius Astr. Obs. Bull.*, No. **35**.

DIGITISED OPTICAL SKY SURVEYS

Part Four:

ASTRONOMY FROM LARGE SCALE DIGITISED SURVEYS

ASTEROID SEARCHES FROM UKST MATERIAL

K.S. Russell [1], *E. Bowell* [2], *S.J. Bus* [2] *and B. Skiff* [2]

[1] Anglo-Australian Observatory, Private Bag, Coonabarabran, NSW 2357, Australia.
[2] Lowell Observatory, 1400 West Mars Road, Flagstaff, Arizona 86001, U.S.A.

1 Introduction

The UK Schmidt Telescope (UKST) has been involved in asteroid work since 1979, initially on the UK-Caltech Asteroid Survey (UCAS) and more recently with the Lowell-UK Asteroid Survey (LUKAS). In addition to these major programs, a number of smaller projects have been undertaken. These include searches for Trojan asteroids of Saturn and Jupiter, distant moons of Neptune and Uranus and distant solar system objects. Most of these are on-going projects.

There are two situations where it is necessary to utilise an advanced automated microdensitometer. The first, as typified by the LUKAS program, is where a large subset of images need to be identified, characterised and extracted from a great many plates. In this case the manual effort would be prohibitive. The second is the 'needle in the haystack' type search, where current visual searches would be unproductive. Many interesting solar system objects cannot be identified from a single plate but show up when compared with an archival plate. Digital blinking is a far more powerful technique than the manual equivalent.

2 UCAS

The UCAS program was conducted in collaboration with Caltech. Starting in 1979, a total of 101 plates were taken over the period 1979-1981. The basic aim was to track a group of asteroids, extending the search area as we went, in order to limit loss of candidates and thus avoid some of the extreme bias present in earlier surveys of this type. The program was largely successful in this aim and has, for example, drastically modified the known distribution of asteroid orbital inclinations. Over 1200 good orbits and a number of Earth approaching asteroids (Shoemaker et al.

237

H. T. MacGillivray and E. B. Thomson (eds.), Digitised Optical Sky Surveys 237–244.

1984) were extracted from the data. UCAS is now regarded as one of the standard references in asteroid research.

The principal problem arising in this survey was the immense effort required to manually search over 100 plates, containing up to four hundred asteroid images each, many of them close to the plate limit. As a result, the full analysis took many years and work on the precise photometry is still in progress.

3 LUKAS

While UCAS had gone some way to removing observational bias there were clearly still some problems in this area, such as the severe clumping of the nodes. LUKAS was conceived to try to overcome as far as possible such effects, present in all other surveys, by selecting widely distributed fields observed over many years.

The experience of the UCAS program taught us also that we would need to look at improved techniques, both to increase productivity and to reduce the manpower implicit in standard measurement and analysis methods.

Productivity was improved by tracking at the mean asteroid rate during exposure. Indeed, some superb early test exposures were estimated to contain in excess of 2000 asteroid images, compared to 300-400 for untracked plates. From a set of eight plates exposed over three lunations, we can derive good orbits for the majority of asteroids on the plates.

Productivity gains of this magnitude have an adverse affect on the practicality of using standard measurement techniques. Even under poor seeing conditions, we could anticipate at least 1000 candidates. We had little choice but to turn to an automated microdensitometer, and in our case COSMOS was conveniently available.

4 Reduction Techniques

4.1 Initial Approach

The fundamental difference between LUKAS plates and normal plates is that all astronomical objects, external to the solar system, are trailed in a characteristic manner. All stars are easily recognised by a more or less constant trail length and orientation. Many galaxies should also be easily recognised using the same criteria.

This is confirmed in the digitised data with a single proviso: faint images near the sky background become fragmented, which disguises both the orientation and trail length in a substantial number of images, as much as 30% in some cases.

Additional information is available to us in the context of this research. We know that each asteroid image should be present on most of the plates in the set. Using this information, plus limits on the anticipated asteroid motion, we are able to reject a large proportion of the remaining images. This can be carried out most successfully where we have a plate pair taken on the same night in similar conditions.

While we are able to identify 85-95% of asteroid images in this manner, it is still far from routine and depends too heavily on the time distribution of plate exposures. We consider this approach only partially successful.

4.2 Archival Identification

At the time when we were attempting to overcome the problem discussed above, a complete digital map of the sky was beginning to emerge from the COSMOS group (see the article by Yentis et al., these proceedings). We therefore decided to approach the problem in a different way.

We obtained, from COSMOS, image lists for all survey fields in the area of sky overlapping our asteroid plates. An image matching exercise was carried out between the archival data and the asteroid plate data and all images which matched were rejected. The remaining real images we expected to be solar system objects.

Note that the archival plates are simple sidereally tracked plates, whereas our asteroid plates were tracked at the mean anticipated asteroid motion. This complicates the image-matching process considerably. Yet again, image fragmentation caused a problem. While the majority of images from the tracked plates matched those from the archival plates, a substantial fraction did not. Extending the error box in the direction of the image trail helped alleviate the problem and we were eventually able to eliminate the majority of astronomical images.

4.3 Problems

One of the major outstanding problems lies with the complex diffraction patterns generated by relatively bright stars. The background following is unable to adequately follow the true sky background around bright stars with the result that fragments of the diffraction pattern generate anomalous images. One can minimise this problem by eliminating the area around bright stars, but this is a significant area and will lead to a loss of asteroid images. Even so, fragments of the diffraction spikes remain to complicate the analysis. The very brightest stars can generate diffraction spikes extending across the entire plate. Contamination from false diffraction images remains at about 15% of the final image list, even after strictly excluding areas around bright stars. It may be that by selecting undeblended images when close to bright stars we might reduce the difficulty.

Satellite trails are unavoidably common on UKST plates. In some circumstances they can generate a multiplicity of images which are difficult to identify and reject automatically. However, this is not a major problem on most plates.

A proportion of the remaining images can be identified visually as dust or flaws, but the majority of false images have been identified as spurious grain clumps which lie very close to the sky background.

4.4 Results

Our automated analysis is currently able to reduce the initial list of 100 000 images from each plate to less than 2000. These particular plates were tracked at the Jupiter

Trojan rate and we expected about 500-600 asteroid images. There is, therefore, substantial contamination from spurious images. After the first plate pair is linked, the contamination has been virtually excluded — a few % at most. After full linking of all eight plates, the contamination is nil but we estimate the loss of real asteroid images at about 5-7%.

5 Jupiter Trojans

This program, an offshoot of LUKAS, was undertaken in collaboration with two other groups, to characterise the Trojan asteroids of Jupiter (Shoemaker et al. 1984).

Four fields were studied using the UKST, only two of which have been analysed so far. Due to the great sensitivity of the UKST and the advantage of tracking at the anticipated velocity of the asteroids, we were able to double the number of known Trojans in the L5 area from these first two fields. A total of fifty Trojans were discovered, only one of which was previously known. A plot of the velocity vectors derived from the residual images in the COSMOS data clearly shows the small group of anomalous vectors representing the Trojan asteroids (Fig. 1).

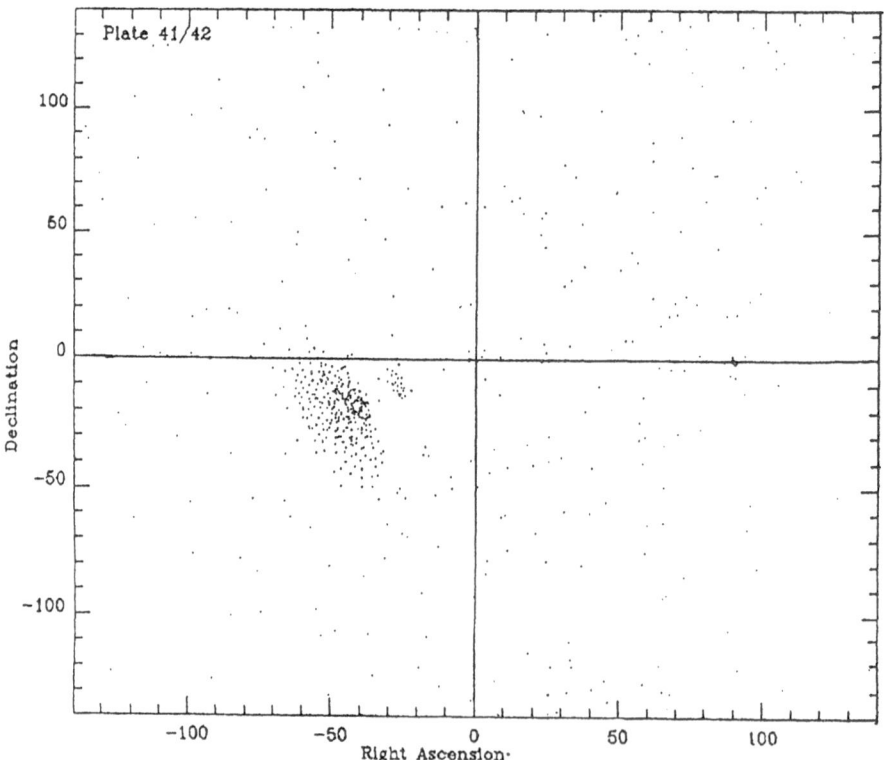

Fig. 1. Asteroid motion vectors from COSMOS data.

6 Isolated Plates

It is commonly supposed that no useful dynamical information can be obtained from single observations of solar system objects. However, this is not strictly true. In a paper by Bowell et al. (1989), it has been shown that a single plate exposed at opposition can be analysed to yield both an approximate inclination and semi-major axis. In Fig. 2, a plot is shown of the sine of the orbital inclination against semi-major axis for several hundred asteroids. These have been taken from accurate orbital data and it is possible to see structure, such as the Kirkwood gaps and some Hildas out at 4au. In Fig. 3 we show a similar plot except that the inclination and semi-major axis have been derived from motion vectors obtained from a single plate. Clearly much of the structure is still visible.

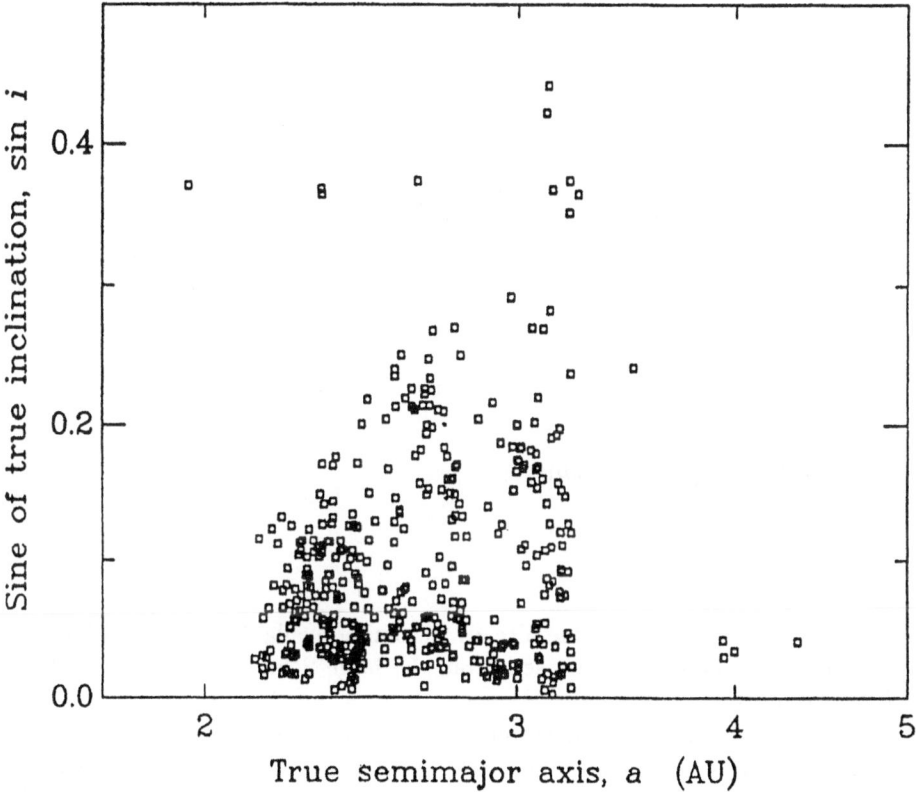

Fig. 2. Inclination versus semi-major axis from orbital data.

A great many plates meet the criteria to allow the derivation of meaningful motion vectors. Many of these plates have been, or will shortly be, measured and it is our intention to develop software to extract these single asteroid observations. The large number of such observations will allow us to use statistical techniques to investigate certain features of the asteroid belt. This program will, in addition,

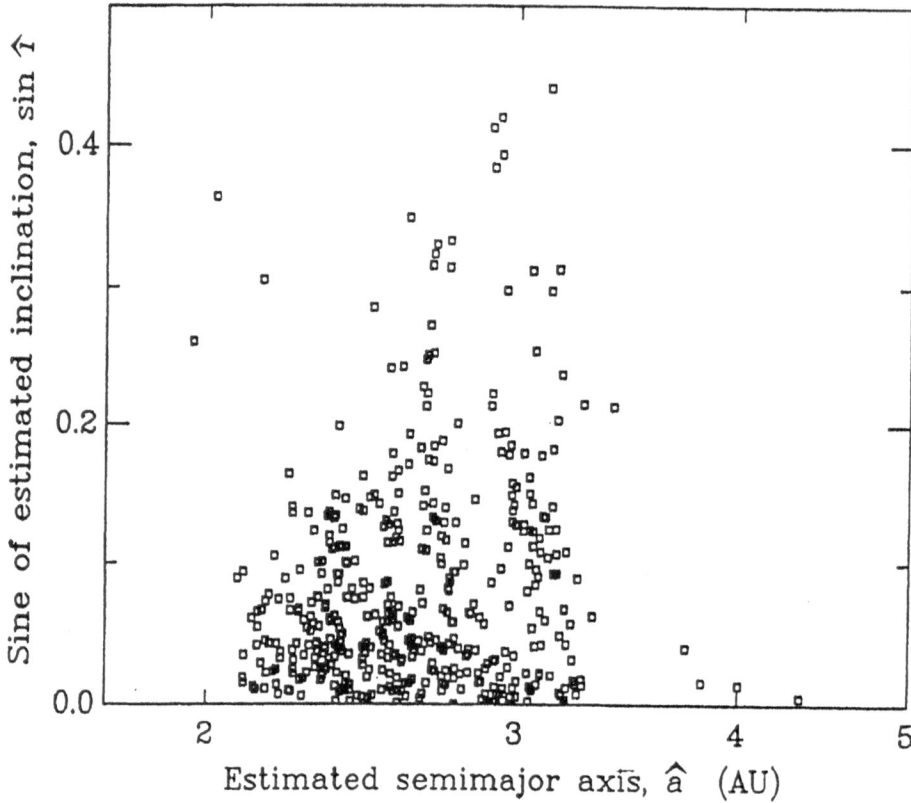

Fig. 3. Estimated inclination versus semi-major axis from a single plate.

search for distant solar system objects.

7 Conclusions

Large, digitised sky catalogues are already becoming available in a useful form. It seems likely that the entire plate sets from the large Schmidt Telescopes will be digitised within the next decade.

At present the astrometric precision of these digitised datasets does not match the inherent accuracy of the original plate material, but the techniques to correct this situation are understood and we can expect to see a steady improvement in this area. Much of the problem lies in the analysis of the data rather than in the measurement process.

A number of formats for storing and presenting this information are being utilized. While there are immediate benefits from treating the data on a plate-by-plate basis, I believe that the greatest utility will be realised from converting the data to an image based catalogue, with multiple entries where an image is identified on several plates. Faster computers and greater on-line storage capacity will ease the path to such databases.

The long term integrity of the original plates is not assured, due to continuing problems with 'gold spot' and it is therefore important that as much measurement information as possible is retained should it become necessary to re-analyse the raw digital information in unexpected ways.

References

Shoemaker, E.M., Bus, S.J., Dunbar, R.S., Helin, E.F., Dawe, J.A., Barrow, J., Hartley, M., Morgan, D., et al., 1984. *'Mars-crossing asteroids discovered in the UK-Caltech Asteroid Survey'*, Bulletin American Astronomical Society, **16**, 691.

Bowell, E., Skiff, B.A., Wasserman, L.H., Russell, K.S., 1989. *'Orbital Information from Asteroid Motion Vectors'*, Proceedings of Asteroids, Comets, Meteors III, Uppsala 1989. Eds. Lagerkvist, Rickman, Lindblad, Lindgren.

Discussion

van Dessel :

Regarding the discussion about the precision of the measurements needed for the determination of orbits and the accuracy of the orbits obtained: did you also treat objects with well known, reliable orbits and compare your own results with the known orbit, and were you satisfied?

Russell :

Yes, we always locate known objects and do check that these agree with the accepted orbit.

Cannon :

Your tracked plate technique clearly lets you detect fainter asteroids, but does it really help you to find more distant objects? It seems to me that you are actually introducing a major bias by using tracked plates, being effectively 'tuned' to finding asteroids at a particular distance. Second, has anyone yet actually found any Saturn 'Trojans'?

Russell :

Yes it does indeed introduce a bias. However, we believe this can be precisely evaluated. Going fainter would certainly allow us to find more distant objects providing they exist. No Saturn Trojans have yet been found to my knowledge.

Dodd :

In your residuals plot, are the objects with very small residuals fainter than say the main asteroid belt and Trojan objects — i.e. are these low residual objects likely to be very distant solar system objects?

Russell :

It is possible, although we have not investigated the low residual group in any serious manner.

Stobie :

In your diagram of RA residuals and DEC residuals you showed two groups of asteroids. These were data from combining two plates. There were many other objects that had paired that appeared on the diagram scattered all over the plot. What are these objects?

Russell :

The objects with very small residuals may be interesting although we have not investigated them visually. The scattered objects are spurious and disappear after further linking.

MacGillivray :

How much of your bright star 'junk' images could be removed by ignoring the deblend parameter in COSMOS data in the region of these bright stars?

Russell :

We checked both the blended and deblended object lists and decided that the deblended list was best. We have not attempted to use the lists selectively but perhaps we should.

Parker :

When doing work with 'single-plate' material you talked about the need for a more precise image classification scheme. After the interesting paper by Dr. Odewahn, do you think it is worth looking at an AI approach such as that of neural networks?

Russell :

It seemed to me that the neural network approach was only another way of approaching the classification problem without obvious major benefits over the others. However, it may be that in a complex situation the neural network approach would be better.

UBV STAR COUNTS IN SELECTED AREA 54, AND GLOBAL STRUCTURE OF THE GALAXY

T. Yamagata [1,2] *and Y. Yoshii* [1]

[1] National Astronomical Observatory, Mitaka, Tokyo 181, Japan.
[2] Royal Observatory, Blackford Hill, Edinburgh EH9 3HJ, U.K.

Abstract

We have obtained new UBV star count data in 16 square degrees centered on Selected Area 54 $[(l, b) = (200°, +59°)]$, based on 10 plates taken with the Kiso Schmidt telescope.

We have constructed a model for the Galaxy based on systematic analysis of wide-area surveyed star counts, including our new data in SA54. Our results are summarized as follows: the sun is situated at 40pc north from the mid-plane of the disc; the scale length of the thin disc is 3.8kpc; the scale height of the thick disc is 900pc; the axial ratio of the spheroid is 0.84; local normalization of the thick disc relative to the thin disc is 0.019 and that of the spheroid is 1/900.

1 Introduction

In order to derive statistically meaningful results, the data of star counts should be based not only on accurate photometry but also on observations over a sufficiently wide area to minimize the effect of local fluctuation in the spatial distribution of stars. Multi-colour star count data which satisfy these criteria have been obtained only in three regions:

1. UBV data in 21.46 square degrees in the direction of SA57 near the NGP $[(l, b) = (81°, +87°)]$ (Yoshii et al. 1987),
2. BV data in 11.5 square degrees at the SGP $[(l, b) = (0°, -90°)]$ (Gilmore & Reid 1983), and
3. BV data in 17 square degrees in a 22^h field $[(l, b) = (37°, -51°)]$ (Gilmore et al. 1985, hereafter GRH).

H. T. MacGillivray and E. B. Thomson (eds.), Digitised Optical Sky Surveys 245–253.

In addition to these data, star counts in other areas are indispensable for constraining structural parameters of the Galaxy, especially those of radial structure. We have been making new UBV star counts in several areas which are appropriate for this purpose.

In the present study, we make new UBV star count data in 16 square degrees in the direction of Selected Area 54 $(l, b) = (200°, +59°)$. We chose SA54 for the following reasons:

1. SA54 is located in almost the opposite direction to the GRH region in the sky. Thus this direction is appropriate to estimate the radial structure when combined with the data in the GRH region.
2. We can use the UBV photometric sequence of Purgathofer (1969) for SA54. It includes 66 stars for $6.6 \leq V \leq 17.6$.

Using the data on wide area surveys, including our new star counts in SA54, we derive the constraints on structural parameters of the Galaxy with the help of a Galaxy model which consists of a thin-disc, thick-disc, and spheroid components.

2 UBV Star Counts in SA54

Observations were undertaken with the Kiso Schmidt telescope which has a corrector mirror of 105cm in diameter and a main mirror of 150cm. We obtained 3 plates in the U band (Kodak IIa-O and Schott UG1), 3 plates in the B band (IIa-O and GG385), and 4 plates in the V band (IIa-D and GG495). Each plate is 14 inches in size covering an area of $6° \times 6°$ with a scale of 62.6 arcmin mm^{-1}. We measured the plates with the PDS2020GMS microdensitometer at Kiso Observatory. The measured area on each plate is 240 mm × 240 mm which corresponds to $4°.2 \times 4°.2$ on the sky.

A cubic spline function was used to describe the sky background in the measured region on each plate. Each image is recognized by the connectivity of the pixels for which the value of relative intensity exceeds an imposed threshold value. Spurious images were removed by plate-pairing. Standard UBV magnitudes were derived from the total intensity using the UBV photometric sequence for SA54 (Purgathofer 1969; Yamagata & Sasaki 1991). Star/galaxy discrimination was performed by means of the diagram shown in Fig. 1. A detailed description of the reduction is given elsewhere (Yamagata 1986; Yamagata & Yoshii 1991). The data consist of a sample of about 7600 stars brighter than $V = 18$mag, and are shown by the filled circles in Figs. 3c and 4.

3 Global Structure of the Galaxy

We assume that the Galaxy consists of three components: a thin disc, thick disc, and spheroid. In this section, we briefly explain our method for analyzing the data and the model. Details of the method of constructing the Galaxy model are described in Yoshii et al. (1987), and those of the analysis and the results are discussed in Yamagata & Yoshii (1991).

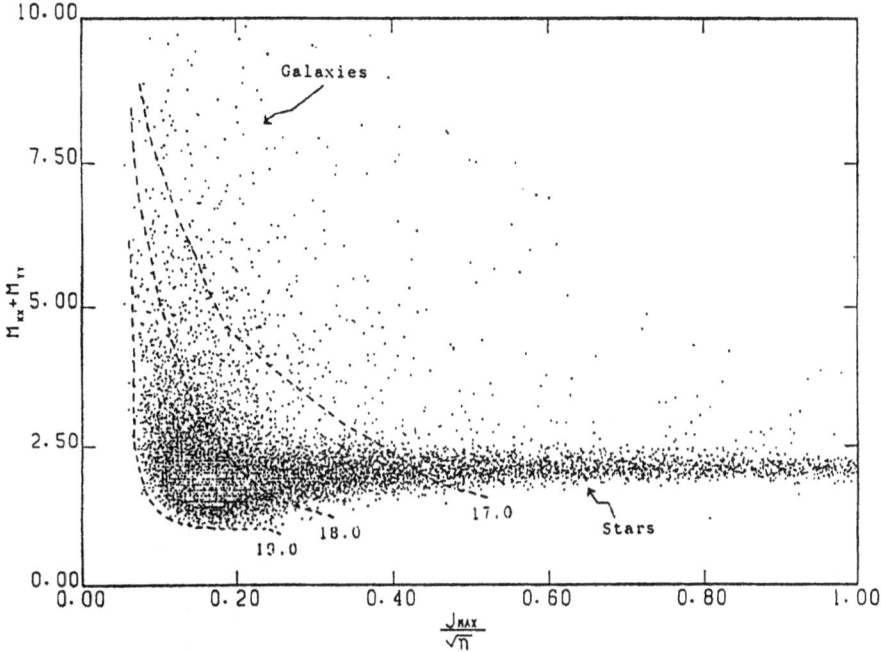

Fig. 1. Star/galaxy discrimination diagram. The abscissa is the peak intensity divided by the number of pixels which form an image $J_{max}/\sqrt{n}$ and the ordinate is the second-order moment of the intensity distribution within an image $M_{xx} + M_{yy}$. Stellar images form a highly elongated distribution along the horizontal axis, whereas galaxian images form a dispersed distribution along the vertical axis. The loci of constant apparent V-magnitude are shown by the dashed lines; the stars become systematically fainter toward the left part of their distribution and the galaxies become so toward the lower part of their distribution. The star/galaxy discrimination is reliable down to $V = 18$ mag.

3.1 The Method of the Analysis

a) Data used in the Analysis

The star count data used in this study are as follows:

1. *UBV* data in 21.46 square degrees at the NGP,
2. *BV* data in 11.5 square degrees at the SGP,
3. *BV* data in 17 square degrees towards the centre (GRH),
4. *UBV* data in 16 square degrees towards the anti-centre (SA54; our new data).

In addition to the data in these four directions, we use $(5)JF$ data in 0.3 square degrees in two directions of SA57 $[(l, b) = (65°, +86°)]$ and SA68 $[(l, b) = (111°, -46°)]$ (Kron 1980). Although the surveyed area of the data (5) is small compared with those of other data, these are the only data available to date which have limiting magnitude deep enough for the analysis of the spheroid component.

b) Separation of the Components

We can separate the contribution of each component on star counts by magnitude and colour. According to the Galaxy model, stars with $12 < V < 17$ and

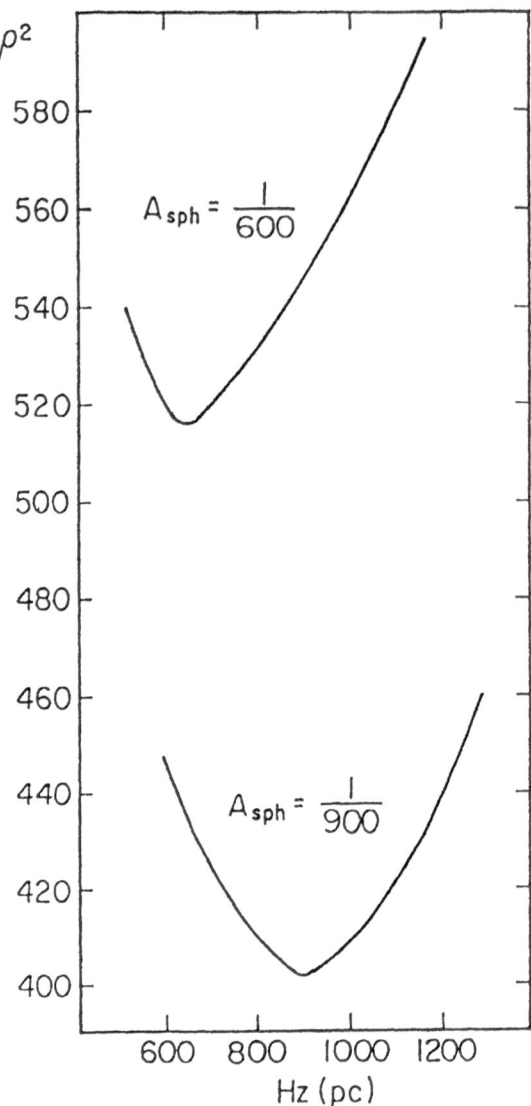

Fig. 2. The least-squares sum p^2 versus the scale height of the thick disc H_Z for each of $A_{sph} = 1/600$ and $1/900$. Min (p^2) in the case of $A_{sph} = 1/900$ is much smaller than that of $A_{sph} = 1/600$.

$B - V > 1.0$ consist of exclusively thin-disc stars. The colour distribution for high-latitude stars with $V \sim 20 - 22$mag have two pronounced peaks; the red peak consists of thin-disc stars and the blue peak of spheroid stars. Therefore, we can estimate the contributions of thin-disc and spheroid components using stars with these ranges of magnitude and colour. However, there is no magnitude or colour range where thick-disc stars manifest themselves. Therefore, we can analyze the contribution of the thick-disc component after those of thin-disc and spheroid com-

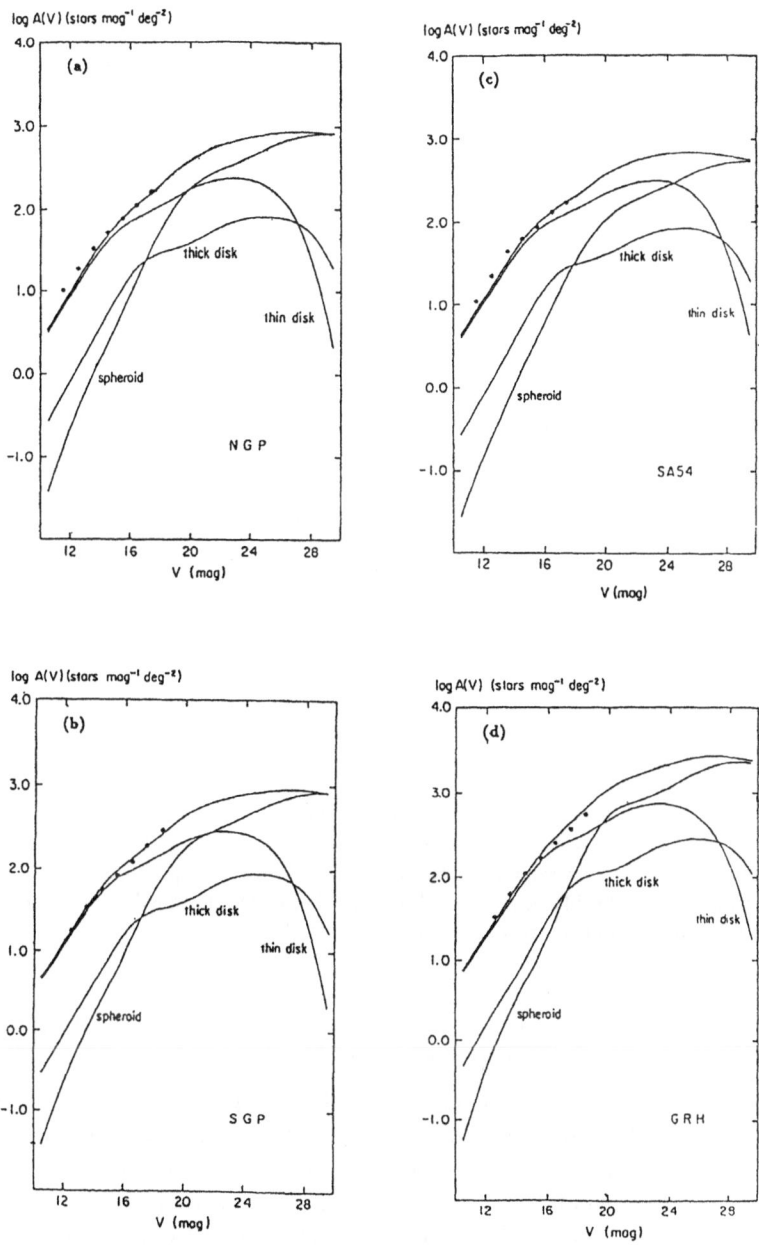

Fig. 3. Differential number count $A(V)$ of stars at intervals of one apparent V magnitude per square degree plotted against V. The thin lines represent predicted contributions of the thin disc, thick disc, and spheroid, and the thick line represents their sum. The observed counts are shown by the circles. **(a)** The NGP region close to SA57 $[(l, b) = (65°, +86°)]$. The data are from Yoshii et al. (1987). **(b)** The SGP region $[(l, b) = (0°, -90°)]$. The data are from Gilmore & Reid (1983). **(c)** The SA54 $[(l, b) = (200°, +59°)]$. The data are from our present study. **(d)** The GRH region $[(l, b) = (37°, -51°)]$. The data are from Gilmore et al. (1985).

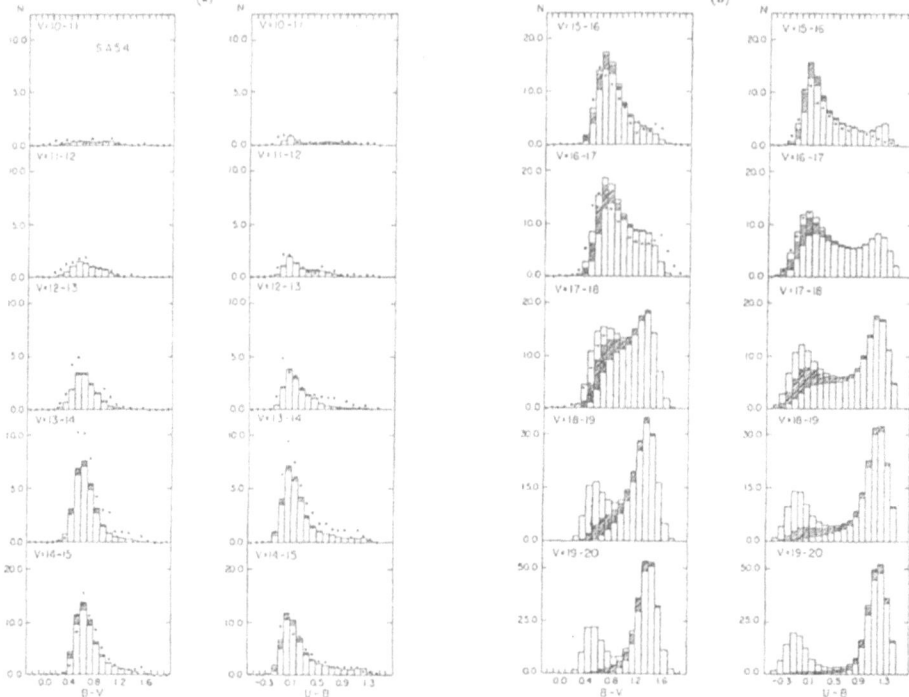

Fig. 4. $B - V$ and $U - B$ colour distributions for SA54 at colour interval of 0.1 mag as a function of **(a)** apparent V magnitude from 10-11 (upper) to 14-15 (lower), **(b)** apparent V magnitude from 15-16 (upper) to 19-20 (lower). The filled circles indicate the observations. The hatched regions in the histograms show the predicted contributions of thick-disc stars, and the blank regions above and below the hatched region correspond to thin-disc and spheroid stars, respectively.

ponents are determined. The contribution of the thick-disc component is maximum in stars with $15 < V < 18$ and $B - V < 1.0$.

c) Determination of Parameters for each Component

We can estimate parameters for each component using the data of specific directions. We can estimate the vertical structure (i.e. distance of the sun from the mid-disc $Z_\odot$, scale height H_Z) using data from the galactic polar regions. In order to estimate radial structure (i.e. scale length H_R), data from a pair of directions towards the centre and anti-centre with the same latitude $|b|$ are appropriate. For the axial ratio of the spheroid q, data from more than two areas in the sectional plane for $l = 90°$ and $270°$ with different latitudes are ideal.

3.2 Results

a) Thin Disk

The observed ratio of the number of thin-disc stars in the SGP N_{SGP} to that in the NGP N_{NGP} is not unity. This implies that the sun is not situated on the mid-plane of the disc. We calculate the ratio N_{SGP}/N_{NGP} as a function of $Z_\odot$ for different values of H_Z. The relation between H_Z and $Z_\odot$ is obtained by comparing the calculated ratio with the observations. If we assume $H_Z = 350 \pm 25$pc which is the most probable value in the literature (e.g. Bahcall & Soneira 1980; Yoshii 1982; Gilmore & Reid 1983), we find that the distance of the sun from the mid-plane of the disc is $Z_\odot = 40 \pm 3$pc north.

Following a similar method, we can derive the relation between H_Z and H_R of the thin disc using the ratio of the number of stars in the GRH region to that in SA54. For the most probable value of $H_Z = 350 \pm 25$pc, the scale length is determined to be $H_R = 3800 \pm 270$pc.

b) Spheroid

We closely follow Bahcall & Soneira's (1980, 1984) analysis and estimate an axial ratio of the spheroid and its local normalization relative to the thin disc in terms of our own Galaxy model using the faint star count data of Kron (1980).

We calculate the number ratio between spheroid stars (i.e. stars in the blue peak) in SA57 and those in SA68 as a function of q. Comparing with the observed values, we obtain $q \approx 0.84$.

We can estimate the local normalization of the spheroid A_{sph} relative to the thin disc from the number ratio between stars in the blue peak and those in the red peak. The derived value of A_{sph} from data in SA57 is 1/900 which is smaller than that from SA68:1/600; however, they should coincide with each other within errors in the observations.

c) Thick Disk

Using the parameters for the thin disc and spheroid derived above, we can finally estimate the contribution for the thick-disc. We calculate the number of NGP and SGP stars in an area of 1 square degree with $15 < V < 18$ and $B - V < 1.0$ as a function of a local normalization A_{thick} of the thick disc, its scale height H_Z, and A_{sph}. Thus the relation between A_{thick} and H_Z can be obtained by comparing the observed number with the calculated one for each of $A_{sph}=1/600$ and 1/900.

In order to obtain the most probable set of H_Z, A_{thick} and A_{sph}, we calculate a sum of residuals

$$\rho^2 = \sum_i (N_i^{obs} - N_i^{model})^2 \qquad (1)$$

and find where ρ^2 is minimized in the relation among H_Z, A_{thick} and A_{sph}. The summation runs over $15 < V < 18$ and $B - V < 1.0$. As shown in Fig. 2, ρ^2 exhibits a sharp minimum with respect to H_Z, and the $\min(\rho^2)$ in the case of $A_{sph}=1/900$ is much smaller than that of $A_{sph}=1/600$. Thus the model with $A_{sph}=1/900$ gives a better solution, and we obtain $(H_Z, A_{thick}, A_{sph})=(900 \text{ pc}, 0.019, 1/900)$.

d) Comparison of the Model with the Observations

Parameters for the three-component Galaxy model are summarized in Table 1 with their values determined in this study.

Figure 3 shows the differential number counts $A(V)$ at intervals of one apparent V magnitude per square degree plotted against V; Fig. 3a to 3d correspond to the NGP, SGP, SA54, and GRH, respectively. The thin lines represent contributions of

Table 1. Parameters for the Galaxy model.

Solar position:		
	Galactocentric distance	8 kpc (given)
	Distance from the disc plane	40 pc north (obtained)
Thin Disk component:		
	Local normalization	0.093 stars pc^{-3} (given)
	Scale height, late-type dwarfs	350 pc (given)
	subgiants	250 pc (given)
	Scale length	3.8 kpc (obtained)
	Luminosity function	McCluskey (dip)
	CM diagram main sequence	Wielen
	giant sequence	open cluster M67
	Metallicity at $(R, Z) = (R_\odot, 0)$	0.0 dex (given)
	Metallicity gradient, Z direction	-0.5 dex kpc^{-1} (given)
	R direction	-0.04 dex kpc^{-1} (given)
Thick disk component:		
	Local normalization to disc	0.019 (obtained)
	Scale height	900 pc (obtained)
	Scale length	3.8 kpc (given)
	Luminosity function	globular clusters
	CM diagram	globular clusters
	Metallicity at $(R, Z) = (R_\odot, 0)$	-0.76 dex (given)
	Metallicity gradient, Z direction	-0.1 dex kpc^{-1} (given)
	R direction	-0.04 dex kpc^{-1} (given)
Spheroid component:		
	Local normalization to disc	0.0011(=1/900) (obtained)
	Density law	deprojected $r^{1/4}$ law
	Effective radius	2.7 kpc (given)
	Axial ratio	0.84 (obtained)
	Luminosity function	globular clusters
	CM diagram	globular clusters
	Mean metallicity	-1.5 dex (given)

thin disc, thick disc, and spheroid. The thick line represents their sum. The observed counts used in this paper are shown by filled circles. In all the four regions, the model counts agree with the observations over a range of observed magnitude.

Figure 4 shows the $B - V$ and $U - B$ colour distributions for SA54 stars at colour intervals of 0.1 mag. Histograms show the model distributions; the hatched regions indicate the contribution of the thick disc, and the blank regions above and below the hatched region correspond to the thin disc and spheroid, respectively.

The observed colour distributions for $10 < V < 18$ are shown by the filled circles. Agreement between the model and observed distributions is satisfactory over a whole range of observed V magnitude, for both of the $B - V$ and $U - B$ colours.

4 Acknowledgements

T.Y. is indebted to the Yamada Science Foundation for financial support for the international exchange program of researchers. This work was supported in part by the Grant-in-Aid for Encouragement of Young Scientists from the Ministry of Education, Science and Culture, Japan (63790140).

References

Bahcall, J.N., Soneira, R.M., 1980. *Astron. J. Supp.*, **44**, 73.
Bahcall, J.N., Soneira, R.M., 1984. *Astron. J. Supp.*, **55**, 67.
Gilmore, G., Reid, N., 1983. *Mon. Not. R. Astron. Soc.*, **202**, 1025.
Gilmore, G., Reid, N., Hewett, P., 1985. *Mon. Not. R. Astron. Soc.*, **213**, 257.
Kron, R.G., 1980. *Astron. J. Supp.*, **43**, 305.
Purgathofer, A.Th., 1969. Lowell Obs. Bull. No. 147.
Yamagata, T., 1986. *Annals Tokyo Astron. Obs.*, 2nd Ser. **21**, 31.
Yamagata, T., Sasaki, T., 1991. In preparation.
Yamagata, T., Yoshii, Y., 1991. *Astron. J.* Received.
Yoshii, Y., 1982. *Pub. Astron. Soc. Japan*, **34**, 365.
Yoshii, Y., Ishida, K., Stobie, R.S., 1987. *Astron. J.*, **93**, 323.

Discussion

Odewahn :
> Can you determine from your star count data whether or not the bulge component in the Galaxy follows a $r^{1/4}$ law distribution?

Yamagata :
> No. Our model doesn't contain the bulge component. The data used in our analysis are those at high galactic latitude. In order to determine the contribution of the bulge component, the data near the galactic centre should be used. Therefore, in this case, the effect of absorption or reddening is an important factor.

A SEARCH FOR MACROSCOPIC DARK MATTER IN THE GALACTIC HALO THROUGH MICROLENSING

M. Moniez

Laboratoire de l'Accélérateur Linéaire, IN2P3-CNRS, Université de Paris-Sud, F-91405 Orsay Cedex, France.

Abstract

We present a status report of our project to search for dark objects in the Galactic Halo via gravitational microlensing of stars in the Large Magellanic Cloud. The search uses a combination of Schmidt photographic plates and CCD images, and is sensitive to lensing by objects in the range $10^{-7}M_{\odot}$ to $10^{-1}M_{\odot}$. An automated analysis software has been successfully tested on CCD test images. Using these preliminary results, we calculate the number of microlensing events expected on plate and CCD programs.

1 Introduction

It is widely believed that the flat rotation curves observed for spiral galaxies like our own indicate that such galaxies are surrounded by a 'halo' of dark matter (Trimble 1987; Primack et al. 1988). While the mass of the dark haloes should be as much as ten times that of the visible parts of galaxies, the composition of the haloes is not known. Candidates range from new, weakly interacting elementary particles to dark astronomical objects, like brown dwarfs or black holes. The elucidation of the nature of the halo inhabitants would have profound implications for cosmology and theories of galaxy formation.

The mass of hydrogeneous compact halo objects could range from 10^{-7} to 10^{-1} solar masses (De Rújula et al. 1991). As Paczynski (1986) pointed out, objects in this range of mass could be detected by monitoring the brightness of individual stars in the Large Magellanic Cloud (LMC). A massive object D passing close enough to the line of sight OS from the observer O to the observed star S induces a relativistic light deflection and magnification (Fig. 1). Only the magnification is measurable,

255

H. T. MacGillivray and E. B. Thomson (eds.), Digitised Optical Sky Surveys 255–263.
© 1992 *Kluwer Academic Publishers.*

and its amplitude is more than 0.3 magnitudes if the impact parameter is less than the Einstein radius R_E, given by:

$$R_E^2 = \frac{4GMD}{c^2}, \quad D = \frac{OD.DS}{OS} \tag{1}$$

Due to the perfect alignment required for a sizeable magnification, the microlensing phenomenon is sensitive to the relative motion of the observer, star and deflector. The result is an achromatic and symmetric light curve. For stars in the LMC, a deflector of mass M and relative speeds of order 200km/s, the mean time τ during which the magnification is greater than 0.3 magnitudes is given by:

$$\tau = (70\text{days})\sqrt{\frac{M}{M_\odot}} \tag{2}$$

Since R_E^2 is proportional to the deflector mass and since the number of deflectors, for a given halo total mass, is inversely proportional to their mass, the probability that a given star is amplified by a given amount at any moment is independent of the deflector mass. For stars in the LMC, a halo with total mass $4 \times 10^{11} M_\odot$ and a microlensing threshold of 0.3 magnitudes, this probability is about 0.5×10^{-6}. And if one assumes that all deflectors have mass M, since τ is proportional to $\sqrt{M}$, the number of microlensing events expected in a fixed observation time is inversely proportional to $\sqrt{M}$.

The selection criteria for microlensing events are light curve achromaticity and symmetry, and uniqueness of the event.

2 Two Experiments

Our group * (Milsztajn 1990; Moniez 1990) has two experiments under way: the first, using photographic plates taken with the ESO Schmidt telescope, will monitor more than one million stars with a time scale of the order of a day, and then will be sensitive to mean lensing durations between 1 and 30 days corresponding to deflector masses between 10^{-4} and 10^{-1} $M_\odot$. Plates are taken in two colours, B and R, with a one hour exposure time, and then digitized on MAMA (Machine Automatique à Mesurer pour l'Astronomie, Observatoire de Paris). Each plate requires 8 hours for digitisation and generates 2 gigabytes of information.

The second, using a large CCD mosaic, will monitor more than 30 000 stars with a time scale of half an hour, and then will probe lensing durations between 1 hour and 3 days, corresponding to deflector masses between 10^{-7} and 10^{-3} $M_\odot$. The CCD camera is under construction in Saclay, and will have a total area of 76mm × 29mm consisting of an array of 2 × 8 Thomson CCDs. Images are taken in two colours with 10 mins. exposure time, assuming a 40-cm telescope with broad-band filters.

* Institut d'Astrophysique de Paris, Observatoires de Paris et Marseille, Laboratoire d'Astronomie Spatiale de Marseille, Département d'Astrophysique et de Physique des Particules Élémentaires du CEN Saclay, Laboratoire de l'Accéléreteur Linéaire.

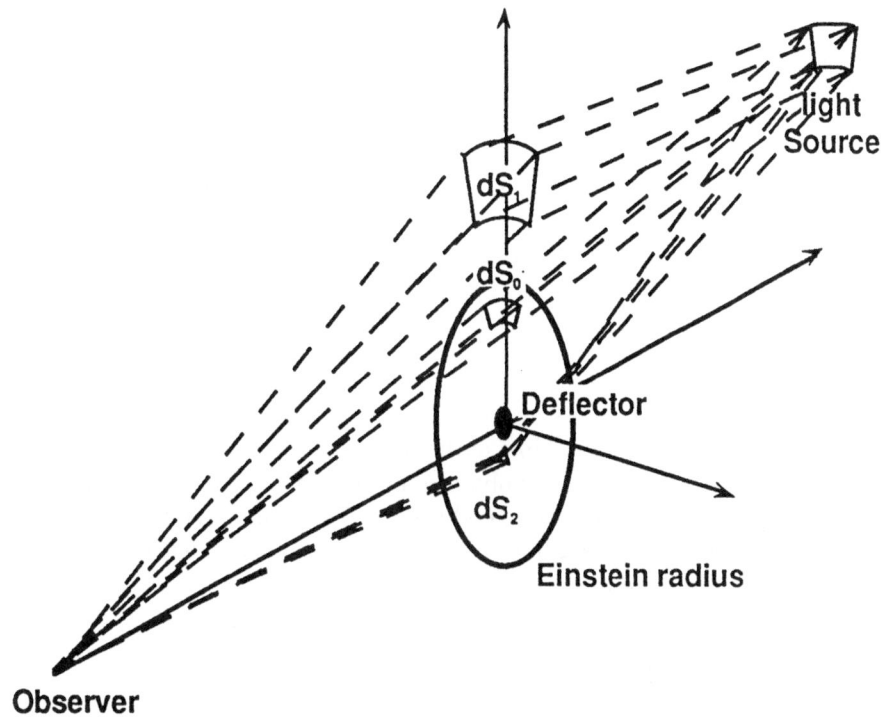

Fig. 1. A microlensing event. Undeflected light coming from light source crosses deflector plane on dS_0. The deflected light gives two images, dS_1 and dS_2 in deflector plane. The surface brightness is the same for all dS, the apparent amplification is then $(dS_1 + dS_2)/(dS_0)$.

3 Data Analysis

The analysis has to be completely automated, given the huge quantity of data to process. An analysis software package has been developed. It works according to the following steps:

- A reference file is created, either from the 'best' couple of images or from a composite image. All stars found on other images will be mapped to the stars found on the reference file, so it has to be of very good quality.
- For each image, a pattern recognition algorithm finds a first order fit to the reference image in a completely automatic way. This algorithm has been successfully used to map images from different telescopes.
- A least squares fit is made on isolated stars to fit more accurately the two images.
- For each star, the algorithm searches for a star at an equivalent position on the reference image.
- A relative calibration then maps all brightnesses to the brightnesses measured on the reference image. The algorithm used is a generalization of the classical

method for variable star observation consisting of the observation of a stable neighbouring star. We make the assumption that the mean brightness of a large sample of neighbouring stars in a given range of brightness is constant. In a small area of an image, we then choose different subranges of brightness, and measure the mean brightness of these samples of stars on this image and on the reference image. We can interpolate the relative calibration curve from these points [†]. If we have an absolute calibration on the reference image, the result is an absolute calibration.

– The final output is a light curve (luminosity vs time) per star and per colour.
– At the present time, the resolution in relative photometry is measured by the rms dispersion of the light curves, assuming that most stars are of constant brightness.

Careful optimization allows an execution time between $O(n)$ and $O(n \text{ Log } n)$, where n is the number of stars. Most of the time is spent in the photometry routine. The results shown below have been obtained using an early version of our algorithm, which provides a relative precision good enough for our minimal needs. The whole set of data has been processed in 6 minutes CPU time per colour [‡], extrapolating to 3 hours for a photographic plate.

4 A CCD Test Run

To test the feasibility of the CCD program, our group observed the LMC with the 40-cm GPO refracting telescope at La Silla in December of 1990. A single 576 pixel × 405 pixel Thomson CCD chip was mounted on the telescope giving a field of $11' \times 8'$. The essential part of the observations consisted of 63 exposures in the middle of the bar of the LMC. The high chromatic aberration of the GPO forced us to use narrow filters($\lambda = 4520\text{Å}$, $\Delta\lambda = 340\text{Å}$, and $\lambda = 6480\text{Å}$, $\Delta\lambda = 110\text{Å}$) and then long exposure time (30 mins.) to reach $m_B = 19$.

We found a relative precision shown in Fig. 2 as a function of magnitude ranging from 4 to 13%.

A casual inspection of stars with large fluctuation in magnitude resulted in the discovery of three candidate cepheid variables, an eclipsing binary and some long-period variable stars (Figs. 3–5). This set includes all the known periodic variable stars of this sky area. One of the candidate cepheids was previously unknown. More careful analysis of this data sample would certainly lead to the discovery of many more variable stars. We note, however, that we found no light curves that resemble that expected for a microlensing event.

We have investigated several possible algorithms for identifying microlensing events from the characteristics of observed light curves. In order to be sensitive to microlensing by objects in our halo, an algorithm should have an inefficiency for the rejection of random fluctuations of the light curves of less than 10^{-6}. The

---

[†] In fact, to take into account the non-linearity of photographic plates, the algorithm maps both sky background and measured brightness, with a multi-dimensional interpolation.
[‡] The computer used was an IBM 3090/600 VF.

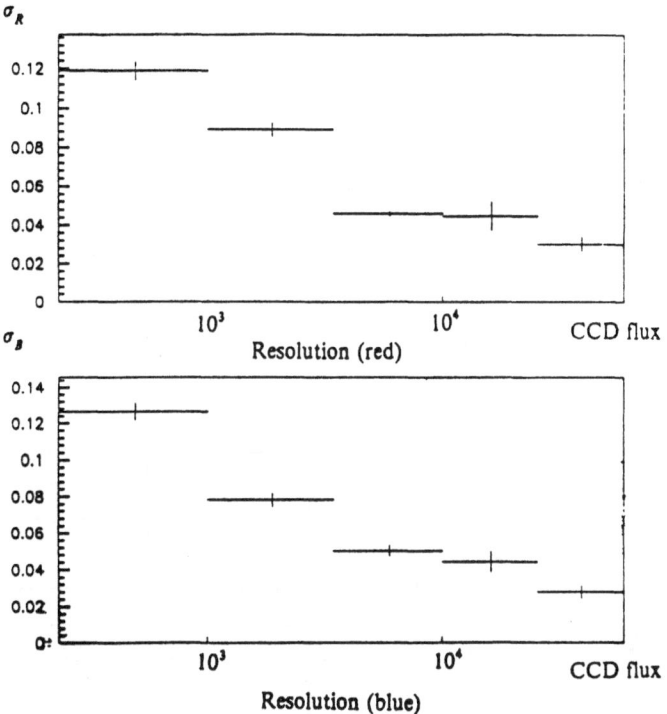

Fig. 2. Magnitude resolution for CCD images using the red and blue filters versus the CCD flux. A flux of 10^3 corresponds to about 17th magnitude.

inefficiency for rejection of variable stars will depend on their type and number. For cepheids, our measurement of their rate indicates that a rejection inefficiency of 10^{-3} is necessary. On the other hand, an algorithm should have a good efficiency for accepting real microlensing events. The efficiency must be estimated via a Monte-Carlo simulation of the observation sequence and detector resolution. The efficiency will depend on the distribution of deflector masses and, to a lesser extent, on the halo model.

In our present algorithm, we search for stars whose light curves exhibit, within any time window, a sufficiently large deviation from the mean of the measurements outside the window. The threshold for accepting a deviation is chosen so as to eliminate random fluctuations of light curves due to the system resolution. Each light curve is tested with window widths and positions varying over the total duration of the observing period. Each time a sufficiently large deviation is found, the light curve is fitted with Paczynski's theoretical light curve separately in the two colours. The event is accepted as a candidate microlensing event if the fit gives compatible amplifications and times (of maximum light) for the two colours.

For our observing sequence of December 1990, we have used the Monte-Carlo technique to study the algorithm's efficiency in accepting events of peak amplification of at least 0.3 magnitudes. The efficiency is zero for masses below $M =$

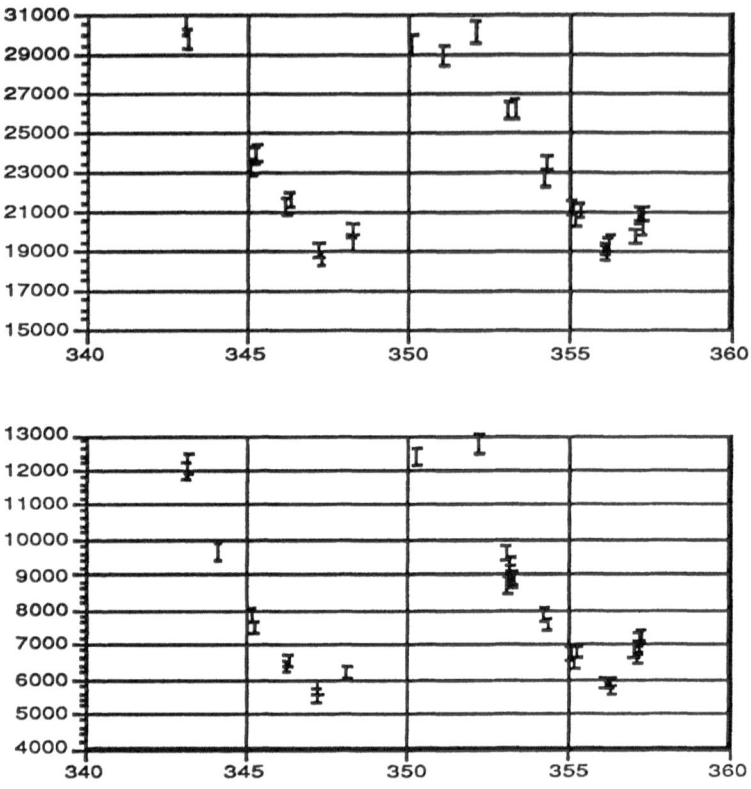

Fig. 3. A Cepheid variable (top: blue filter, bottom: red filter). The abscissa is the day of the year 1990, the ordinate the measured CCD flux.

$10^{-8}M_{\odot}$, where the mean lensing time is shorter than the exposure time. It increases with mass above this value, reaching a value of 20% for deflector masses of $M = 10^{-5}M_{\odot}$. It falls to zero for $M = 10^{-1}M_{\odot}$ where the mean lensing time is of order the total observation time (1 month). The number of events expected for the 1990 sequence is about 0.01 for deflector masses in the range $10^{-7} - 10^{-3}M_{\odot}$. The number of events is nearly independent of deflector mass in this range because the rising efficiency is compensated by the falling total number of microlensing events.

For the next observing season, we plan to use the CCD camera now under construction, multiplying by 16 the number of monitored stars. We also hope to increase the number of observing nights to 100. The number of microlensing events that can be expected with this program as a function of the assumed deflector mass is shown in Table 1. Since the number of expected events is of order one, we will be able to definitively test the hypothesis that the Halo is made of dark astronomical objects only with a global improvement in light collection, given the fact that refinement of our algorithms could yield only a limited increase in acceptance.

It is hoped to replace the GPO with a 40-cm reflector, allowing wider-band

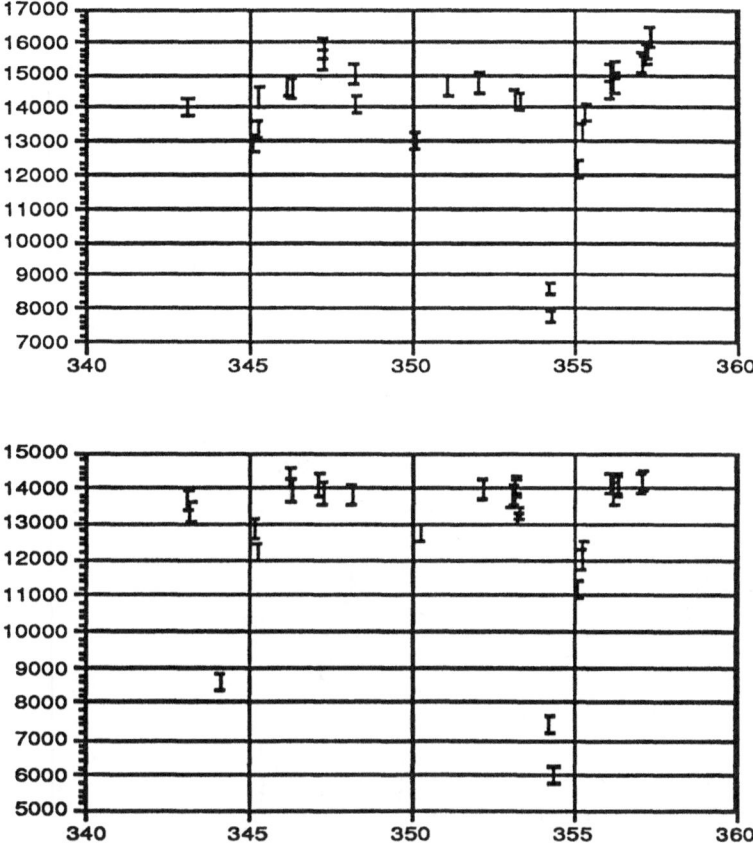

Fig. 4. An eclipsing binary.

filters and shorter exposure times. The number of microlensing events expected as a function of the assumed deflector mass is shown in Table 1.

5 Photographic Plates: Preparatory Analysis

During the winter of 1990-1991, 28 pairs of blue and red Schmidt plates were taken at La Silla over a three month period. Exposure times were typically one hour, allowing us to easily find isolated stars of magnitude 20 in the red and blue. The plates are to be digitized on the MAMA machine at the Observatoire de Paris.

As a preliminary step, we have digitized two entire plates and five 4-cm^2 zones on 20 plates, 10 of each colour. The first measurements give a precision of 0.14 magnitudes for relative photometry on the 1000 brightest stars per cm^2, corresponding to at least one million stars per plate. Using this result, we can compute the expected number of events for the plate program. The figures are shown in Table 1 for 1990-1991 observations and for a proposed 1991-1992 program. If no events are found, such a program would allow us to place significant limits on the fraction of the halo that is contained in objects in the range $10^{-4} - 10^{-1}$ M$_\odot$.

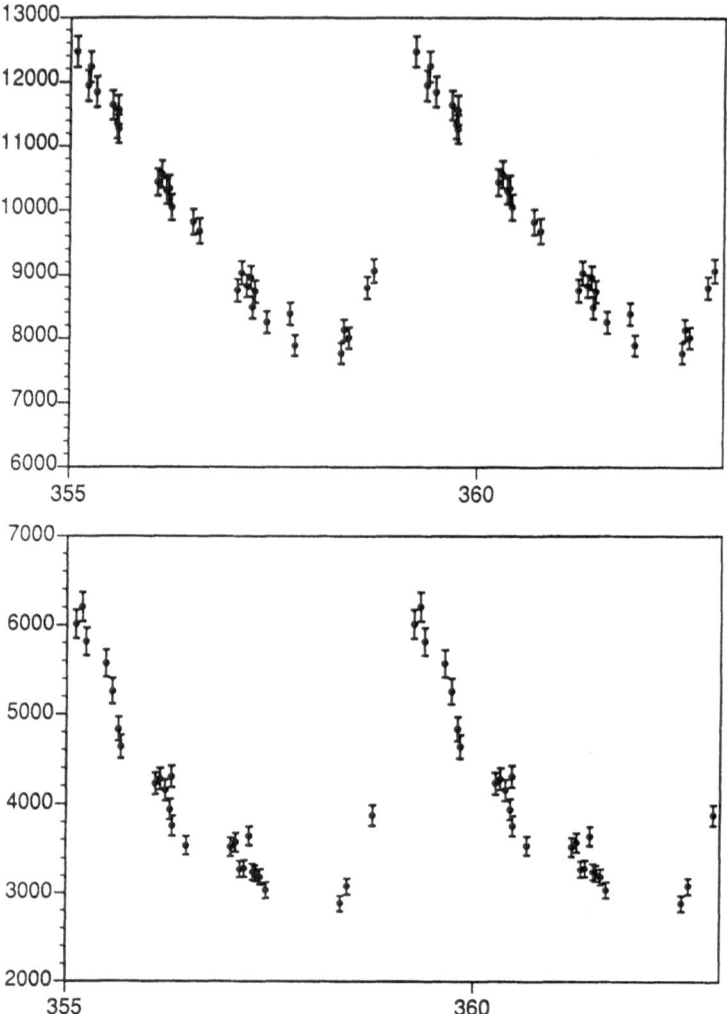

Fig. 5. A Cepheid, with all points on one period. We show here twice the period.

6 Acknowledgements

We have benefitted greatly from discussions with C. Alcock, A. Baranne, D. Benett, A. Bijaoui, A. de Rújula, P. Felenbok, B. Fort, K. Greist, Y. Mellier, C. Penny-packer, B. Sadoulet, C. Stubbs and A. Terzan. We thank J.-P. Berger, R. Burnage and Ph. Veron of the Observatoire de Haute Provence for the help with the GPO telescope and CCD tests. We thank S. D'Odorico and H. Van der Laan of ESO for encouragement and help with the CCD camera. Finally, we thank the ESO staff at La Silla, especially O. and G. Pizarro, D. Hoffstadt and H.E. Schuster, for the observations with the Schmidt telescope and for help with the CCD observations.

Table 1. The expected number of observed microlensing events as a function of assumed deflector mass, M, for the Schmidt program (1990-91 alone and combined with proposed 1991-92 observations), for the proposed 1991-92 CCD program, on the GPO and on the 40-cm reflector. The estimates are based on an isothermal halo model with a total mass of 4.0×10^{11} $M_\odot$ between the galactic core and the LMC. For the 1991-92 Schmidt program, we assume 100 pairs of plates spread over 5 months, 10^6 stars per plate with a mean relative photometric resolution of 0.14 magnitudes in red and blue. For the CCD program, we assume 150 consecutive nights from which we subtract the weeks centred on the full moon and 20% randomly selected unusable nights. We assume 30 000 monitored stars with a mean resolution of 0.1 magnitudes.

$M(M_\odot)$	Schmidt 1990-91	Schmidt 1990-92	GPO 1991-92	40 cm 1991-92
10^{-0}	0.3	2.6		
10^{-1}	1.0	4.8		
10^{-2}	1.9	8.2		
10^{-3}	2.2	10.0	1.2	3.0
10^{-4}	1.5	7.5	2.2	5.7
10^{-5}			2.8	10.0
10^{-6}			2.3	11.0
10^{-7}			1.3	6.6

References

De Rújula, A., Jetzer, Ph., Massó, E., 1991. CERN preprint CERN-TH 5787/91.

Milsztajn, A., January 1990. 10th Moriond Workshop on new and exotic phenomena, Les Arcs. Published by Editions Frontières, 1991.

Moniez, M., January 1990. 10th Moriond Astrophysics Meeting, Les Arcs. Published by Editions Frontières, 1991.

Paczynski, B., 1986. *Astrophys. J.*, **304**, 1-5.

Primack, J.R., Seckel, D., Sadoulet, B., 1988. *Ann. Rev. Nucl. Sci.*

Trimble, V., 1987. *'General Reviews of Dark Matter'*, Ann. Rev. Astron. Astrophys.

A DEEP PROPER MOTION SURVEY OF THE PLEIADES FOR VERY LOW MASS STARS AND BROWN DWARFS

N.C. Hambly

Department of Astronomy, University of Leicester, University Road, Leicester
LE1 7RH, U.K.

1 Introduction

The nature of the Very Low Mass (VLM) main sequence remains a fascinating but elusive topic, despite the many searches and surveys for faint red stars. Here we will follow the usual definitions of objects of mass m in the range $0.3 > m > 0.08$ solar masses being called very low mass, and those with $m < 0.08 M_\odot$ being labelled brown dwarfs (BDs). Theoretical stellar models predict that objects having $m < 0.08 M_\odot$ will be unable to support stable hydrogen fusion reactions in their cores — such stars become degenerate cooling brown dwarfs.

Observational surveys have shown VLM stars of $m = 0.2$ solar masses to be most common — searches for lower mass objects are hampered by their intrinsic faintness and extremely red spectral distribution. Exceptional VLM stars/BD candidates discovered recently include GD165B, a companion to a white dwarf (Becklin & Zuckermann 1988); Wolf 424AB, an apparent brown dwarf binary system with component masses 0.06 and 0.05 solar masses (Heintz 1989 — an understandably controversial result); and MH2115-4518, probably the faintest star yet discovered (Hawkins, private communication). Major effort has been expended in searching for binary companions and field stars, yet the relatively younger open clusters provide a rich source of lower main sequence stars in a relatively small area of the sky. This report describes a deep proper motion survey of the Pleiades for these objects — a more detailed description and analysis will appear in *Mon. Not. R. Astron. Soc.* (Hambly et al., in press).

2 Observations and Data Reduction

We used COSMOS measures of POSS R plates (first epoch, 1951) and an UKST OR plate (second epoch, 1989) to provide proper motions. Using software described

H. T. MacGillivray and E. B. Thomson (eds.), Digitised Optical Sky Surveys 265–270.

by Hawkins (1986), the plates are divided into small areas and mapped onto a co-ordinate system defined by the second epoch plate, thus eliminating systematic positional errors due to emulsion shifts and plate/telescope distortions. The combination of COSMOS internal positional accuracy, software sophistication and large time base-line gives the requisite accuracy for us to pick out Pleiades members in the proper motion diagram (Fig. 1) — the cluster main sequence stars are selected outwith the SW circle, while an estimate of the 'background' contamination in the diagram is provided by the NE circle. This leads us directly to a list of members along with associated membership probabilities in a given luminosity bin from the background estimator. The COSMAG photometric parameter was calibrated using accurate photoelectric photometry of stars in the field, and a small sample of members were further measured in J,H and K at the UK Infra-Red Telescope (UKIRT).

3 Discussion

Figure 2 shows the Pleiades infrared main sequence. Open circles are higher mass members, numbers denoting fainter stars from our proper motion survey. Open squares are Hyades members shifted to the Pleiades distance (Leggett & Hawkins 1988) and open triangles with the dotted line show the empirical old–dwarf main sequence from Bessell (1991). The solid line and squares are a theoretical VLM model from Stringfellow (1989); mass points are 0.2, 0.15, 0.1, 0.09, 0.08, 0.07, 0.06 and 0.05 solar masses from top to bottom. The three solid squares at the faint end are unpaired 'I-only' images from Hambly & Jameson, (1991 — see also Leggett & Hawkins 1989) and the solid triangle is the CCD BD·candidate of Jameson & Skillen (1989). These last four objects have no proper motion measurements, and also appear to be background non-members in Fig. 2, as is the proper motion member star 1. With the exception of star 28, the rest of the proper motion selected sample form a well-defined main sequence that is cooler than the old dwarf dotted line as expected for the young Pleiades pre-main sequence (PMS). The theoretical isochrone of Stringfellow does not coincide exactly with the PMS, but given the uncertainties in theoretical parameters and bolometric corrections/temperature scales for VLM dwarfs, agreement is fair. One intriguing question that realistically remains unanswered is that of the nature of the faintest object, star 5. It lies between the 0.08 and 0.07 mass points in the theoretical curve, implying it is one of the least massive stars identified. Whether it is a BD or a VLM main sequence star is impossible to say given uncertainties of at least 0.5 magnitudes in the computation and conversion of the theoretical points.

Figure 3 shows the Pleiades luminosity function (LF) from the proper motion members. The data are rebinned by half a magnitude between squares and crosses, square root errorbars are shown. The open circles and solid line are the field LF from Hawkins & Bessell (1988). No normalisation has been done — the agreement between the two LFs is remarkable (the same agreement is found for the field and Hyades LFs in Leggett & Hawkins 1988) and implies a similar underlying mass function. Age effects are difficult to see at the faint end where the Pleiades sample is incomplete.

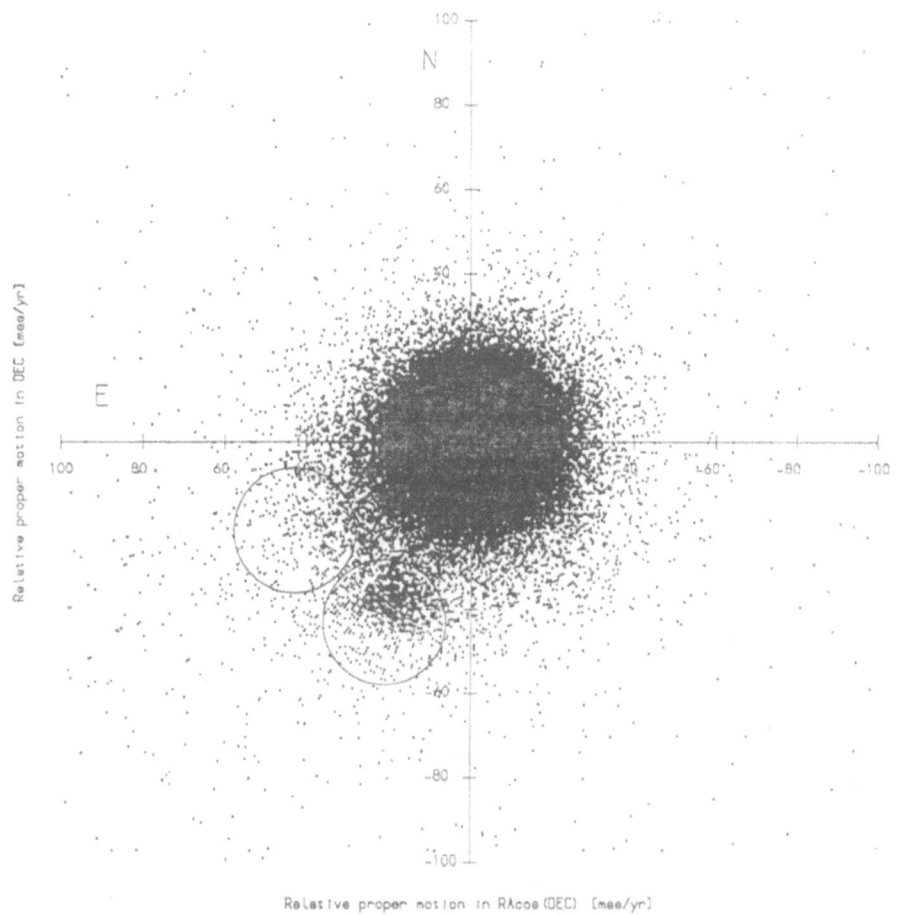

Fig. 1. The proper motion diagram for the Pleiades field — the cluster members stand out clearly from the general field scatter at (20 mas/yr, -40 mas/yr). The main sequence is selected out by the SW circle; an estimate of the 'background' contamination is made from the NE circle.

4 Conclusions and Future Work

We have used the combination of Schmidt plates and the powerful measuring machine COSMOS to produce a deep proper motion survey of the Pleiades cluster. Follow–up infrared photometry is essential to correctly identify cool, faint main sequence members. The cluster LF is the same as that of the field within the errors, and implies a single, global underlying mass function of star formation. As future work, we plan to examine more of the list of proper motion members with RIJHK photometry and infrared spectra, and we hope to use the superior accuracy of SuperCOSMOS to push the survey fainter using I plates spanning 4 or 5 years.

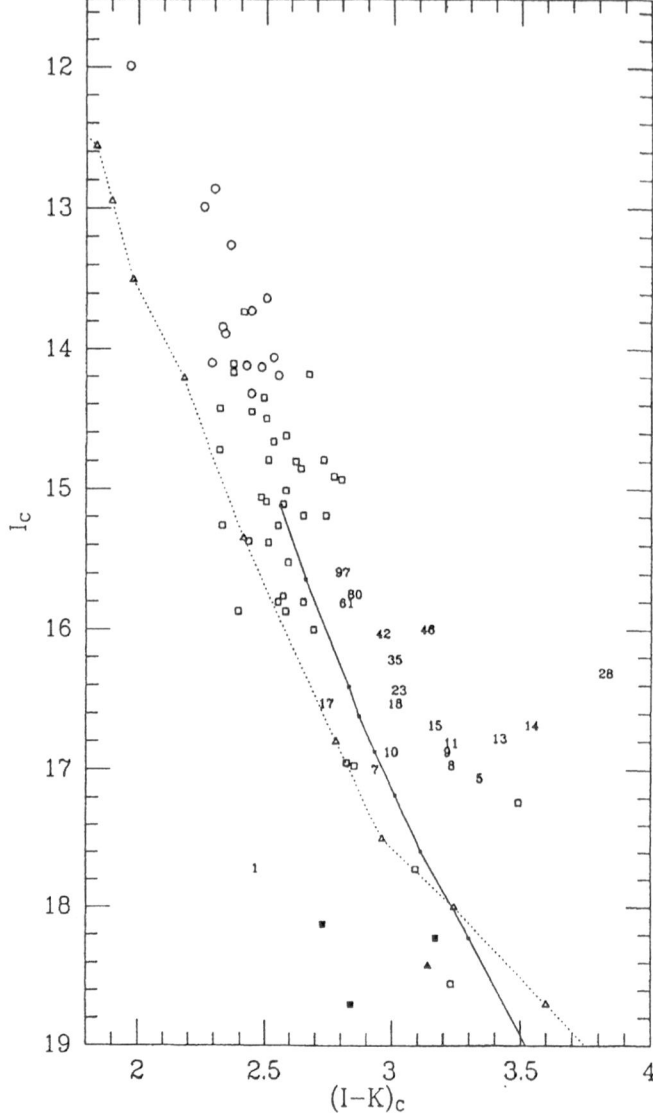

Fig. 2. The infrared main sequence of the Pleiades cluster. Numbers denote stars from our proper motion list; open circles are higher mass members from Stauffer (1982). Open squares are Hyades members from Leggett & Hawkins (1988) shifted to the Pleiades distance; open triangles plus the dotted line define the old disk dwarf main sequence (Bessell 1991). The solid squares and triangles are faint candidates that have no proper motions — the triangle is JS1 from the CCD survey of Jameson & Skillen (1989), the three solid squares are 'I-only' images from Hambly & Jameson (1991). Small solid squares plus the solid line are a 70 Myr theoretical isochrone for VLM stars/BDs from Stringfellow (1989) — mass points are 0.2, 0.15, 0.10, 0.09, 0.07, 0.06, and 0.05 solar masses from top to bottom.

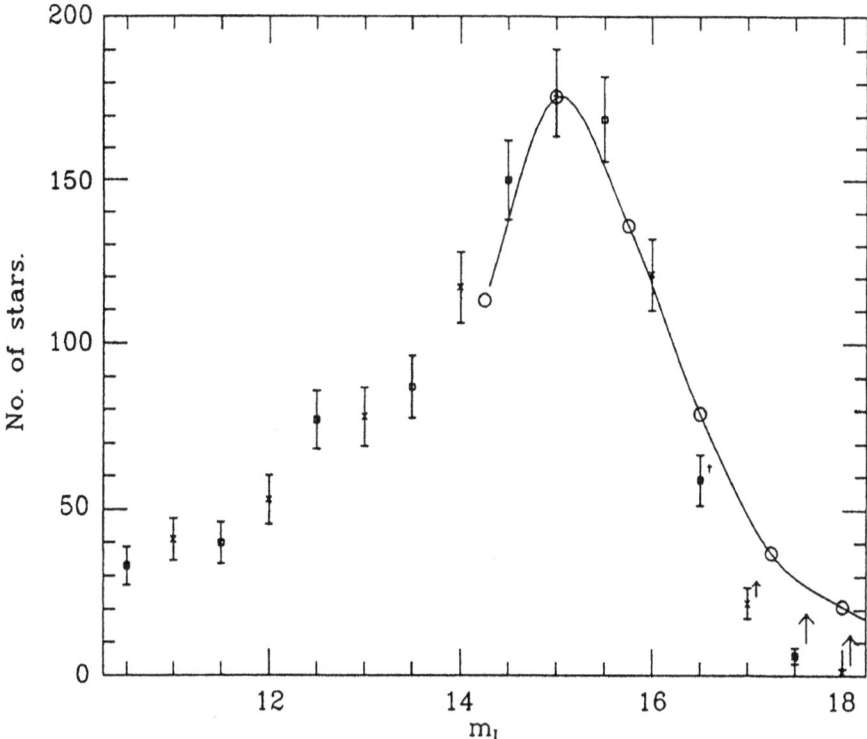

Fig. 3. The Pleiades luminosity function shown in 1 magnitude bins and rebinned by a 0.5 magnitude bin shift. Open circles and the solid line are the field from Hawkins & Bessell (1988).

5 Acknowledgements

The author gratefully acknowledges the UKST and COSMOS groups for plate material and measures; the close supervision and collaboration of Drs R.F. Jameson and M.R.S. Hawkins; a SERC studentship and STARLINK computer facilities.

References

Becklin, E., Zuckermann, B., 1988. *Nature*, **336**, 656.
Bessell, M.S., 1991. *Astron.J.*, **101**, 662.
Hambly, N.C., Jameson, R.F., 1991. *Mon. Not. R. Astron. Soc.*, **249**, 137.
Hawkins, M.R.S., 1986. *Mon. Not. R. Astron. Soc.*, **223**, 845.
Hawkins, M.R.S., Bessell, M.S., 1988. *Mon. Not. R. Astron. Soc.*, **234**, 117.
Heintz, W., 1989. *Astr. Astrophys.*, **217**, 145.
Jameson, R.F., Skillen, W.J.I., 1989. *Mon. Not. R. Astron. Soc.*, **239**, 247.
Leggett, S., Hawkins, M.R.S., 1988. *Mon. Not. R. Astron. Soc.*, **234**, 1065.
Leggett, S., Hawkins, M.R.S., 1989. *Mon. Not. R. Astron. Soc.*, **238**, 145.
Stauffer, J.R., 1982. *Astron. J.*, **87**, 1507.
Stringfellow, G.S., 1989. PhD thesis, University of California at Santa Cruz.

Discussion

Schachter :

What are the predictions of different theoretical models below the threshold of hydrogen burning?

Hambly :

Different models agree well in logL (better than 0.2) but not well in T_{eff} (15% variation typically). Since near infrared bolometric corrections are strongly dependent on T_{eff}, this causes major uncertainties in the comparison of theory and observation.

Leggett :

1. Did you use the Burrows et al. models?

2. On your I:I-K diagram is there any difference between the Hyades and Pleiades, as you would expect due to their age difference?

Hambly :

1. We used the Stringfellow models as they fit the ZAMS better than any others.

2. There is some evidence that the Pleiades stars are younger than the Hyades but this is debatable. However, both Pleiades and Hyades appear younger than the old field M-dwarfs as expected.

THE ABSOLUTE PROPER MOTION OF
THE GLOBULAR CLUSTER M3

M.J. Irwin [1], *N. Argue* [2] *and R.D. Scholz* [3]

[1] Royal Greenwich Observatory, Madingley Road, Cambridge CB3 0EZ, U.K.
[2] Institute of Astronomy, Madingley Road, Cambridge CB3 0HA, U.K.
[3] Zentralinstitut für Astrophysik, Pottsdam-Babelsberg, DDR-1502, Germany.

Abstract

The combination of Schmidt plates and an automatic measuring machine provides a powerful means to investigate the proper motions of the Galactic halo globular clusters and the more nearby dwarf spheroidal satellites with respect to a reference frame of thousands of background galaxies. Using Tautenburg Schmidt plates measured on the APM facility we have determined the absolute proper motion of the globular cluster M3 to be $\mu_x = -0.13 \pm 0.07$ and $\mu_y = -0.14 \pm 0.05''$/cent.

1 Introduction

Our present knowledge of the kinematics of the Galactic globular cluster system is mainly based on radial velocity studies. The key to further progress lies in the determination of absolute proper motions. Webbink (1988) reviewed the efforts in both directions and encouraged new attempts to obtain absolute proper motions of globular clusters directly with respect to background galaxies and quasars.

Proper motions of a whole series of globular clusters have been investigated by Cudworth and coworkers by measuring the relative motion of cluster stars with respect to field stars. This required various assumptions concerning the direction of the solar apex and the secular parallaxes of the chosen field stars around each cluster. Not only were the results affected by the choice of solar apex and motion but they were also afflicted by large errors due to the small number of field stars used (~ 50) and their inherent proper motion dispersion (Cudworth 1979). Although a very high accuracy ($\pm 0.02''$/cent) in the relative proper motion of the cluster stars was achieved, the uncertainty in the absolute proper motion of the whole cluster

H. T. MacGillivray and E. B. Thomson (eds.), Digitised Optical Sky Surveys 271–279.
© 1992 *Kluwer Academic Publishers.*

remained of the order of 0.1 to 0.4″/cent. Brosche et al. (1983, 1985) used Lick stars with known absolute proper motion as reference stars to connect globular cluster motions to an extragalactic reference frame, whilst Tucholke (1989) derived the proper motion of globular clusters relative to the background of the SMC using automated measurements on ESO Schmidt plates.

This work describes the first measurement of the absolute proper motion of a globular cluster directly with respect to a large number ($\sim$ 2000) of background galaxies. The Tautenburg Schmidt telescope has been taking photographic plates of globular clusters since the early 1960s. With a 25 year base-line, it is now possible to use this material to study the proper motions of these globular clusters. This 2m Schmidt telescope has a plate scale of 51.4″/mm. With a useful plate size of 24 cm, each plate covers some 3° × 3° of sky, providing sufficient background galaxies to define an absolute inertial reference frame. Until recently the plates were measured on a semi-automated measuring machine, ASCORECORD. Because of the relatively slow measuring rate, the number of reference galaxies and stars that could be included were severely restricted (Schilbach 1982; Scholtz & Rybka 1988). Among the fields for which first epoch plates are available are several centred on the globular cluster M3 (Scholtz & Hirte 1991). Seven pairs of Tautenburg Schmidt plates of M3 were processed on the APM facility in Cambridge in order to fully extract all the information available on the plates. This system has already been extensively used for astrometric work using UK Schmidt and Palomar Schmidt photographic plates (Kibblewhite et al. 1982; Evans 1988). Here we present preliminary results of our determination of the absolute motion of M3 using two plate pairs measured on the APM.

2 Observations and Measurements

All plates used in this project were taken on the Tautenburg Schmidt telescope. Table 1 lists the relevant parameters of all the useful plate material available for M3. The plates were measured by placing them on an optically flat 1/4″ thick glass plate using nonane (a liquid of comparable refractive index to glass) to stick the plate onto the glass surface. This procedure ensures that the plates are orthogonal to the measuring beam over the whole area, removes the need for autofocussing of the laser spot and prevents fringes occurring from the coherent illuminating laser light. The total available field of view per plate is 10.5 square degrees and the plate pair overlaps are given in Table 1. In all cases the cluster field centre is within ±7′ of the centre of the overlap zone, giving a good distribution of field stars and reference galaxies about the cluster. Images within about 2′ of the cluster centre were too crowded to be reliably measured and were excluded from subsequent processing.

One of the plates (6226) was measured twice, with a time interval between measurements of several days, to assess the repeatability of the APM measurements of Tautenburg Schmidt plates. The positional repeatability for different magnitude classes and types of object (stars and galaxies) is shown in Table 2. The stellar magnitudes are only indicative since they were based on an approximate linear relationship between diameter and magnitude noted by Schilbach (1986) in a general investigation of Tautenburg Schmidt plates. Further assessment of the plate-to-plate

Table 1. M3 Plate Material.

Plate No. (pairs)	Epoch	Emulsion	Limiting Magnitude	Plate centre (1950.0)	Overlap (□°)
2167	1966.2	103a-O	B=21.0	$13^h39.6 + 28°36$	9.2
6226	1986.4	ZU-21	B=20.5	$13^h39.3 + 28°52$	
2174	1966.2	103a-O	B=21.0	$13^h39.5 + 28°37$	9.2
6232	1986.4	ZU-21	B=20.0	$13^h39.5 + 28°54$	
2175	1966.2	103a-O	B=21.0	$13^h39.3 + 28°35$	10.0
6999	1989.2	ZU-21	B=20.5	$13^h39.1 + 28°38$	
2176	1966.2	103a-O	B=21.0	$13^h39.6 + 28°36$	10.0
7000	1989.2	ZU-21	B=20.5	$13^h39.9 + 28°32$	
2873	1969.3	ZU-21	B=20.0	$13^h39.8 + 28°35$	10.0
7002	1989.2	ZU-21	B=20.5	$13^h39.7 + 28°34$	
2504	1967.4	103a-D	V=20.5	$13^h39.4 + 28°36$	10.0
7016	1989.3	103a-D	V=19.0	$13^h39.6 + 28°37$	
2883	1969.4	103a-D	V=20.5	$13^h39.5 + 28°35$	10.0
7019	1989.3	103a-D	V=19.5	$13^h39.6 + 28°37$	

Notes:

1. The second epoch V plates do not go as deep as the first epoch plates because only old Kodak 103a-D emulsion was available.
2. The variation in overlap area was partly due to the calibration wedge in the NE corner of all the second epoch plates and partly due to a shift of field centre in the case of #6226 and #6232.
3. All exposure times were between 20 and 30 mins.

positional errors was made by pairing up all plates taken at the same epoch (Scholz 1990). For these plate pairs a positional accuracy between 1μm and 2μm ($0.05''$ and $0.10''$) was achieved for stars in the magnitude range B = 10 to 17, indicating that the main source of error in any astrometric comparison would be governed by the plate material, not the APM. For fainter objects the plate-to-plate errors are completely dominated by random errors in the recording process — poisson and photographic grain noise (see for example Irwin 19,85). So for B = 19 the stellar positional accuracy is 3μm whilst for galaxies it rises to 5μm.

Both field-dependent and magnitude-dependent positional errors were looked for also. Whereas coma-like effects could not be found, magnitude-dependent system-atic shifts of up to a few μm were apparent for some of the plate pairs. This appears to be caused by a combination of asymmetric image profiles and photographic sat-uration. Basically, in good seeing, some 3 to 4 magnitudes above the plate limit all the stellar images begin to saturate. Any centering algorithm, whether it be a centre of gravity method or even a profile fitting method, sees a varying fraction

Table 2. APM positional repeatability on Tautenburg Schmidt plate # 6226.

Diameter (μm)	B Magnitude	No. of Stars	Error (μm)	No. of Galaxies	Error (μm)
35 - 70	> 18.5	4057	1.0	3142	2.0
70 - 110	17.0 - 18.5	4231	0.6	2200	1.4
110 - 175	14.0 - 17.0	3156	0.5	1458	1.3
175 - 260	12.0 - 14.0	1466	0.4	557	1.3
260 - 350	10.0 - 12.0	288	0.4	141	1.4
350 - 610	8.5 - 10.0	110	0.5	63	1.5
610 - 990	< 8.5	11	0.5	9	1.8

of the image profile as a function of magnitude. If the underlying profile is asymmetric, the derived centre will depend in a subtle way on how much of the image is lost to saturation. Fortunately, the vast majority of stars in the cluster are not saturated, and the same is true of the background galaxies that are used for an inertial reference frame. Indeed, no statistically significant magnitude effect was seen in the galaxy measurements. So, providing the brightest cluster stars are not used in determining the cluster proper motion, this magnitude dependence is not a problem.

3 Reduction Methods

Using different samples of reference galaxies (see Table 3), the plate-to-plate solutions for the determinations of the absolute proper motion of all stars was done with complete 1st, 2nd and 3rd order polynomials and with the method of stepwise regression described in Hirte et al. (1990). All terms were only coordinate dependent. With the stepwise regression, a global plate-to-plate transformation is computed with terms up to and including cubic polynomials. Only statistically significant terms are retained and any plate-solution is based on the minimum number of polynomial coefficients required to define the least-squares solution. The global polynomial approach was used for plate pairs #2167/6226 and #2174/6232. The proper motion of the cluster sample is then simply defined as the mean difference in positions between the two epochs. Although the other plate material has been partially reduced, we will restrict the remainder of the paper to the results obtained from the two plate pairs mentioned previously.

3.1 Plate Matching

After automatically classifying all objects measured on each plate into the categories stellar, non-stellar, merged and noise, plate pairs taken at the same epoch were

Table 3. Systematic Error of the Absolute Proper Motion of M3.

No. of Galaxies	Plates 2167/6226 $\mu_x, \mu_y('' / \text{cent})$	Plates 2174/6232 $\mu_x, \mu_y('' / \text{cent})$	Mean Systematic Error using both pairs
434[1]	0.055 0.052	0.078 0.076	0.048 0.046
877[2]	0.049 0.051	0.073 0.069	0.044 0.043
1037[3]	0.039 0.037	0.053 0.051	0.033 0.032
1234[4]	0.044 0.043	0.061 0.056	0.038 0.035

Notes:

1. Selected galaxies were outside a cluster radius of 8', had diameters between 110 and 260μm and were classified as galaxies on at least 3 out of the 4 plates.
2. Selected galaxies were outside a cluster radius of 8', had diameters between 70 and 260μm and were classified as galaxies on at least 3 out of the 4 plates.
3. Selected galaxies were outside a cluster radius of 16', had diameters between 70 and 260μm, were classified as galaxies on at least 2 out of the 4 plates and had $< 3\sigma$ positional errors.
4. Selected galaxies were outside a cluster radius of 16', had diameters between 70 and 260μm, were classified as galaxies on at least 2 out of the 4 plates and had $< 4\sigma$ positional errors.

matched in order to exclude noise images and images only appearing on one plate. In general, plate matching consists in pairing up objects between a reference plate and a comparison plate using several passes of ever decreasing search radius, starting with bright objects and large search radii (up to several mm), finally iterating down to faint objects within a target search radius of 30μm. This corresponds to a proper motion of 7.5''/cent, so that only very high proper motion objects were excluded. Only astrometric criteria were used in the matching, since with a 30μm final search radius the probability of generating a spurious mismatch is negligible ($< 0.01\%$ at these number densities). In order to exclude objects that were unreliably recorded or measured (due to, say, excessive image crowding, dirt or emulsion defects) only those images classified as stars on all plates or as galaxies on at least three plates were used in subsequent analyses. Finally the faintest stars and galaxies (B > 19) were not used both because of their rapidly increasing positional errors and the uncertainty in the image classification of galaxies near the plate limit.

3.2 Determination of Proper Motion — Polynomial Method

Up to 3rd order polynomial plate models were used without any magnitude- dependent terms. Magnitude-dependent terms were unnecessary because the magnitude interval for the objects used had already been restricted to essentially eliminate this problem. The mean proper motion error for an individual star was $\sim 0.7''$/cent for faint stars ($17 < B < 18.5$) and $\sim 0.3''$/cent for brighter stars ($B < 17$). Assuming all stars within $8'$ radius are cluster stars, there were 550 well measured (i.e. stellar on all plates) faint and 125 bright stars. Therefore, the random error for the cluster motion, obtained by simply averaging the mean individual stellar proper motions, is of order $0.03''$/cent. We decided not to use the brighter cluster subset in the estimate, since even after allowing for the extra information present in the membership probabilities obtained by Cudworth (1979) for stars with $V < 16$, the possible magnitude-dependent systematic errors could begin to dominate the solution.

An estimate of the systematic error of the mean absolute proper motion of the cluster can be obtained directly by dividing the unit weight error of the plate-to-plate solution by the epoch difference and the square root of the number of galaxies used in the solution. This error represents the error in defining an inertial reference frame in the field centre where the cluster is situated. In Table 3 the systematic errors in the cluster motion obtained independently from each pair of plates and for the mean absolute proper motion of the cluster are shown using different samples of reference galaxies. The final two categories include galaxies in the magnitude range $B = 16$ to 19 and should be directly compatible with the magnitudes of the cluster stars used from $B = 17$ to 18.5. This restriction is important because of the possibility of magnitude-dependent systematic errors on at least some of the plates. For example a systematic error of 1μm would contribute a fictitious $0.025''$/cent proper motion to the cluster. The worse results for the second pair of plates are due to the lower limiting magnitude and poorer image quality on plate 6232.

4 The Proper Motion of M3

Using the same reference galaxies and stars on both pairs of plates gives two independent results for the mean absolute proper motion of the star sample which can be used to help estimate the internal accuracy. From the different polynomial models applied in the plate solutions, the linear model generally gave greater differences (up to a factor of 2) between the two plate pairs. For the complete second and third order polynomials the difference between the two pairs of plates was typically less than $0.10''$/cent; whilst for the third order stepwise regression this figure dropped to $0.05''$/cent. Given that there is no unique 'correct' reduction model the stepwise regression approach appears to give the more reliable results from among the polynomial models. Figure 1 illustrates the sensitivity of the cluster motion results to the reference galaxy sample used, different choices for the cluster radius and of using stars within differing magnitude ranges. Taking only the faint stars (B=17 to 18.5) and a radius of $8'$ we obtain an absolute cluster proper motion of $\mu_x = -0.13 \pm 0.07''$/cent and $\mu_y = -0.14 \pm 0.05''$/cent for a different choice of reference galaxies. The derived cluster proper motion does not depend on the value used for the cluster radius until the contamination by field stars becomes significant. If stars brighter than B=17 are used, magnitude-dependent systematic effects become noticeable. The proper motion of the field star data lies between the old Lick antapex (Klemola & Vasilevskis 1971) and the new Lick antapex of Hanson (1987) for stars with B=16. Given the uncertainty of our magnitude scale, this agreement is excellent. Also shown in Fig. 1 is the antapex for the Galactic globular cluster system taken from Mihalas & Binney (1981). The derived proper motion of M3 lies along this direction, indicating a space motion in the same direction as the sun with roughly half of the solar orbital velocity.

5 Discussion

We have shown that it is possible to obtain the absolute proper motion of Galactic globular clusters directly with respect to a large number of background galaxies using Tautenburg Schmidt plates measured on the APM facility. In comparison with Cudworth (1979), the epoch difference is only moderate and the accuracy in measuring the position of any individual object is much worse because of the smaller epoch difference and our much smaller plate scale. However, the advantages in being able to use a reference frame of thousands of background galaxies outweigh these drawbacks. Indeed the reference frame can be established to a few ''/cent accuracy. Most of the magnitude-dependent effects noted in determining the cluster motion using stars brighter than B=17 are due to contamination of the cluster stars with field stars having a relatively high proper motion. However, in spite of the similar direction of the solar apex with respect to faint stars and the globular cluster system, we found a significant difference between the field and cluster proper motions for stars of the same magnitude interval. The agreement of the field star proper motion with the Lick results shows that the overall reduction strategy is capable of producing very reliable absolute proper motions for Galactic halo systems.

In future work we intend to address the problem of determining the proper motions of some of the Galactic dwarf spheroidal satellites; in addition extending the M3 work to fully exploit all of the Tautenburg plate material.

6 Acknowledgements

We wish to thank the Tautenburg staff, particularly Dr. F. Börngen and Dr. Ziener for taking the plates. R. Scholz would like to thank the Institute of Astronomy, Cambridge for hospitality and financial support during some of this work. The APM is a national astronomy facility financed by the Science and Engineering Research Council.

References

Brosche, P., Geffert, M., Ninkovic, S., 1983. *Publ. Astron. Inst. Czech. Acad. Sci.*, **56**, 145.
Brosche, P., Geffert, M., Klemola, A.R., Ninkovic, S., 1985. *Astron. J.*, **90**, 2033.
Cudworth, K.M., 1976. *Astron. J.*, **81**, 519.
Cudworth, K.M., 1979. *Astron. J.*, **84**, 1312.
Eichorn, H., Jefferys, W.H., 1979. *Publ. Leander McCormick Obs.*, **XVI**, 267.
Evans, D., 1988. PhD Thesis Cambridge.
Hanson, R.B., 1987. *Astron. J.*, **94**, 409.

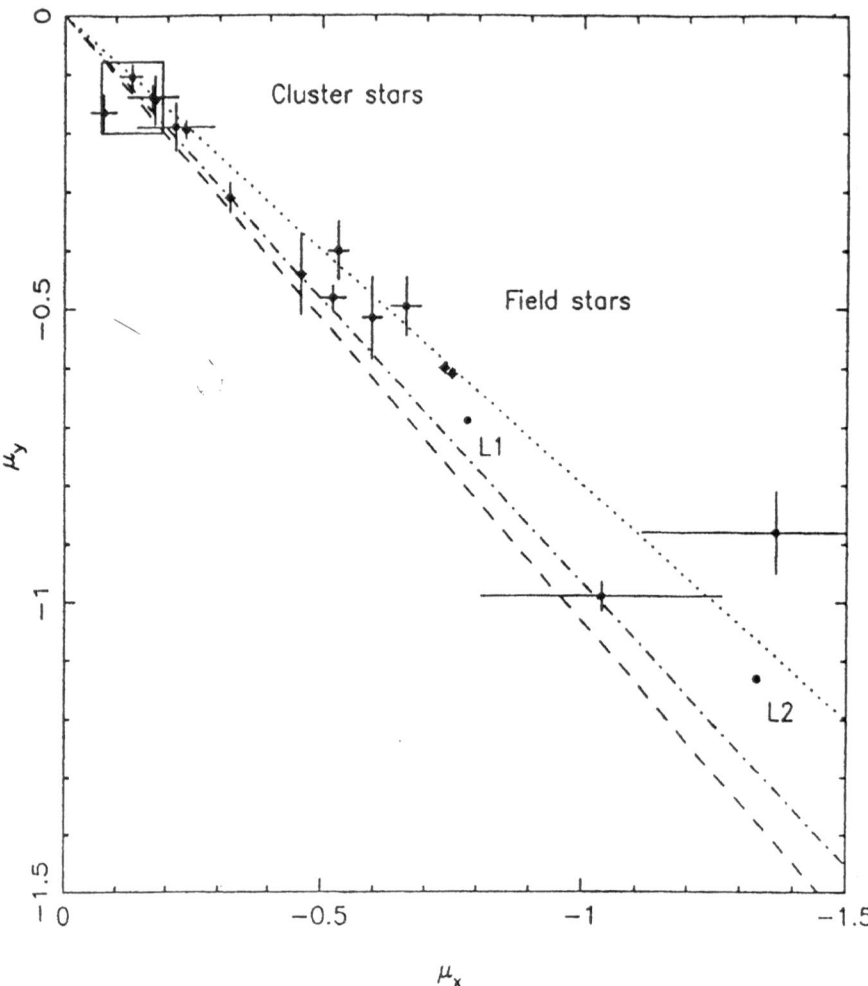

Fig. 1. Mean absolute proper motion of cluster stars and field stars ($''$/cent) as a function of the apparent magnitude. As expected there is a general trend of increasing proper motion of the field stars for brighter samples. The cluster stars show no strong correlation of proper motion with magnitude. Points corresponding to field and cluster stars together with their 1σ error bars are marked. The square box centred on the cluster stars represents the $\pm 1\sigma$ error limits for the cluster motion. The points labelled L1 and L2 are Lick proper motion measurements for faint and bright stars respectively taken from Klemola & Vasilevskis (1971). The three lines represent: dashed – solar antapex with respect to globular clusters; dot-dash – old Lick solar antapex; dotted – new Lick solar antapex.

Hirte, S., Dick, W.R., Schilbach, E., Scholz, R.D., 1990. In *'Errors, Bias and Uncertainties in Astronomy'*, Proc. *IAU Colloq. xx*, ed. C. Jaschek. In press.
Irwin, M.J., 1985. *Mon. Not. R. Astron. Soc.*, **214**, 575.
Kibblewhite, E.J., Irwin, M.J., Bridgeland, M.T., Bunclark, P.S., 1982. *Occ. Rep. R. Obs. Edinburgh*, **10**, 79.
Klemola, A.R., Vasilevskis, S., 1971. *Publ. Lick Obs.*, **XXII**, part III.
Mihalas, D., Binney, J., 1981. *Galactic Astronomy*, Freeman, San Francisco.
Schilbach, E., 1982. *Astron. Nach.*, **303**, 335.
Schilbach, E., 1986. *Astron. Nach.*, **307**, 61.
Scholz, R.D., 1990. Dissertation (A), Potsdam.
Scholz, R.D., Rybka, S.P., 1988. *Astron. Nach.*, **309**, 47.
Scholz, R.D., Hirte, S., 1991. *Astron. Nach.*, **312**. In press.
Tucholke, H.J., 1989. PhD Thesis, Münster.
Webbink, R.F, 1988. In *'The Harlow Shapley Symposium on Globular Cluster Systems in Galaxies'*, *IAU Symp.* No. 126, p. 49, eds. J.E. Grindley & A.G. Davis Philip.

THE STRUCTURE AND STELLAR CONTENT OF THE SMALL MAGELLANIC CLOUD

L.T. Gardiner [1], M.R.S. Hawkins [2] and D. Hatzidimitriou [3]

[1] Department of Astronomy, University of Edinburgh, Blackford Hill, Edinburgh EH9 3HJ, U.K.
[2] The Royal Observatory, Blackford Hill, Edinburgh EH9 3HJ, U.K.
[3] The Anglo-Australian Observatory, P.O. Box 296, NSW 2121, Australia.

1 Introduction

The combination of wide field UKST photographic plates and the COSMOS measuring machine at ROE is the ideal tool for studying the structure and stellar content of the outer regions of the Small Magellanic Cloud (SMC), where the image density is low enough for reliable measurement of stellar images. This project was initiated several years ago and has resulted in several papers by Hatzidimitriou et al. (1989), Hatzidimitriou & Hawkins (1989), Gardiner & Hawkins (1991), Gardiner & Hatzidimitriou (1991). Our fundamental interest in the structure of the SMC lies in determining the effects of past interactions with the Galaxy and with the Large Magellanic Cloud (LMC) which are known to have produced the Magellanic Stream, H I Bridge and the young stellar link between the Magellanic Clouds (Irwin et al. 1990) and thought to be responsible for the complicated kinematics and geometrical structure of the central SMC regions (see e.g. Mathewson & Ford 1984). The study of the stellar populations in the field regions of the SMC gives us an insight into the evolution of dwarf irregular galaxies and the processes responsible for star formation in previous epochs under very different conditions from those prevailing today.

The observational data for the project are derived from 38 blue and red UKST plates in 6 fields, with approximately three plates per filter per field in order to effect a $\sqrt{n}$ improvement in the photometric accuracy. The plates in each field were digitised by COSMOS, paired up and calibrated by CCD sequences which were obtained at the Danish 1.5m telescope at ESO. Colour, magnitude and positional data were obtained for a total of 1.1 million images in an area comprising 130 square degrees representing virtually the entire outer area of the SMC beyond 2° from the optical centre ($RA = 00^h51^m, DEC = -73°$, 1950, de Vaucouleurs & Freeman

281

H. T. MacGillivray and E. B. Thomson (eds.), Digitised Optical Sky Surveys 281–289.
© 1992 Kluwer Academic Publishers.

1972). From the comparison between photometry derived for overlapping areas in adjacent fields, it was found that the random photometric accuracy of the magnitudes in each filter is less than 0.1 mag, while systematic discrepancies between the photometry of neighbouring fields did not exceed 0.2 mag. This discrepancy is due mainly to field effects as a result of variable desensitisation towards the edges of the plates. For the fainter regime of magnitudes ($R > 18$), where photometric accuracy is more important, these discrepancies were generally less than 0.1 mag.

Two main methods were adopted in order to analyse the data:-

1. Each field was divided into square grid regions 0.87 square degrees in area, and the colour-magnitude diagram (CMD) for each region was studied separately;
2. Different types of stellar population were selected on the basis of their colours and magnitudes, and the surface distribution of these populations was analysed.

A representative colour-magnitude diagram of a 0.87 square degree region in the northern area is presented in Fig. 1, with the main features labelled. The HB/clump is the most prominent feature and is due to core-helium burning stars aged from a few 10^8 yr up to about 10 Gyr. They are the more massive equivalents of the horizontal branch (HB) stars found in Galactic globular clusters and are a characteristic feature of the CMDs of both intermediate-age open Galactic clusters and Magellanic Cloud clusters. The main sequence comprises hydrogen-burning stars with ages up to a maximum of 2 Gyr, older main sequence stars being too faint to be detected above the effective magnitude limit at $R = 20$.

2 The 3-Dimensional Structure of the SMC Outer Regions

A smoothed contour plot of the surface distribution of HB/clump stars is presented in Fig. 2. This plot is a reliable guide to the morphology of the SMC structure projected onto the plane of the sky, since the clump stars represent the dominant intermediate-age (≤ 10 Gyr) population in the outer regions. There is apparently an inner elliptical structure extending to about 4° from the SMC centre in the SW and to about 3° in the NE. This distribution is similar to that found for carbon stars by Azzopardi & Rebeirot (1991) and may represent a component resembling a Fornax-type elliptical galaxy with spheroidal kinematics which the carbon stars seem to possess (Hardy et al. 1989). The outer contours appear to be more radially symmetric with respect to the SMC centre, but an elongation along the N-S axis is also apparent.

In order to obtain information on the extension of the SMC along the line-of-sight direction, we have made use of the luminosity distribution of clump stars in each grid region. A number-magnitude relation for the clump stars was constructed from which the contribution of Galactic foreground stars, derived from the outermost regions possessing negligible numbers of stars belonging to the SMC, was subtracted. A Gaussian profile was then fitted to the data with a fixed offset to allow for the contribution of red giant branch stars, since these stars could not be separated from true clump stars on the basis of their colours and magnitudes. The dispersion of the fitted Gaussian was then corrected for the intrinsic range

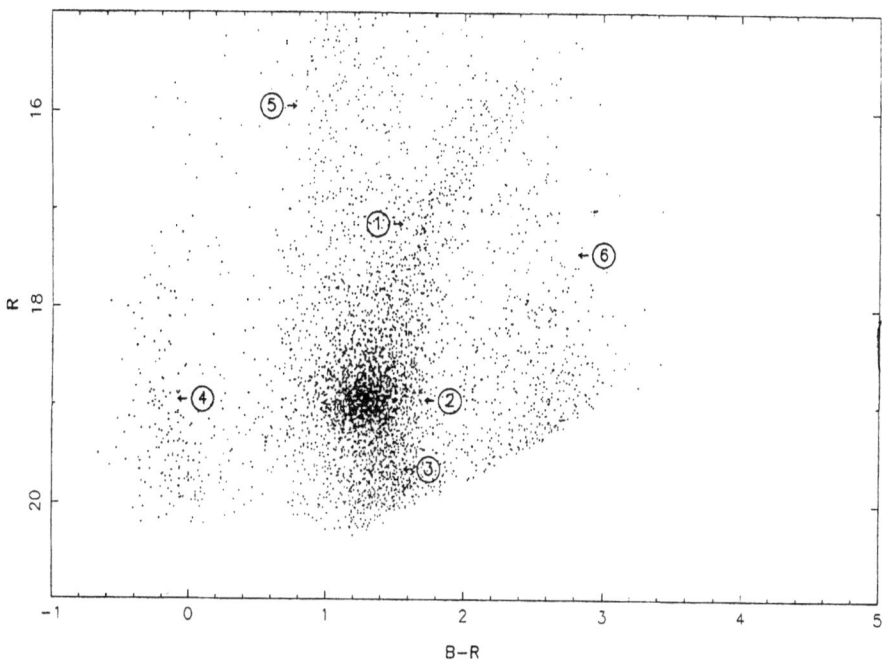

Fig. 1. An example of colour-magnitude diagram from Field 51 showing the various stellar populations: (1) Red Giant Branch (2) HB/Clump (3) Subgiant Branch (4) Main Sequence (5) Galactic Foreground — 'blue peak' (6) Galactic Foreground — 'red peak'.

in luminosity of the clump feature in order to give an estimate of the geometrical dispersion of distance moduli of the clump stars. The intrinsic clump dispersion is due mainly to the evolution of clump stars from their zero-age HB locations (see the evolutionary tracks reproduced in Seidel et al. 1987), but smaller contributions result from a dispersion in age and metallicity of the stellar populations as well as from random photometric errors. The depths derived in this manner for each grid region are plotted in Fig. 3. The parameters of the Gaussian fit also give the mean clump magnitude from which the mean distance modulus of each region can be obtained.

An examination of Fig. 3 shows that the depths in the eastern half of the SMC are systematically greater than in the western half but become smaller in the extreme eastern area. Excluding the very large depths above 20 kpc in the south-east, the mean depth of the eastern area is 10.5 ± 3.3 kpc compared to 4.9 ± 3.5 kpc for the western area. The model which best describes the structure revealed by the depth and distance modulus information is a 2-component model consisting of a far component of small thickness covering most of the outer area of the SMC and a near component in the eastern area located some 7-10 kpc in front. Large depths occur along lines of sight where both components are present, while the trend towards smaller depths (and in fact lower distance moduli) in the extreme east can be explained by the dominance of the near component in these areas. The very

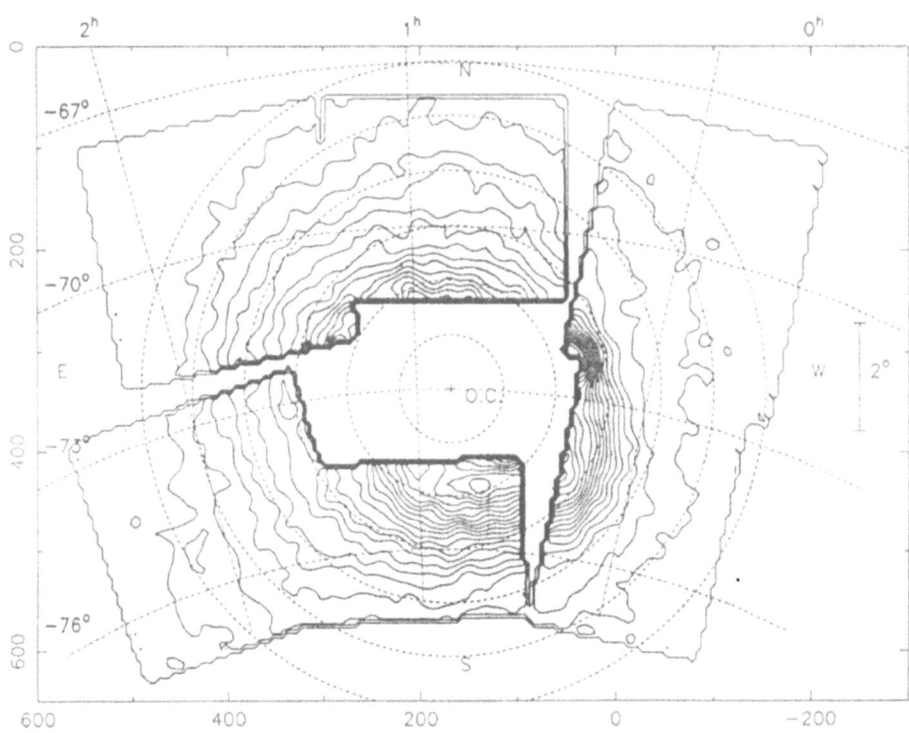

Fig. 2. Contour plot of the surface distribution of HB/Clump stars defined by the colour-magnitude limits $0.6 < B - R < 1.8$, $18.2 < R < 19.5$ for all six survey fields. Concentric radii at 1,2,3,4,5,6 degrees from the SMC optical centre are shown. The contour levels increase towards the SMC centre and consist of levels of 9,13,17,21 etc. stars per 5.6×5.6 arcmin pixel. The contribution from Galactic foreground stars is about 7.5 counts per pixel. To obtain the number of stars per square degree, multiply by 115.

large depths (> 20 kpc) in a few regions in the south-east do not fit satisfactorily into this picture and appear to be due to a different feature altogether. Study of the kinematics of clump stars in the eastern outer regions is essential to determine whether the two components can be separated based on their kinematics. Preliminary results obtained by Hatzidimitriou et al. (1991a) (see also Hatzidimitriou et al. 1991b) indicate that the more distant stars possess higher radial velocities, but a larger sample is required to give conclusive results.

The tidal models of the Galaxy-LMC-SMC interaction of Fujimoto & Murai (1984) (see also Murai & Fujimoto 1980) predict the existence of several past encounters between the LMC and the SMC. The proposed near collision about 200 million years ago is able to reproduce the main features of the Magellanic Stream and also predicts depths as great as 30 kpc for the SMC as a result of tidal disruption by the LMC. The observational depths we have derived are based on the 2σ dispersion of distance moduli in a given grid region, but if the overall depth is as

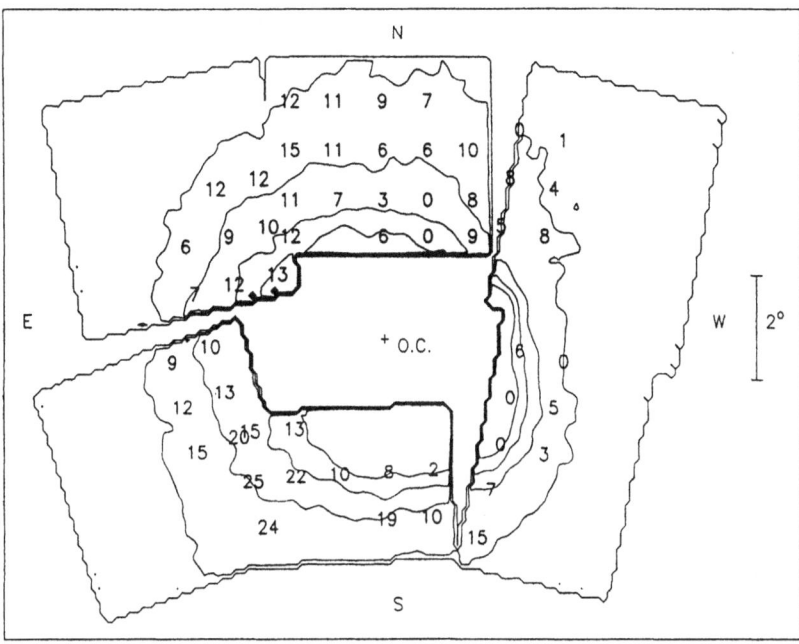

Fig. 3. Depths determined for each grid region in kpc shown as a function of position in the SMC. The contours for the HB/Clump population are superimposed consisting of levels of 12,24,36,48 stars per 5.6 × 5.6 arcmin pixel.

large as 4σ, as suggested by Feast (1989), then much of the eastern area possesses depths between 20 and 30 kpc. Mathewson & Ford (1984) suggested that this recent encounter has led to the creation of two separate components, on the basis H I velocity maps and distances of young cepheid variables located in the central SMC area. However, our two-component picture differs from their model in that their nearer component (the SMC Remnant) lies to the SW, while ours lies to the east. Our results confirm that the outer regions of the SMC are disrupted with the main disturbance acting along our line of sight, whereas in contrast no broad clump feature has been observed in the LMC (Irwin 1991).

3 The Stellar Age Distribution

We present here some of the results we have derived on the age composition in the outer regions of the SMC. Detailed age resolution for the stellar populations of the SMC is limited to populations younger than about 2 Gyr due to the faint magnitude limit of our data at around $R = 20$. To obtain an overview of how these younger populations (compared to the HB/clump which consists largely of populations older than 2 Gyr) are distributed over the surface of the SMC, we

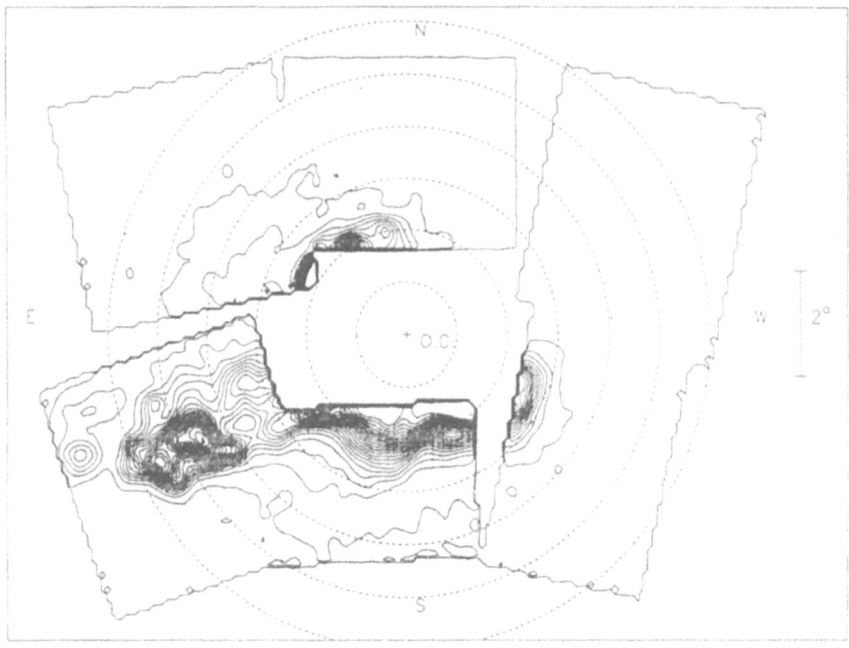

Fig. 4. Contour plot of the surface distribution of main sequence stars with $B - R < 0.1$, $R < 20$ for all six survey fields. The contour levels generally increase towards the SMC centre and consist of levels of 1,3,5,7 etc. stars per 5.6 × 5.6 arcmin pixel.

have produced a contour plot of their surface distribution in Fig. 4. The difference between the distributions of the main sequence populations and the clump stars is striking, and suggests that the role of external interactions may be important in stimulating star formation in the past 2 Gyr. Global star formation activity in the SMC outer regions has apparently ceased sometime before 2 Gyr ago, as shown by the absence of main sequence stars in large parts of the north-western outer area. However, in the east, the Wing feature comprising large numbers of young blue stars is very prominent, and there is a bar-like distribution of main sequence stars along the NE-SW direction corresponding to the major axis of the SMC (de Vaucouleurs & Freeman 1972). The presence of main sequence stars out to beyond 4° from the SMC centre in the NE and SSE directions is also a notable feature of the distribution.

An increase in the age resolution can be achieved by plotting the ratios of main sequence stars in different magnitude ranges. In Fig. 5 we have plotted the ratio of main sequence stars with $16.5 < R < 17.5$ to that with $18.5 < R < 19.5$, giving us an idea of the relative contribution of populations aged younger than about 0.6 Gyr to the total population younger than about 1.3 Gyr (from consideration of the Yale

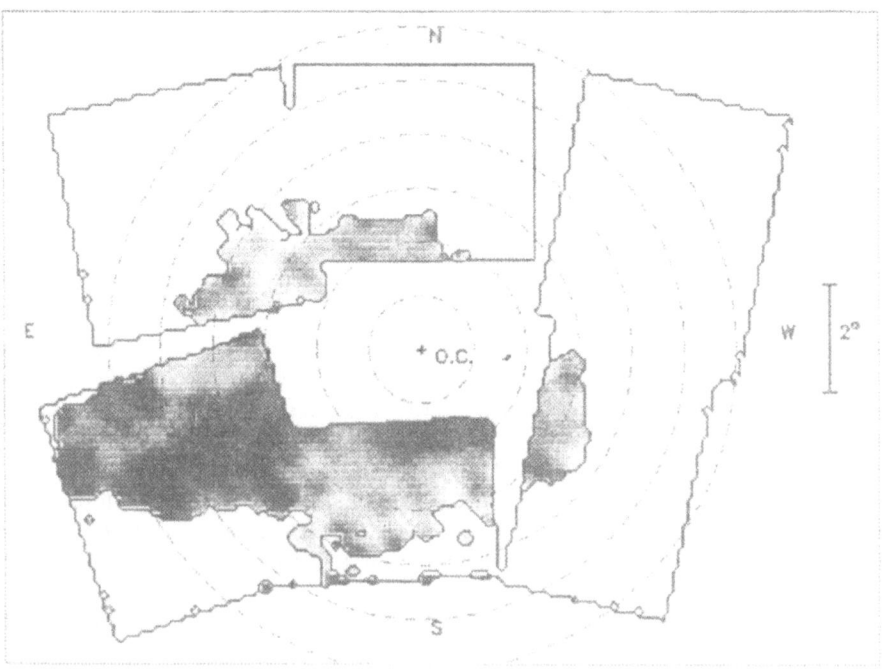

Fig. 5. Age distribution contour map for main sequence populations. The ratio of main sequence stars with $16.5 < R < 17.5$ to stars with $18.5 < R < 19.5$ is plotted using greyscale representation. This ratio is related to the relative contribution of very young (< 0.3 Gyr) main sequence populations defined by $16.5 < R < 17.5$ to the total population younger than about 1.3 Gyr (according to the Yale isochrones). The areas shaded darkest correspond to areas dominated by very young stars.

isochrones), with the areas with the largest contribution shaded darkest. The Wing region obviously consists almost entirely of very young stars while the NE and SSE outlying areas have an older average age than the Wing and the more central area. The fact that these outlying areas have an older age composition with respect to the inner areas indicates that the presence of stars younger than 2 Gyr in these areas did not result from tidal forces pulling young stars outwards from the central regions. It is probable that external interactions have triggered star formation in these areas — ram pressure between the gaseous halos of the Magellanic Clouds may have led to condensations in the H I distribution creating new sites of star formation — but processes internal to the SMC cannot yet be discounted, since dwarf irregular galaxies often show signs of irregular star formation patterns (e.g. the infrared study of dwarf galaxies in the Virgo cluster by James 1991).

In conclusion, the large areal coverage obtained by combining the use of automated plate measuring machines and high-quality photographic material have enabled the systematic investigation of the structure and stellar content of the SMC outer regions, greatly enhancing our understanding of our second nearest galaxian neighbour.

References

Azzopardi, M., Rebeirot, E., 1991. *IAU Symp. No. 148*, p. 71, eds. Haynes, R., Milne, D., Kluwer, Dordrecht, Holland.

Feast, M.W., 1989. *Recent Developments of Magellanic Cloud Research*, p. 75, eds. de Boer, K.S., Spite, F., Stasinka, G., Observatoire de Paris.

Fujimoto, M., Murai, T., 1984. *IAU Symp. No. 108*, p. 115, eds. van den Berge, S., de Boer, K.S., Reidel, Dordrecht, Holland.

Gardiner, L.T., Hatzidimitriou, D.H., 1991. *Mon. Not. R. Astron. Soc.* In preparation.

Gardiner, L.T., Hawkins, M.R.S., 1991. *Mon. Not. R. Astron. Soc.* In press.

Hardy, E., Suntzeff, N.B., Azzopardi, M., 1989. *Astrophys. J.*, **344**, 210.

Hatzidimitriou, D., Cannon, R.D., Hawkins, M.R.S., 1991a. *IAU Symp. No. 148*, p. 107, eds. R. Haynes, D. Milne, Kluwer, Dordrecht, Holland.

Hatzidimitriou, D., Cannon, R.D., Hawkins, M.R.S., Teo, A.C.Y., 1991b. In preparation.

Hatzidimitriou, D., Hawkins, M.R.S., 1989. *Mon. Not. R. Astron. Soc.*, **241**, 667.

Hatzidimitriou, D., Hawkins, M.R.S., Gyldenkerne, K., 1989. *Mon. Not. R. Astron. Soc.*, **241**, 645.

Irwin, M.J., 1991. *IAU Symp. No. 148*, p. 453, eds. Haynes, R., Milne, D., Kluwer, Dordrecht, Holland.

Irwin, M.J., Demers, S., Kunkel, W.E., 1990. *Astron. J.*, **99**,191.

James, P., 1991. *Mon. Not. R. Astron. Soc.*, **250**, 544.

Mathewson, D.S., Ford, V.L., 1984. *IAU Symp. No. 108*, p. 125, eds. van den Bergh, S., de Boer, K.S., Reidel, Dordrecht, Holland.

Murai, T., Fujimoto, M., 1980. *Publs. Astr. Soc. Japan*, **32**, 581.

Seidel, E., Da Costa, G.S., Demarque, P., 1987. *Astrophys. J.*, **313**, 192.

de Vaucouleurs, G., Freeman, K.C., 1972. *Vistas Astr.*, **14**, 163.

Discussion

Humphreys :

1. Did you have a control field for determining the foreground contribution?

2. Have you determined the luminosity function?

Gardiner :

1. The control areas used for the determination of the foreground contribution lay at distances of $7° - 8°$ from the SMC centre in our existing fields where the numbers of SMC stars are very small, leaving no visible signs on the colour-magnitude diagram. The method we use to determine the clump luminosity dispersion includes an offset parameter which renders the results largely insensitive to small fluctuations in the foreground number counts.

2. Luminosity functions were determined for several regions and are due to be published in a forthcoming paper by Gardiner & Hatzidimitriou.

Odewahn :

It is common for morphological features in paired Magellanic type galaxies to be present in both systems. A prominent outer ring of late type population I stars is present in the LMC. Do you find evidence for such a structure in the SMC?

Gardiner :

From our surface count data we do not find features of this kind for the SMC.

Cannon :

How long will a feature like the SMC Wing last, before it becomes smeared out more or less uniformly around the SMC?

Gardiner :

I do not know the answer to this question precisely, but it appears that a non-uniform distribution can last at least 1 Gyr from the observed asymmetry in the distribution of the < 1 Gyr main sequence.

Raychaudhury :

Your surface density-age distribution appears very interesting, and seems to have interesting dynamical consequences. How does your plot correlate with similar plots that can be made with the ages of other tracers, eg. globular clusters/carbon stars?

Gardiner :

The number of studied clusters lying beyond 2° from the optical centre is very small, so a meaningful comparison is difficult to make. Young objects (cepheid variables, blue and red supergiants, small nebulae) are concentrated in the bar of the SMC, while we do observe a bar-like distribution of < 0.2 Gyr main-sequence stars. The more dispersed distribution of carbon stars compares well with the distribution of the largely intermediate age HB/clump population.

THE VARIABLE STAR SYSTEM OF THE SCULPTOR DWARF GALAXY

C. Goldsmith

Department of Physics, University of York, 4700 Keele Street, North York, Ontario M3J 1P3, Canada.

1 Introduction

The Sculptor dwarf galaxy, discovered by Shapley in 1939 (Shapley 1939), contains the richest collection of variable stars of any dwarf galaxy in the Local Group. At present, there are 603 known variables in Sculptor (Van Agt 1978). This makes Sculptor the ideal 'laboratory' for the study of variable stars in population II systems.

This project has several goals:

- Confirm the variability of, and determine precise periods for, the known variable stars in Sculptor;
- search for new variable stars. Counting statistics predict that there are ~1050 variables in Sculptor;
- search for double-mode RR-Lyrae stars;
- determine the period-frequency distribution for Sculptor.

The current status of the project is described herein.

2 Data and Photometry

The data came from the following sources: Thirty-two 20″ plates of Sculptor were obtained in 1985 and 1986 using the 100″ Dupont telescope at Las Campanas. These plates have been digitized with the COSMOS facility of the ROE. For the purpose of photometric calibration, images of several SAAO E regions were obtained with the 1.5m telescope at Cerro Tololo. Various regions of the Sculptor field were also imaged with the same instrument.

The photometric reductions were carried out using the technique of Harris et al. (1981). From this, σ_{residual} for the standard stars was 0.01 in V and 0.02 in B-V. A

H. T. MacGillivray and E. B. Thomson (eds.), Digitised Optical Sky Surveys 291–294.

second order polynomial fitting function was then used to calibrate the COSMOS magnitudes of the program stars to B in this photometric system. The mean scatter about these fits for all plates was 0.1 magnitudes in B.

3 Search and Discovery

Mean B magnitudes were determined for every object which appeared on 25 or more of the 32 plates. Standard deviations for these means were also calculated. Figure 1 is a plot of σ_B vs B for the south-east quadrant of Sculptor. The presence of RR-Lyrae stars is confirmed by the existence of the 'plume' centered at B=20.5 in the contour plot. RR-Lyrae candidates were chosen for further study based on their position in this plot. All members of the 'plume' were chosen, resulting in 319 candidates for the SE quadrant. Additional candidates were chosen from the region $17 < B < 19.5$ and $\sigma_B > 0.15$ in order to look for stars in the BL-Herculis or W-Virginis class of variables. The mean B magnitude for the RR-Lyrae of 20.5 combined with an assumed $M_B = 1.0$ (Hoffmeister et al. 1985) gives an approximate distance modulus of 19.5, or d~79 kpc.

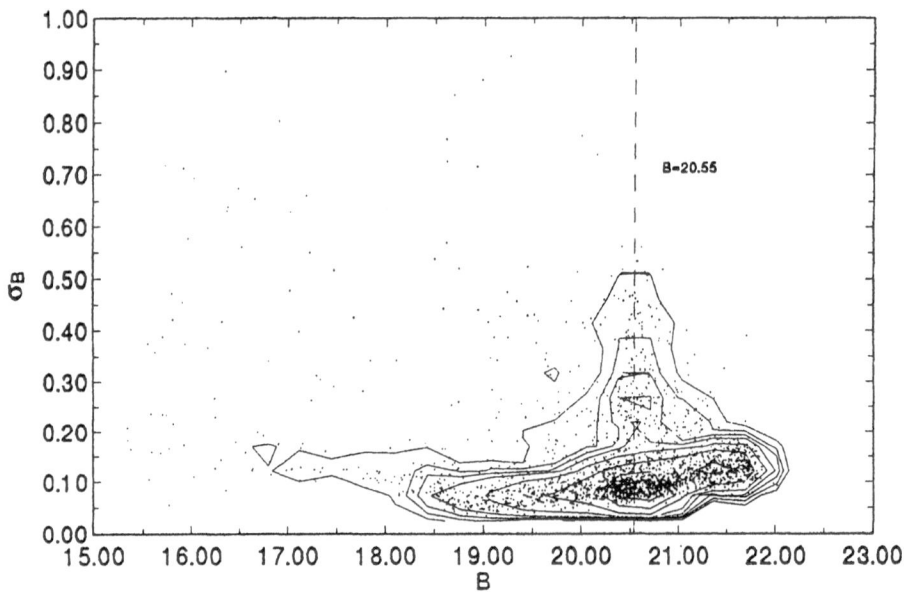

Fig. 1. Magnitude error distribution.

4 Analysis

Period searches are being carried out using the modified fouriergram procedure of Scargle (1982). This technique corrects the traditional method's inability to handle unevenly spaced data. It has been shown that the Fourier technique is the best choice for the analysis of periodic data (Swingler 1989).

Figure 2 shows a typical 'periodogram' produced by the period finding software. The most likely period present in the data occurs where the periodogram power is greatest, in this case at 0.517 days. Figure 3 shows the lightcurve for the same star. Va008 (Van Agt variable #8) is a type ab RR-Lyrae star displaying all of the characteristics of this type (amplitude ~ 1 mag, period ~ 0.5 days).

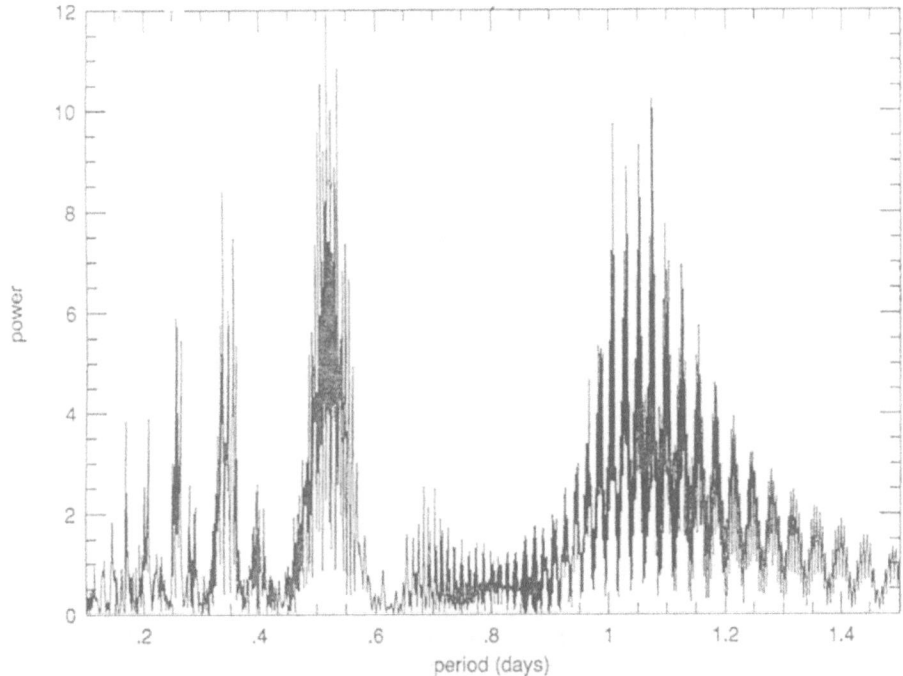

Fig. 2. va008 periodogram.

5 Future Work

Analysis of the variables in Sculptor is ongoing, with the following still to be done:

- A search for double-mode RR-Lyrae. This will involve the 'pre-whitening' (removal of periods from the data) of variables whose primary period falls in the expected range of 0.35 - 0.45 days. Searches for secondary periods will then be made in the range 0.5 to 0.6 days.
- Derivation of the period-frequency histogram for Sculptor. This will indicate how, if at all, Sculptor fits into the Oosterhoff dichotomy (apparent division of mean P_{ab} into two groups; centered at $P_{ab} = 0.55d$ and $P_{ab} = 0.65d$ respectively). It appears that Sculptor *may* belong to the type II (short period) group.

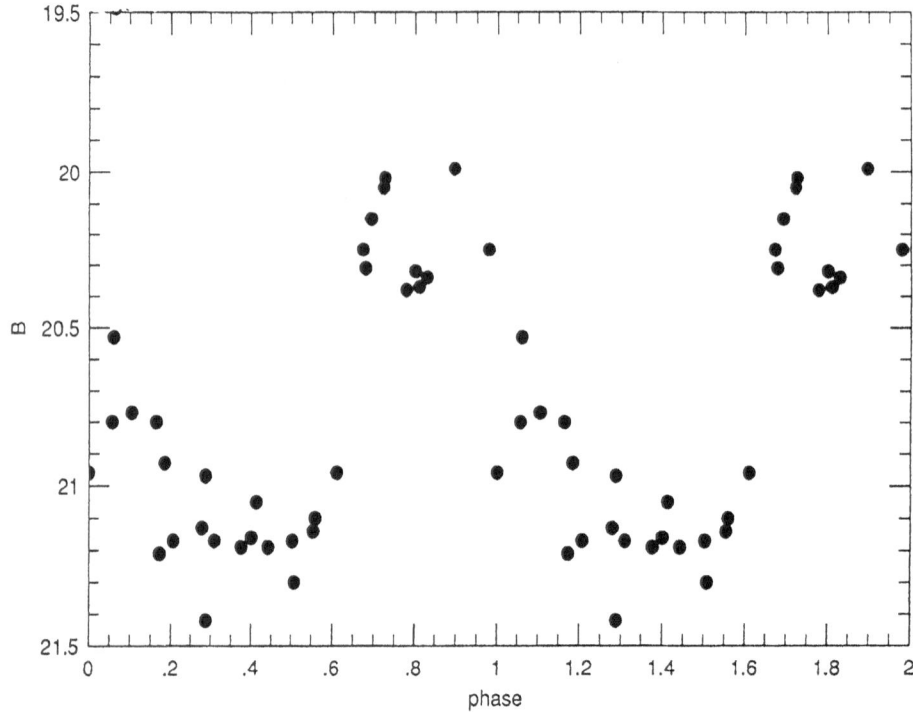

Fig. 3. va008 light curve.

6 Acknowledgements

I wish to thank Dr. J. Nemec and Dr. N. Suntzeff for their initiation of the project in 1985 and for their collection of the plate data. I would also like to thank Dr. K. Innanen for his continued supervision and Dr. H. MacGillivray for overseeing the COSMOS scanning of the plates.

References

Harris, W.E., FitzGerald, M.P., Reed, B.C., 1981. *Publ. Astron. Soc. Pacific*, **93**, 507.

Hoffmeister, C., Richter, G., Wenzel, W., 1985. *Variable Stars* (Berlin Heidelberg: Springer-Verlag).

Scargle, J.D., 1982. *Astrophys. J.*, **263**, 835.

Shapley, H., 1939. *Proc. Natl. Acad. Sci. U.S.*, **25**, 565.

Swingler, D.N., 1989. *Astron. J.*, **97**, 280.

Van Agt, S., 1978. *Publ. David Dunlap Obs.*, **3**, 7.

NORMAL SURFACE BRIGHTNESS DWARF GALAXIES IN THE FORNAX CLUSTER

S. Phillipps and J. Davies

Department of Physics, University of Wales College of Cardiff, P.O. Box 913, Cardiff CF1 3TH, Wales, U.K.

1 Introduction

There are strong selection effects in favour of selecting galaxies of certain surface brightnesses (Disney & Phillipps 1983; Davies 1990), and against finding galaxies of other surface brightnesses. Both low and high surface brightness objects of a given total luminosity will have faint isophotal magnitudes and/or small isophotal sizes. This means that low surface brightness galaxies, in particular, can generally only be seen if they are quite nearby. Thus if we look in the direction of a cluster, almost all the low surface brightness galaxies we see will be genuine cluster members (Phillipps, these proceedings). 'Normal' surface brightness galaxies, on the other hand, are easy to see out to great distances, so they swamp other types of galaxy in any magnitude- or size-limited catalogue (which is why these surface brightnesses are considered 'normal', of course). Thus even if we look in a cluster area, apparently faint normal surface brightness objects will be mostly background objects. It is therefore still controversial whether intrinsically low luminosity but normal surface brightness dwarf galaxies exist at all (e.g. Davies et al. 1988; Ferguson & Sandage 1988; Irwin et al. 1990).

2 The Data

We have therefore selected a set of normal surface brightness yet apparently faint galaxies from the APM scans of UKST plates used in our overall survey of the Fornax Cluster (Phillipps, these proceedings). The galaxies chosen have intensity profiles which are reasonably well fitted by exponentials and have extrapolated central surface brightnesses between 21 and 23 B magnitudes per square arc second (Bμ). Magnitudes calculated from the exponential fits are 16.5 to 19.5. We then observed two Fornax fields with the AUTOFIB multi-object fibre system on the AAT, obtaining a total of 64 redshifts.

H. T. MacGillivray and E. B. Thomson (eds.), Digitised Optical Sky Surveys 295–298.
© 1992 *Kluwer Academic Publishers.*

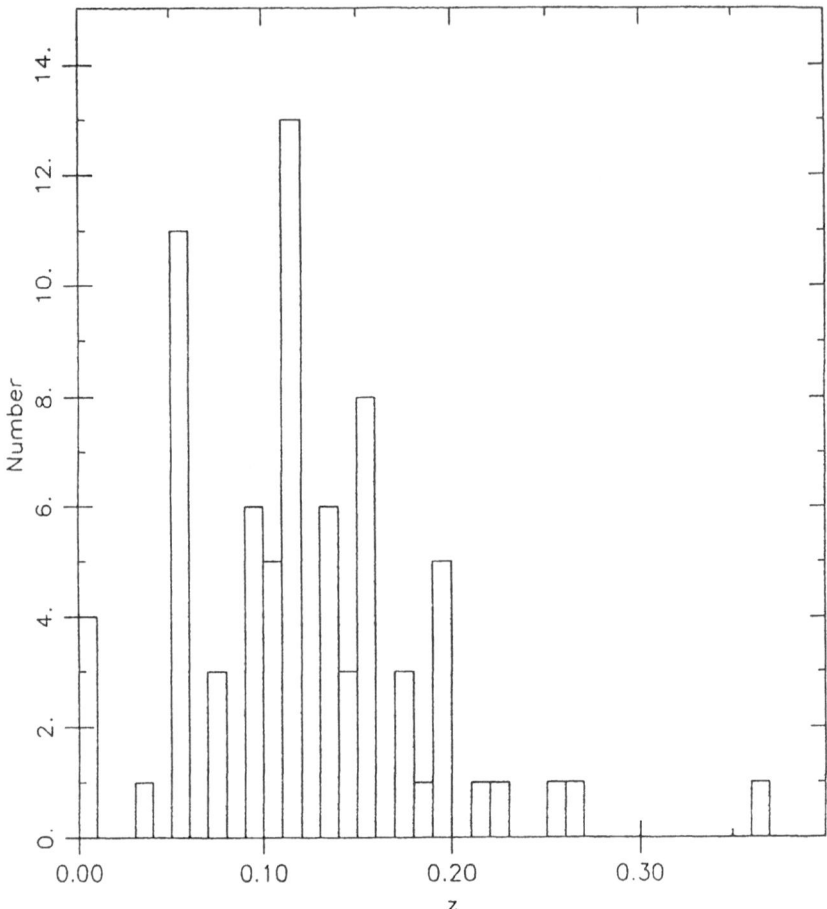

Fig. 1. Redshift distribution for the sample galaxies. Note particularly the four low redshift cluster members around ($z < 0.01$). Note also the redshift 'spikes' around $z = 0.055$ and $z = 0.115$ among the background galaxies. These correspond remarkably well to the 'spikes' seen in redshift surveys at the SGP (35 degrees away) by Broadhurst et al. (these proceedings). If the structures are continuous, their implied sizes are over 100 and 200 Mpc respectively.

3 Results

The results are shown in Figs. 1 and 2. Four of the objects are found to have redshifts indicative of cluster membership. They have surface brightnesses of 21.8 to 22.1 Bμ, quite typical of many giant spirals or bright Magellanic irregulars, yet have total magnitudes around -13.5 to -11.5 (assuming a Fornax distance modulus of 31.0). (In one case, the object may possibly be a bright patch in a somewhat larger irregular diffuse galaxy, though). Note that these objects lie a long way off the

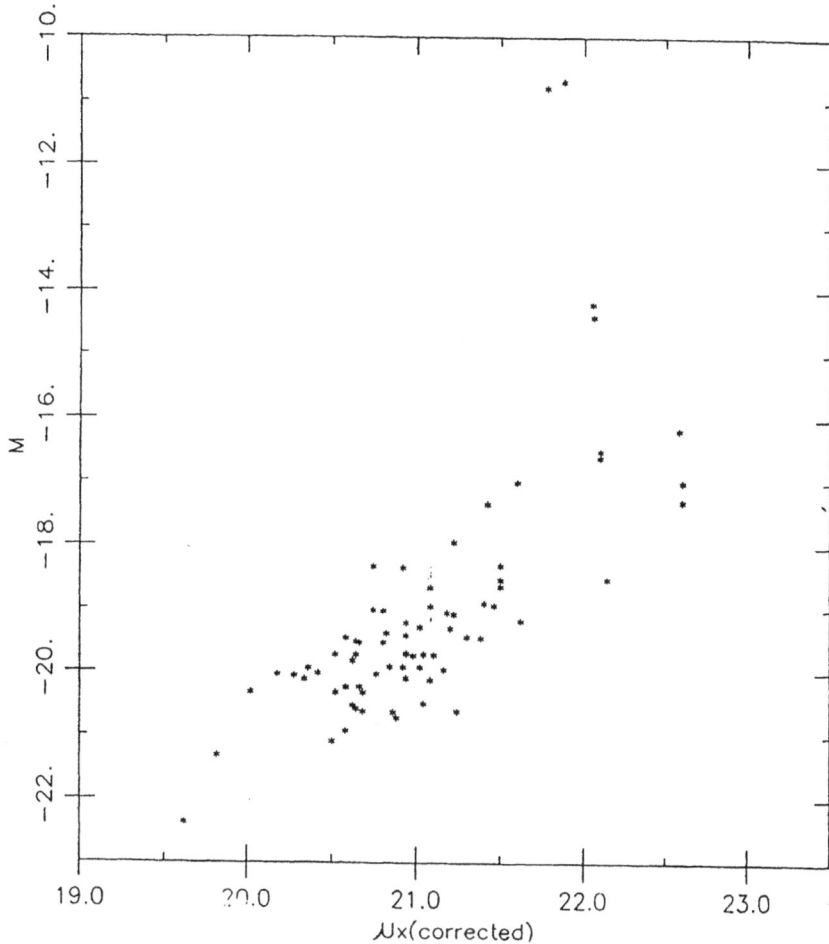

Fig. 2. Intrinsic surface brightness — absolute magnitude plot for the sample galaxies. The cluster galaxies are the four objects fainter than $m = -15$. Note the spurious surface brightness luminosity relation for the background galaxies induced by distance effects.

canonical luminosity-surface brightness relation advocated by Bingelli et al. (1984) and others, which would suggest absolute magnitudes (for our distance scale) of about -15.5 at these surface brightnesses. Note too that the background galaxies do show a clear luminosity-surface brightness relation, but this is entirely spurious and arises from the larger cosmological and k-corrections to the surface brightnesses of the more distant, and hence intrinsically more luminous, objects (since all galaxies in the sample have roughly the same apparent surface brightness).

References

Binggelli, B., Sandage, A., Tarenghi, M., 1984. *Astron. J.*, **89**, 64.

Davies, J.I., 1990. *Mon. Not. R. Astron. Soc.*, **244**, 8.

Davies, J.I., Phillipps, S., Cawson, M.G.M., Disney, M.J., Kibblewhite, E.J., 1988. *Mon. Not. R. Astron. Soc.*, **232**, 239.

Disney, M.J., Phillipps, S., 1983. *Mon. Not. R. Astron. Soc.*, **205**, 1253.

Ferguson, H., Sandage, A., 1988. *Astron. J.*, **96**, 1520.

Irwin, M.J., Davies, J.I., Disney, M.J., Phillipps, S., 1990. *Mon. Not. R. Astron. Soc.*, **245**, 289.

SURVEYS FOR LOW SURFACE BRIGHTNESS GALAXIES

S. Phillipps

Department of Physics, University of Wales College of Cardiff, P.O. Box 913,
Cardiff CF1 3TH, Wales, U.K.

1 The Survey

In recent years it has become apparent that there exist substantial numbers of Low
Surface Brightness Galaxies (LSBGs) in clusters. These were not seen in earlier
surveys because of technical limitations and severe selection biases. A group of us
at Cardiff (Phillipps, Disney, Davies, Evans) and the APM group at Cambridge
(Kibblewhite, Cawson, Irwin) have therefore attempted to make the first objective
(machine selected) survey for LSBGs by using the APM machine to scan UKST
plates of cluster areas, primarily the Fornax Cluster area. As mentioned elsewhere
in these proceedings, the advantage of using measuring machine data is not that
selection effects can be avoided, but that they can be quantified. This point is
particularly important for LSBG samples.

Technically, the Fornax survey involves several different, though linked, stages.
Images are detected in the usual way as sets of connected pixels above some threshold (for our 'standard' data set this was set at a limit corresponding to 25.5 B
magnitudes per square arc second, henceforth Bμ). This then provides a parameterized image list. All images brighter than a convenient limit in integrated intensity
(corresponding to B = 19.1) were then raster scanned individually using 0.5 arc second pixels (as in the original images scan) to give digitised maps. Calibration was
done directly on the maps by comparison with absolutely calibrated CCD images of
a number of the brighter galaxies. Radial intensity profiles are obtained by ellipse
fitting using the GASP package. The final surface brightness accuracy is 0.1 magnitudes over the range 22.5 to 26.5 Bμ. LSBGs were then defined as galaxies with
approximately exponential intensity profiles (this removed a minority of galaxies
with 'curved' surface brightness profiles suggestive of background giant ellipticals
or cluster compact M32-like dwarfs) and with extrapolated central surface brightnesses, from an exponential fit, fainter than 22.5 Bμ ('standard' disc galaxies are
claimed to have a narrow range of surface brightnesses around 21.5 Bμ). We are left

299

H. T. MacGillivray and E. B. Thomson (eds.), Digitised Optical Sky Surveys 299–303.
© 1992 *Kluwer Academic Publishers.*

with around 250 LSBGs, the most diffuse of which have central surface brightnesses around 24.5 Bμ (see Davies et al. 1988).

To go deeper still, we have obtained data at coarser resolution (2″ pixels). This involved rescanning the plate and making four separate 4000 × 4000 pixel maps, which cover approximately the same area as the original scans. This raw pixel data was then median filtered to an effective resolution of 8″. The main advantage of the median filter is that it significantly reduces the noise without blurring the images, and it also effectively removes small contaminating star images. The low background noise so obtained then allows us to run the image detection algorithm with a very low intensity threshold (27.3 Bμ). Selecting images with a given minimum size (diameter 16″) then gives us our very low surface brightness (VLSBG) sample. The most extreme of these (e.g. Fig. 1) have central surface brightnesses around 26.5 Bμ, the same as the classic giant LSBG Malin 1 (Irwin et al. 1990).

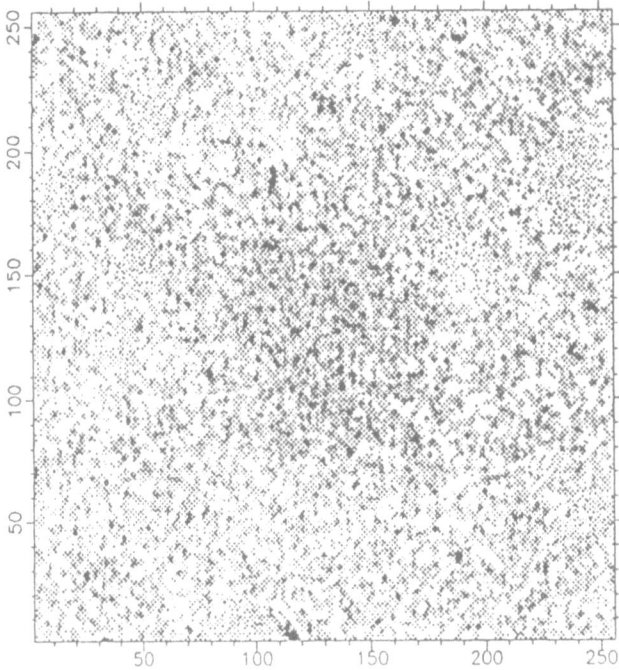

Fig. 1. Very low surface brightness galaxy in Fornax with central surface brightness close to 26 Bμ.

2 Photometric Parameters

The combined data set contains about 320 galaxies. Because of strong selection effects against finding LSBGs at large distances (they have small isophotal sizes), the majority will be genuine cluster members, not background objects. This is confirmed by their spatial distribution about the cluster centre. Note that the lowest SB objects have the most compact distribution. The overall sample confirms and extends results found from an earlier sample.

1. After allowing for selection effects (it is clear that at each SB there is a smallest exponential scale size which allows a galaxy to exceed the magnitude or size limit of a sample), the number of galaxies per magnitude interval in SB is roughly constant (see Fig. 2), or in terms of intensities $n(I) \propto I^{-1}$. There is no sign of a cutoff at the lowest SBs (and CCD surveys confirm the existence of yet more diffuse objects).
2. There is no correlation between scale size and SB except that forced by the selection limits (Fig. 3).
3. There is a maximum scale size for LSBGs (about 2kpc) and below this numbers increase as $n(a) \propto a^{-2}$. Together with the distribution of central intensities, this implies that LSBGs have a steep luminosity function with $n(L) \propto L^{-3/2}$.

Note that large LSBGs dominate the covering factor in the cluster while small LSBGs completely dominate the galaxy numbers.

Additional current and planned surveys are aimed at determining whether these conclusions still hold in different types of cluster (Fornax is a compact cluster dominated by early type galaxies) and in the general field.

3 LSBG Colours

In an additional study (Evans et al. 1990) we have scanned corresponding red plates to determine galaxy colours. Unsurprisingly, galaxies which we classified as dIs because of their irregular appearance or profiles turned out to be bluer than the objects of smooth appearance which we classified as dEs. In addition, nucleated dEs are redder than ordinary dEs at the same magnitude. There are reasonably clear trends for fainter dIs to be bluer but fainter dEs to be redder. There is a marginal trend for lower SB objects to be redder but the interpretation of the data in terms of a simple fading model is not obvious.

References

Davies, J.I., Phillipps, S., Cawson, M.G.M., Disney, M.J., Kibblewhite, E.J., 1988. *Mon. Not. R. Astron. Soc.*, **232**, 289.
Evans, Rh., Davies, J.I., Phillipps, S., 1990. *Mon. Not. R. Astron. Soc.*, **245**, 164.
Irwin, M.J., Davies, J.I., Disney, M.J., Phillipps, S., 1990. *Mon. Not. R. Astron. Soc.*, **245**, 289.

Discussion

Cannon :
  What fraction of your dEs are nucleated dENs?

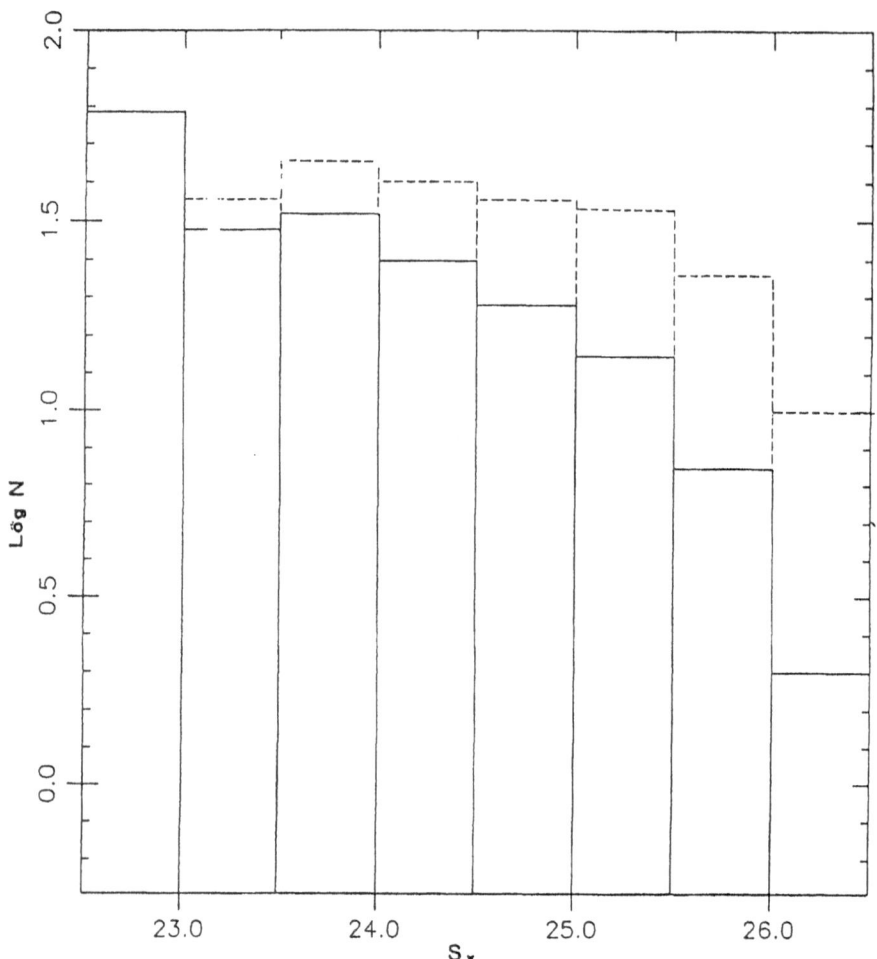

Fig. 2. Numbers of LSBGs of different central surface brightness as observed (solid line) and as corrected for selection limits (dashed line).

Phillipps :

The APM profiles are not particularly suitable for finding nuclei. The standard result is that perhaps half the higher luminosity dEs are nucleated, with the fraction decreasing as you go fainter.

Hawkins :

How would a galaxy like the Carina dwarf fit into your sample?

Phillipps :

Carina would have a central surface brightness even fainter than any of our sample, around 28 Bμ I think. It's a good example of only being able to see very low surface brightness galaxies if they are very nearby.

Hambly :

Naive question: is there significant mass in LSBGs/VLSBGs — i.e. could missing mass in galaxy clusters be partly accounted for by these objects?

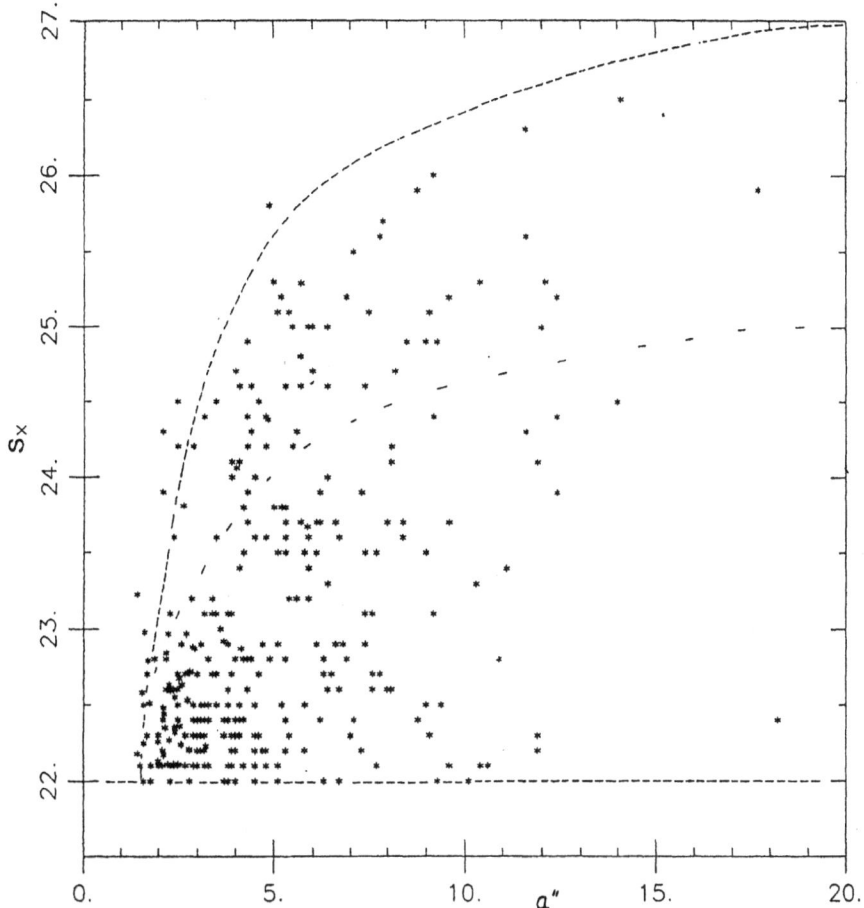

Fig. 3. Distribution of LSBGs in scale size and central surface brightness showing the selection limits of the samples. (Inner curve is for the standard sample, outer curve for the median filtered sample; see text for details).

Phillipps :
If they have 'normal' mass-to-light ratios they probably contribute a few percent, perhaps 10%. If they have high M/L then you can get any number you like!

Raychaudhury :
Does the distribution of low surface brightness galaxies in Fornax follow the underlying distribution of the normal spirals? Is there any indication as to whether the LSB galaxies are bound to the cluster in general or to the individual bright galaxies?

Phillipps :
The low surface brightness galaxies do seem to be distributed throughout the cluster like the 'normal' galaxies, though the very diffuse objects are more concentrated towards the centre. The LSBGs do not seem to be associated with individual bright cluster members.

THE APM BRIGHT GALAXY SURVEYS

S. Raychaudhury

Institute of Astronomy, University of Cambridge, Madingley Road, Cambridge
CB3 0HA, U.K.

Abstract

Three surveys of galaxies brighter than $m_B \simeq 17^m$, together covering more
than a fifth of the sky, are described. In each of these surveys, the Automated Photographic Measuring (APM) facility in Cambridge has been
used to identify and measure images on UKST IIIa-J Sky Survey plates.
All galaxy candidates have subsequently been checked by eye and morphological types assigned. Magnitudes and diameters have been calibrated using
overlap matching and CCD photometry. The three surveys have been used
for scientific projects ranging from the study of large-scale distribution of
galaxies, the determination of the angular and spatial correlation function,
the investigation of large-scale motions in the Local Universe (the 'Great
Attractor'), studies of clusters of galaxies and a compilation of galaxies in
the 'missing strip' between the boundaries of the UGC and ESO catalogues.

1 Motivation and Strategy

In recent years, the use of automated star/galaxy separation algorithms has been
very successful in producing wide-field surveys of galaxies as faint as $m_B = 20.5$
from machine-measured scans of Sky Survey Schmidt plates (see Table 1). Such algorithms produce $\sim$95% complete catalogues of galaxies with $\sim$10–15% stellar contamination in the magnitude range m_B=17.0–20.5. However, for galaxies brighter
than m_B=17, such automated algorithms, which rely principally on image profiles,
turn out to be inefficient (on the APM facility) since the images of stars of comparable magnitude are saturated on Sky Survey plates, and therefore lead to $\gtrsim$50%
stellar contamination.

We therefore decided to use a combination of automated algorithms and human
intervention while preparing catalogues of galaxies brighter than m_B=17 (which we
will refer to as 'bright galaxies' in this paper). The APM facility in Cambridge was

H. T. MacGillivray and E. B. Thomson (eds.), Digitised Optical Sky Surveys 305–311.

Table 1. Wide-Field galaxy surveys: past and present.

Catalogue	Year of Completion	Ω (Sr)	N	$m_{\lim}$ (B/b_J)	$D_{\lim}$ (')	Ref
Shapley-Ames	1932	4π	1246	13.4		1
UGC	1968	$4\pi/2.1$	12921		1	2
SRCBG	1976	$4\pi/2.8$	5481		2.5	3
ESO/Uppsala	1982	$4\pi/2.9$	15455		1	4
APM Bright	1989	$4\pi/9.4$	14681	16.4		5
Hydra-Centaurus	1989	$4\pi/15.2$	19272	17.0		6
Equatorial	1991	$4\pi/8.3$	~ 40000	17.0	0.4	7
APM Faint	1988	$4\pi/9.4$	$\sim 2,000,000$	20.5		8
Caltech	1990	$\sim 4\pi/100$	$\sim 150,000$	$19.0(R_F)$		9
EDSGC	1990	$\sim 4\pi/30$	$\sim 1,500,000$	20.5		10

The columns refer to:

Ω: the solid angle covered by the catalogue in radians; N: the total number of galaxies in the catalogue down to the magnitude limit $m_{\lim}$ and/or diameter limit $D_{\lim}$.

References: (1) Sandage & Tammann 1987; (2) Nilson 1973; (3) de Vaucouleurs, G., de Vaucoulours, A. & Corwin, H,G., 1976; (4) Lauberts 1982; (5) Loveday 1989 (APM); (6) Raychaudhury 1990 (APM); (7) Raychaudhury, Scharf, Lynden-Bell, Sutherland, Maddox & Hudson, ongoing (APM); (8) Maddox 1988, Maddox et al. 1990 (APM); (9) Picard 1991 (COSMOS, POSS II); (10) Nichol et al. this conference (COSMOS).

used to generate a conservative list of 'galaxy candidates' from scans of UKST Sky Survey plates, using a preliminary star/galaxy separation technique. Each candidate was subsequently inspected on film copies and the identified galaxies were assigned a rough morphological classification. The magnitudes and diameters of these galaxies were converted to a consistent system by matching the overlapping edges of plates, and calibrated with the help of CCD photometry.

In this paper, I will describe the three major APM 'bright galaxy' surveys that have been carried out in the last five years. Table 1 compares these surveys with a few other surveys of brighter galaxies (where galaxies were found and measured manually) and of fainter galaxies (compiled from APM/COSMOS scans using automated techniques).

2 The SGP Bright Galaxy Survey

The SGP bright galaxy survey (Loveday 1989) serves as the bright end of the Maddox et al. (1990) SGP survey, and was carried out in the semi-automated fashion described above. The survey area of 4300 deg^2 is approximately defined in Equatorial coordinates by $21^h \lesssim \alpha \lesssim 5^h$, $-72°.5 \lesssim \delta \lesssim -17°.5$, and consists of $\sim$15 000 galaxies complete to $b_j = 16.4$, all of which have been inspected by eye and assigned morphological types. The magnitude system was calibrated using CCD photometry for 252 galaxies obtained at Mount Stromlo.

Loveday et al. (1991) have completed a redshift survey of a 1 in 20 randomly-selected subsample of galaxies brighter than $b_j = 17$ from the SGP bright and faint catalogues. Their sample of 1440 galaxies covers a volume that is 30 times larger than the first CfA ($m_B \leq 14.5$) survey, and provides a very good determination of the luminosity function and mean space density of galaxies ($M_{b_j} \leq -15$), and the spatial two-point correlation function.

3 The 'Great Attractor' Survey in Hydra-Centaurus-Corvus

The discovery of better secondary distance indicators to galaxies have in recent years resulted in large surveys of galaxies designed to find significant departures from the Hubble Flow in the Local Universe. One such peculiar motion survey (Lynden-Bell et al. 1988) found that the Local Group of galaxies is streaming towards the constellations of Hydra-Centaurus at a speed of $\sim$600 km s over and above the Hubble Flow. It has been suggested that the cause might be a 'Great Attractor', a large conglomeration of galaxies in that direction at a distance corresponding to $\sim$4200 km s in the Hubble Flow. Maps of the sky drawn from existing catalogues of galaxies were found not to be deep enough to reveal the presence of a significant concentration of galaxies at that distance.

Therefore, a catalogue of about 19 000 galaxies brighter than $b_j = 17^m$ (for which positions, diameters and magnitudes have been measured) was compiled, using APM scans of 110 UKST IIIa-J plates (Raychaudhury 1990). The magnitude system has been calibrated using CCD photometry obtained at the SAAO and VBO, using 51 galaxies. The scatter about the linear APM–b_j magnitude relation was $0^m.19$. We have obtained more calibration photometry at CTIO and SAAO in April this year, and are currently trying to improve the magnitude calibration.

Figure 1 shows the distribution on the sky of the galaxies belonging to this 'Hydra-Centaurus' catalogue. The general overdensity of galaxies in this region was estimated by analyzing the number counts of galaxies as functions of diameter and magnitude. Simple cluster-finding algorithms were used to identify clusters of galaxies in the Hydra-Centaurus catalogue. From redshifts collected from the literature, these clusters could be shown (Raychaudhury 1989) to constitute two distinct concentrations, one closer than 5000 km s (the 'Hydra-Centaurus Supercluster') and the other between 10000–16000 km s (Fig. 2). This further concentration was named the 'Shapley Supercluster', in honour of Harlow Shapley, who in the early 1930s had noted the existence of this 'remarkable concentration' in his faint galaxy survey in the southern Hemisphere.

The analysis of the distribution of Abell clusters in the sky, together with the X-ray luminosities of the component clusters and the galaxy counts in this region of the sky (Raychaudhury et al. 1991) seem to indicate that the 'Shapley Supercluster' is the most massive galaxy concentration known, and represents a remarkable over-density in terms of the primordial fluctuation that could have led to its formation.

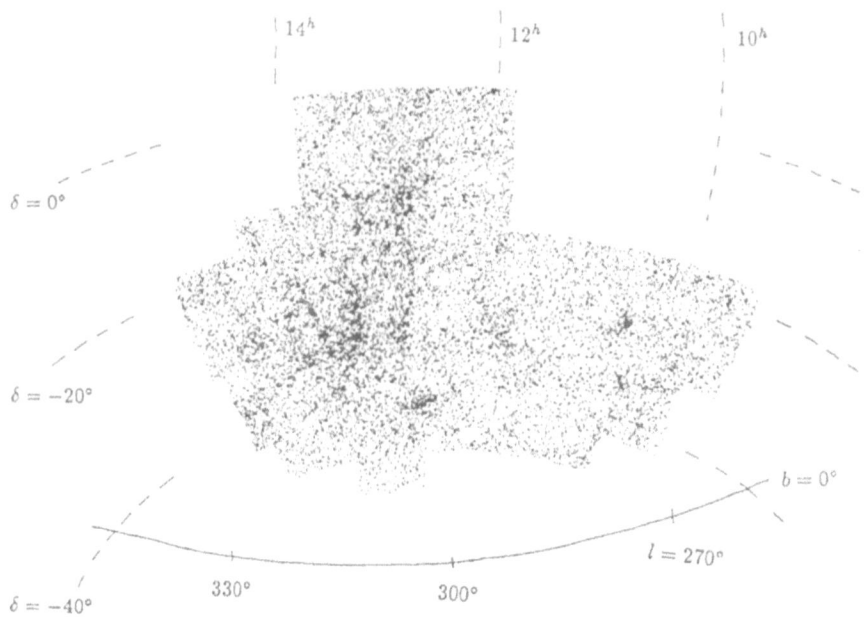

Fig. 1. An equal-area plot of the $\sim$ 19 000 galaxies in the Hydra-Centaurus catalogue down to $b_j = 17^m$. Each point represents a galaxy. An R.A., Dec grid is superposed, along with some Galactic coordinates.

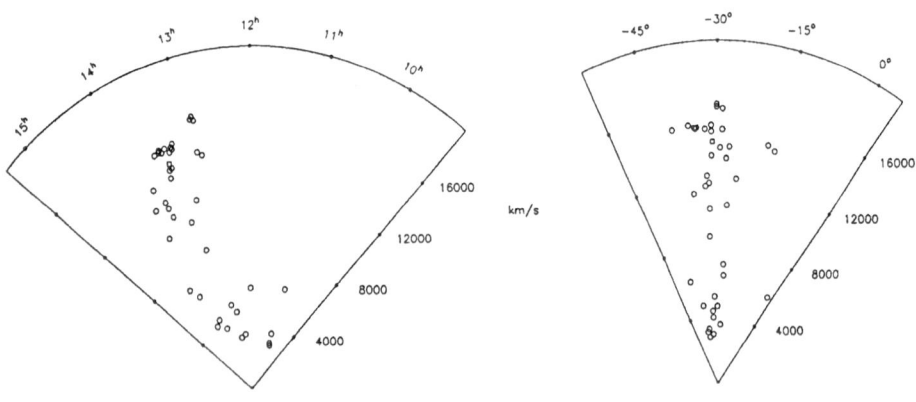

Fig. 2. Galaxy clusters with known redshifts in the direction of Hydra-Centaurus, showing the position on the sky [(a) Right Ascension, (b) Declination] against mean redshift. The plots clearly show the existence of two distinct groups of clusters: the nearer group, between 2800–5000 km s, is known as the 'Hydra-Centaurus Supercluster', and the further group, between 11 000 $< v <$ 16 000 km s is the 'Shapley Supercluster'.

In spite of this, it seems that it is just too far away to be the 'Great Attractor', and cannot account for more than ~10–15% of the peculiar motion of the Local Group. If there has to be a 'Great Attractor' in the form of a large visible concentration of galaxies, it has to be the well-known nearby Hydra-Centaurus Supercluster (Raychaudhury 1989). On the other hand, similar deeper surveys in other parts of the sky could be used to estimate the contribution of other concentrations like the Perseus-Pisces Supercluster to the local peculiar velocity field.

The Hydra-Centaurus catalogue is currently being used as the source of a wide-field redshift survey in the direction of the centre of the 'Great Attractor', and in studies of the luminosity function and distance measures of clusters in the Hydra-Centaurus and Shapley Superclusters.

4 The APM Equatorial Survey: Filling up the 'Missing Strip'

Studies of the large-scale distribution of galaxies that require a homogenous all-sky catalogue of galaxies (e.g. determination of the optical dipole) are hindered considerably by the absence of reasonable catalogues in two substantial patches of the sky: (a) the region of the sky on and near the Galactic plane ($|b| < 15°$); and (b) the Equatorial strip ($-17°.5 < \delta < -2°.5$), a region that was not surveyed by either the ESO and/or UGC catalogues, due to the lack of Sky Survey material at the time of these surveys. While compiling catalogues near the Galactic plane remains a major problem, filling in the 'missing' Equatorial strip is now feasible with the completion of the UKST IIIa-J survey in the region.

We have used the procedure described above to compile a catalogue of all galaxies brighter than $b_j = 17^m$ and larger than $D = 0.4$ arcmin in this so-called 'missing strip' ($|b| \geq 20°$, $-17°.5 \leq \delta \leq +2°.5$), covering about 200 UKST Survey Plates (see Fig. 3). This survey has a substantial overlap with the UGC and ESO catalogues so that direct comparisons can be made. The people involved in this survey are D. Lynden-Bell, C. Scharf, M. Hudson and myself in Cambridge and S. Maddox and W. Sutherland in Oxford. This project is expected to be complete by the end of 1991, and will produce a catalogue of over 40 000 galaxies, with positions, magnitudes, diameters, position angles and morphological information for all of them. The magnitude system will be calibrated by using CCD photometry obtained at CTIO and SAAO: we have another CTIO run in October this year.

5 General Conclusions

Catalogues of galaxies brighter than $B = 17^m$ can be produced very efficiently from machine scans of Schmidt plates by a combination of automated star/galaxy separation and subsequent inspection of candidates. As a result of overlap matching and CCD calibration, the measure of the magnitude limit is objective and reliable. Therefore, such catalogues can be used very effectively for studies of large-scale structure and as sources of redshift surveys.

The two major problems of compiling such catalogues from APM scans are: (a) measures of galaxies brighter than $b_j \simeq 14.5$ and (b) images of merged objects. Since the local background is determined over pixels of size 1 arcmin^2, the

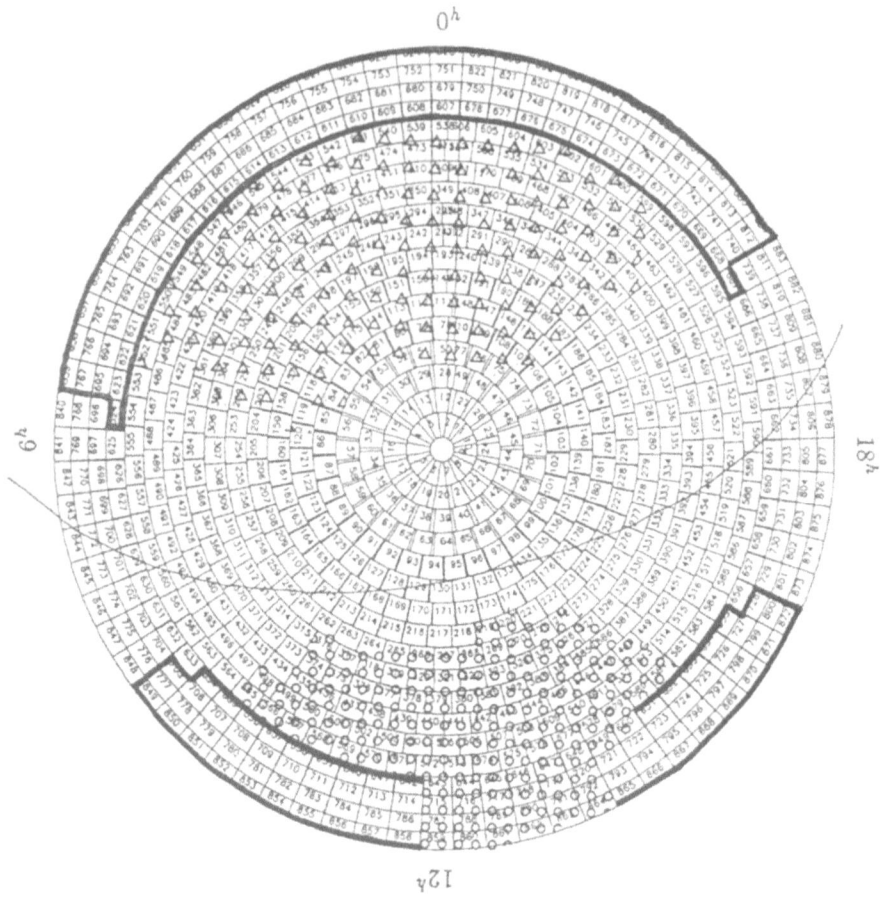

Fig. 3. All fields in the UK Schmidt survey in the southern hemisphere, plotted in equal-area projection. The shaded fields belong to the three surveys described in this paper.

background-following sky-subtraction algorithm makes measures of galaxies with major diameters $\simeq 1.5$ arcmin very unreliable. Also, we have not used any deblending software for merged images in the image detection routine. In order to produce a complete homogenous catalogue, we are in the process of scanning in raster mode all galaxies bigger than 1 arcmin and all merged galaxy images on the plates involved in the Hydra-Centaurus and Equatorial projects.

6 Acknowledgements

Thanks to the APM team, particularly Mike Irwin for help with scanning and analysis, and to Steve Maddox, Jon Loveday and Will Sutherland for loan of experience and software, and Gerry Williger for observational collaboration. Indebtedness to the UKST group at ROE, namely Sue Tritton and Mike Read, for their excellent support. The author is an SERC research assistant at the IoA and a research fellow of St Edmund's College, Cambridge.

References

de Vaucouleurs, G., de Vaucouleurs, A., Corwin, H.G., 1976. *Second Reference Catalog of Bright Galaxies*, University of Texas Press, Austin, Texas, U.S.A.

Lauberts, A., 1982. *The ESO/Uppsala Survey of the ESO(B) Atlas.* Published by ESO.

Loveday, J., 1989. Ph.D. Thesis, University of Cambridge.

Loveday, J., Peterson, B.A., Efstathiou, G., Maddox, S.J., 1991. *Astrophys. J.* Submitted.

Lynden-Bell, D. et al. 1988. *Astrophys. J.*, **326**, 19.

Maddox, S.J., 1988. Ph.D. Thesis, University of Cambridge.

Maddox, S.J., Sutherland, W.J., Efstathiou, G., Loveday, J., 1990. *Mon. Not. R. Astron. Soc.*, **243**, 692.

Nilson, P., 1973. *Uppsala General Catalogue of Galaxies*, Nova Acta Regiae Soc. Scient. Uppsala, Ser. V:A, Vol. 1.

Picard, A., 1991. *Astrophys. J. Lett.*, **368**, L7.

Raychaudhury, S., 1989. *Nature*, **342**, 251.

Raychaudhury, S., 1990. Ph.D. Thesis, University of Cambridge.

Raychaudhury, S., Fabian, A.C., Edge, A.C., Jones, C., Forman, W., 1991. *Mon. Not. R. Astron. Soc.*, **248**, 101.

Sandage, A., Tamman, G.A., 1987. *A Revised Shapley-Ames Catalog of Bright Galaxies*, Carnegie Institution of Washington, Washington, D.C., U.S.A.

Discussion

Monet :
You said that the visual examination of the ~ 500 images on a typical plate indicated that ~ 200 were actually galaxies. What were the other images that had APM parameters that indicated that they were galaxies?

Raychaudhury :
Typically they would be stars with asymmetric haloes or merged images of binary stars. Occasionally there would be diffraction spikes of very bright stars or bright patches of satellite trails that would have to be eliminated by inspection.

Oewahn :
Do you have a value for your threshold surface brightness and are you able to transform diameters and axis ratios to the standard $\mu_B = 25\text{mag}/\square''$ isophotal level?

Raychaudhury :
The threshold surface brightness is about $\mu_B \approx 24.5\text{mag}/\square''$. Correlation with ESO-LV D_{25} values is generally very good for galaxies in common. Absolute calibration is now in progress.

THE APM CLUSTER REDSHIFT SURVEY

S.J. Maddox, G.B. Dalton, G. Efstathiou and W.J. Sutherland

Department of Physics, Keble Road, Oxford OX1 3RH, U.K.

Abstract

Catalogues of rich galaxy clusters selected from the APM Galaxy Survey show that the amplitude of the cluster autocorrelation function, $w_{cc}(\theta)$, depends strongly on the algorithm used to identify the clusters. However, by using a small effective radius to estimate the cluster richness and distance parameters we find that the catalogues are more stable and the resulting $w_{cc}(\theta)$ measurements less sensitive to varying other selection parameters. We have completed a redshift survey of the richest ~ 200 clusters and find that the amplitude of the three dimensional clustering corresponds to $s_0 \sim 13$ h^{-1}Mpc, consistent with the amplitude of $w_{cc}(\theta)$. For scales $s \lesssim 15$ h^{-1}Mpc, the amplitude is compatible with the predictions of the standard CDM model, but on scales $s \gtrsim 20 - 30$ h^{-1} Mpc our measurements suggest more large-scale clustering than predicted.

1 The APM Galaxy Survey

The original APM Galaxy Survey is based on 185 UKST J survey plates centred with declination $\delta < -20°$ and galactic latitude $b \lesssim -40°$. We have recently extended the survey to include 84 extra plates near the celestial equator, covering the south galactic cap up to $\delta = +2.5°$. The complete survey now contains about 5×10^7 images with b$_j \lesssim 22$. The construction of the survey is described in detail by Maddox et al. (1990a, b), and a brief summary is presented here.

The Automatic Plate Measuring (APM) system in Cambridge (Kibblewhite et al. 1984) was used to digitise and measure the magnitude, position and shape for typically 200 000 images on each plate. The magnitude of each image was corrected for the effects of vignetting and differential desensitisation on each plate to ensure a uniform magnitude limit across each field. Then the magnitude zero point for each plate was adjusted to match the neighbouring plates and so make the magnitudes uniform across the whole survey area. Finally the overall magnitude zero point

H. T. MacGillivray and E. B. Thomson (eds.), Digitised Optical Sky Surveys 313–321.

and linearity were determined from CCD photometry of 339 calibrating galaxies, observed at the South African Astronomical Observatory, and the Mt. Stromlo Observatory. After applying these corrections, the survey contains 2×10^7 images to a uniform magnitude limit of $b_j = 21.5$.

The surface brightness profiles and shape parameters of the images were used to automatically distinguish between galaxies, stars and overlapping images. Visual inspection of over 4000 images on the Schmidt plates and 4-meter plates has shown that, over the magnitude range $16 \lesssim b_j \lesssim 20$, the galaxy catalogue is $\sim 95\%$ complete with $\sim 5\%$ contaminating non-galaxy images. For images fainter than $b_j \sim 20.5$, the classification reliability decreases rapidly, so to ensure an acceptably uniform galaxy sample, we use only galaxies brighter than $b_j = 20.5$. We have used the 100μm emission maps from the IRAS survey (Rowan-Robinson et al. 1991) to correct our magnitudes for the extinction caused by dust in our galaxy. This leads to a catalogue of about 3×10^6 galaxies with $b_j \leq 20.5$, as shown in Fig. 1.

All extended objects with $b_j \lesssim 16.5$ have been visually inspected on film copies of the plates, giving a sample of $\sim 20\,000$ galaxies with morphological classifications. These form the APM Bright Galaxy Survey (Loveday 1989), and have been used as the basis of the Stromlo/APM redshift survey (Loveday et al. 1991).

2 APM Galaxy Cluster Catalogues

We have selected many catalogues of galaxy clusters from the basic galaxy survey. Our cluster selection procedures involve two steps: first the identification of potential cluster sites; and second, the estimation of the cluster parameters for each position. We tried three methods for finding the candidate cluster sites: selecting percolated groups of galaxies; selecting peaks in the smoothed galaxy density map; and stepping over a rectangular grid of positions. To ensure that the final cluster catalogues are complete, we selected many more candidate cluster sites than the final number of clusters: typically $\sim 10\,000$ sites were tested to produce a catalogue of ~ 500 clusters. In practice, there is little difference between the large-scale structure in the catalogues based on the different candidate lists. We decided to use the centres from the percolation method because it does not bias against the selection of close pairs of clusters. The initial groups were selected to have at least 20 galaxies linked with a percolation radius of 0.7 times the mean galaxy separation. Although the large-scale structure in the catalogues is not strongly dependent on this first step in the selection, the next step of selecting clusters by richness and distance, can lead to very different catalogues.

To estimate each cluster distance and richness we used an iterative routine which essentially mimics Abell's procedure (Abell 1958). A first estimate of the cluster distance is used to define a circle with metric radius r_A, and all galaxies within this circle are sorted in magnitude. Then we count down the list of galaxies until there are 10 galaxies more than the expected field galaxy count. The magnitude of this galaxy is used to give a new estimate of the distance to the cluster, and hence a new counting radius. These steps are repeated until the distance estimate is consistent with the radius used; typically 4–6 iterations are required for this to converge. If the candidate centre does not correspond to a rich cluster, the iterations do not

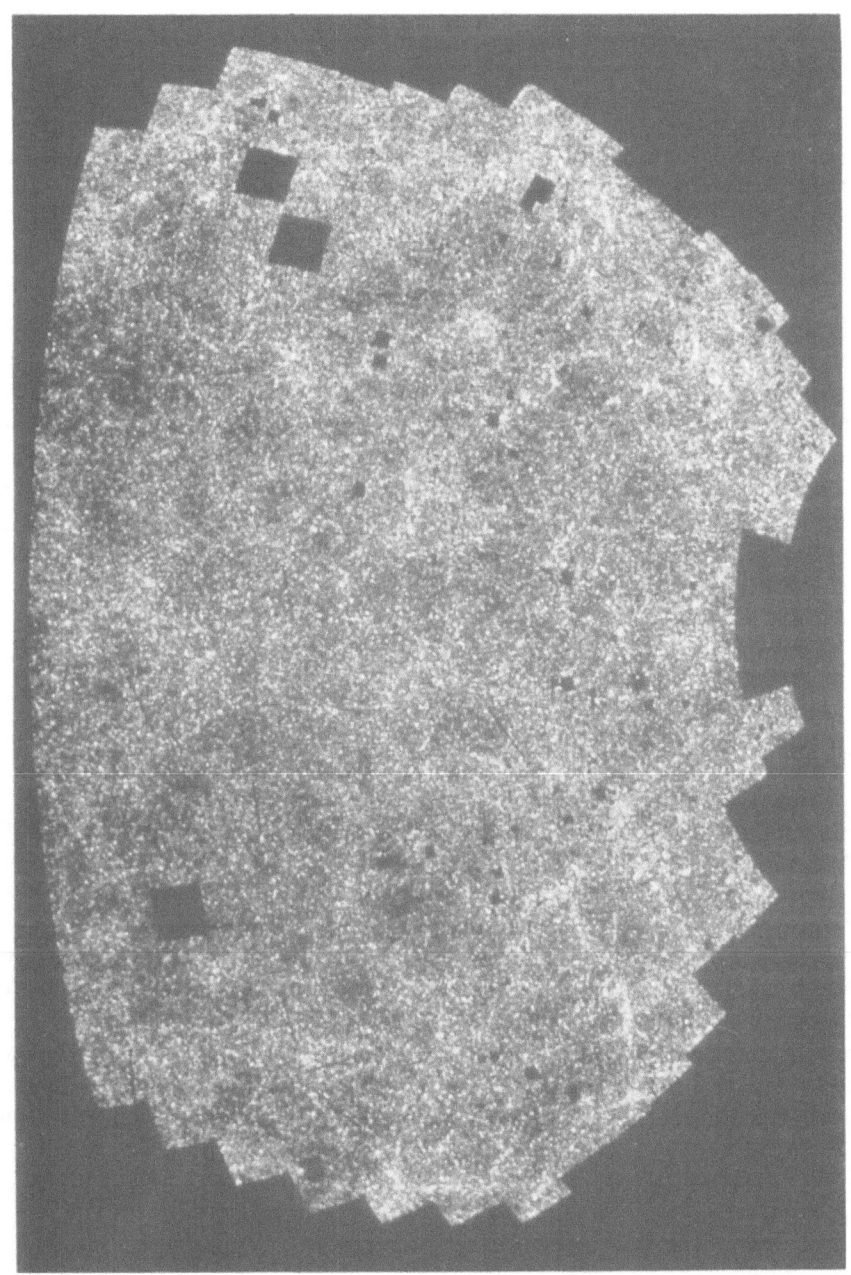

Fig. 1. The distribution of galaxy density in the APM Galaxy Survey shown in an equal area projection centred on the south galactic pole. The map shows a plot of $\sim 3 \times 10^6$ galaxies with the greyscale proportional to the galaxy density.

converge; the radius tends to decrease until there are less than ten galaxies within the circle. At each iteration we estimate the richness, $\mathcal{R}$, by counting the number of galaxies with magnitudes in the range $[m_3, m_3 + 2]$ and subtracting the expected field galaxy count within the circle. The value at the final iteration is taken as the best richness estimate.

Several modifications can be made to the algorithm described above and we made catalogues using combinations of the following options: fixing the centre to the initial estimate, or allowing it to be updated after each iteration; estimating the distance using m_{10}, or m_{20}, or m_x, where $x = \mathcal{R}/4$; estimating the richness by counting galaxies in $[m_3, m_3 + 2]$ or $[m_{10} - 0.5, m_{10} + 1.5]$ or $[m_x - 0.5, m_x + 1.5]$ or $[m_x - 0.5, m_x + 1]$; using various radii and radial weighting schemes in counting the galaxies; for close pairs of clusters either assigning galaxies to only one cluster, or to both; using a fixed global normalisation to estimate the background field counts or using a normalisation defined locally for each cluster.

As seen in Fig. 2, the large-scale distribution of clusters in the catalogues follow the same general features seen in the galaxy map shown in Fig. 1. However we find that the relative contrast of the features is different in different catalogues. The differences are shown quantitatively by the amplitude of the cluster-cluster correlation function, $w_{cc}(\theta)$, as plotted in Fig. 3a. The catalogues with multi-counted galaxies and a global background estimate have an amplitude about four times higher than the single-counted local background catalogues.

The large differences in clustering amplitude arise because some selection procedures give catalogues which contain spurious angular clustering not related to the true spatial clustering (Sutherland 1988). The apparent richness of a cluster can be enhanced by the galaxies from another cluster that is nearby on the sky, even if the pairs of clusters are not spatially correlated. A more subtle effect that is much harder to correct for is that we observe clusters through the generally clustered field of galaxies, so that clusters behind foreground clumps may also be boosted into the sample. Such richness enhancements can bias a catalogue to include more apparently close pairs of clusters than would be seen in a catalogue with a truly uniform richness selection. These spurious close pairs increase the amplitude of $w_{cc}(\theta)$, giving a biased estimate of the clustering. Since the number of clusters in a catalogue increases very rapidly towards poorer clusters, even relatively small projection biases can produce a large effect on the apparent clustering in a catalogue.

In order to reduce such biases, we gave more weight to galaxies near the centre of each cluster, where the contrast relative to the background is highest. We weighted the counts according to $W \propto r_A/(2r + r_A)$ for galaxies with $r < r_A$ and $W = 0$ for $r > r_A$, and used $r_A = 0.75$ h^{-1}Mpc, half the standard Abell radius. With this scheme, the changes in $w_{cc}(\theta)$ caused by multi-counting or single counting galaxies, or using a globally or locally defined background estimate, are much reduced, as shown in Fig. 3b. Of these catalogues the single counted, local background catalogue is least biased by projection effects, and so has one of the lowest amplitudes for $w_{cc}(\theta)$.

Also plotted in Fig. 3b is $w_{cc}(\theta)$ for the Abell clusters in our area (Abell et al. 1989). The higher amplitude at large angles suggests that the Abell catalogue may be biased by projection effects, and this interpretation is confirmed by the

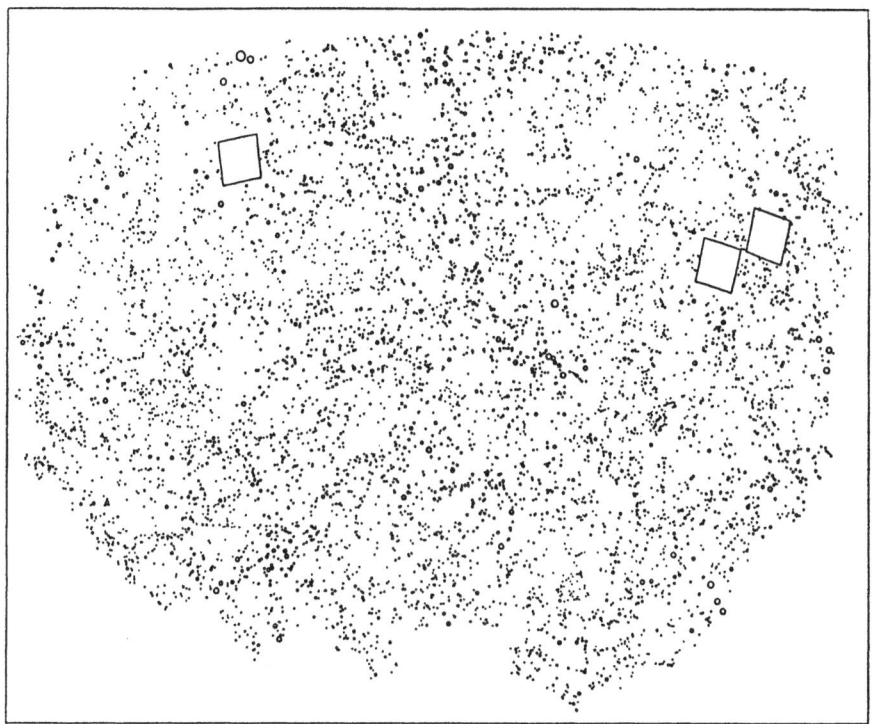

Fig. 2. The distribution of galaxy clusters in a catalogue selected from the APM survey. The size of each circle shows the counting circle used to measure the richness and m_x for each cluster. The squares show the boundaries of missing fields in the APM survey.

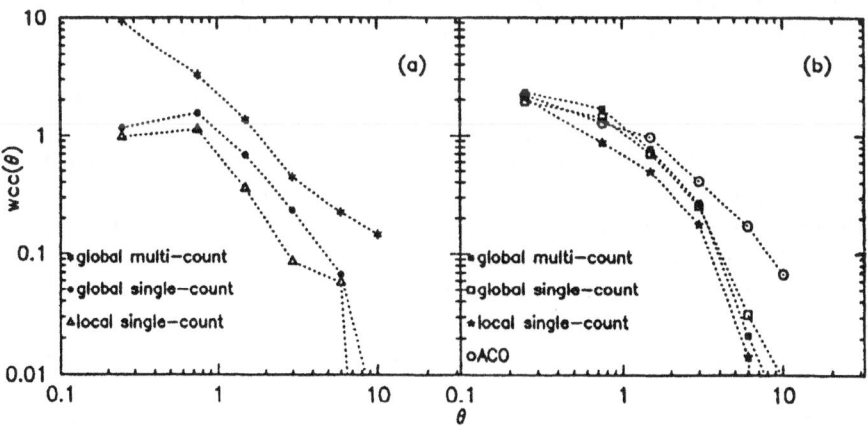

Fig. 3. (a) The $w_{cc}(\theta)$ measurements for several different APM cluster catalogues as labelled. **(b)** $w_{cc}(\theta)$ for the more stable, less biased APM cluster catalogues. Also plotted is $w_{cc}(\theta)$ for the relevant area of the Southern Abell catalogue (Abell et al. 1989).

highly anisotropic clustering in redshift space (Sutherland 1988). The projection bias introduces spurious cluster pairs with very large redshift separation but small angular separation, thus distorting the apparent clustering in redshift space. With no corrections, this bias leads to a high amplitude, $s_0 = 25$ h^{-1}Mpc and very anisotropic clustering (Bahcall & Soneira 1983). However applying a correction (Sutherland 1988) makes the clustering more isotropic, and leads to a smaller amplitude $s_0 = 14$ h^{-1}Mpc.

3 The Redshift Survey

We have recently completed a redshift survey of the richest 243 clusters from our best catalogue (Dalton et al. 1991). The redshifts were obtained from two observing runs with the AAT. For the first run in 1989 we used the automatic fibre system, AUTOFIB, to measure redshifts of the brightest $\gtrsim 20$ members in each of 29 clusters. Subsequent analysis showed that for $\sim 90\%$ of the clusters, using the redshifts of just the brightest two galaxies within the 0.75 h^{-1}Mpc radius would have given a redshift estimate correct to ~ 500kms^{-1}. We then used a maximum likelihood technique to estimate the probability that the 2-galaxy redshift is consistent with the apparent magnitude distribution for each cluster. For all of the 10% of clusters where the brightest 2 galaxies give a poor redshift, we find a very low probability of the 2-galaxy redshift being consistent with the observed magnitude distribution.

This suggested a new observing strategy: we measured redshifts for the brightest 2 galaxies in each cluster that would fit on the 4′ slit of the AAT. Then after the observation of each cluster we immediately estimated the galaxy redshifts, and computed the likelihood that they were consistent with the apparent magnitude distribution. If the likelihood was high we assumed that the initial 2-galaxy redshift estimate is correct, but if the likelihood was low, we observed more galaxies in the cluster to improve the redshift estimate. In this way we managed to measure redshifts for 140 clusters in one 5 night run. The distribution of clusters as a function of redshift and right ascension in the survey is shown in Fig. 4, and it can be seen that the mean redshift is $\overline{z} \simeq 0.08$. The mean space density of the clusters is $\overline{n} \simeq 2.4 \pm 0.2 \times 10^{-5}$ h^{3}Mpc^{-3} which is a little less than the density of Abell richness class $R = 0$ clusters, i.e. on average these clusters are slightly richer than $R = 0$ Abell clusters.

We have measured the spatial correlation function ξ_{cc} from this sample as shown in Fig. 5. The amplitude of ξ_{cc} fitted over the range $2 \lesssim s \lesssim 30$ h^{-1}Mpc corresponds to $s_0 = 13$ h^{-1}Mpc, which is consistent with the angular measurements. Analysis of ξ_{cc} as a function of redshift separation π and projected separation σ shows that there is a slight distortion along the redshift direction. The distortion is consistent with a Gaussian velocity dispersion with $\sigma \sim 700$kms^{-1}, which is mainly due to our redshift errors. The fact that the clustering is nearly isotropic in redshift space, is evidence that the ξ_{cc} from our catalogue represents the true spatial clustering of clusters, not strongly biased by projection effects. Also shown in Fig. 5 is the prediction of the standard CDM model (White et al. 1987; Dalton et al. 1991). Although the amplitude on small scales is consistent with the standard CDM prediction, there appears to be a slight excess of clustering on scales $s \simeq 20 - 30$ h^{-1}Mpc. However,

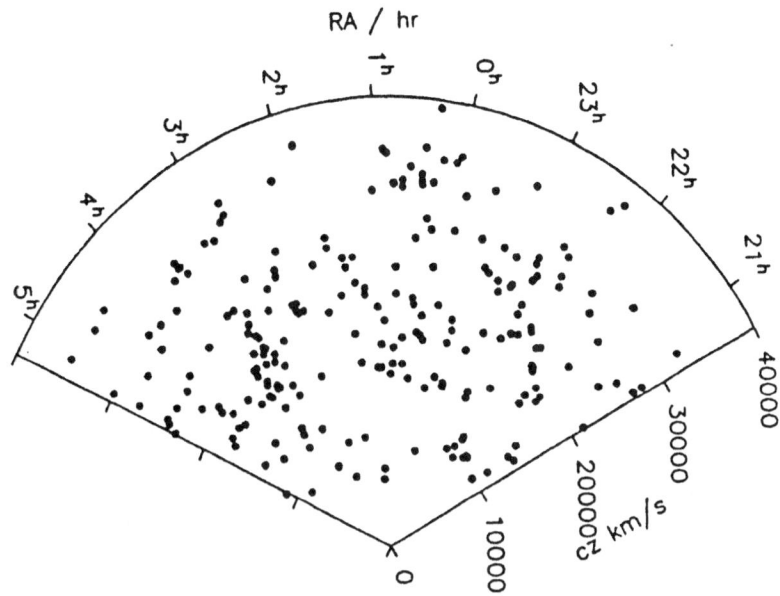

Fig. 4. A cone plot showing the distribution of clusters as a function of redshift and right ascension.

the statistical errors in our current survey are rather large, and this is currently only a marginal detection of excess clustering. We hope to measure redshifts for a further 200 clusters and so confirm this measurement.

References

Abell, G.O., 1958. *Astrophys. J. Suppl.*, **3**, 211.

Abell, G.O., Corwin, H.C., Olowin, R.P., 1989. *Astrophys. J. Suppl.*, **70**, 1.

Bahcall, N.A., Soneira, R.M., 1983. *Astrophys. J.*, **270**, 20.

Dalton, G.B., Efstathiou, G., Maddox, S.J., Sutherland, W.J., 1991. *Astrophys. J.* Submitted.

Kibblewhite, E.J., Bridgeland, M.T., Bunclark, P., Irwin, M.J., 1984. *Astronomical Microdensitometry Conference, NASA Conf. Pub.*, No. 2317, p. 277.

Loveday, J., 1989. PhD Thesis, University of Cambridge.

Loveday, J., Efstathiou, G., Peterson, B.A., 1991. *Astrophys. J.* Submitted.

Maddox, S.J., Sutherland, W.J., Efstathiou, G., Loveday, J., 1990a. *Mon. Not. R. Astron. Soc.*, **243**, 692.

Maddox, S.J., Efstathiou, G., Sutherland, W.J., 1990b. *Mon. Not. R. Astron. Soc.*, **246**, 433.

Rowan-Robinson, M., Hughes, J., Jones, M., Leech, K., Vedi, K., Wallke, K., 1991. *Mon. Not. R. Astron. Soc.*, **249**, 729.

Sutherland, W.J., 1988. *Mon. Not. R. Astron. Soc.*, **234**, 159.

White, S.D.M., Frenk, C.S., Davis, M., Efstathiou, G., 1987. *Astrophys. J.*, **313**, 505.

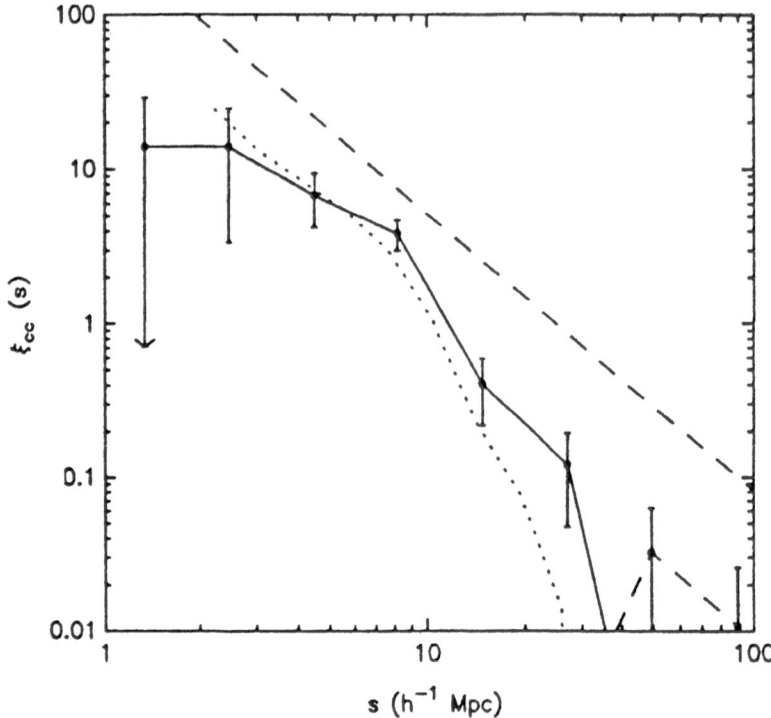

Fig. 5. The spatial correlation function for clusters in the redshift survey is shown by the filled points joined by the solid line where ξ_{cc} is positive and a dashed line where the measurements are negative. The error bars are estimated from the Poisson noise in the cluster pair counts. The dotted line shows the function predicted from the standard CDM model. The straight dashed line shows $(s/25\ h^{-1}\mathrm{Mpc})^{-1.8}$.

Discussion

Hale-Sutton :

You showed the galaxy correlation function for Loveday's 1-in-20 redshift survey and this has a correlation length $r_0 \sim 6.5\ h^{-1}\mathrm{Mpc}$. It is generally accepted that $r_0 \sim 5\ h^{-1}\mathrm{Mpc}$, would you like to comment on this?

Maddox :

The plot is from a preliminary sub-sample of the survey and so the errors on ξ are fairly large. The discrepancy in r_0 depends on the pair-weighting scheme used in the ξ measurements, but even the largest discrepancy is only $\sim 1\sigma$.

MacGillivray :

Could you comment on how the presence of overlapping images at the centres of rich clusters affects:-

1. Your cluster detection.
2. The richness estimate.

Maddox :

We used a fairly high detection isophote ($B \sim 24.5$ mags per arcsec^2) and so image blending is not a major problem in most clusters. Since we test $\sim$ 10 000 candidate clusters to generate the final sample, it is unlikely that we miss a significant number of clusters. We have also tried stepping over a grid of sites covering each field, rather than using the candidate sites. There are no significant differences in either catalogues produced. The effect of blended images of galaxies on the richness estimates are more significant, but still not important. For most clusters, no more than a couple of galaxies have merged, and the worst case would lose maybe $\sim$ 5 galaxies. This is smaller than the random noise expected in the Poisson sampling of the underlying density field, since there are only $\sim$ 50 galaxies per cluster.

Parker :

Have you compared the selection of clusters in the APM catalogue with the EDSG cluster catalogue in the region of overlap?

Maddox :

No, but I would like to make the comparison.

GALAXY CORRELATION FUNCTION CONSTRAINTS ON LARGE-SCALE STRUCTURE

D. Hale-Sutton, T. Shanks and R. Fong.

Physics Department, University of Durham, South Road, Durham DH1 3LE, U.K.

Abstract

We consider whether photometric errors may have induced the apparent large-scale, $\geq 10h^{-1}$ Mpc (where h is the Hubble constant in units of 100 km s^{-1} Mpc^{-1}), power in the angular correlation function $\omega(\theta)$ estimated from the APM galaxy catalogue. Using the galaxy CCD sequences on 39 of the plates in this catalogue we have made an independent estimate of the random and systematic errors in the APM magnitude scale. In particular we have found that the rms error in the plate zero-points is $0.^m084$, a factor of ~ 2 larger than the estimate found by Maddox et al. (1990b). Further, the distribution of the zero-point residuals over the APM survey area suggests that much of this extra error could arise from large-scale magnitude correlations. Thus, taking these observations into account in the estimate of $\omega(\theta)$, it is shown that the excess power in the galaxy correlation function may be due to photometric calibration errors and that the standard CDM galaxy formation model may not be in contradiction with the observations.

1 Introduction

Although large volume redshift surveys have hinted at the presence of large-scale ($\geq 10h^{-1}$ Mpc) structures, such as 'voids', 'filaments', 'sheets' etc., in the galaxy distribution, there has, until recently, been little firm evidence for large-scale power in the objective statistics of such distributions. In particular, the popular standard Cold Dark Matter (CDM) model seemed reasonably able to reproduce such structures with little or no power in the important two-point correlation function at large ($\geq 10h^{-1}$ Mpc) scales (see e.g. White et al. 1987). With the advent of large area digitised optical sky surveys (Heydon-Dumbleton et al. 1989; Maddox et al. 1990a)

H. T. MacGillivray and E. B. Thomson (eds.), Digitised Optical Sky Surveys 323–334.
© 1992 *Kluwer Academic Publishers.*

and the all sky survey of infrared galaxies by IRAS (see e.g. Rowan-Robinson et al. 1990) there has been some indication that there may be statistically reproducible structure in the galaxy distribution on scales of up to $\sim 20h^{-1}$ Mpc (Maddox et al. 1990b; Efstathiou et al. 1990; Collins et al. 1991). The observations suggest that the spatial two-point correlation function has the power-law form $\xi = (r_0/r)^{1.8}$ out to separations $r = 20h^{-1}$ Mpc with a correlation length r_0 of between 5 and $7h^{-1}$ Mpc. Whilst the absolute size of this statistic is quite small at these separations (ranging from 0.4 at $10h^{-1}$ Mpc to 0.1 at $20h^{-1}$ Mpc assuming $r_0 = 6h^{-1}$ Mpc), it leads to large number density fluctuations in volumes with this average dimension and to possibly excessively large structures in the galaxy distribution. Such large-scale power in the correlation function is, however, not predicted in the standard CDM model of galaxy formation, and since the CDM model incorporates certain basic assumptions about the form of initial conditions in the early universe, this result has led to some speculation about the validity of the standard Big Bang model itself.

However, before such far reaching conclusions are accepted, it is important that, in view of the small size of the quantity being measured, a critical assessment of the statistical significance of the correlation function results are made, taking into account both random and systematic errors. This is even more essential in the case where galaxy correlation functions are measured on the sky, as there is typically an order of magnitude decrease in the signal arising from the effects of projection. For example, Fig. 1 shows the angular correlation function $\omega(\theta)$ (solid symbols) estimated from the APM survey scaled to the depth of the Lick catalogue. The excess power of the APM results over the prediction of CDM (solid line) and the $\omega(\theta)$ estimated from Lick (open symbols; Groth & Peebles 1977) is clearly shown. However, if there are number density fluctuations on the sky of order $x\%$, on average, arising from random or systematic errors, then this will induce artificial 'clustering' in the angular correlation function of $\sim (x/100)^2$. Thus it can be seen that 10% fluctuations can give apparent angular correlations of ~ 0.01, i.e. this is at a level which is important for these studies. One way in which such number density variations can be induced is through uncertainties in galaxy apparent magnitudes. If there are on average fluctuations of order δm magnitudes over the sky, this will lead to percentage fluctuations in number density of $x \approx 230 s \delta m\%$ where s is the slope of the logarithmic number counts at the apparent magnitude limit of the galaxy catalogue. The slope at $20.^m0$ is $s \approx 0.45$ and so errors in the magnitudes of $\delta m \approx 0.^m1$ are sufficient to give the 10% fluctuations in the surface number density.

In this presentation we explore, in particular, how photometric errors may have percolated through into the angular clustering analysis of one very large area galaxy catalogue: the APM Galaxy Survey. Thus, in Section 2 we briefly describe our methods and results for making independent estimates of the random and systematic errors in the APM magnitude scale and how these estimates affect the observations of galaxy clustering. In Section 3 we assess our interpretation of the results and conclude on the significance of the rejection of the CDM model.

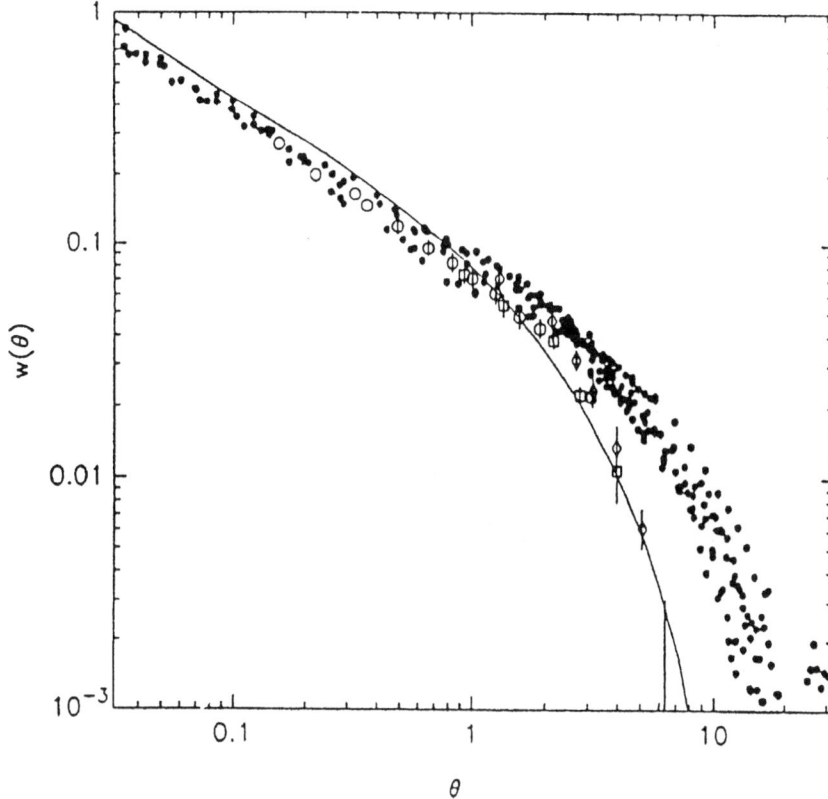

Fig. 1. The APM galaxy angular correlation function scaled to the depth of the Lick catalogue. This figure is reproduced from Fig. 3 of Maddox et al. 1990b. The solid and open symbols are the $\omega(\theta)$ estimates from the APM and Lick (Groth & Peebles 1977) catalogues respectively. The solid line is the prediction from the CDM model with Hubble constant $h = 0.4$.

2 Application to a Large Area Digital Sky Survey: the APM Survey

The APM Galaxy Survey (Maddox et al. 1990a) consists of approximately 2 million galaxies gathered from the automatic scanning of 185 Schmidt J-band photographic plates covering a contiguous area of some 4300 square degrees in the southern Galactic cap. In their photometric calibration of the survey, Maddox et al. (1990c, hereafter MES) almost exclusively used a magnitude matching technique for the galaxies in the plate overlaps to provide a consistent magnitude scale over the entire catalogue area. Whilst the 'internal' estimate of the error in the measured difference in the plate zero-points is quite small, as there are, on average, 460 galaxies in these plate overlaps to $b_j = 20.^{m}5$, the 'external' estimates could be considerably larger due to the propagation of errors through the matched network and the possibly remaining systematic errors that may not have been successfully removed during the matching process.

To provide an external check (but note not a direct external calibration) of their galaxy magnitudes, MES obtained B and V CCD galaxy sequences on 39 of their 185 plates. In this section we describe how these sequences can be used to provide a more direct external estimate of the random and systematic errors in the 'matched' APM galaxy magnitudes. We will also discuss in this context some of the tests that MES themselves have carried out.

2.1 RMS Errors in the Plate Zero-Points

The routines to convert the 'unmatched' APM magnitudes for the galaxies in the 39 sequences in Table 3 of MES to 'matched' APM magnitudes were kindly provided by Dr. Maddox. On plotting matched photographic magnitudes versus CCD magnitudes for the galaxies in these sequences, it was noticed that there were quite often cases where the dispersion of the points around a best correlation fit for individual sequences was considerably smaller than the dispersion for all sequences plotted together. As this type of effect could quite simply be produced by errors in the APM matched zero-point offset for each plate, we attempted to quantify this by using an analysis of variance (full details of the method are given in Fong et al. 1991). The zero-point offset c_{ij} as given by the jth galaxy in the CCD sequence on the ith plate is

$$c_{ij} = (m_A)_{ij} + (b_j)_{ij}/1.097 \qquad (1)$$

where m_A is the matched APM magnitude of the galaxy and b_j is its CCD calibration (this correlation is a good representation of the overall data). The method then estimates a mean within-sequence variance s_W^2 of the zero-point c_{ij} about a mean fit to each sequence $\overline{c_i}$ and a between-sequence (or between-plate) variance s_B^2 of $\overline{c_i}$ about the overall zero-point for all the sequences considered together ($\overline{c}$).

Using these measures of variance it is then possible to objectively test the hypothesis that the variations in the mean zero-point for each sequence $\overline{c_i}$ have arisen from the noise associated with the dispersion of c_{ij} about $\overline{c_i}$ (i.e. that there are no real plate-to-plate errors in the zero-points). Under this hypothesis s_W^2 and s_B^2 should be roughly equal and so the ratio F given by

$$F = s_B^2/s_W^2 \qquad (2)$$

should be close to unity. The significance of deviations from unity can then be obtained from the F-test. Using all 267 galaxies in the sequences on 39 plates we found that $F = 1.99$ and the F-test rejected the hypothesis at the 99.89% confidence limit.

This result clearly seems to indicate that there are indeed real plate-to-plate errors in the zero-points of the APM matched magnitudes. The size of this mean rms plate-to-plate error σ can be estimated from s_B^2 given that this quantity partly contains the jitter of c_{ij} about $\overline{c_i}$

$$\sigma = (s_B^2 - s_W^2)^{\frac{1}{2}} \qquad (3)$$

Again for the 267 galaxies we found $\sigma = 0.^{m}084$ and this is a factor of ~ 2 larger than the internal value of $0.^{m}04$ rms estimated by MES on the basis of the propagation of errors in their network of intermeshed plates. The 95% upper and lower confidence limits we found for σ were $0.^{m}129$ and $0.^{m}046$, respectively, with the value of $0.^{m}04$ lying just outside this range.

Some question over the validity of using the zero-point for the sequence on field 152 was raised by MES since the value for this sequence deviates fairly widely from $\bar{c}$. If we omit this sequence from our analyses (or omit the two galaxies that have been claimed to be causing the discrepancy) we find $\sigma = 0.^{m}069$ with 95% upper and lower limits of $0.^{m}113$ and $0.^{m}023$. This value for σ is still ~ 1.7 times larger than the value claimed by MES but now the lower limit less strongly rejects their estimate. We emphasise, however, that poorly determined sequence magnitudes *will* be accounted for in the analyses (they are incorporated into s_W^2) and thus there does not seem any justifiable reason why certain sequences should be, a posteriori, omitted from the reductions.

2.2 Systematic Errors in the Plate Zero-Points

Having obtained an estimate of the rms error in the plate zero-points which was a factor of 2 larger than the internal MES estimate of $0.^{m}04$ based on the assumption that the plate overlaps were independent, it occurred to us that, as suggested above, the larger error may arise through correlated errors that have not been removed successfully by the matching process. In Fig. 2 is shown the spatial distribution of the zero-point residuals Δ_i for the difference between the zero-point for each sequence $\overline{c_i}$ and the zero-point for all the sequences put together $\bar{c}$ ($\Delta_i = \overline{c_i} - \bar{c}$). As can be seen there is evidence for some overall trend of these residuals from East (at the left) to West (at the right) in this figure which is suggestive of correlated zero-point errors.

Again to quantify this, we divided the RA range (which runs from West to East) into four equal zones and then estimated the mean residual for each zone. The results for this are shown as solid circles in Fig. 3 where the errors are based on the assumption that each sequence zero-point is independent and that the error on the mean residual arises solely from the jitter of c_{ij} about $\overline{c_i}$. From this figure the large-scale correlations in the residuals are more clearly seen. If we make the hypothesis that there is no gradient in this data with RA, we found that the value of χ^2 is 27.3 (where we have included the extra error of $0.^{m}04$ per sequence claimed by MES) and this is rejected at a confidence limit which is greater than the 99.99% with three degrees of freedom (one degree of freedom being lost since the weighted mean of Δ_i over all the sequences is equal to zero). On omitting the sequence for field 152 (open circles in Fig. 3) the hypothesis is still significantly rejected (here $\chi^2 = 18.3$ which is significant at the 99.95% confidence limit).

Thus, these observations do in fact suggest that the plate zero-points are correlated over the APM survey area. With the small number of sequences available, it is clearly impossible to form an accurate description of these correlations. However, as a simple model of the mean residuals in one dimension, we fitted the linear form

$$\overline{\Delta} = c_1(\alpha + c_2) \tag{4}$$

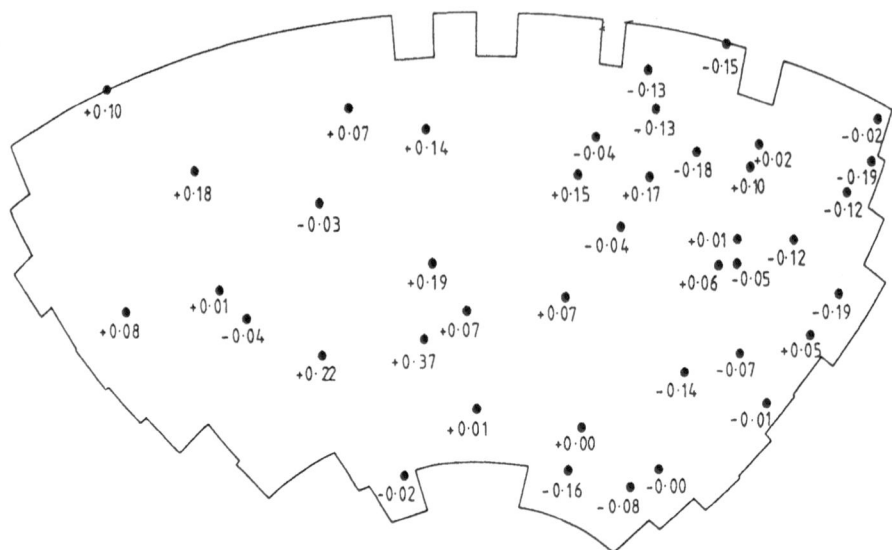

Fig. 2. The CCD zero-point residuals (in magnitudes) for the galaxy sequences in the APM survey. The solid line shows an equal area projection of the APM survey boundary as in Fig. 9 of MES.

to the solid circles in Fig. 3. Here α is the RA in hours where the RA lies in the range $0 \leq \text{RA} < 12\text{hrs}$ and $\alpha = \text{RA} - 24$ where the RA lies in the range $12 \leq \text{RA} < 24\text{hrs}$. The constant c_2 is fixed by the requirement that the weighted mean of $\overline{\Delta}$ is zero and c_1 (the slope) by the observations. By minimising the χ^2 with this model we found

$$\overline{\Delta} = 0.030(\alpha + 0.049) \tag{5}$$

with $\chi^2_{min} = 7.1$ and this fit is shown as a solid line in Fig. 3. With three degrees of freedom, this value of χ^2 shows that the model is marginally rejected by the data, indicating that, as we had guessed, a more complicated model may be required. However, this model will be sufficient to illustrate the effects due to such systematic errors. Excluding the sequence on field 152 gave similar results for the gradient.

2.3 Other Tests for Photometric Errors

In their assessment of the photometric errors associated with the galaxy magnitudes in the APM survey, MES carried out various tests of their data. One such test was to calculate the correlation function $C_{CCD}(\theta)$ of the CCD residuals at different angular separations θ. In Fig. 4 we show this correlation function as calculated from both the actual observed CCD residuals (solid points) and the residuals predicted at the positions of the CCD sequences (solid line) by the model in (5). As can be seen from this figure there is very good agreement between the observations and the model (the open points exclude the sequence on field 152).

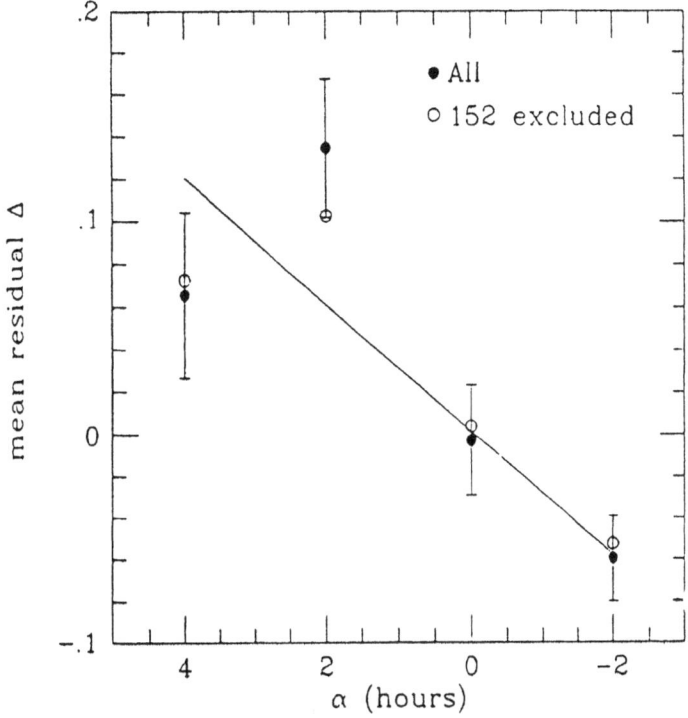

Fig. 3. The RA dependence of the mean CCD zero-point residuals (in magnitudes). The solid and open symbols are the mean residuals which include and exclude the sequence on field 152 respectively. The solid line is a weighted least squares linear fit to the solid points.

As the absolute value of the correlation function in Fig. 4 is small at separations of $0° < \theta < 60°$ (the upper value being the limit of their plot), MES took these observations to imply (and this was reasonably confirmed by their estimates of the error) that this correlation function was consistent with no correlations in the matched APM zero-point residuals. However, our model of such residuals seems to suggest that this figure is not a very good discriminant of such correlations, since the solid line *is* based on a simple gradient in the magnitudes. The lack of an effect in this figure at small angles (except at $\theta = 0°$) as opposed to that in Fig. 3 may partly be explained by the fact that the correlation function, being weighted to pairs of residuals, is dominated by where the sequences in Fig. 2 are most densely situated. As this occurs in a small area in the north western half of the APM survey, the full gradient of the model may not be being sampled uniformly. We suggest that if the CCD sequences had been placed more uniformly over the APM survey area, $C_{CCD}(\theta)$ would have been a better measure of the correlations in the APM magnitudes.

A second feature of Fig. 4 is the agreement between the observed and model correlation functions at $\theta = 0°$. As $C_{CCD}(0)$ is effectively the variance in the plate

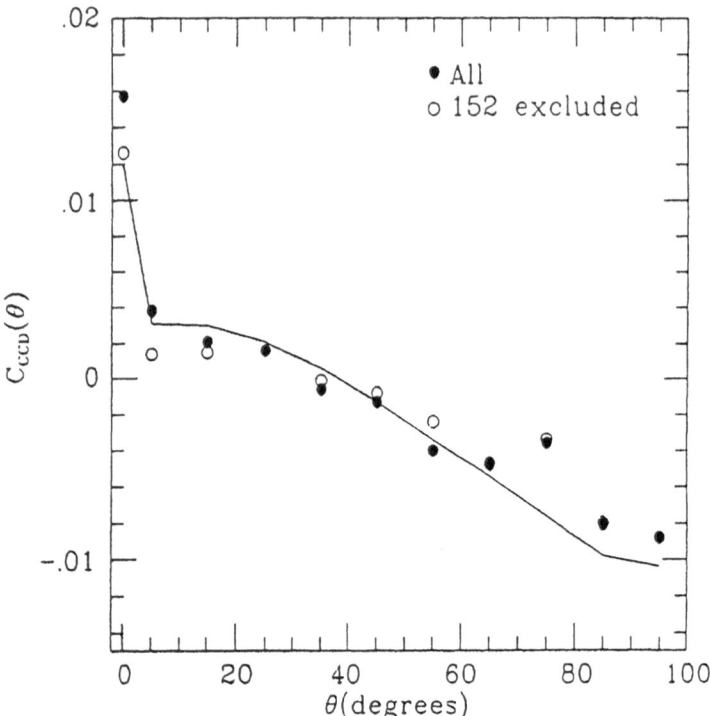

Fig. 4. The correlation function of the residuals in the CCD zero-points. The solid and open symbols are the correlation functions for the residuals which include and exclude the sequence on field 152 respectively. The solid line uses the predicted residuals from the model in (9).

zero-points, this agreement implies that the larger rms error seen in Section 2.1 *can* be explained by correlated zero-points as was anticipated above (the CCD error per sequence has been added in quadrature to the model for $C_{CCD}(0)$ to allow a direct comparison between the data points and the model).

A further test of the photometric errors in the APM magnitudes is based on the difference between ω_{intra} and ω_{inter}, the intra- and inter-plate estimates of the galaxy angular correlation function $\omega(\theta)$. As was stated in Section 1, magnitude errors of order δm can lead to errors in this galaxy correlation function of $\sim (2.3s\delta m)^2$, where s is the slope of the logarithmic number counts at the depth of the catalogue. In the case of random magnitude errors from plate-to-plate, such effects in the correlation function show up strongly in ω_{intra} but not so much in ω_{inter}, since the latter necessarily averages over two or more plates. Maddox et al. (1990b) estimated that $\omega_{intra} - \omega_{inter} = 1.7 \times 10^{-3}$ and this is consistent with the value of 1.71×10^{-3} expected on the basis of rms plate-to-plate errors in the magnitudes of $0.^m04$ and a slope $s \approx 0.45$ for the number counts at a depth of $20.^m0$.

This conclusion, however, relies on the assumption that the errors in the plate zero-points are uncorrelated. To demonstrate the effects of correlated errors we carried out a simulation of the APM survey magnitudes which included the simple gradient in the zero-points estimated in Section 2.2 above. We considered an array of 19×10 Schmidt plates with an underlying systematic East-West error of $0.^m013$ per plate [consistent with (5)] together with a random plate-to-plate error of $0.^m035$ rms. Using a slope of $s = 0.45$ to predict the cell counts in each plate we found, using the Peebles estimator, that

$$\omega_{intra} - \omega_{inter} = 1.7 \times 10^{-3} \tag{6}$$

as expected (a small part of the correlated error contributes to this), but here the correction to ω_{inter} is not negligible

$$\omega_{inter} = 5.0 \times 10^{-3} \tag{7}$$

with the total correction to the correlation function estimated from the real catalogue lying in the range 6.5×10^{-3} at $\theta = 0.5°$ to 5.1×10^{-3} at $\theta = 5°$. This clearly demonstrates that $\omega_{intra} - \omega_{inter}$ is not a good discriminator of correlated zero-point errors.

Thus, in conclusion, we have shown that some of the main tests for photometric errors considered by MES are fairly insensitive to the sort of correlated errors in the magnitudes discussed in Section 2.2 above and that the uncertainties in the galaxy correlation function could be considerably larger than were previously anticipated.

2.4 A 'Corrected' APM Galaxy Correlation Function

In the previous section it was demonstrated that correlated errors in the APM magnitude scale could lead to significant errors in the estimated galaxy correlation function. However, as it is difficult to specify exactly what form these correlated errors take (see Section 2.2), we will rely on estimates of the rms plate-to-plate error in the zero-points σ given in Section 2.1 to provide an estimate of the correction to ω_{intra}

$$\omega_{intra} = (2.3s\sigma)^2 \tag{8}$$

It then follows that ω_{inter} can be found knowing both the slope of the counts at the depth of the catalogue and the difference $\omega_{intra} - \omega_{inter}$ (which we take to be the observed value of 1.7×10^{-3}). At an angular separation of $3°$ there is about an equal contribution to the total correlation function from ω_{inter} and ω_{intra}, and so with the best estimate of $\sigma = 0.^m084^{+0.^m107}_{-0.^m065}$ (the limits being approximately ± 1 st. dev.) we obtain as a correction to $\omega(\theta)$

$$\omega_{total} = 0.007^{+0.004}_{-0.003} \tag{9}$$

at the $b_j = 20.^m0$ limit of the catalogue.

In Fig. 5 the dashed and dotted lines represent, respectively, the best estimate and ± 1 st. dev. envelope of the corrected correlation function based on the $17.^m0 < b_j < 20.^m0$ $\omega(\theta)$ in Fig. 1 of Maddox et al. (1990b) and the values given in (9). As in Fig. 1 above, the solid points are the uncorrected APM estimates of $\omega(\theta)$ but it can be seen that the new, corrected estimate of this function is in much better agreement with the original $\omega(\theta)$ from the Lick catalogue (shown as open symbols); here the 1 st. dev. lower bound passes quite well through these points at the largest scales. Also, the solid line is the prediction from the CDM galaxy formation scenario (with Hubble constant $H_0 = 40$ km s^{-1} Mpc^{-1}) and, contrary to the conclusions of Maddox et al., this prediction no longer appears to be significantly excluded by our new estimation of the errors associated with the photometry of the APM catalogue.

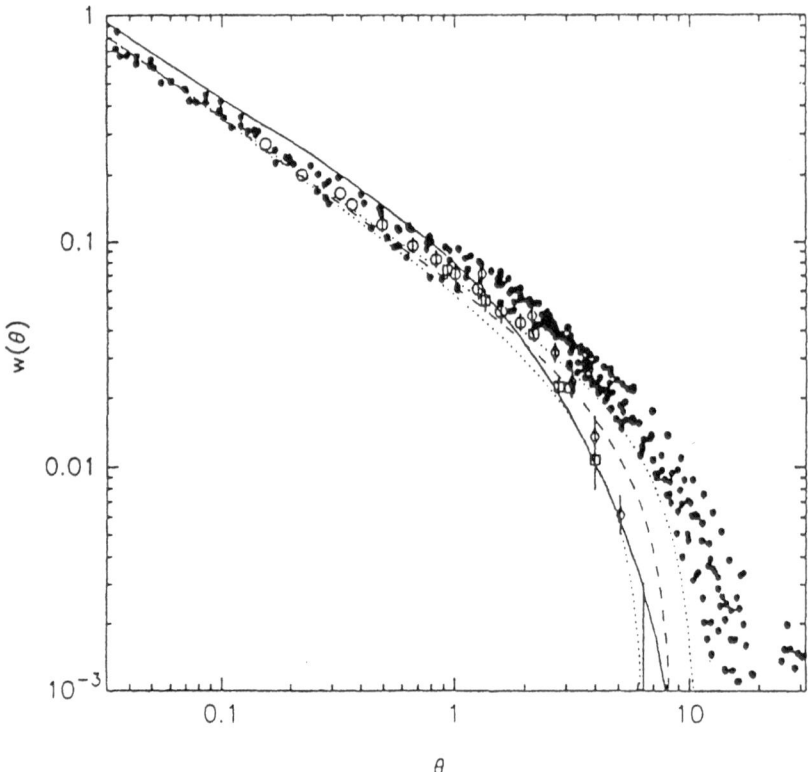

Fig. 5. The corrected APM angular correlation function. The solid and open symbols and the solid line have the same meaning as in Fig. 1. The dashed and dotted lines represent, respectively, the best estimate and ±1 st. dev. envelope of the corrections to the APM $\omega(\theta)$.

3 Discussion

Our conclusion that the CDM scenario may not be excluded by the correlation function estimated from the APM catalogue has been criticised by Maddox (see the discussion at the end of this contribution) on a number of issues and it is worthwhile addressing some of these points as part of this section. Firstly, it has been suggested that the larger estimate of the rms error of $0.^{m}084$ in the sequence zero-points in Section 2.1 arises from small-scale ($\sim$arc minute) 'ripples' in the emulsion sensitivity within a plate. Although no such evidence has been presented for such an effect on these scales, such small-scale ripples would result in larger random variations in the CCD zero-points for sequences taken from different places on the same plate, as the CCD sequences cover areas of order this dimension. This result would clearly be of importance for a wide variety of situations in which the photometry of all objects on a plate are calibrated using a single CCD sequence. However, it would be expected that part of this extra variation would again be seen in the estimate of s_W^2 and thus be taken care of in our analysis of variance. In contrast, the evidence from our work suggests that the source of the larger rms error from sequence-to-sequence is adequately explained by the observed correlations in the plate zero-points; such correlations from plate-to-plate would be difficult to explain on the basis of random errors in the zero-points.

Maddox has also suggested that correcting $\omega(\theta)$ as in Section 2.4 by subtracting a constant from the estimates is not an adequate way of describing such errors. Whilst this may be true on scales much larger than $10°$, our simulations indicate that even if $\omega_{total} = \omega_{inter}$, our best estimate of the correction at $5°$ should be ~ 0.006 (rather than 0.007) and this makes little difference to our conclusions.

Finally, Maddox et al. (these proceedings) have suggested that if the $\omega(\theta)$ estimated from the faintest magnitude slice ($20.^{m}0 < b_j < 20.^{m}5$) contains all the uncertainties associated with the photometry errors, then this estimate subtracted from the five other brighter $\omega(\theta)$ slices does not affect the large-scale power for the brightest three. Whilst this may be true, it presupposes that the correction to $\omega(\theta)$ for each magnitude interval is the same and this is not necessarily so. For example, even for the same rms error in the plate zero-points, the correction to $\omega(\theta)$ increases to brighter magnitudes due to the increase in number-count slope. It also should be noted that the matching procedure is biased to giving accurate magnitudes for the fainter galaxies as these dominate the numbers in the plate overlaps. It is possible to see how the zero-points for the brighter galaxies could be in error (if there are scale errors in the magnitudes) and how the corrections to $\omega(\theta)$ could be worse for the brighter slices.

In conclusion, then, we believe that we have obtained a fair and independent estimate of the errors associated with the estimate of the APM angular galaxy correlation function and, in particular, we find that our best estimate of this function does not exclude the standard theory of Cold Dark Matter.

4 Acknowledgements

We would especially like to thank Steve Maddox for providing much help and information concerning the APM survey and for useful discussions. We also acknowledge useful discussions with Nigel Metcalfe, Richard Bower and John Lucey. TS and DHS respectively thank the Royal Society and the SERC for financial support.

References

Collins, C.A., Nichol, R.C., Lumsden, S.L., 1991. *Mon. Not. R. Astron. Soc.* Submitted.

Efstathiou, G., Kaiser, N., Saunders, W., Lawrence, A., Rowan-Robinson, M., Ellis, R.S., Frenk, C.S., 1990. *Mon. Not. R. Astron. Soc.*, **247**, 10p.

Fong, R., Hale-Sutton, D., Shanks, T., 1991. *Mon. Not. R. Astron. Soc.* Submitted.

Groth, E.J., Peebles, P.J.E., 1977. *Astrophys. J.*, **217**, 385.

Heydon-Dumbleton, N.H., Collins, C.A., MacGillivray, H.T., 1989. *Mon. Not. R. Astron. Soc.*, **238**, 379.

Maddox, S.J., Sutherland, W.J., Efstathiou, G., Loveday, J., 1990a. *Mon. Not. R. Astron. Soc.*, **243**, 692.

Maddox, S.J., Efstathiou, G., Sutherland, W.J., Loveday, J., 1990b. *Mon. Not. R. Astron. Soc.*, **242**, 43p.

Maddox, S.J., Efstathiou, G., Sutherland, W.J., 1990c. *Mon. Not. R. Astron. Soc.*, **246**, 433 (MES).

Rowan-Robinson, M., Lawrence, A., Saunders, W., Crawford, J., Ellis, R., Frenk, C.S., Parry, I., Xiaoyang, X., Allington-Smith, J., Efstathiou, G., Kaiser, N., 1990. *Mon. Not. R. Astron. Soc.*, **247**, 1.

White, S.D.M., Frenk, C.S., Davis, M., Efstathiou, G., 1987. *Astrophys. J.*, **313**, 505.

Discussion

Maddox :

(1) The excess variance in the CCD magnitudes is probably caused by small-scale (few arcmin) ripples in the emulsion sensitivity and so will not propagate across the survey, and will not significantly affect the $\omega(\theta)$ measurement.

(2) Even if the variance is due to a gradient in the matching errors, you should subtract the error-correlation function, not just a constant. If we make the most extreme assumption that all of the signal in the faintest ω measurement is due to errors in the catalogue, then we should subtract this from the brighter measurements.

The resulting measurements for the brightest 3 slices are hardly affected and still show more large scale clustering than CDM and Lick.

Hale-Sutton :

I draw your attention to section 3 of this contribution where I discuss these points.

THE EDINBURGH/DURHAM SOUTHERN GALAXY CATALOGUE

R.C. Nichol [1], *C.A. Collins* [2], *L. Guzzo* [3] *and S. Lumsden* [2,4,5]

[1] Department of Astronomy, University of Edinburgh, Blackford Hill, Edinburgh EH9 3HJ, U.K.
[2] Royal Observatory, Blackford Hill, Edinburgh EH9 3HJ, U.K.
[3] Osservatorio Astronomico di Brera, Milan, Italy.
[4] Astrophysics Group, Imperial College of Science, Technology and Medicine, Blackett Laboratories, Prince Consort Road, London SW7 2BZ, U.K.
[5] Astrophysics Department *, University of Oxford, Keble Road, Oxford OX1 3RH, U.K.

1 Introduction

1.1 Motivation

The study of the large-scale structure in the universe has been of prime importance in our understanding of how galaxies form. On scales greater than 10 h^{-1}Mpc, the underlying mass distribution of the universe is a relic of the primordial density fluctuations in the early universe. If light traces the mass, then the large-scale distribution of galaxies and clusters of galaxies we see today must be predicted by theories of the early universe and galaxy formation.

Over the last ten years, much time has been invested in constructing redshift surveys of galaxies. Although these surveys measure the 3 dimensional distribution of galaxies, they still only sample the relatively nearby universe. Large redshift surveys now planned should rectify this, although they are still some years from completion. The only other ways of probing the clustering distribution on scales greater than 10 h^{-1}Mpc are to use either the projected 2 dimensional distribution of galaxies or the distribution of clusters of galaxies. Results from just such studies are the topic of this paper.

Large area 2-D catalogues of galaxies and clusters of galaxies have been invaluable tools in studying the galaxy distribution and constraining galaxy formation

---

* Present address of S. Lumsden.

H. T. MacGillivray and E. B. Thomson (eds.), Digitised Optical Sky Surveys 335–344.

models. The Lick astrographic survey of galaxies (Shane & Wirtanen 1967) was for many years the largest survey in existence, with $\sim 10^6$ galaxies covering $\sim \frac{2}{3}$rds of the sky. The Abell Cluster Catalogue (Abell 1958), recently extended to cover the whole sky (Abell et al. 1989), was until recently the only large-area cluster catalogue available. Both these catalogues were constructed from visual searches of photographic plates, which has prompted many authors to distrust the structure seen in these catalogues (Geller et al. 1984; de Lapparent et al. 1986; Sutherland 1988).

With the advent of high-speed microdensitometers like COSMOS and APM, it is now possible to construct objectively large-angle galaxy catalogues out to cosmologically interesting depths. This article details recent results obtained from a large digitised galaxy catalogue centred on the SGP.

1.2 The Edinburgh/Durham Southern Galaxy Catalogue

The Edinburgh-Durham Southern Galaxy Catalogue (hereafter EDSGC) consists of COSMOS data for 1.5 million galaxies, covering 60 Schmidt fields centred on the SGP (21h $< \alpha <$ 3h, $-45° < \delta < -20°$, $\sim$ 0.5 steradians). The fully automated star/galaxy classification techniques we developed to construct the catalogue result in a 95% completeness to a limit of $b_j \sim$ 20.5 and a residual stellar contamination of 10% brighter than this magnitude (Heydon-Dumbleton et al. 1989). One of the most difficult aspects to the construction of a digitised galaxy survey from individual photographic plates is the photometric matching of these fields. The photometric calibration of the survey has been motivated by the stringent limits in plate-plate errors which are necessary in order that spurious structures are not introduced (Geller et al. 1984; de Lapparent et al. 1986). Over our 60-plate mosaic we have obtained accurate CCD galaxy calibration sequences for 30 fields. For each of these fields we have obtained B and V CCD images of **two** loose clusters. Typically, this results in $10 - 15$ galaxies per field which enables the zero point of each field to be calculated to a magnitude uncertainty of 0.05. For each field without CCD sequences, the surrounding zero-pointed fields are used to determine the plate background magnitude. This procedure minimises the photometric uncertainties that would be introduced if the field overlaps were used to bootstrap calibrations across many fields. The final calibration uncertainty on each plate is $0.^m05$. The consistency between this result and the calibration uncertainty from the CCD sequences shows the stability of the calibration procedure and indicates that there is no significant systematic variation of the photometric zero point across the area of individual fields (Collins et al. 1991b).

2 The Galaxy Angular Correlation Function

The two-point angular correlation function, $\omega(\theta)$, and its spatial equivalent, $\xi(r)$, are the two statistics which have been most commonly used to quantify the distribution of galaxies and clusters on large scales. These correlation functions measure the excess number of galaxies over that expected from a random distribution at a given spatial or angular separation. The main advantage of this function as a measure of large-scale structure is that it provides a direct measure of the amplitude of density fluctuations on different scales, under the assumption that galaxies trace the mass. In this way, the correlation function can provide a direct constraint on the amplitude of fluctuations in the early universe and the shape of the initial fluctuation spectrum (Peebles 1980; Bond & Couchman 1988). In addition, any putative feature in the correlation function may well reflect a particular preferred scale of galaxy clustering.

A convenient definition for $\omega(\theta)$ is in terms of the probability of finding a galaxy, with areal number density N_g, in a solid angle element $\delta\Omega$, distance θ from a randomly chosen galaxy:

$$\delta P = N_g[1 + \omega(\theta)]\delta\Omega. \tag{1}$$

In a more useable form $\omega(\theta)$ can be expressed as

$$1 + \omega(\theta) = \frac{n_{gg}}{n_{gr}}\frac{2N_r}{N_g}, \tag{2}$$

where n_{gg} is the number of galaxy pairs with a separation θ and n_{gr} is the number of cross pairs between the galaxy data and a random catalogue of galaxies, inside the same boundary, with a separation θ. For a full review of $\omega(\theta)$ see Peebles (1980) or Nichol (1991).

Figure 1 shows $\omega(\theta)$ estimated over the full area of the EDSGC. We have estimated the correlation function at 4 different magnitude limits (b_j = 17.5, 18.5, 19.5, 20.3) and scaled them independently to the depth of the Lick survey. The error bars on the EDSGC $\omega(\theta)$ points have been empirically determined from analysis over smaller regions of the survey. It is clear that our $\omega(\theta)$ provides strong evidence for large-scale power on scales $\simeq 10°$ or $30h^{-1}$Mpc. Also shown are the correlation functions estimated from the Lick Survey (Groth & Peebles 1977), the APM survey (Maddox et al., these proceedings) and the POSS II Survey (Picard 1991). There is substantial agreement on the amplitude of clustering from all the the machine-based galaxy catalogues. Each find significantly more power on large scales than is seen in the Groth & Peebles (1977) analysis of the Lick catalogue. For a full discussion of the EDSGC $\omega(\theta)$ see Collins et al. (1991).

This has serious repercussions for the popular biased Cold Dark Matter (CDM) model of galaxy formation (Frenk 1986), as this model is unable to predict the large-scale structure seen in these automated catalogues (Maddox 1990). However, recent work on low biased CDM models may be able to explain the large-scale power seen in the angular correlation function (Couchman & Carlberg 1991).

3 The Cluster Catalogue

The EDSGC is an ideal database for constructing an objective galaxy cluster catalogue as it contains accurate positions and photometry for over one million galaxies. During the construction of the EDSGC, all images detected by COSMOS were deblended to check for merged objects (Beard et al. 1990). This is vital for the construction of any automated cluster catalogue, since otherwise the number of

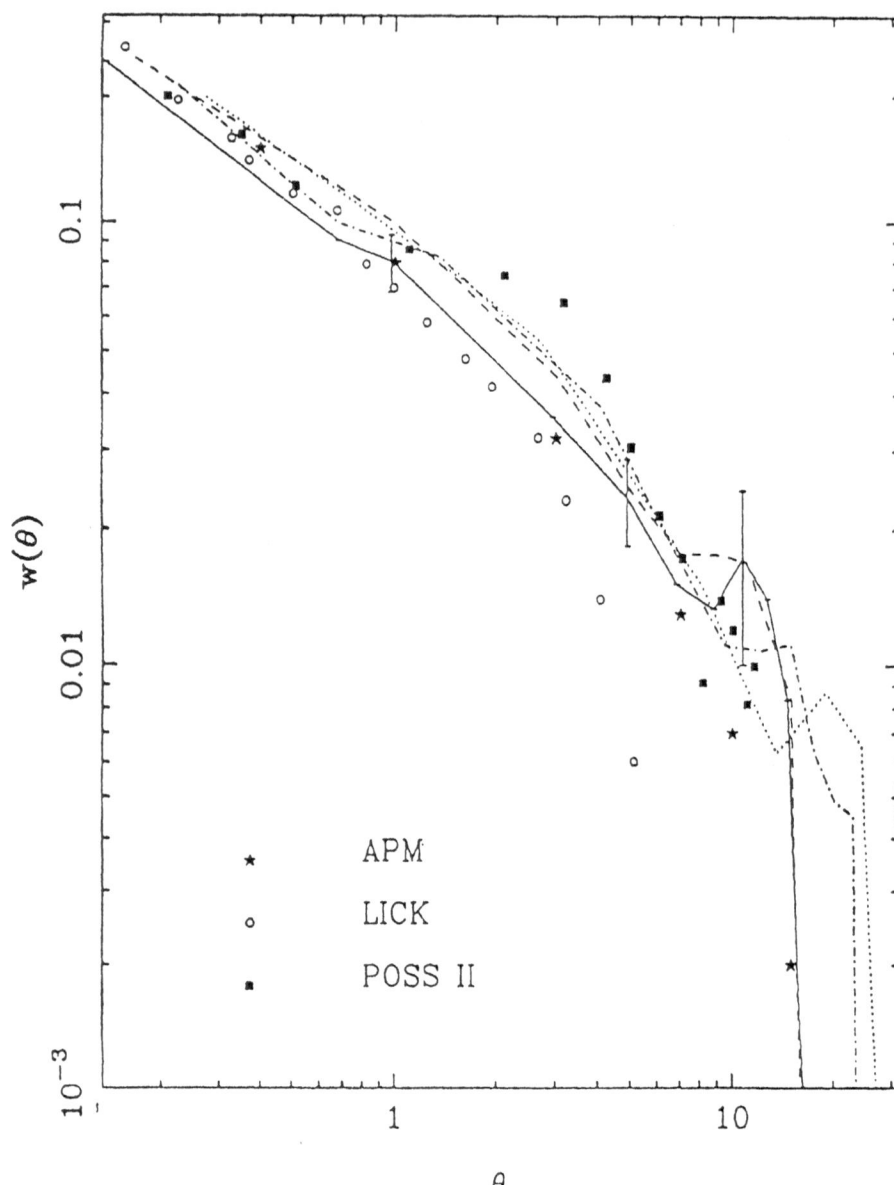

Fig. 1. The angular two-point correlation function for the EDSGC, APM, POSS II and Lick surveys. The lines are the EDSGC estimates of $\omega(\theta)$ at the magnitude limits $b_j = 17.5$ (solid), 18.5 (dashed), 19.5 (dot-dashed) and 20.3 (dots). The results have been scaled to the depth of the Lick survey ($b_j = 18.6$). The scaling of $\omega(\theta)$ with magnitude depth (number density) is an important test on the reality of the form of $\omega(\theta)$ (Peebles 1980). The error bars shown were calculated internally. All the machine-based galaxy catalogues find more power on larger scales than the visually constructed Lick catalogue.

galaxies in the centres of rich clusters would be grossly underestimated. A full description of the construction of the Edinburgh/Durham Cluster Catalogue (EDCC) and a comparison with the Abell catalogue can be found in Lumsden et al. (1991). The essential details of this paper are highlighted below:

- The galaxy data were binned at three different magnitude limits, $b_j = 18.5$, 19.5, 20.5.
- The binned data were smoothed (scale $\sim 1°$) to establish the large-scale variations in the galaxy data. This *sky* frame was subtracted from the data.
- A peak-finding algorithm was employed to find overdensities above a given threshold. These candidate clusters were then deblended in search of daughter images.
- An Abell style analysis was performed on the candidate clusters.
- The 3 selections of clusters (according to the different magnitude ranges) were combined to give a catalogue of clusters with the Abell statistics of richness, m_{10} and m_3. These values were corrected for local sky contamination.
- For overlapping clusters, a two-sided Kolmogorov–Smirnov test was used to compare the magnitude distributions of the clusters. Clusters with the same distribution were combined and new Abell statistics were evaluated.

The reason for performing an Abell analysis on the candidate clusters was to present the EDCC in a familiar framework and to compare it with the Abell catalogue. This provides qualitative estimates of the completeness and accuracy of the Abell catalogue.

- Within the EDCC area, 731 clusters or groups were detected. The EDCC completeness limit is $m_{10}(b_j) = 18.75$.
- Over 80% of the clusters that were in the all-sky Abell catalogue, brighter than our completeness limit, were identified within the EDCC area.
- 70% of the EDCC clusters, brighter than the EDCC completeness limit, are previously undetected.
- There is a large scatter in the richness estimates of clusters in common between the two catalogues, with Abell's estimate of the clusters richness tending to be higher than that of the EDCC estimate. Also Abell's estimate of m_{10} is systematically brighter than the EDCC estimate.
- We estimate that the EDCC is 90% complete and can be used for statistical studies of galaxy clusters.

4 The EDCC Redshift Survey

One of the most important results in recent observational cosmology has been the form of the spatial two-point correlation function of clusters, $\xi_{cc}(r)$. The amplitude of $\xi_{cc}(r)$ is approximately 10 times higher than that of the galaxies with power on scales out to $\sim 50h^{-1}$Mpc (Bahcall & Soneira 1983; Ling et al. 1986; Sutherland 1988). The importance of this result is that both the clusters and galaxies cannot be fair tracers of the underlying mass distribution. The main motive behind the

EDCC was to study the distribution of clusters, via $\xi_{cc}(r)$, using a sample free from projection effects and other systematic errors that have troubled previous investigations of the cluster distribution.

We have now completed the observations of a subset of the 731 clusters within the EDCC. The EDCC Redshift Survey is one of the largest cluster redshift surveys in existence and is an excellent database to re-estimate $\xi_{cc}(r)$. We have obtained spectra for several galaxies per cluster for $\sim$ 100 clusters, thus completing the subset of richness greater than 35 members and $m_{10} < 18.75$. It is vital to obtain several redshifts per cluster to address the problems of possible projections and to be fully confident of the redshift of the cluster. Figure 2 shows the distribution of the number of galaxy redshifts per cluster.

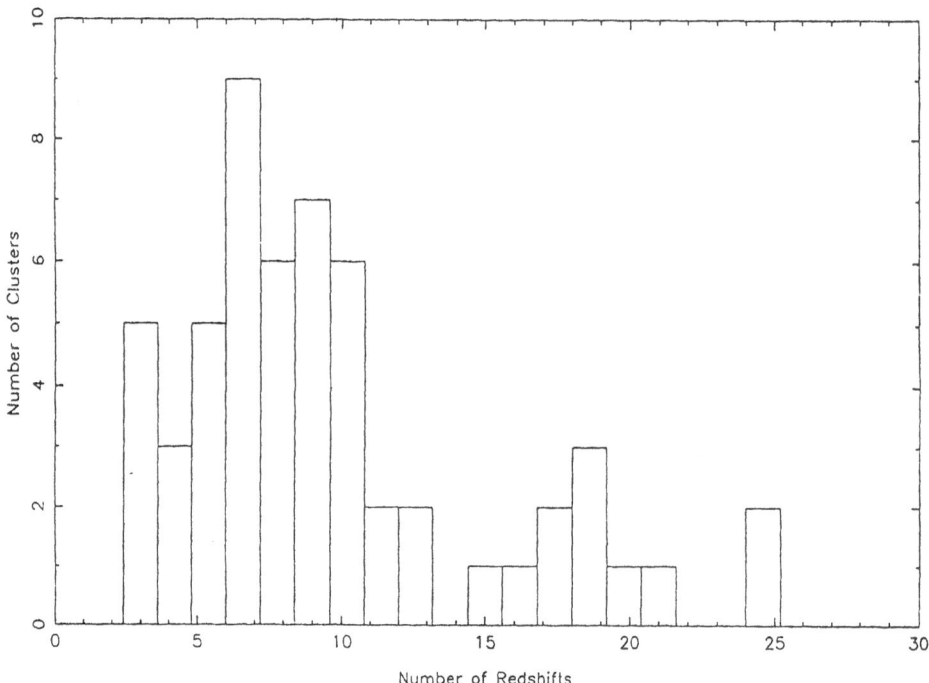

Fig. 2. The distribution of cluster galaxy redshifts. All the interlopers have been removed from this diagram, so this represents the number of true cluster redshifts obtained for a cluster.

The cone diagram in Fig. 3 is for the 53 clusters with data already reduced (less than half the data we possess). This is the deepest large-angle redshift survey to be carried out in the southern hemisphere, probing structures out to $z\sim$ 0.1. From the m_{10}-log(cz) relationship for these clusters this corresponds to an $m_{10}(b_j) = 18.5$, the completeness limit for the EDCC. Even though this is not the complete sample, large scale structures within the cluster distribution are observed. The supercluster between 0 and 1 hour at z=0.11 has not been seen before so prominently.

RA

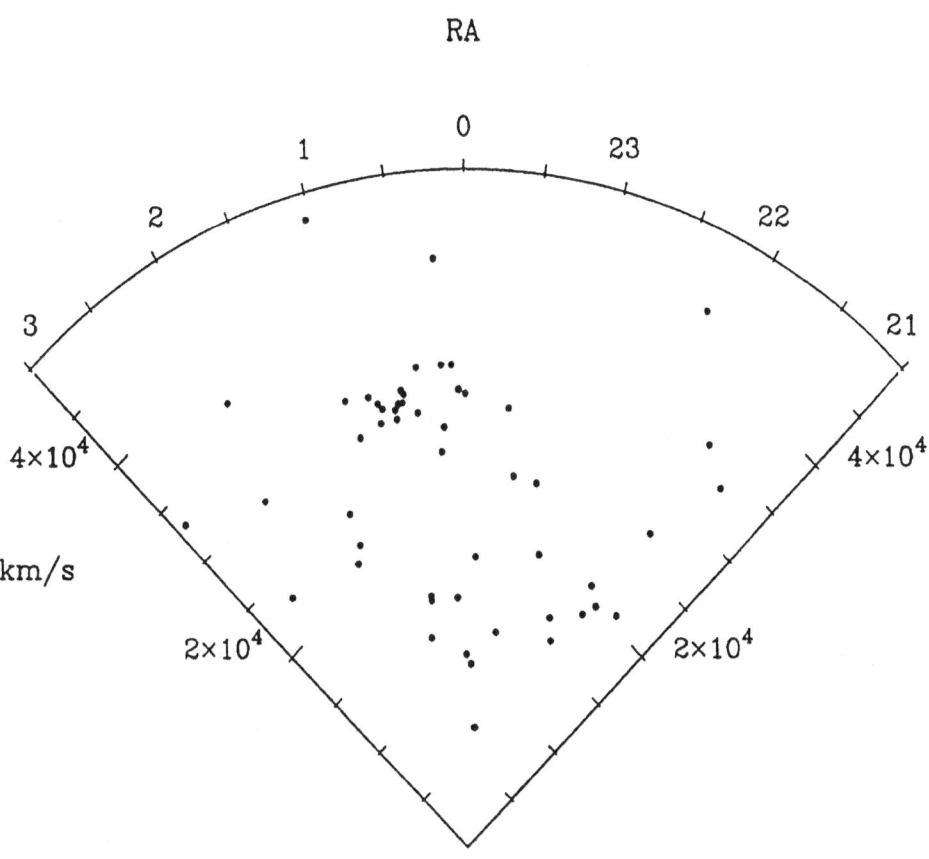

Fig. 3. The cone plot in right ascension for the 53 clusters whose data has been reduced. This is for the declination range $-45° < \delta < -25°$.

The redshift distribution for these clusters is shown in Fig. 4. The arrow marks are the positions of the peaks in the galaxy redshift distribution, on the scale of $128\ h^{-1}$Mpc, claimed by Broadhurst et al. (1990). There is very good agreement between the structure seen in the EDCC redshift survey and the position of the Broadhurst peaks. For further discussion of this agreement see the paper by Broadhurst et al. in this volume.

5 Acknowledgements

We would like to thank Harvey MacGillivray, the COSMOS group and the UKSTU for their continued support during this long–term project, and Tom Shanks, Dick Fong and the Durham group for many useful discussions. We are grateful to Tom Broadhurst for his help during the writing of this article.

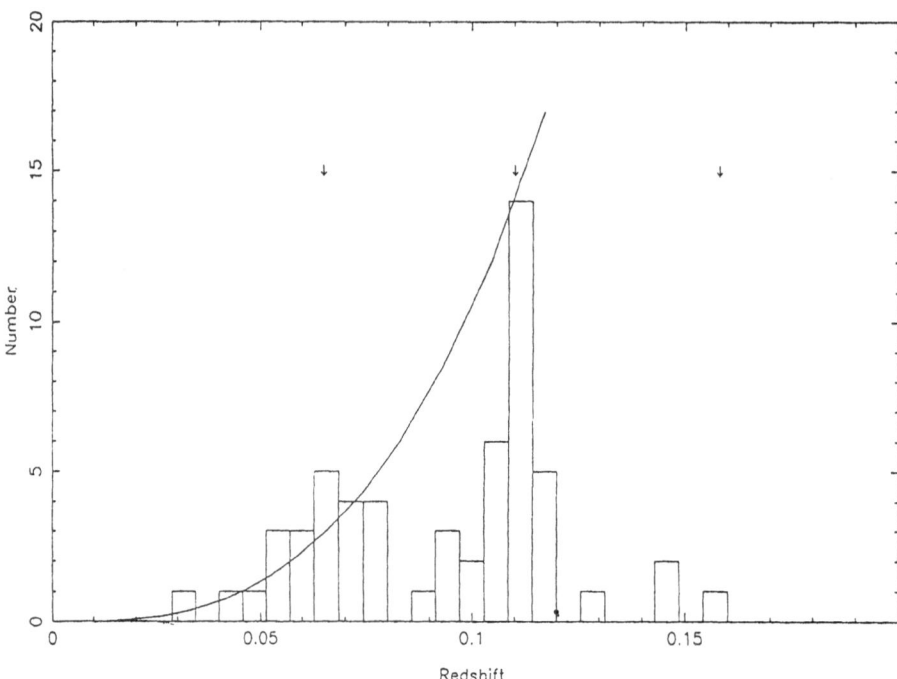

Fig. 4. The redshift distribution for the clusters in Fig. 3. The arrow marks are the positions of the Broadhurst peaks. The solid line is the selection function for our cluster redshift survey (no luminosity dependence has been included).

References

Abell, G.O., 1958. *Astrophys. J. Suppl.*, **3**, 211.

Abell, G.O., Corwin, H.G., Olowin, R.P., 1989. *Astrophys. J.Suppl.*, **70**, 1.

Bahcall, N.A., Soneira, R.M., 1983. *Astrophys. J.*, **270**, 20.

Beard, S.M., MacGillivray, H.M., Thanisch, P.F., 1990. *Mon. Not. R. Astron. Soc.*, **247**, 311.

Bond, J.R., Couchman, H., 1988. *Proc. Second Canadian Conference on General Relativity and Relativistic Astrophysics*, p.385, eds. Coly, A. & Dyer, C., World Scientific, Singapore.

Broadhurst, T.J., Ellis, R.S., Koo, D.C., Szalay, A.S., 1990. *Nature*, **343**, 726.

Collins, C.A., Nichol, R.C., Lumsden, S.L., 1991a. *Mon. Not. R. Astron. Soc.* In press.

Collins, C.A., Nichol, R. C., Shanks, T., 1991b. *Mon. Not. R. Astron. Soc.* In preparation.

Couchman, H.M.P., Carlberg, R.G., 1991. Preprint.

de Lapparent, V., Kurtz, M.J., Geller, M.J., 1986. *Astrophys. J.*, **304**, 585.

Frenk, C.S., 1986. *Phil. Trans. R. Soc. Lond.*, **330**, 517.

Geller, M.J., de Lapparent, V., Kurtz, M.J., 1984. *Astrophys. J.*, **287**, L55.

Groth, E.J., Peebles, P.J.E., 1977. *Astrophys. J.*, **217**, 385.

Heydon-Dumbleton, N.H., Collins, C.A., MacGillivray, H.T., 1989. *Mon. Not. R. Astron. Soc.*, **238**, 379.

Ling, E.N., Frenk, C.S., Barrow, J.D., 1986. *Mon. Not. R. Astron. Soc.*, **223**, 21P.

Lumsden, S.L., Nichol, R.C., Collins,C.A., 1991. *Mon. Not. R. Astron. Soc.* Submitted.

Maddox, S.J., Efstathiou, G., Sutherland, W.J., Loveday, J., 1990. *Mon. Not. R. Astron. Soc.*, **242**, 43p.

Nichol, R. C., 1991. Ph.D. Thesis. University of Edinburgh.

Peebles, P.J.E., 1975. *Astrophys. J.*, **196**, 647.

Peebles, P.J.E., 1980. *The Large-Scale Structure of the Universe,* Princeton University Press, Princeton, New Jersey.

Peebles, P.J.E., Hauser, M.G., 1974. *Astrophys. J. Suppl.*, **192**, 239.

Picard, A., 1991. *Astrophys. J.*, **368**, L7.

Shane, C.D., Wirtanen, C.A., 1967.*Publs. Lick Obs.*, **192**, 209.

Sutherland, W., 1988. *Mon. Not. R. Astron. Soc.*, **234**, 159.

Williams, B., Peacock, J.A., Heavens, A., 1991. *Mon. Not. R. Astron. Soc.* In press.

Discussion

Deul :

Now that you have a mapped 3-D distribution of clusters of galaxies would you care to comment on how well this fits the Voronoi foam (tesselations)?

Nichol :

Williams, Peacock & Heavens (1991) find that the Voronoi foam model cannot simultaneously account for the observed ξ_{gg}, ξ_{cc} and periodicity (Broadhurst et al. 1990). Also recent work by Williams has shown the most severe constraint on the Voronoi model is by the angular correlation function, which it fails to reproduce.

Maddox :

Have you measured the angular correlation function of the cluster catalogue?

Nichol :

Not yet. It is something we plan to look at soon in conjunction with the spatial correlation function from the redshift survey.

MacGillivray :

To what precision do you require the cluster redshifts in order to investigate the spatial correlation function for clusters?

Nichol :

I think the main question here is not the accuracy of the redshift, but is whether there is a cluster there to take a redshift and if so, how confident you are that you have the cluster Z, not an interloper. Even with ten redshifts of a real cluster it is hard to get the cluster Z to an accuracy of a few 100 km s^{-1}, due to the peculiar motions in the cluster. But this is fine, an error of 300 km s^{-1} corresponds to an error of 1/100 in the redshift at a redshift of 0.1. This is the error in the distance, so if you wanted to measure the cluster-cluster correlation function and found $r_o = 20h^{-1}$Mpc, the error is tiny compared to other errors like fitting the data points and the error on those points.

As we are always dealing with small number statistics, the main worry is how confident you are of the cluster Z. If you haven't removed projection effects, or if you have measured the Z of an interloper, you introduce spurious clusters. This is the main concern.

Parker :

(1) Of the 50 cluster 'candidates' observed spectroscopically, how many are true clusters, i.e. not projections along the line of sight or sheets + less rich clusters, and if interlopers etc. are removed would the 'cluster' members that remain still have been picked up by your selection criteria?

(2) How much does the stellar contamination of $\sim 10\%$ in the EDSGC change as a function of magnitude? I have eyeballed and classified all galaxies down to $b_j \leq 16.8$ in 7 UKST fields in the EDSGC and find 15% contamination in 1 field and only 2–5% in all the others.

Nichol :

(1) The clusters I showed have a secure redshift since most of their galaxy redshifts agree. Clusters that didn't have a reliable redshift are not shown. As for removing interlopers, this would have little effect and I am not sure that is what you would want to do. During the selection in 2-D of these clusters, each cluster is corrected for background (or foreground) contamination. For each cluster a 4° square is used around the cluster to estimate the contamination in the magnitude range $M_3 \rightarrow M_3 + 2$ for the cluster.

(2) Heydon-Dumbleton et al. (1989) discuss fully the star/galaxy separation procedure. They eyeballed 5 plates across the EDSGC of varying quality. They also carried out simulations to test their S–G separation. Their findings were (i) $> 95\%$ complete with $< 10\%$ stellar contamination, (ii) only $\sim 3\%$ RMS residual variation in the number of objects classed as galaxies. This was carried out above $b_j = 20$ and was not found to be a function of magnitude. You have eyeballed 7 plates only to a magnitude of $b_j = 16.8$, 6 of those have contamination of 2–5%, not bad!

THE ROE/NRL CLUSTER CATALOG: I. CORRELATION WITH SOUTHERN ABELL CLUSTERS

H. Gursky [1], *R.G. Cruddace* [1], *B.V. Stuart* [1], *D.J. Yentis* [1], *H.T. MacGillivray* [2]
and C.A. Collins [2]

[1] Naval Research Laboratory, Code 4100, 4555 Overlook Ave. SW, Washington DC 20375, U.S.A.
[2] Royal Observatory, Blackford Hill, Edinburgh EH9 3HJ, U.K.

1 Introduction

The Royal Observatory, Edinburgh, (ROE) and the Naval Research Laboratory (NRL) have collaborated to produce a catalog of clusters of galaxies from the galaxies listed in an object catalog produced by ROE from COSMOS scans of the UK Schmidt Telescope survey (UKST) material, namely the collection of IIIa-J survey plates of fields south of 0 degrees declination. The preparation and characteristics of the object and cluster catalogs are described in this volume (Yentis et al. these proceedings). The project of scanning the UKST plates has been ongoing at ROE for the past 4 years and was recently completed. The techniques used in the compilation of the ROE/NRL object catalogue are similar to those used in providing the Edinburgh/Durham Southern Galaxy Catalogue (Collins et al. 1988; Heydon-Dumbleton et al. 1989; Nichol et al. these proceedings). However, the latter catalogue is based on only 60 Southern sky survey fields, while the ROE/NRL study (on the IIIa-J data at least) encompasses all Southern sky survey fields outside the Galactic plane (numbering over 700 fields). The ROE/NRL object catalogue contains several tens of millions of galaxies down to the magnitude limit (20.5m-21m) determined by the star/galaxy separation algorithm. A catalog of 731 rich clusters has already been obtained from the Edinburgh/Durham catalog as described in this volume (Nichol et al., these proceedings), which has been derived from the same plate and scan material but over the much smaller area of sky.

The intent with the ROE/NRL cluster catalog is to be complete to statistical limits set by the material, and most of the constituent members, approximately 67 000 clusters, have few galaxies. At the present time, the catalog is more

H. T. MacGillivray and E. B. Thomson (eds.), Digitised Optical Sky Surveys 345–351.
© 1992 *Kluwer Academic Publishers.*

a database of astronomical objects and we are in the process of establishing its validity as a cluster catalog. In this paper we examine its contents compared to rich clusters listed by Abell et al. (1989: ACO) which presents both the clusters found by Abell on the Palomar Schmidt (POSS) plates and southern clusters found on the UKST plates using the criteria established by Abell in his initial paper (Abell 1958). Other studies of the Abell clusters have been restricted to a small subset of the clusters (cf. Scaramella et al. 1991). In the present paper, we make a comparison with the entire ACO catalog.

The finite size of clusters and the recognized difficulty of defining their centers make precise positional coincidence unreliable for establishing the correspondence of a celestial object with a cluster. Because we have a large data set, we can use a separation analysis to determine which entries of our data set correspond to the ACO clusters. The separation analysis consists of finding sensible correlations for pairs of clusters (one from each data set) as a function of the separation of the pair members.

2 Characteristics of the ROE/NRL Clusters

As discussed in Yentis et al. (these proceedings), clusters were found in the galaxy distribution using both percolation and binning techniques. Briefly, percolation uses a 'nearest neighbor' criterion to define a cluster; binning uses the local overdensity. These techniques are used to find statistically significant groups, which are then characterized and filtered using various techniques, including moments analysis, an Abell analysis, and a maximum likelihood fit to a King model to the galaxy distribution.

As an example of the characteristics of the catalog as a data set, Fig. 1 is a scatter diagram of the number of galaxies (corrected for the local background) found within a fitted radius, for a random sample of catalog entries. As noted, most entries correspond to sparsely populated clusters with $N < 10$. The right edge of the distribution, toward lower overdensity, is determined by the statistical criteria used for generating entries into the cluster data base, since these are functions of the local overdensity. The background galaxy density varies significantly over the plate material, but $0.25\,\mathrm{min}^{-2}$ is a good average and is used to calculate the curve showing where clusters with unit overdensity would fall. The average overdensity of selected clusters varies from about 2.5 at large radii to more than 10 at the low end. An overdensity of 10 implies average galaxy separations of about half an arc minute and is probably close to the confusion limit of the material. The left side of the distribution, toward higher overdensity, is in the direction of higher signal-to-noise compared to the background galaxies and should not be limited by statistical considerations imposed by the processing algorithms. Since large numbers of clusters are revealed at high overdensity for the $N < 10$ portion of the distribution, the absence of such high overdensity clusters at larger N is most likely intrinsic to the nature of clusters.

Figure 2 provides the same information for the sample of ROE/NRL clusters found to be coincident with ACO clusters within 5'. (The choice of 5' will be discussed below.) Comparison of Figs. 1 and 2 reveals an excess of clusters for

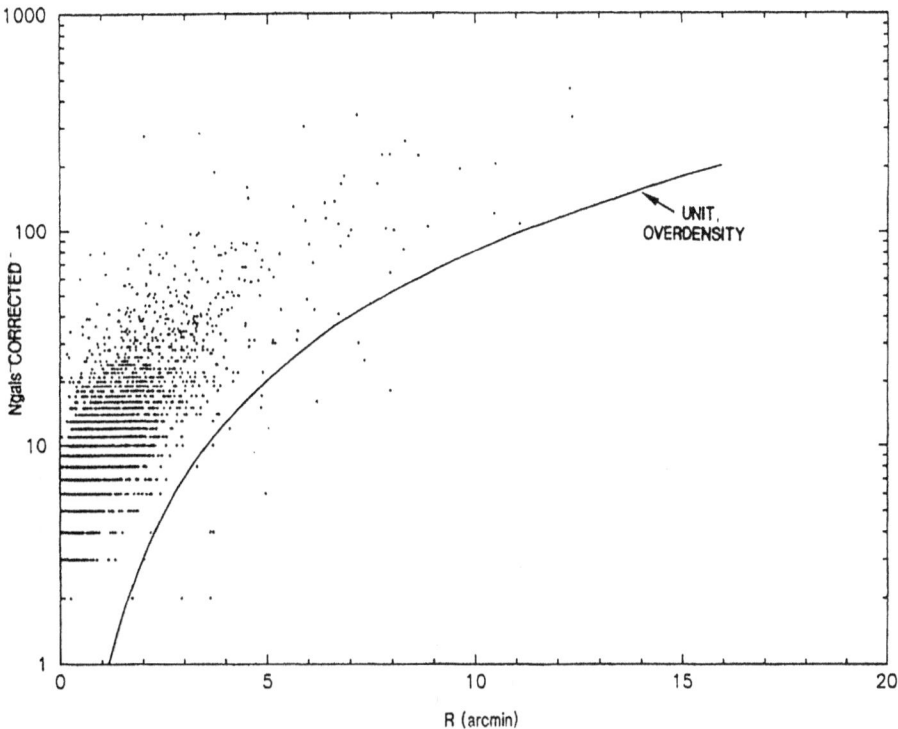

Fig. 1. Scatter diagram showing the number of galaxies over background, N, and the radius, r, for a random set of NRL/ROE clusters.

values of N greater than about 10. These clusters, of course, must be the same as those found by Abell and his colleagues. The same information is presented in Fig. 3 which gives the differential number of clusters as a function of N for the whole catalog and for the clusters found to be coincident with ACO clusters. Not unexpectedly, the fraction of our clusters revealed as ACO clusters rises with N; however, a large number of high-N clusters do not appear to be present in the ACO lists. The discrepancy is a factor of two at $N = 100$, three at $N = 50$ and ten at $N = 20$.

The upper curve in Fig. 3 seems to reveal that our clusters represent two independent distributions since there is a distinct break in the curve at around $N = 10$. The high N distribution overlaps the lower curve that contains mostly ACO clusters. As noted, however, the numbers are highly discrepant.

3 Direct Correlation with Abell Clusters

We have performed a separation analysis to correlate our cluster catalog with the ACO catalog. For each ACO cluster in their tables 4 and 5 we find all ROE/NRL catalog entries within 20′. Because of the high space density of our clusters, about 4 deg^{-2}, there will generally be one or two clusters present in the search radius. Figure 4 is a histogram of all separations and reveals a substantial peak at $2 - 3'$

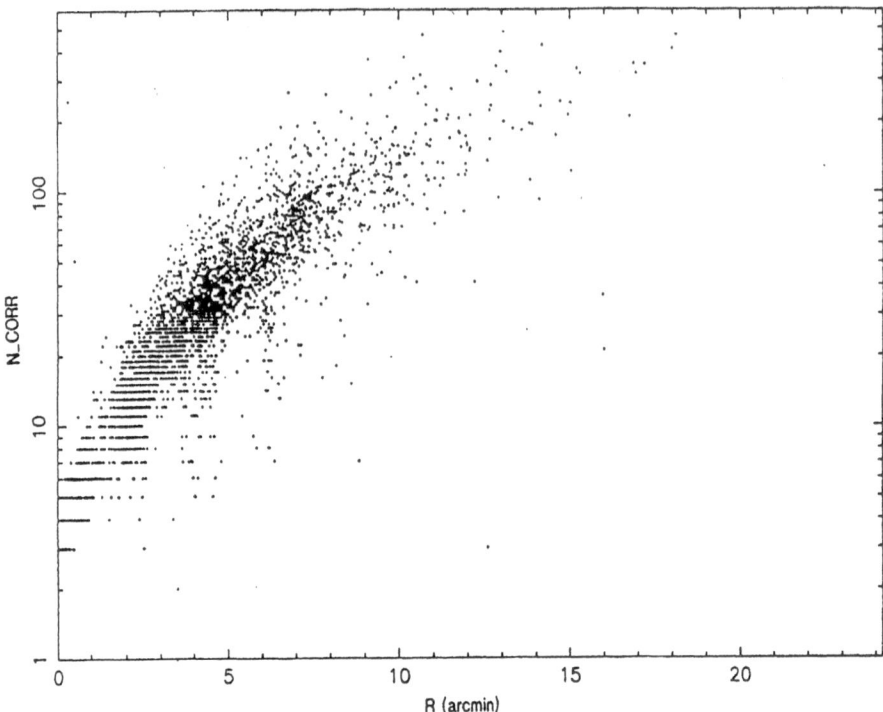

Fig. 2. $N - r$ diagram for ROE/NRL clusters found within 5' of an ACO cluster.

corresponding to real matches between the two data sets. The gradual increase at larger separation corresponds to the increasing likelihood of accidental matches since the annular search area increases linearly with separation. With this data set, 5' separation appears to be a boundary between where real matches dominate and accidental matches dominate. This plot alone makes it clear that we are making matches with most of the Abell clusters. The actual number of matches within 5' is about 2300, compared to the 2099 entries in the ACO catalog. The excess significantly exceeds the difference since a number of ACO clusters are not found as matches within 5', and is accounted for by the instances in which there are two or more ROE/NRL clusters that match an ACO cluster. There is also a tail to the distribution, an excess above the straight line expected from just the contribution from accidental matches. This may relate to the fact, noted below, that some cluster centers are poorly determined. We may also be seeing cluster-cluster correlations.

To determine better the comparison between the two catalogs, we examined the fate of every ACO cluster and generated finding charts for all those that did not reveal a matching ROE/NRL cluster within 5'. This amounted to 18% of all ACO clusters for which we had scan material. A casual examination of the finding charts reveals that the principal cause for these absences is discrepant positions between ourselves and ACO. A more detailed examination by one of the authors (H.G.) reveals the following four categories for the fields with approximately equal numbers:

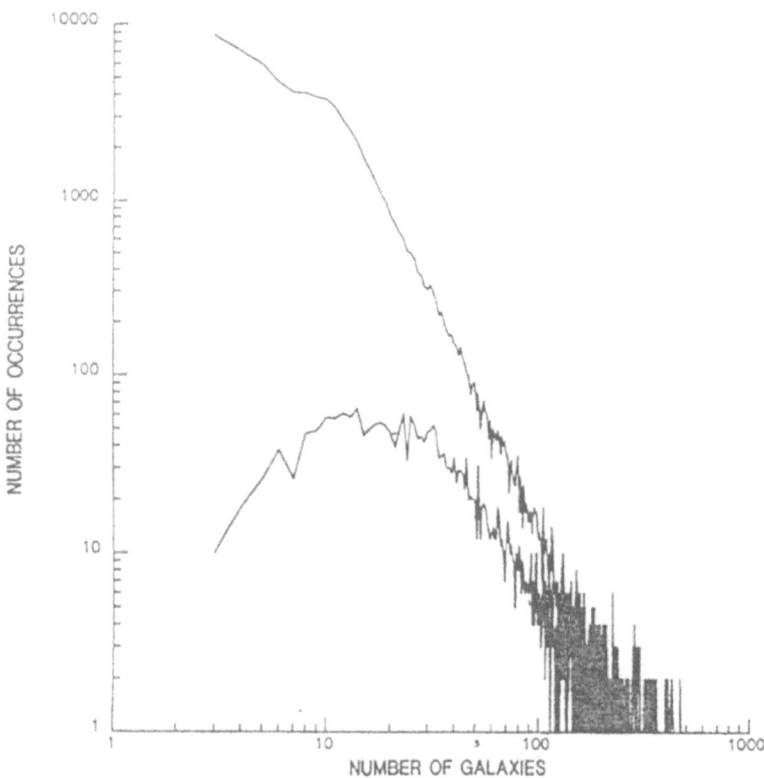

Fig. 3. Number distribution of ROE/NRL clusters versus N. The upper curve is for the data set, the lower curve, those clusters within $5'$ of an ACO cluster.

a. A complex field where ACO seem to have made a better choice of the cluster center or where we appear to have missed a cluster.

b. A complex field where we seem to have made a better choice of the cluster center.

c. A complex field where it was not obvious which of the two cluster centers is correct.

d. A field with no obvious cluster.

To further check on the difficulty of finding cluster centers, we examined the ACO catalog itself which lists 275 objects in the overlap zone between the POSS and UKST plates for which two independently determined positions are provided. For this sample, approximately 15% of the clusters show a $5'$ or greater discrepancy between the two positions. Thus it appears that the mismatch in positions for some clusters is a reflection of the problem of finding their centers rather than a failure of our analysis.

In at least some of the cases where the ACO cluster is missing from the ROE/NRL

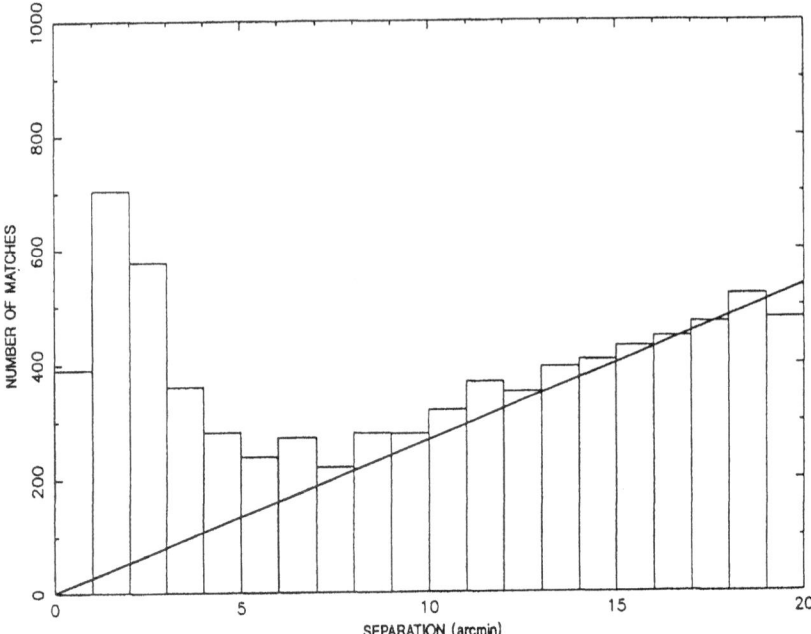

Fig. 4. Number distribution of pairs of ROE/NRL-ACO clusters versus the separation between the pair members.

catalog, there appear to be bright galaxies present in the field, indicative of a relatively nearby cluster. It does appear that our cluster finding algorithm fails to find nearby clusters characterised by widely separated galaxies without correspondingly larger numbers of faint galaxies.

4 Conclusions

Taking into account the fact that centers are difficult to define for many clusters, we find more than 90% of the entries in the the ACO catalog. The residue is composed of ACO clusters that we seem to have missed and those that simply do not seem to be present. The most significant defect that we have found with our algorithm has to do with finding nearby clusters and those in very crowded fields. We expect to correct this defect by regenerating the cluster catalog, starting with galaxies brighter than the star/galaxy separation limit. This will have the effect of preferentially selecting clusters with brighter members. Also, with a lower density of galaxies we can use larger percolation radii and lower overdensities as selection criteria for clusters, both of which will allow the selection of geometrically larger and nearer clusters. The catalog as it now exists selects clusters based on numbers of galaxies and not with their brightness.

As a whole, when we look at the distribution of the number of galaxies in our clusters a dual distribution appears to be present, with a break at about $N = 10$. The ACO clusters belong to the higher N value clusters, but are much fewer in number.

References

Abell, G.O., 1958. *Astrophys. J. Suppl.*, **3**, 211.

Abell, G.O., Corwin, H.G., Olowin, R.P., 1989. *Astrophys. J. Suppl.*, **70**, 1.

Collins, C.A., Heydon-Dumbleton, N.H., MacGillivray, H.T., 1988. *Mon. Not. R. Astron. Soc.*, **236**, 7p.

Heydon-Dumbleton, N.H., Collins, C.A., MacGillivray, H.T., 1989. *Mon. Not. R. Astron. Soc.*, **238**, 379.

Scaramella, R., Zamorani, G., Vettolani, G., Chincarini, G., 1991. *Astron. J.* Submitted for publication.

Discussion

Raychaudhury :

Since your cluster-finding algorithm has no faint cutoff in magnitude, it would be prone to projection effects, perhaps more so than the Abell cluster catalog. So it is not surprising that you find so many 'clusters' not listed by Abell.

Gursky :

It is certainly correct that Abell was very selective in incorporating clusters in his catalog and we are less so. However, observers generally agree that the Abell lists are incomplete in that there are many fewer clusters at large distances than expected from the numbers found at small distances.

MacGillivray :

I would like to point out that the ROE/NRL cluster-seeking algorithm is at the present time optimised for the detection of compact, distant clusters. For this reason it is to be expected that there may be some difficulties with the more amorphous low-redshift clusters. It is our intention to reprocess the COS-MOS/UKST object catalogue to detect the lower surface brightness clusters.

Nichol :

Do you have any statistics about the number of clusters missed in the Abell catalogue?

Gursky :

There appear to be a large number, perhaps 60% more, of additional clusters of comparable richness in the South compared to those listed by Abell, Corwin and Olowin. However, Abell used fairly stringent requirements on brightness, radius and number compared to us for inclusion in our catalog.

THE ROE/NRL CLUSTER CATALOG: II. CORRELATION WITH SERENDIPITOUS EINSTEIN X-RAY SOURCES

H. Gursky [1], R.G. Cruddace [1], B.V. Stuart [1], D.J. Yentis [1], H.T. MacGillivray [2] and C.A. Collins [2]

[1] Naval Research Laboratory, Code 4100, 4555 Overlook Ave. SW, Washington DC 20375, U.S.A.
[2] Royal Observatory, Blackford Hill, Edinburgh EH9 3HJ, U.K.

1 Introduction

A large number of x-ray sources were discovered serendipitously during the Einstein Mission. One set, brought together as the medium sensitivity survey (Maccacaro et al. 1982) and the extended medium sensitivity surveys (Stocke et al. 1991), comprised sources found in the field of the imaging proportional counter during observations of specific objects. The other comprised sources found when the observatory maneuvered from one target to the next and the imaging proportional counter was functioning. During such maneuvers, a half degree swath of sky was scanned and occasionally a bright source would be detected. Such sources were catalogued as the slew survey (Elvis et al. 1991). For both surveys, substantial effort has gone into obtaining optical and radio counterparts to the x-ray sources. We have correlated these catalogs with the ROE/NRL cluster catalog obtained from the UK Schmidt Survey (UKSS) of the southern hemisphere in order to determine independently which x-ray sources in the Einstein Mission intensity range originate in clusters of galaxies to the limit of the UKSS material. The details of the correlation analysis and a description of our catalog is given in the previous paper (Gursky et al., paper I, this volume).

2 The Slew Survey

The slew survey revealed 1067 sources with a typical flux of 5×10^{-12} ergs/cm^2-sec and positional errors of about $2'$. Slews covered almost the entire sky. Elvis et al. (1991) have made correlations with certain published catalogs of objects that are

H.T. MacGillivray and E. B. Thomson (eds.), Digitised Optical Sky Surveys 353–359.

known to exhibit significant x-ray emission and find optical and radio counterparts for 626 of their entries. Of 80 galaxy cluster counterparts, 32 are in the southern hemisphere and in the region covered by the ROE/NRL catalog — 29 from the Abell et al. (1990) catalog and 3 from other lists of clusters.

We used a separation analysis to correlate our cluster catalog with the slew sources by finding all ROE/NRL clusters within 20′ of each entry in the slew catalog. Figure 1 shows the number of matches at each separation out to 20′. The linear increase at S greater than about 5′ is just what is expected from random matches since the annular search area increases as S. The excess number of matches at low S can be attributed to real coincidences between clusters and x-ray sources; in particular, there are 83 matches within 5′ whereas only about 30 are expected based on random associations. Thus, this analysis reveals that there should be about 50 real associations between slew x-ray sources and clusters in the region comprising the UKSS plates. These numbers include duplications, cases where more than one of our clusters comes within 5′ of an x-ray source. Since the space density of the x-ray sources is very low, there is no case of the reverse, where more than a single x-ray source correlates with a given cluster. As discussed in paper I, the choice of 5′ is somewhat arbitrary; indeed, Fig. 1 indicates that a significant excess may exist out to 6′.

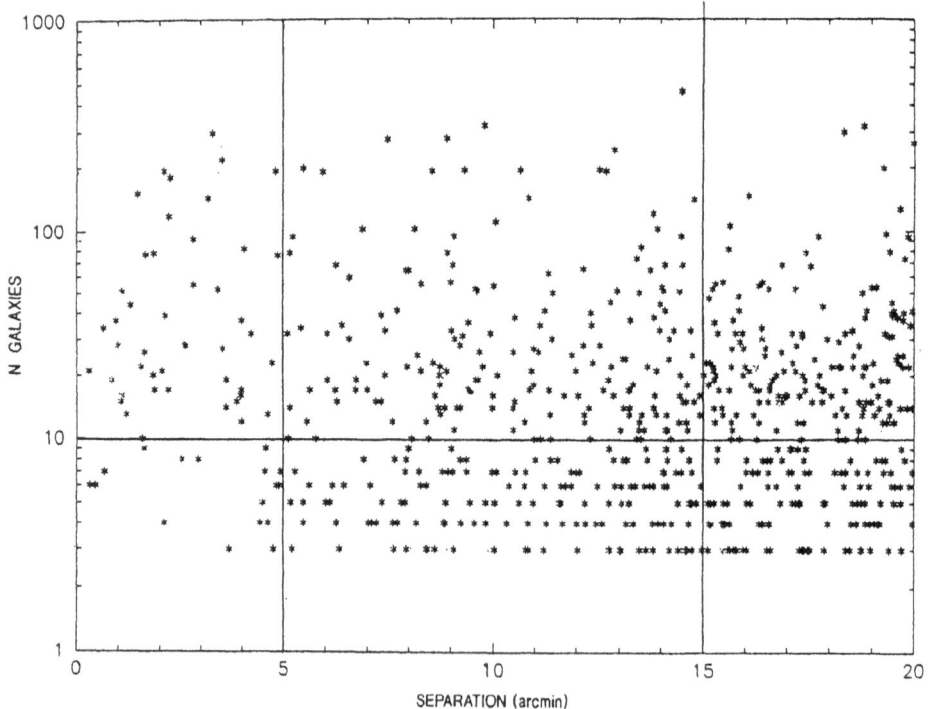

Fig. 1. Number distribution of matching pairs of ROE/NRL clusters — slew sources versus the separation between the pair members.

Figure 2 shows the same data, but in the form of a scatter diagram with the number of galaxies in the cluster, N, corrected for the local background, and the separation of the matching pair as the two parameters plotted for each match. These data reveal that a distinct break appears at $N = 10$, which, as was noted in paper I, also appears for both the cluster catalog as a whole and the ACO clusters. Of the 83 matches, 70 are in the block $N > 10$, $S < 5'$, but only 20 are expected based on random associations. The remaining 13, in the block $N < 10$ and $S < 5'$, are consistent with being random associations. Again we find an excess of 50, but concentrated in the $N > 10$, $S < 5'$ block.

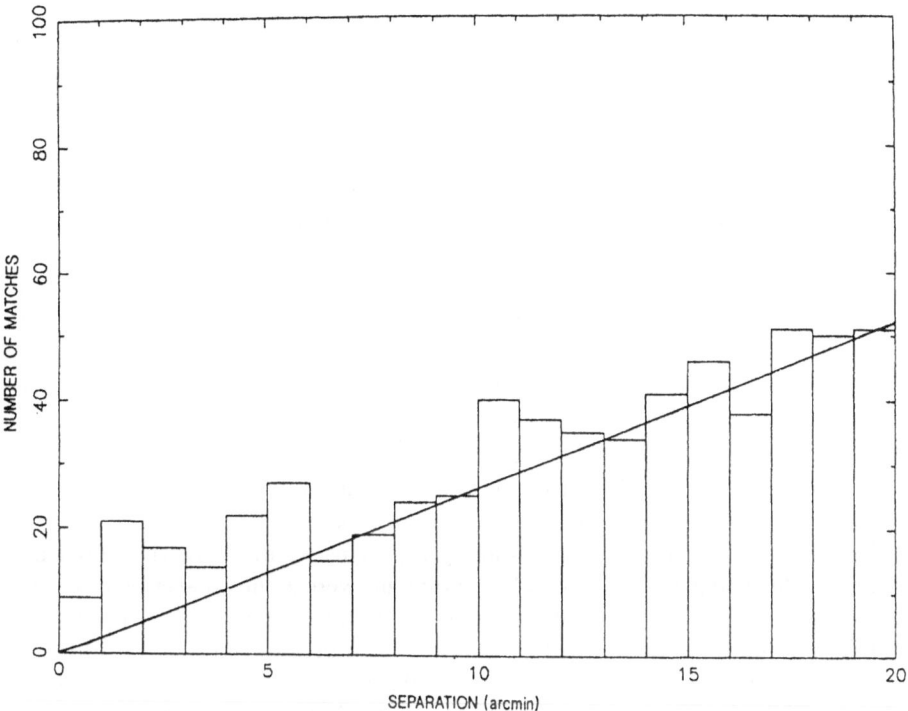

Fig. 2. Scatter diagram showing the number of galaxies in the matching ROE/NRL cluster and separation, for each of the pairs shown in Fig. 1.

Our next step was to search through the list of 83 matches, comparing our cluster candidates to candidates proposed by Elvis et al. (1991). There are 17 duplicate ROE/NRL cluster candidates; thus, the 83 matches correspond to only 56 individual x-ray sources. Of these, we find all 29 southern ACO clusters and 1 of 3 non-ACO clusters. These clusters range in N from 9 to 467 with a median value of about 120 galaxies. Finding all the ACO clusters is a small surprise, since in paper I we show that we 'miss' about 18% due to a variety of reasons. We find 17 other matches with $N > 10$ ranging in N from 10 to 467, with a median value of about 22. Of these Elvis et al. list non-cluster identifications for 7. Thus our analysis has selected credible cluster candidates for about 40 slew sources, of which 30 were previously catalogued clusters. The discrepancy between 40 and the expected 50 can be accounted for by duplications and statistical fluctuations.

The principal conclusion of this analysis is that the NRL/ROE catalog is a reliable source of cluster candidates for the brighter Einstein sources using the techniques outlined here. Also, a very high percentage (about 75%) of cluster x-ray sources at the intensities recorded by the slew survey are rich clusters as originally defined by Abell (1958) and listed in the ACO catalog. We find all of these, plus others that, on average, are of lower N values than the ACO clusters. In paper I we noted that the ROE/NRL cluster catalog was dominated by clusters that were not listed by ACO, with the exception of clusters with the highest values of N.

3 The Medium Survey

The medium survey has gone through several iterations. The extended medium survey, comprising 835 sources found in a net search area of 780 deg^2 at galactic latitudes greater than 20 degrees, was subject to a thorough study by Stocke et al. (1991) with a goal of finding optical candidates for all entries. The group achieved remarkable success, having found candidates for 95% of them; in particular, they found 104 clusters of galaxies. Fifty-seven of these were more distant than $z = 0.2$, nominally the limit of the Schmidt survey plates and the ACO survey. Of the remainder, only 15, or 30%, were ACO clusters. By comparison with the slew survey, where 75% were ACO clusters, we see clearly the sharp change in fraction of sources showing up as Abell clusters with decreasing x-ray intensity. This change reflects the much fainter sources found in the medium survey, in the range 5×10^{-14} to 3×10^{-12} ergs/cm^{2}sec, significantly lower than those in the slew.

We proceeded with a separation analysis just as we did with the slew survey. Figures 3, the histogram of the number of matches versus the separation of the pair, and 4, the scatter diagram showing the excess number of galaxies of the matched ROE/NRL clusters and the separation of the pair, present our basic results. From Fig. 3 we estimate that the sample, about 400 sources, shows an excess of about 25 sources within the 5' separation limit used above; however, there does not appear to be any significant excess at separations beyond 3'. Figure 4 reveals that there is no excess in the block $N < 10$, $S < 5'$ (17 matches predicted on chance, 18 observed); in the block $N > 10$, $S < 5'$ there are 45 matches shown and only 23 predicted to be present on chance. Thus, as with the slew sources, almost all the cluster candidates should appear with $N > 10$ and most of these should appear at $S < 3'$.

To go further, we compared, one-by-one, the 65 matches that appear at $S < 5'$, with their optical identifications as reported by Stocke et al. (1991). First we found that there were 7 duplications, reducing the number of unique matches to 57. We then sorted the matches into those with $N > 10$, $S < 3'$, where, based on Fig. 3, almost all the real matches should appear, and the 'Remainder', and counted the number of identifications of the various categories listed by Stocke et al. The result is shown in Table 1. (The total number of entries in the table is 58 since there is one case of two candidates for a given source.)

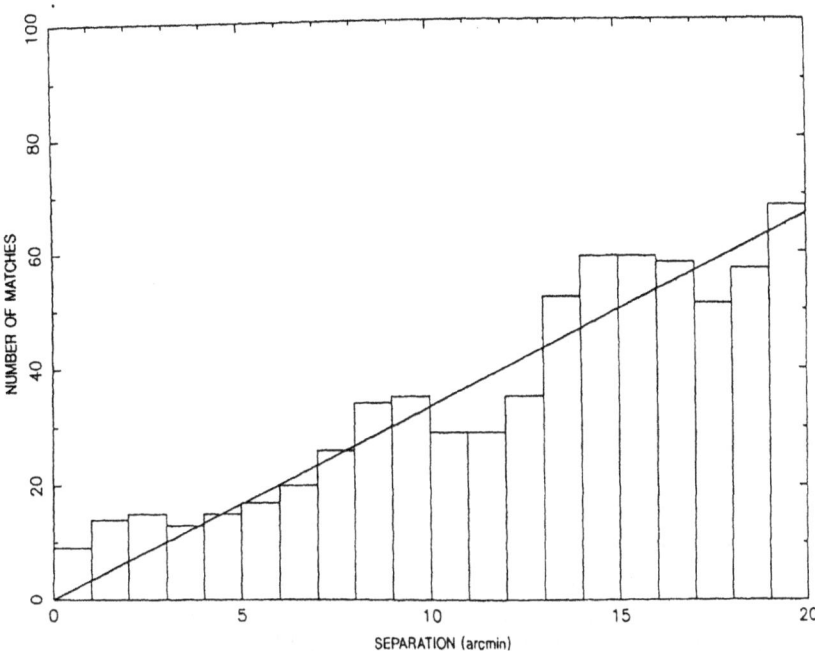

Fig. 3. Number distribution of matching pairs of ROE/NRL clusters — medium sensitivity sources versus the separation between the pair members.

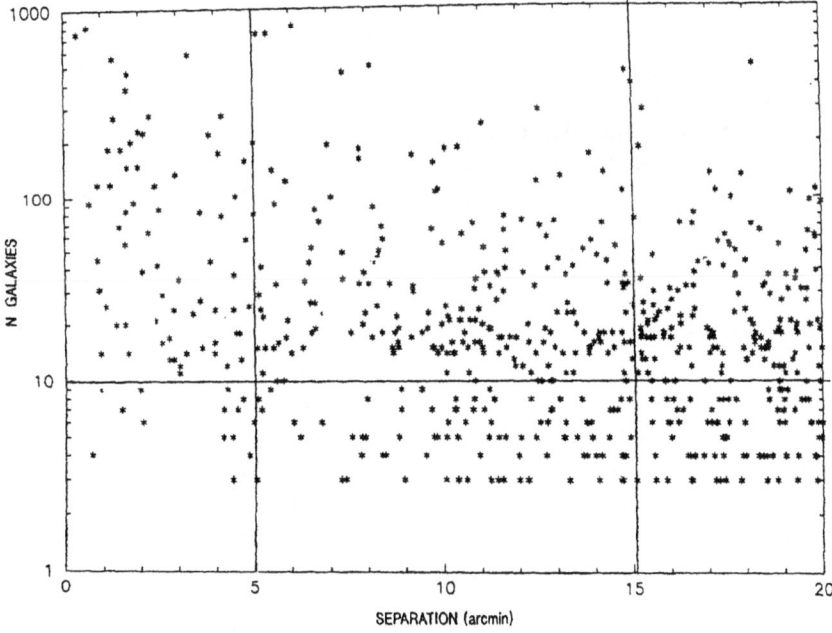

Fig. 4. Scatter diagram showing the number of galaxies in the matching ROE/NRL cluster catalog and separation, for each of the pairs shown in Fig. 3.

Table 1.

	$N > 10,\ S < 3'$	Remainder
Clusters	18	4
AGN	4	18
Bl Lac	0	3
Stars	2	7
Unidentified	1	1

The number of clusters found by Stocke et al. is essentially the same as what we estimate should be present. However, we note the following:

a. Stocke et al. cite cluster counterparts (with $z < 0.2$) for 7 other sources, including two ACO clusters, which do not appear in our catalog.

b. Three of the four clusters in the Remainder column are large N clusters (65, 121 and 159) with radii much larger than their separation from the matching x-ray source. These are all ACO clusters and probably represent genuine matches with x-ray sources, especially in view of the results in paper I regarding the difficulties in defining cluster centers.

c. The remaining cluster in the Remainder column, MS0450.6-5602, may represent a different kind of object compared to the other cluster X-ray sources. It is characterized by a small, closely packed group of galaxies. We find one other, MS0905.6-0817,2, identified by Stocke et al. (1991) as an AGN at z=0.071, that may be in the same category.

d. There may be a few misidentifications in the above table, especially among the 4 AGNs in $N > 10,\ S < 3'$ column. For example, MS2118.4-1050, is identified as a z=0.092 AGN. Our catalog shows a 14-member cluster to be present there as well.

4 Conclusions

The principal conclusions from this paper and paper I are technical in nature, having to do with the efficiency of our analysis for finding certain kinds of clusters. In paper I we demonstrated that we could find over 90% of ACO clusters. Thus, it was not entirely surprising that we find all 29 of the ACO clusters listed in the slew survey, although we failed to find 2 of 5 ACO clusters within a 5′ separation for the medium survey. We find at least 10 uncatalogued clusters among the slew x-ray sources.

In the case of the medium survey, we find 18 (or 22 counting the 4 in the Remainder column) of the 29 low redshift clusters identified by Stocke et al. (1991) in the South. Since Stocke et al. confirm the presence of their clusters by measuring redshifts, this figure provides a lower limit to our ability to find low richness clusters.

Thus, even in its present state the cluster catalog provides a useful and quantitative starting point for identifying clusters and making identifications with other objects, such as x-ray sources. We should be able to improve the catalog and the identification process by making use of the magnitude of the galaxies and the radii of the clusters. The confidence in x-ray source counterparts can be further improved by using known characteristics of the class of objects; for example, the finite size of cluster x-ray sources.

References

Abell, G.O., 1958. *Astrophys. J. Suppl.*, **3**, 211.
Elvis, M., Plummer, D., Schachter, J., Fabbiano, G., 1991. *Astrophys. J. Suppl.* Submitted.
Maccacaro, T. et al., 1982. *Astrophys. J.*, **253**, 504.
Stocke, J.T, Morris, S.L., Fleming, T.A., Gioia, I.M., Maccacaro, T., Schild, R., Wolter, A., Henry, J.P., 1991. *Astrophys. J. Suppl.* Submitted.

AN AUTOMATED SEARCH FOR COMPACT GROUPS OF GALAXIES

I. Prandoni [1], A. Iovino [1], R.K. Bhatia [1], H.T. MacGillivray [2], G. Palumbo [3] and P. Hickson [4]

[1] Osservatorio Astronómico de Brera, Via Brera 28, 20121 Milano, Italy.
[2] Royal Observatory, Blackford Hill, Edinburgh EH9 3HJ, U.K.
[3] Osservatorio Astronomico di Bologna, Via Zamboni 33, I-40126 Bologna, Italy.
[4] University of British Columbia, Vancouver B.C., V6T 1W5, Canada.

1 Why an Automated Search for Compact Groups?

Compact groups (hereafter CGs) are small, relatively isolated systems of galaxies with projected separations comparable to the diameter of the galaxies themselves.

Stephan's quintet is a classical example, where interactions are clearly taking place between member galaxies, suggesting a true physical association of group members (Fig. 1). However, in general, redshift information is needed to assess the reality in space of what can be just a projection effect.

The most recent and largest sample of CGs available is the one obtained through visual searches on red POSS prints by Hickson in 1982 (covering nearly 67% of the sky). It contains 100 CGs, on which extensive work has been done, generating controversy and debate regarding the true nature of such systems and their implications [the possible absence of precursors and descendants, the discrepant red-shifts, the unexpected population properties (see Hickson 1990; White 1990)]. It is therefore of great interest to get a larger and deeper sample of CGs.

Nowadays, large galaxy catalogues are becoming available using fast measuring machines and star-galaxy separation algorithms on plates like the UKST, whose quality is superior to the original POSS plates. A natural choice is to search in such machine-based catalogues using an automated procedure. This makes it possible to produce a sample of CGs, which is homogeneous within well-specified and uniformly applied selection criteria. The advantage of such an approach is evident, being not only more reliable and faster, but also allowing the exploration of the parameter

H. T. MacGillivray and E. B. Thomson (eds.), Digitised Optical Sky Surveys 361–365.

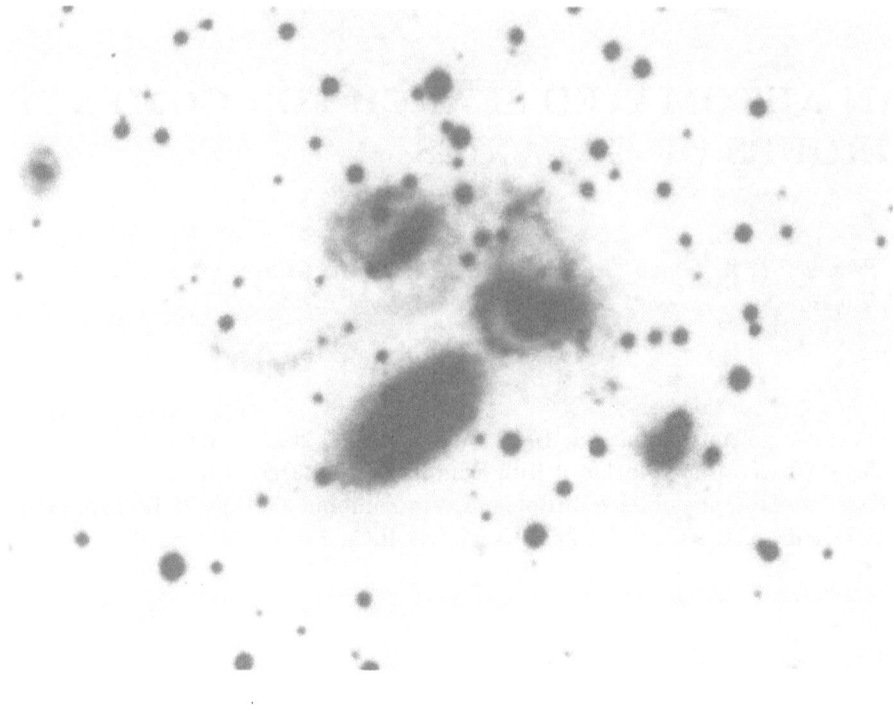

Fig. 1. Stephan's quintet.

space for the extraction of different samples, varying the original Hickson selection criteria.

We are currently undertaking a programme aimed at identifying CGs in the COSMOS data of the UKST Southern Sky survey.

2 Algorithm Used

The selection criteria used in the algorithm (Hickson 1982) are:

$$n \geq 4 \text{ with } m \leq m_B + 3 \qquad \text{(richness)}$$
$$R_N \geq 3R_G \qquad \text{(isolation)}$$
$$\mu_G < 26 \qquad \text{(compactness)}$$

where m_B is the estimated red magnitude of the brightest group member, R_G is the radius of the smallest circle containing the centers of the group members, μ_G is the mean surface brightness contained by this circle, and R_N is the distance from the center of the circle to the nearest non-member galaxy satisfying the same magnitude criterion. According to the above criteria, an obvious choice is to restrict the search for CGs around the galaxies brighter than $m_{lim} - 3$, if m_{lim} is the limiting magnitude of the galaxy catalogue available. These galaxies are the candidates for the brightest group member, and therefore are the starting point for the search of a possible CG.

The next step is to define the maximum size of the area up to which it is 'reasonable' to extend this search.

Using the isolation constraint and the compactness constraint, it is possible to define an upper limit R_{max} for the size of the CG, in the sense that the probability of having a compact group with radius $R_G > R_{max}$ is less than e.g. 10^{-10}.

The isolation constraint easily gives $R_{max\ isol}$ using the relationship:

$$P = \exp -8\pi R^2 N \tag{1}$$

for the probability of having no galaxies between R and $3R$. N is the surface density of galaxies between m_B and $m_B + 3$, obtained from number counts available in the literature (e.g. Shanks et al. 1984).

The compactness constraint also gives an $R_{max\ comp}$ using the relationship:

$$P = 1 - \exp(-\pi R^2 N) \sum_{i=0}^{n-1} (\pi R^2 N)^i / i! \tag{2}$$

for the probability of having n or more galaxies within a radius R and the relationship between R and n imposed by the limit on μ_G:

$$\mu_G \geq -2.5 \log n + 2.5 \log(\pi R^2) + \tilde{m}. \tag{3}$$

$\tilde{m}$ was imposed to be the same for all the group members and equal to $m_B + 1.5$.

The richness constraint, on the other hand, defines a lower limit R_{min} for the radius of the CG, using the expression:

$$P = 1 - \exp(-\pi R^2 N) \sum_{i=0}^{3} (\pi R^2 N)^i / i! \tag{4}$$

for the probability of having 4 or more galaxies within R.

Fixing a cut-off for each of the above probabilities P, it is possible to define an area in the plane m_B and R_G such that the groups inside it form a statistically homogeneous sample. Figure 2 has been obtained choosing $P = 10^{-10}$.

It is interesting to note that the list of groups originally selected by Hickson (points in Fig. 2) fit well within this area (a possible problem exists for the line expressing the richness constraint, where having neglected clustering could lead to an underestimation of probabilities). The dashed lines are obtained varying the cut-off probability to 10^{-11} and 10^{-9}.

R_{max} is defined for each m_B as the minimum between $R_{max\ comp}$ and $R_{max\ isol}$, the compactness constraint being the more stringent at brighter magnitudes, the isolation one at fainter magnitudes.

The size L of the box around each starting galaxy where the CG search has to be performed is an appropriate multiple of R_{max}: we must be sure to be able to

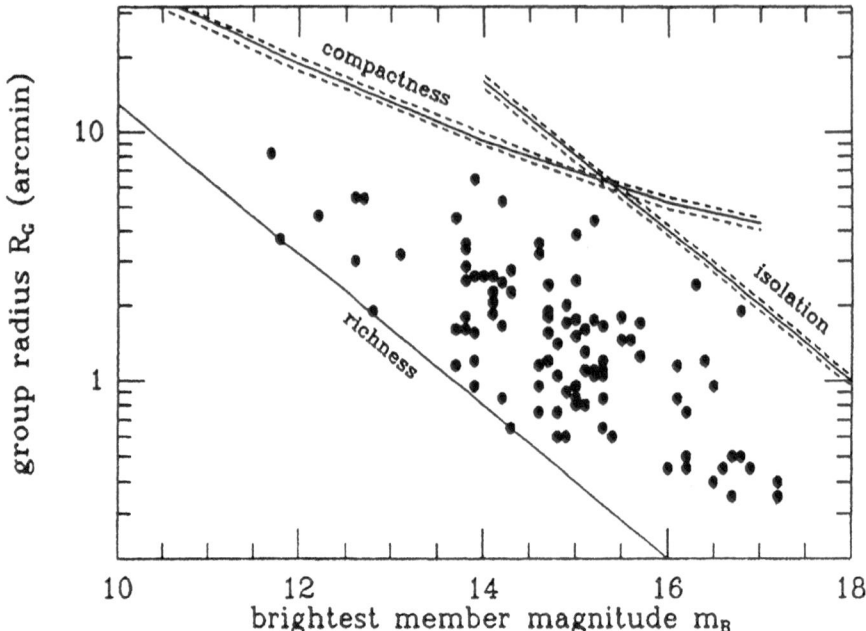

Fig. 2. Area defined by Hickson's criteria in the plane m_B and R_G. Solid lines correspond to a probability cut-off $P = 10^{-10}$, while dashed ones correspond to $P = 10^{-11}$; points represent Hickson's groups.

fully check the isolation constraint also in the 'worst' geometrical position of the starting galaxy with respect to the center of the CG candidate ($L = 8R_{max}$).

Indexing and ranking tables were used to speed up the extraction of the galaxies in this box.

Such an approach turns out to be very efficient, even in comparison to classical techniques (e.g. percolation). The main advantages are the direct implementation of the selection criteria, and therefore the possibility of varying them easily, and the production of a statistically homogeneous sample in terms of the criteria themselves.

3 First Results and Future Work

The algorithm described above was applied to the catalogue of galaxies obtained using COSMOS scans (MacGillivray & Stobie 1984) of 8 UKST b_j plates in an area near to the SGP. Figure 3 shows two of the three CGs found. On relaxing the compactness criterion, two other groups were found.

We plan to apply our algorithm to the whole of the southern sky using the COSMOS/UKST Southern Sky database (see Yentis et al., these proceedings), going deeper than Hickson did, and to explore the parameter space for the extraction of different samples varying the original Hickson selection criteria.

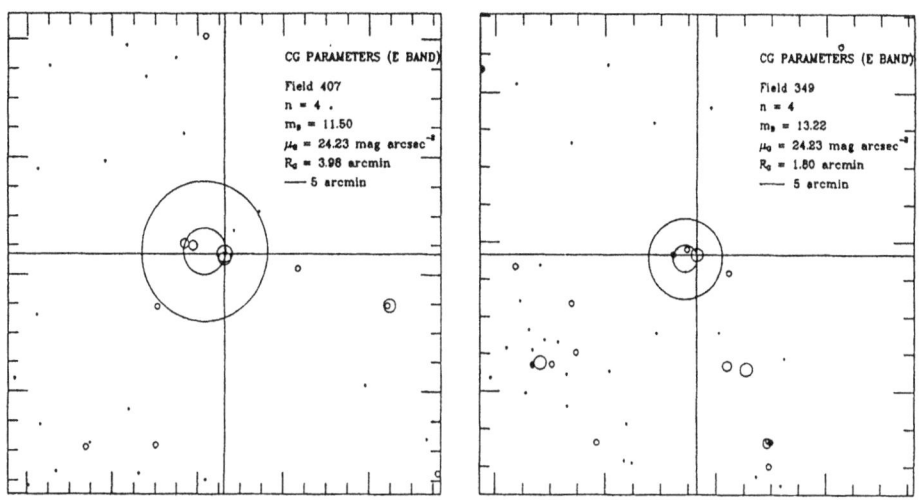

Fig. 3. Two compact groups found in the COSMOS Southern Sky data. The plot shows the box of size L (see text) centred on the brightest group member. The two circles of radius R_G and R_N are also shown.

References

Hickson, P., 1982. *Astrophys. J.*, **255**, 382.

Hickson, P., 1990. *'Paired and Interacting Galaxies'*, *IAU Colloquium* No. 124, eds J.W. Sulentic and W.C. Keel, in press.

MacGillivray, H.T., Stobie, R.S., 1984. *Vistas in Astronomy*, **27**, 433.

Shanks, T., Stevenson, P.R.F., Fong, R., MacGillivray, H.T., 1984. *Mon. Not. R. Astron. Soc.*, **206**, 767.

White, S.D.M., 1990. In *Dynamics and Interactions of Galaxies*, p. 380. Ed. R. Wielen, Heidelberg, Springer-Werlag.

THE MUENSTER REDSHIFT PROJECT — MRSP

W.C. Seitter

Astronomisches Institut der Universität Münster, Domagstraße 75, D-4400 Münster, Germany.

Abstract

MRSP is the attempt to provide a three-dimensional survey of galaxies and quasars to be employed as a uniform data set for studying strongly interdependent cosmological properties: large-scale structure and larger-scale homogeneity and isotropy, parameters of the Friedmann-Lemaître universe, and predictions from hierarchical and inflationary models. So far, the survey covers about 360 square degrees. Data from the first 180 square degrees were used to develop methods and to obtain preliminary results, presented here to illustrate the above applications.

1 Introduction

MRSP is designed to provide galaxy and quasar data for problem-solving in observational cosmology. Since it appears that on large scales (better: in large volumes) the criteria defining a suitable sample are somewhat different from those used for the detection and description of relatively nearby objects, high spatial resolution may be sacrificed for large numbers of data, from which complete and unbiased subsamples with well-defined properties can be drawn.

Observational material are photographs, taken with the UK-Schmidt and ESO-Schmidt telescopes, directly and with objective prisms, a reduction system including two high accuracy measuring machines PDS2020GM^{PLUS}, RISC-computers designed for fast handling of large numbers of in- and out-put data, and dedicated software for automatic reductions at all levels. In most cases, film copies of the original plates are used; some glass copies and original direct and objective prism plates are employed for special projects, e.g. superposition of scans from direct plates or red objective prism plates to gain depth (gain: 1.4 magnitudes from 6

H. T. MacGillivray and E. B. Thomson (eds.), Digitised Optical Sky Surveys 367–381.

plates, larger redshifts for galaxies up to $z = 0.7$, and for quasars up to $z = 4.7$). General information concerning the MRSP is found in Seitter et al. (1988, 1989), Duerbeck (1989), Horstmann et al. (1989).

2 Two-Dimensional Data

Objects on the direct plates are classified as star or galaxy or are rejected. Their right ascensions and declinations (m.e. $\Delta\alpha$ and $\Delta\delta < 1''$; measuring accuracy: 0.7μm for 15μm step width) and their magnitudes [for galaxies both isophotal and aperture magnitudes ($m_J \leq 20.4$ with measuring accuracies $\Delta m_J = 0.1$)] are determined, together with simple parameters (e.g. moments) describing shape and orientation. The measured positions are used to locate the object and to determine astrometric wavelength zero points for the objective prism plates (positional accuracy in the direction of dispersion: $0.49''$). From the magnitudes we derive corrected total galaxy magnitudes B_T^0 ($\Delta B_T^0 = 0.27$ at $z = 0.1$, the large scatter being mainly due to the uncertain K-correction). The image parameters are used to determine morphological classes (for $14.0 \leq m_J \leq 17.0$: 90% within one Hubble-type (E, S0, Sa, Sb, Sc-Ir); for $17 < m_J \leq 18$: 85%; for $18.0 < m_J \leq 19.0$: $\approx 80\%$). About 22 500 galaxies per plate are found up to $m_J = 20.4$ at high galactic latitudes. Misclassifications of objects amount to $< 10\%$ for $12 \leq m_J \leq 18$ and $< 5\%$ for $18 < m_J \leq 20.4$. Details are given in Horstmann et al. (1989), Cunow (1989), Aniol & Horstmann (1991), Spiekermann (1991a,b,c), Tucholke et al. (1991).

3 Redshifts from Low-Dispersion Objective Prism Spectra

Redshifts z are measured on objective prism IIIa-J plates (reciprocal linear dispersion at $H_\gamma = 246$ nm/mm) up to $z = 0.3$ for about 12 000 objects per plate in the range $14 \leq m \leq 19.7$. Five methods, presented in Table 1, are employed; z-errors are checked internally ($\Delta z = 0.01$) from all data of a given plate and externally from 202 galaxy redshifts obtained with slit spectra. The random error is $\Delta z = 0.02$ for 85% of the measured redshifts, 15% show systematic deviations. For early-type galaxies the random error is slightly smaller, while the systematic deviations reduce to 6%. Final checks are made using distribution functions: Gaussian error distribution, (m, z)-diagram compared with Monte-Carlo (MC)-simulations, (M, z)-diagram compared with MC-simulations, $N(m)$-diagram of galaxies with redshifts compared to all galaxies measured two-dimensionally. Sources of incompleteness are determined and statistical corrections derived.

Spectral features of star-like objects are used to separate stars and quasars. The stellar spectral types, together with colour indices from J- and R-plates, are used to study galactic structure. Up to several hundred quasar candidates with $15 \leq m \leq 19$ are found per plate; for about half of them, automatic z-determination is possible; the results are checked interactively. Details are given in Schuecker (1988, 1990, 1992) and Meijer (1990).

Table 1. Redshift measurements from low dispersion objective prism plates.

Wavelength zero point:	
astrometric	accuracy: 1000 kms^{-1} at $z = 0$
	1500 kms^{-1} at $z = 0.1$

Measuring Procedure:	
automatic	range: $14.0 < m_J < 19.7$

Methods:		
correlation method		line spectra
	I.	(after continuum subtraction with fuzzy filter)
		comparison with redshifted G/K spectra
least squares methods		total spectrum
	II.	comparison with redshifted galaxy spectra of different morphological type
	III.	comparison with redshifted G/K type spectra
break methods		absorption breaks
	IV.	separation-amplitude criterion
	V.	separation-shape criterion
additional criteria		
		redshifts and absolute magnitudes
		outside the range covered by observations
		lead to rejections

4 Statistical Properties of Galaxies

Information contained in the data is summarized in the basic observational diagram m versus z with isopleths (numbers of objects in each (m, z)-bin) for 12 plates. They are adjacent to each other (along at least one side) and (partially) cover a range in declination and right ascension of $22° \times 22°$ near the SGP. This 'five-dimensional' $[\alpha, \delta, m, z, N(m, z)]$ view of more than 100 000 galaxies is shown in Fig. 1.

Morphological classification is required, both for the interpretation of structures (e.g. normal clusters of galaxies dominated by E-galaxies, and rare clusters dominated by S-galaxies) and for the derivation of correction factors (e.g. internal absorption in galaxies for the determination of B_T^0).

Mean true axial ratios Q_0 are obtained from apparent axial ratios, separately for the major morphological types, assuming Gaussian distribution of true axial ratios

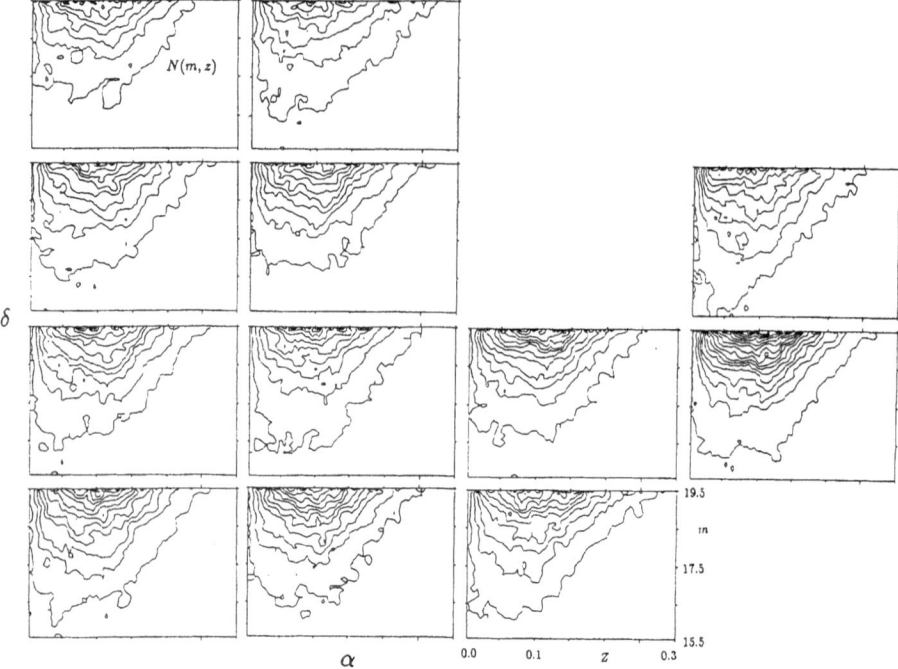

Fig. 1. 'Five-dimensional distribution' of more than 100 000 MRSP-galaxies. Note that vertical lines within each field represent uncalibrated luminosity functions, lines parallel to an isopleth which covers sufficiently large m- and z-intervals are lines of constant absolute magnitude and thus imply constant values for H_0. Lines of constant q_0 and constant K-correction are nearly parallel to lines of constant Hubble values, underlining the difficulty of deriving these corrections.

and random orientations. The results are: $Q_{0,\mathrm{E(oblate)}} = 0.66 \pm 0.18$, $Q_{0,\mathrm{E(prolate)}} = 0.74 \pm 0.18$, $Q_{0,\mathrm{S}} = 0.13 \pm 0.06$.

Internal absorption and optical depth for face-on view were determined for 15 000 S-type galaxies using three models discussed by Disney et al. (1989), see Fig. 2.

Both the (universal) luminosity function (LF) and the (universal) diameter function (DF) of galaxies up to $z = 0.3$ were determined. The different parts of the LF and DF were obtained after transformation to the present hypersphere from galaxy counts (corrected with a magnitude-dependent selection function) in given M_{BT}^0 and diameter intervals, respectively, z-intervals and solid angles. The results are in good agreement with Peterson et al. (1986), Kirshner et al. (1983), Huchra et al. (1983) and others. For details see Cunow (1991) and Schuecker (1990).

5 Clusters and Superclusters of Galaxies

Among the objectives of MRSP are the search for and study of clusters of galaxies and superclusters. About 50 bright clusters of galaxies per plate are readily seen,

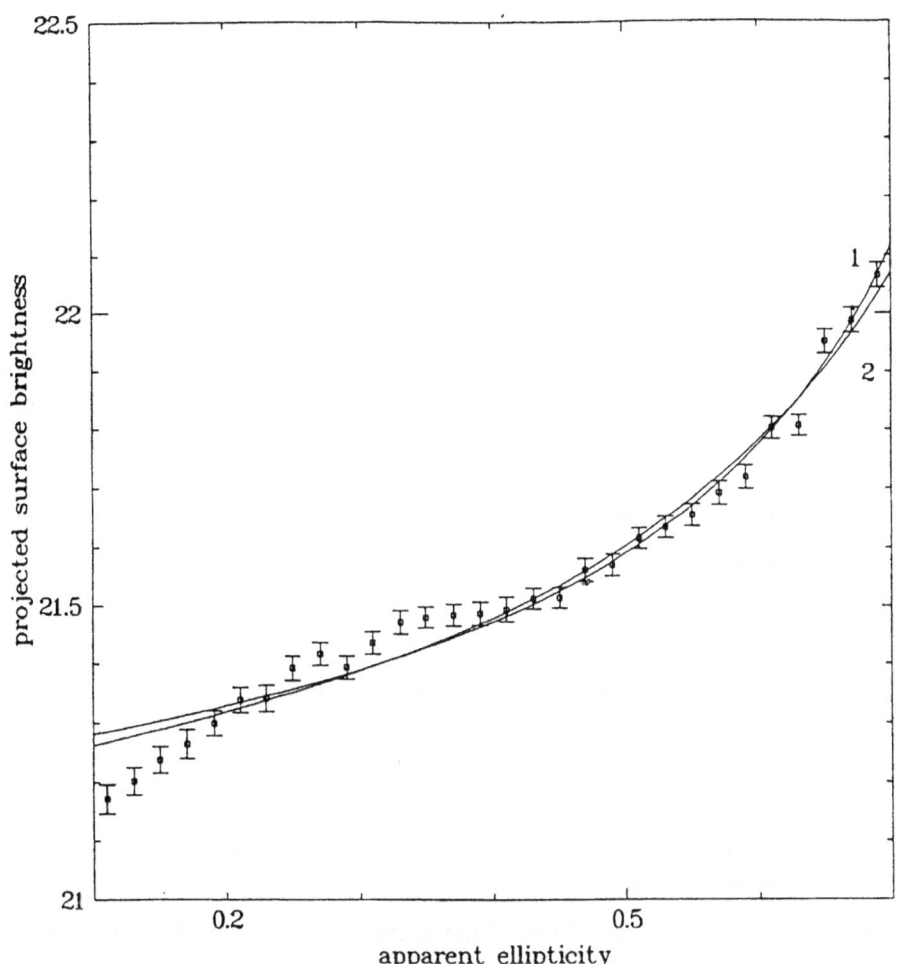

MRSP-data for 15 000 S-type galaxies with $Q_\circ = 0.13$ and $J_T < 18$(crosses) and best fitting models.

Model	face-on absorption $A_\circ$	face-on optical depth $\tau_\circ$	ζ
slab	0.43	0.86	$(1.0)^*$
screen	0.33	0.30	—
sandwich	0.43	0.86	1.0

* property of the model.

Fig. 2. Projected surface brightness of 15 000 S-type galaxies with $Q_0 = 0.13$ and $m_{\mathrm{J,T}} \leq 18$ versus apparent mean ellipticity. The measured data (crosses, vertical lengths = error bars) are compared with Disney et al.'s (1989) three models: slab (1), screen (2) and sandwich (3). (1) and (3) have the same best-fit parameters.

both in two-dimensional projection and in z-histograms. Their redshifts generally lie between $0.05 \leq z \leq 0.15$. Faint clusters become apparent on the basis of automated percolation analysis. So far, 181 faint clusters have been found in 6 fields. Their redshifts are estimated to be $0.1 \leq z \leq 0.4$, with median $z = 0.26$.

Two outstanding superclusters were found: the Sculptor supercluster at $z = 0.11$ and the Cetus-Sculptor supercluster at $z = 0.14$. The overdensity factor of the former is $f \approx 74$, its richness $N_{cl}^{sc} = 5$. Since according to Bahcall and Soneira (1984), 70% of the superclusters with $40 < f < 100$ are of richness 2, the Sculptor cluster is unusually dense. Filaments and bridges are observed between the member clusters. The Sculptor supercluster is an accepted target for pointed ROSAT observations. We hope to be able to find evidence of hot intra-supercluster gas whose presence in any supercluster has not yet been established. The properties of the Cetus-Sculptor supercluster are $f \approx 12$ and $N_{cl}^{sc} = 3$. Details are given in Naumann (1990), Ungruhe (1990), Schuecker (1990).

6 Parameters for the Friedmann-Lemaître-Universe from Statistical Methods

MRSP data have been used to develop methods for the statistical derivation of cosmological parameters. Table 2 lists the definitions, methods, preliminary values obtained from the MRSP-data, and references. It should be emphasized that the present results, based on 46 000 galaxies, are used to illustrate the methods. With the set of 144 000 galaxy redshifts now available, we expect more reliable and internally consistent numerical values. Eventually, all parameters will be derived simultaneously in order to obtain an internally consistent parameter system.

Two methods, the Schechter fit and the Kolmogorov-Smirnov fit, are used to derive the Hubble constant:

- Superposition of the uncalibrated luminosity function of MRSP galaxies, transformed to $z = 0.1$ and the apparent Virgo cluster LF, both represented by the same Schechter function, yields the *differential* distance modulus. With the known distance modulus of Virgo, the Hubble constant is obtained using the Mattig (1958) formula.
- Superposition of the normalized distribution functions of apparent magnitudes of MRSP and Virgo cluster galaxies yields the differential distance modulus. The best *fit* of the two distribution functions is determined by the Kolmogorov-Smirnov *test*.

The mean values of the Hubble constant determined by the two methods in six adjacent ESO/SERC fields are given in Table 2. It should be noted that H_0 relies only on the assumed distance modulus of the Virgo cluster, not, however, on the velocity of the Local Group relative to the Virgo cluster. Quite to the contrary, the latter can be determined when the heliocentric velocity of the Virgo cluster is known. Assuming 1247 km s^{-1} (Zhao et al. 1990), the infall velocity is 223 km s^{-1}.

The density parameter Ω_0 is determined from the galaxy luminosity function and M/L as given by Faber & Gallagher (1979). Higher values are obtained from the fluctuation diagram, discussed below. The results are given in Table 2.

Table 2. Cosmological parameters (preliminary and not internally consistent)

Name	Definition	Method	Value	References
Hubble constant	$H_0 \equiv \dfrac{\dot{R}}{R_0}$	Schechter-Fit Kolmogorov-Smirnov-Fit (distance modulus of Virgo $m - M = 31.7$)	$H_0 = 57.1 \pm 2.2$ km s^{-1} Mpc^{-1} $H_0 = 63.8 \pm 2.7$ km s^{-1} Mpc^{-1}	Duemmler 1991
density parameter	$\Omega_0 = \dfrac{8\pi G \rho_0}{3H_0^2}$	Galaxy LF $< L_B > = (2.51 \ldots 6.21) \cdot 10^8 L_B^\odot h$ Mpc^{-3} and $M/L = 70 \ldots 100\, h\, M_\odot/L_\odot$	$\Omega_{0,min} : 0.07$ ($H_0 = 100$), 0.19 ($H_0 = 60$) $\Omega_{0,max} : 0.23$ ($H_0 = 100$), 0.64 ($H_0 = 60$)	Schuecker 1990
		from turn-off mass in fluctuation diagram	$\Omega_0 \approx 1$	Seiter 1991
Lambda parameter	$\lambda_0 = \dfrac{\Lambda}{3H_0^2}$	galaxy distributions compared with MC-simulations	$\Lambda \geq 0$	Feige 1991
		quasar distribution compared with MC-simulations	$\lambda_0 : 1.33$	Feige 1991
pressure parameter	$\Pi_0 = \dfrac{8\pi G p_0}{3H_0^2}$	gravithermodynamic theory (Saslaw and Hamilton 1984)	not yet determined, assumed to be 0	—
acceleration parameter	$q_0 = \dfrac{1}{2}\Omega_0 + \Pi_0 - \lambda_0$	redshift-volume test	$\left. \begin{array}{l} q_0 < 0.25 \\ 0.3 \leq q_0 \leq 0.6 \end{array} \right\}$ ($\Lambda = 0$)	Ott 1988 Ott 1991
		computation from contributing parameters	$q_0 = 0.12$ ($\lambda_0 = 0$, $\Omega_0 = 0.23$) $q_0 = -1.21$ ($\lambda_0 = 1.33$, $\Omega_0 = 0.23$)	
curvature	$K = \dfrac{k}{R_0^2}$ $= \left(\dfrac{H_0}{c}\right)^2 (\Omega_0 + \lambda_0 - 1)$	computation from contributing parameters	ranges of K from the above data $K = 4.4 \cdot 10^{-7} \ldots 1.5 \cdot 10^{-7}$ Mpc^{-2} $K = 0$ Mpc^{-2} $K = -8.5 \cdot 10^{-8} \ldots -1.0 \cdot 10^{-7}$ Mpc^{-2}	—

The Λ-parameter was tested with galaxy and quasar samples (Feige 1991). Figure 3 compares the observed $N(z)$ histogram of 500 MRSP quasars with simulations. The simulation parameters are given in the graphs, other assumptions are $H_0 = 100$ km s^{-1} Mpc^{-1}, $q_0 = 0.5$ and no evolution of the quasar luminosity function. In this particular application, simulations using $\lambda_0 = \Lambda/3H_0^2 = 1.33$ give better results than those with $\Lambda = 0$. The value for λ_0 is, however, not consistent with other parameters and no tested parameter combination without evolution gives a satisfactory fit to the $N(m)$ histogram.

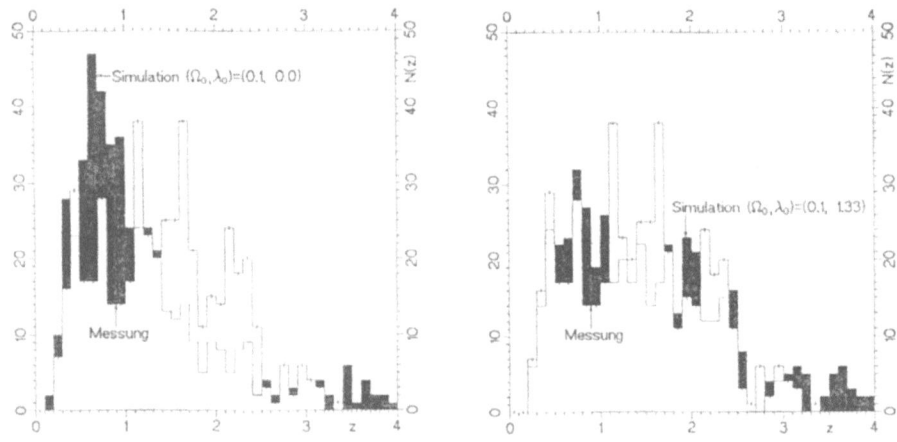

Fig. 3. Observed $N(z)$ histogram of 500 MRSP quasars compared with simulations. The simulation parameters are given in the graphs, other assumptions are $H_0 = 100$ km s^{-1} Mpc^{-1}, $q_0 = 0.5$ and no evolution of the luminosity function.

The pressure parameter Π_0 is generally assumed to be 0. Concerning pressure due to radiation energy, this appears to be justified in the present universe. The possible contribution of pressure associated with mass involves the determination of the gravithermodynamic behaviour of galaxies (referred to below in another context). The negative pressure associated with vacuum energy is equal to the vacuum energy density.

The deceleration parameter q_0 is determined directly from the redshift-volume test using galaxies in well-defined (M, z)-bins with no Malmquist bias. The deceleration parameter can, of course, be computed when λ_0 and Ω_0 are known. The comparison of the two numbers thus obtained may be used to determine errors in both derivations. Details are given in Duemmler (1991), Schuecker (1990, 1992), Feige (1991), Ott (1991).

7 Homogeneity in the Universe

Two methods have been employed to measure the degree of homogeneity within the survey depth of $z = 0.3$. The first method compares the measured number density of galaxies n at a given z with the expected number density $\bar{n}$ in a homogeneous universe. The resulting normalized number densities for galaxies $-20.8 \leq M \leq -19.2$ are nearly equal to 1 over the interval $0.11 \leq z \leq 0.21$. With the Malmquist effect at both the faint end (plate limit) and the bright one (overexposure), the lowest and highest z-ranges require large and unreliable incompleteness corrections. The small deviations from $n/\bar{n} = $ const in the optimal range are an intrinsic property of the sample, which shows that at most variations of a few percent occur on scales of a few percent of the distance covered by reliable data. Variations on scales larger than those covered by the survey cannot be ruled out, because the $n/\bar{n}$ distribution may have a constant slope over the observed depth, depending on the choices of K-correction and q_0. It seems, however, quite artificial to assume the presence of only two kinds of wavelengths for inhomogeneity: the observed small wavelengths and those much larger than $z = 0.1$. If this point of view is accepted, one may, in the reverse, use the disappearance of the slope as indicator of the proper choice for the combined values of K and q_0.

The second homogeneity test uses a method analogous to the fluctuation parameter ϵ introduced by Stoeger et al. (1987). The ϵ-parameter is defined as

$$\epsilon = |n_{\text{galaxy}} - \bar{n}_{\text{galaxy}}|_{\max} / \bar{n}_{\text{galaxy}}. \tag{1}$$

With $\bar{n}$ determined over the total observed volume, n is measured in increasingly larger subvolumes. The maximum differences between n and $\bar{n}$, normalized to $\bar{n}$, are found by shifting the subvolumes in small steps over the total volume. In the present sample, subvolumes are more or less elongated because the total depth in z is larger than in α, δ. This may be neglected in volumes with a relatively smooth distribution of galaxies, while the presence of clustering leads to differences of up to 30%. The overall decrease of fluctuation amplitudes with volume size leaves values of 10% to 30% for volumes of $50\,000\ h^{-3}$ Mpc^3 shifted over $10^5\ h^{-3}$ Mpc^3. This points to the fact that homogeneity within 1% may be found with test volumes $10^6\ h^{-3}$ Mpc^3, and suggests the tendency towards small deviations from homogeneity for the largest available scales of observation. Details are given in Schuecker & Ott (1991).

8 Test for a Hierarchical Universe

The gravithermodynamic theory of Saslaw & Hamilton (1984) provides the parameters for studying hierarchies of matter condensations in the universe. The quantities

$$\beta = <N> / <\delta N^2 > \qquad \text{and} \qquad b = 1 - \sqrt{\beta} \tag{2}$$

measure the deviation from Poisson distribution averaged over a given volume of space. b is zero for the exact Poisson case, $-\infty$ for perfect homogeneity and 1 for complete hierarchical clustering. Partial hierarchical clustering of galaxies on scales of the order 20 Mpc is suggested by $b = 0.6 = $ const found in survey volumes

$\geq 5\,000h^{-3}$ Mpc^3. Clustering evolution with decreasing redshift can be traced from the CMWB through the (still uncertain) X-ray background and quasars to galaxies. Details are given in Schuecker (1988), Seitter et al. (1989).

9 Density Fluctuations and Tests for the Inflationary Universe

The fluctuation diagram (Fig. 4) is a tool for studying primeval density fluctuations and their evolution. To construct the diagram, the ϵ-parameter obtained in the second homogeneity test (Section 7) is used with the assumptions that:

- light traces matter;
- density fluctuations are equal to the ϵ-parameter;
- galaxies (with $-20.2 \geq M \geq -20.8$) which determine ϵ and which contribute approximately 12% to the total mass are representative for all (baryonic) matter.

The linear part of the fluctuation diagram can be interpreted either as the simultaneous view of evolving scale-independent density fluctuations, or as the expected evolutionary track of a single perturbation between horizon crossing and collapse. Evolution on the linear track is predicted by theories, among them inflation, which leads to a nearly constant adiabatic fluctuation value $\delta\rho/\rho$ at horizon crossing, a Harrison-Zel'dovich spectrum during linear evolution inside the horizon and a break-off into the non-linear regime at $\delta\rho/\rho \approx 1$. This obviously implies a sequence of break-off sizes and masses, which, for given redshifts, can be determined observationally.

The present MRSP data may not yet permit reliable conclusions, because they cover only a small range of wavelengths λ. In addition, the wavelengths are the third roots of volumes which have differently elongated shapes (Section 7), a circumstance which may cause unwanted smoothing of the matter distribution. In the worst case, this can affect the break-off position. The fluctuations currently observed at break-off have the sizes of superclusters of galaxies. Since the data cover essentially the hypersphere at $z \approx 0$, superclusters are just beginning to collapse. With the mass of superclusters known from other investigations, various parameters are derived and summarized in Table 3.

Each object in collapse follows, of course, a strictly non-linear path. Several objects in collapse up to clusters of galaxies at $\log \delta\rho/\rho \approx 1.3$ and even fully collapsed galaxies at $\log \delta\rho/\rho \approx 2.5$ can, however, be connected by an almost straight line. This is an artifact which may be explained by the scale-(time)-dependence of non-linear evolution and the fact that time-scales are not displayed in the diagram.

10 Summary

Applications of MRSP galaxy and quasar samples to cosmological problems have been presented. The statistical methods used for deriving various quantities and

Table 3. Questions and answers in the fluctuation diagram.

Questions	Answers
Scale-invariant fluctuation spectrum of matter (test for inflationary models)	in $\dfrac{\delta g}{g} \propto \dfrac{\delta\rho}{\rho} \cdot \left(\dfrac{\lambda}{cH^{-1}}\right)^n$ (λ =scale length) $n = 2.7 \pm 0.1$ for the scale-invariant Harrison-Zel'dovich spectrum, expected to follow inflation
Amplitude of horizon-size density fluctuations	$(\delta\rho/\rho)_{\mathrm{HOR}} = 10^{-6}$ (extrapolated)
Fluctuation type (adiabatic or isothermal) from comparison of horizon values of $\delta T/T$ and $\delta\rho/\rho$ (test for inflationary models)	from COBE upper limit ($\delta T/T < 2 \cdot 10^{-5}$) and extrapolated horizon-size density fluctuations ($\delta\rho/\rho = 10^{-6}$), and $\delta T/T = \mathrm{const}\ \delta n_B/n_B$ (n_B = number density of baryonic matter) $= \delta\rho/\rho$ const < 20, which seems to approach the adiabatic value const $= 3$
Linear scale and amplitude at onset of gravitational instability	observed scale $= 25\,h^{-1}$ Mpc at $\delta\rho/\rho \approx 0.5$ (supercluster scale)
Mass of horizon-size fluctuations (when turn-off mass is known)	from scale of instability and associated supercluster mass $1.5 \cdot 10^{16}\,M_\odot$ $\left(\dfrac{\delta\rho}{\rho}\right)_t = \left(\dfrac{M}{M_{\mathrm{HOR}}}\right)_t^{-\frac{n}{3}} \left(\dfrac{\delta\rho}{\rho}\right)_{\mathrm{HOR}}$ for $t = t_0$ $M_{\mathrm{HOR}} = 2.3 \cdot 10^{23}\,M_\odot$
Mean density of the universe and density parameter from masses at turn-off from the linear regime	with $H_0 = 60$, $M_{\mathrm{SUPERCLUSTER}} = 1.5 \cdot 10^{16}\,M_\odot$, $\lambda_{\mathrm{SUPERCLUSTER}} = 25\mathrm{Mpc}$ at $\delta\rho/\rho \approx 0.5$ $\rho(25\mathrm{Mpc}) = 7.5 \cdot 10^{-27}$ kg m^{-3} $\Omega_0 \approx 1$. The parameter combination lies within the limits of our present knowledge (Bahcall 1986) and shows that $\Omega_0 = 1$ can be obtained from matter associated with clusters of galaxies (possibly baryonic).
Nature of dark matter	clustered and thus probably baryonic

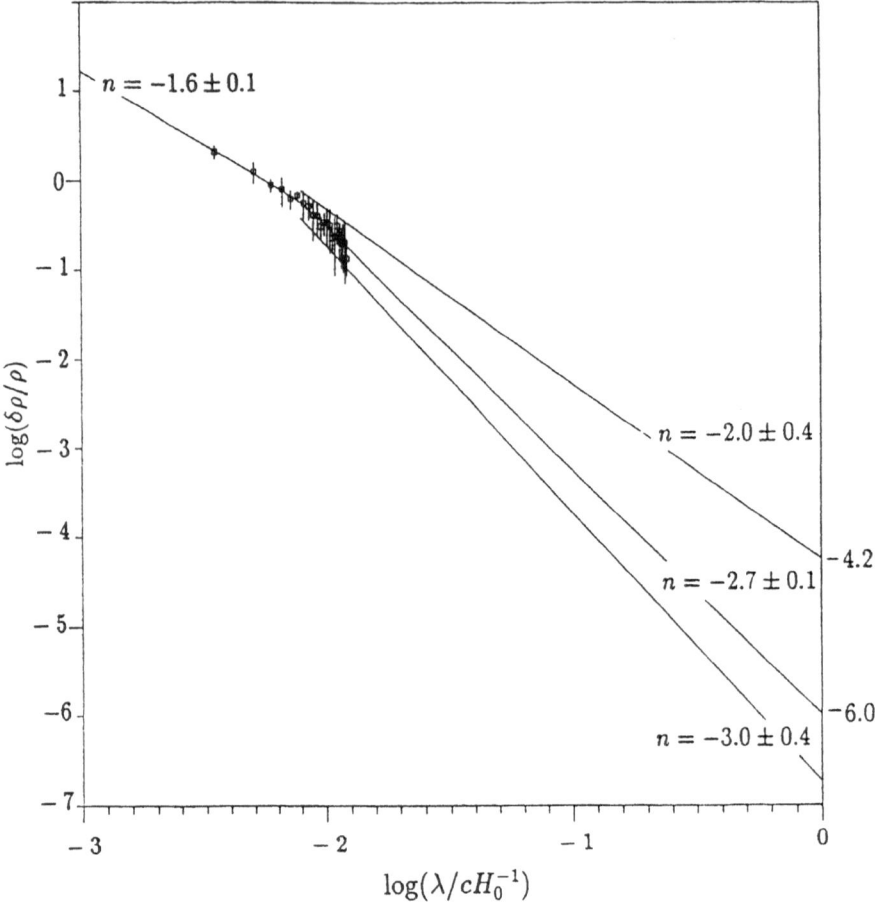

Fig. 4. The fluctuation diagram giving a simultaneous view of scale-independent fluctuations which have crossed the particle horizon but have not yet reached gravitational collapse (linear evolution, slope: -2.7 ± 0.1) together with fluctuations which have entered the non-linear regime (fluctuations in or after collapse, approximated by a line with slope -1.6). Properties derived from the diagram and additional input data are summarized in Table 3.

parameters give values within the ranges expected from published results, based on smaller samples with higher measuring accuracy. We thus consider our preliminary results as indicators that MRSP data can be used to address cosmological questions. The reliability of the results will of course depend on the size of the samples which, thanks to the pleasant cooperation with our hosts from the UK Schmidt-Telescope Unit, is growing at a satisfactory rate and hopefully will include more than 50 fields with 1500 square degrees, a total volume of $10^8 \ h^{-3}\mathrm{Mpc}^3$ and more than half a million redshifts in the foreseeable future.

11 Acknowledgements

Support by the Deutsche Forschungsgemeinschaft (Se 345/14-1,2,3) is gratefully acknowledged.

References

Preprints and unpublished theses of the authors can be obtained from the above address upon request.

Aniol, R., Horstmann H., 1991. In *'Large-Scale Structure, Observations and Instrumentation', Proceedings 2nd DAEC-Workshop*, Publ. Observatoire de Paris at Meudon, in press.

Bahcall, N.A., 1988. *Ann. R. Astron. Astrophys.*, **26**, 631.

Bahcall, N.A., Soneira, R.M., 1984. *Astrophys. J.*, **277**, 20.

Cunow, B., 1989. Diploma Thesis, Universitaet Muenster.

Cunow, B., 1991. In preparation.

Disney, M., Davies, J., Phillips, S., 1989. *Mon. Not. R. Astron. Soc.*, **239**, 939.

Duemmler, R., 1991. *Astron. Astrophys.* In press.

Duerbeck, H.W., Duemmler, R., Horstmann, H., Ott, H.A., Schuecker, P., Seitter, W.C., Teuber, D., Tucholke, H.-J., 1989. *Rev. Mex. Astron. Astrophys.*, **19**, 92.

Faber, S.M., Gallagher, J.S., 1979. *Ann. R. Astron. Astrophys.*, **17**, 135.

Feige, B., 1991. Diploma Thesis, Universitaet Muenster.

Horstmann, H., Schuecker, P., Seitter, W., Aniol, R., Budell, R., Cunow, B., Meijer, J., Teuber, D., Tucholke, H.-J., 1989. *CDS Bulletin d'Information* No. **37**, 43.

Huchra, J., Davis, M., Latham, D., Tonry, J., 1983. *Astrophys. J. Supp.*, **52**, 89.

Kirshner, R.P., Oemler, A., Schechter, P.L., Shectman, S.A., 1983. *Astron. J.*, **88**, 1285.

Mattig, W., 1958. *Astronomische Nachrichten*, **284**, 109.

Meijer, J., 1990. Diploma Thesis, Universitaet Muenster.

Naumann, M., 1990. Diploma Thesis, Universitaet Muenster.

Ott, H.-A., 1988. In *Lecture Notes in Physics*, **310**, p. 274, Springer, Berlin.

Ott, H.A., 1991. To be published.

Peterson, B.A., Ellis, R.S., Efsthathiou, G., Shanks, T., Bean, A.J., Fong, R., Zeng-Long, Z., 1986. *Mon. Not. R. Astron. Soc.*, **221**, 233.

Saslaw, W.C., Hamilton, A.J.S., 1984. *Astrophys. J.*, **276**, 13.

Schuecker, P., 1988. In *Lecture Notes in Physics*, **310**, p. 142, Springer, Berlin.

Schuecker, P., 1990. Doctoral Thesis, Universitaet Muenster.

Schuecker, P., 1992. In preparation.

Schuecker, P., Ott, H.-A., 1991. *Astrophys. J. Lett.* In press.

Schuecker, P., Ott, H.-A., Horstmann, H., Gericke, V., Seitter, W.C., 1989. In *'Astronomy, Cosmology and Fundamental Physics'*, Astrophysics and Space Science Library, Vol. 155, p. 468, eds. Caffo, M., Fanti, R., Giacomelli, G., Renzini A., Kluwer, Dordrecht.

Seitter, W.C., Duerbeck, H.W., Tacke, M. (eds.), 1988. *'Large-Scale Structures in the Universe – Observational and Analytical Methods'*, Lecture Notes in Physics *310*, Springer, Berlin.

Seitter, W.C., Ott, H.-A., Duemmler, R., Schuecker, P., Horstmann, H. 1989. In *'Morphological Cosmology'*, Lecture Notes in Physics *332*, p. 3, eds. Flin, P., Duerbeck, H.W., Springer, Berlin.

Spiekermann, G., 1991a. Diploma Thesis, Universitaet Muenster.

Spiekermann, G., 1991b. This volume.

Spiekermann, G., 1991c. To be published.

Stoeger, W.R., Ellis, G.F.R., Hellaby, C., 1987. *Mon. Not. R. Astron. Soc.*, **226**, 373.

Tucholke, H.-J., Schuecker, P., Horstmann, H., Seitter, W.C., 1991. *Astrophys. Sp. Sc.*, **177**, 209.

Ungruhe, R., 1990. Diploma Thesis, Universitaet Muenster.

Zhao, J., Pan, R., Huang, S., He, Y., 1990. *Acta Ap. Sinica*, **10**, 315.

Discussion

Beard :

In your star/galaxy separation, do you have an estimate for the fraction of stars contaminating the galaxy sample? Late-type stars can appear to be galaxies at a redshift of $Z \sim 0.1$ on the objective-prism plates. When you determine redshift by the least squares method, do you include some masks at zero redshift to eliminate these contaminating stars?

Seitter :

Star/galaxy separation is performed on the direct plates and the classification error is estimated to be 5% at the faint end where most of the galaxies are. We have not carried out the test you mention, but thank you for your suggestion. An indirect (though weak) test is the distribution of late-type stars. All objects classified as stars with sufficiently good spectra are assigned to their spectral classes for Galactic studies. We find no densities which differ much from those expected. This may indicate that there are no gross errors.

Parker :

There are 2 crucial aspects to the whole of the work on galaxy redshifts from UKST low dispersion prism spectra:-

a) The internal calibration with slit spectra — this must be performed and demonstrated over the full magnitude range ($15 \leq b_j \leq 19.5$) of applicability if the technique is valid.

b) The UKST material employed, i.e. use of 'film' copies of 'original' objective-prism plates as opposed to the original plates themselves. Tests and comparisons MUST be performed because of the very low S/N objective-prism spectra at $b_j \geq 18.0$ and the severe contrast reduction of film copies with $b_j \leq 17.0$ etc.

Seitter :

a) I agree that calibration with slit spectra is essential. Unfortunately, most of the slit spectra at faint magnitudes and $z > 0.1$ are unpublished. Many applications for observing time at the ESO 3.6m-telescope have been turned down and personal requests to observers with relevant data have remained unanswered. We were grateful to have your 60 standards from 1986. Additional data which have recently been published are now being used and, thanks to your comment, partially included in this paper.

b) Concerning the higher quality of original plates there are two answers:

1. Original plates are simply not available in the required numbers.

2. our error determinations have been made with film copies and since they are satisfactory, we expect only small improvements from using original plates.

Odewahn :

David Burstein has shown in a recent paper that tests of the magnitude corrections depending on inclination are highly affected by whether you use a magnitude-limited or a diameter-limited sample. He proposes a method based on the Malmquist bias effect. I would suggest that your large catalog of redshifts, diameters and axis ratios would be well suited for such tests.

Seitter :

Thank you for your suggestion.

A ~700 GALAXY MAGNITUDE-LIMITED REDSHIFT SURVEY NEAR THE SGP

Q.A. Parker

Royal Observatory, Blackford Hill, Edinburgh EH9 3HJ, U.K.

Abstract

A description of a magnitude-limited redshift survey covering 7 adjacent UKST fields is presented together with some preliminary results. The survey is $\sim 91\%$ complete to $b_j \leq 16.5$ but does extend to $b_j \leq 16.8$. The survey of ~ 700 redshifts will form one of the largest, independent, contiguous, well sampled and most homogeneous redshift catalogues ever compiled. The special features of this catalogue are its coverage (~ 200 deg^2), mean sampling depth ($\sim 125h^{-1}$ Mpc) and field-to-field photometric accuracy ($\Delta b_j \pm 0.05$). Morphological classifications have been determined for all 937 galaxies in the survey fields down to $b_j \leq 16.8$ and COSMOS image parameters are also available. This extensive catalogue should prove an extremely valuable resource for many applications, but particularly in characterising the precise form and nature of the features in the large-scale galaxy distribution with which any theoretical model must be necessarily reconciled.

1 Introduction

The nature of the large-scale structure provides one of the most important observational constraints for models of the evolution and formation of such features in the galaxy distribution. In particular, the existence and persistence of the 'sponge-like' topological detail found in the latest redshift surveys (de Lapparent et al. 1986; da Costa & Pellegrini 1988; Haynes & Giovanelli 1988; and this survey) present a serious challenge to such models on $\sim 50 - 100h^{-1}$ Mpc scales. The form of the 2-pt spatial correlation function shows little evidence of clustering at these scales and so is difficult to tie in with the observed structure. With Cold Dark Matter (CDM), biased galaxy formation models have been necessary (Kaiser 1984) to try to account for the observed 3-D distribution of galaxies and the prevalence of large

H. T. MacGillivray and E. B. Thomson (eds.), Digitised Optical Sky Surveys 383–387.

$\sim 20 - 50h^{-1}$ Mpc semi-symmetric 'voids'. With such structures, if luminous matter traces mass, then an $\Omega = 1$ Universe is difficult to account for. Furthermore, when deeper, more complete redshift surveys are made, voids and 'cells' hinted at in the shallower surveys are not 'filled in' but become more clearly resolved (de Lapparent et al. 1986). The persistence of such features to higher redshifts has yet to be seriously addressed. Wide angle, sparsely sampled redshift surveys do not reveal such features clearly. Deep pencil beam surveys (e.g. Koo et al. 1987), reveal narrow galaxy peaks and wide $100h^{-1}$ Mpc troughs but only sample a very limited window which may not be representative and says little about the precise forms of the underlying structures. Indeed the apparent striking periodicity in these peaks found by Broadhurst et al. (1990; and in these proceedings), demands the compilation of deeper, more complete and wider-angle redshift surveys such as that presented here.

2 The Survey

A large redshift survey near the SGP to $b_j \leq 16.8$ has been steadfastly compiled using the SAAO 1.9m telescope and the multi-object spectroscopy (MOS) system, 'FLAIR', on the UKST (Parker & Watson 1990), which offers a more efficient means of performing surveys of this nature (e.g. Watson 1988). So far, ~ 700 redshifts have been compiled with a general completeness level of $\sim 95\%$ to $b_j \leq 16.5$. Of these ~ 260 have been obtained with FLAIR and the rest with the 1.9m at SAAO. A contiguous 2.5 hour strip in RA at a constant DEC of $-35°$ has been sampled covering the UKST SERC survey fields 408, 349, 350, 351, 352, 353 and 354. The 'continuous strip' observing strategy is not only an observationally efficient slice to adopt but should also be more sensitive to large-scale feature detection than a simple square area of the same solid angle. The mean depth of the current survey is $\sim 125h^{-1}$ Mpc but sampling to a depth $\geq 250h^{-1}$ Mpc is achieved. The object database has been obtained from COSMOS machine measures of UKST/SERC J survey plates of the 7 fields of interest. Star/galaxy separation of the COSMOS data is performed using a geometric criterion which is highly successful ($\sim 95\%$) at these magnitudes (e.g. MacGillivray & Stobie 1984). Indeed, visual examination of the images of all COSMOS-'classified' galaxies to $b_j = 16.8$ in each field has revealed only $\sim 7\%$ contamination by multiple stars and spurious images for the entire survey area. COSMOS image parameters are available for all survey galaxies. Furthermore the well calibrated Edinburgh Durham Southern Galaxy Catalogue (EDSGC) magnitudes are used (Heydon-Dumbleton et al. 1989) which ensure the consistency of field-to-field magnitude limits to ± 0.05, giving a homogeneous sample. Effects on the contribution to $\xi(r)$ from different fields due to magnitude-limit inconsistencies should thus be small. Morphological classifications for all galaxies in the survey have also been obtained from eyeball scrutiny of the UKST plate material. Independent classifications of the galaxy samples in each field were made by different observers based on the de Vaucouleurs extension to the Hubble scheme. The homogeneity and completeness of the survey should enable a complete statistical description, determination of proper group membership and allow following of the inter-connected features seen in adjacent fields.

The adopted magnitude limit of this survey is compatible not only with the Durham/AAT/SAAO redshift survey (Shanks et al. 1989) with which this independent and contiguous sample can be rigorously compared but also with the new 'redshift map of 4000 galaxies' currently being constructed with the FLAIR System on the UKST which covers the 60 fields of the EDSGC at a 1 in 3 sampling rate to $b_j \leq 17.0$ (Broadbent et al., these proceedings).

3 Survey Aims

The overall aims of this survey are quite broad but include:

- **To obtain** a 3-D description of the galaxy distribution sufficient in depth and across the line of sight to intersect many 'scale-lengths' of typical void dimensions so that their properties can be characterised and so that a more representative 'fair' sample of the Universe is obtained.
- **To determine** the 2-pt spatial correlation function for our sample of ~ 700 galaxies in contiguous fields out to $\sim 200h^{-1}$Mpc to determine both the slope of the power law and the correlation length as these constrain the possible spectrum of primordial density fluctuations on large scales (Peebles 1980). At these depths the influence of the LSC on the form of the correlation function should also be small. Confirmation of the 'shoulder' feature seen in $\xi(r)$ between $1 - 7h^{-1}$ Mpc (Shanks et al. 1989) should be possible. The verification of such a feature could imply non-scale-free clustering at small scales.
- **To form** an intermediate comparison sample between the shallower but wide-angle coverage of the extended CfA survey and the narrow but very deep pencil beam surveys in the hope of learning something about galactic evolution.
- **To investigate** the relative space densities of galaxies of different morphological types, variations in their luminosity functions and possible segregation in galaxy groups and the occurrence of emission lines in their spectra as these too have a bearing on galactic evolution and environmental influences.
- **To obtain** a quantitative description of the observed galaxy sample into groups, associations and field galaxies using techniques of hierarchical clustering for studying the properties of small scale structure.

The results for this deep survey can also be compared with the results from planned deeper surveys such as the ESO key program of de Lapparent et al. who aim to obtain ~ 700 galaxy redshifts in a 0.4 deg^2 area to $R \leq 20.5$.

4 Preliminary Results

Redshifts for 107 galaxies in field 349 are given by Parker et al. (1986), whilst 116 redshifts in field 352 are given by Parker & Watson (1990). Significant voids are revealed along with various well defined galaxy groups and 'sheets' which have only a limited velocity dispersion (typically $\leq 200 \mathrm{Kms}^{-1}$). The expanded 'near complete' contiguous survey of 7 UKST fields presented here **for the first time** has revealed the interconnected nature and extent of these features, providing a valuable database for characterising these structures on larger ($\sim 200h^{-1}$Mpc) scales. The detailed study of the galaxy distribution in the survey area awaits the satisfactory completion of the full catalogue.

A cone diagram of right ascension and redshift for the current sample is shown in Fig. 1. The wedge is $\sim 6°$ thick in declination space. The cone covers all 7 fields in the survey, and ~ 680 galaxy redshifts are plotted (the rest have velocities > 25000Kms^{-1}). The remarkable cellular structure in this plot is self evident. About a dozen significant voids and various well defined galaxy features are apparent. Structure is traced out well to $\sim 200h^{-1}$Mpc, $\sim 2\times$ the depth of the CfA survey. At this depth, this survey represents a continuous strip $\sim 130 \times 20$Mpc in size across the line of sight. Typical void dimensions are $\sim 20 - 50h^{-1}$Mpc, in good agreement with the current wider angle, though shallower, redshift surveys.

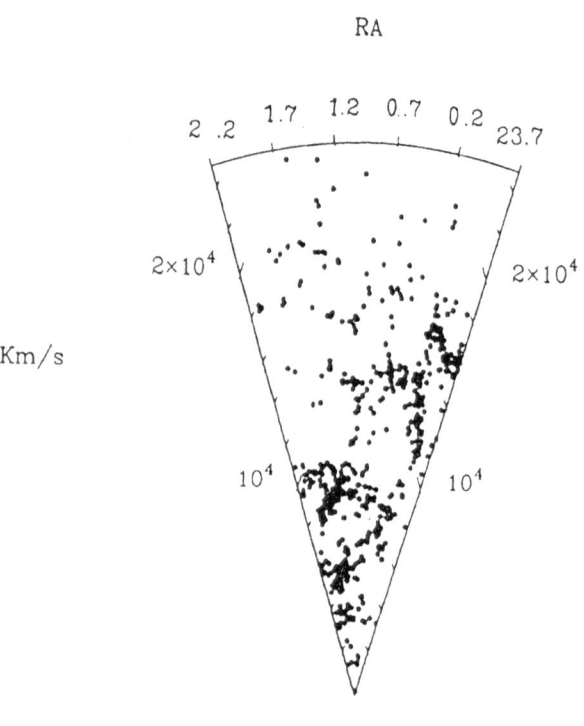

Fig. 1.

The morphological mix of galaxies between fields in the survey is fairly consistent ($\sim 55\%$ S/Irr and 45% E/S0 with a 5% variance) except for field 408 which exhibits a marked ($\sim 60\%$) preponderance for ellipticals. The overall percentages compares well with the results from the Durham/AAT redshift survey (DARS), e.g. Shanks et al. (1984), which found 43% E/S0's and 57% S/Irr types. Their survey of ~ 330 field galaxies, also based on COSMOS data, has a similar magnitude limit ($b_j \leq 16.75$) though consists of 5 discrete fields. The percentage of galaxies exhibiting detectable emission lines in this survey of $\sim 47\%$ also compares well with $\sim 53\%$ for the DARS survey.

References

Broadhurst, T.J., Ellis, R.S., Koo, D.C., Szalay, A.S., 1990. *Nature*, **343**, 726.

da Costa, L.N., Pellegrini, P.S., 1988. *Astrophys. J.*, **327**, L1-5.

de Lapparent, V., Geller, M.J., Huchra, J.P., 1986. *Astrophys. J.*, **302**, L1-5.

Haynes, M.P., Giovanelli, R., 1988. *'Large-Scale motions in the Universe'*, p. 31-70, Princeton University Press, Princeton.

Heydon-Dumbleton, N.H., Collins, C.A., MacGillivray, H.T., 1989. *Mon. Not. R. Astron. Soc.*, **238**, 379.

Kaiser, N., 1984. *Astrophys. J.*, **284**, L9.

Koo, D.C., Kron, R.G., Szalay, A.S., 1987. XIIIth Texas Symposium on Relativistic Astrophysics, p. 28, ed. M.P. Ulmer.

MacGillivray, H.T., Stobie, R.S., 1984. *Vistas Astron.*, **27**, 433.

Parker, Q.A., MacGillivray, H.T., Hill, P.W., Dodd, R.J., 1986. *Mon. Not. R. Astron. Soc.*, **220**, 901.

Parker, Q.A, Watson, F.G., 1990. *Astron. Astrophys. Suppl. Ser.*, **84**, 445.

Peebles, P.J.E., 1980. *'The Large-Scale Structure of the Universe'*, Princeton University Press.

Shanks, T., Hale-Sutton, D., Fong, R., Metcalfe, N., 1989. *Mon. Not. R. Astron. Soc.*, **237**, 589.

Shanks. T., Stevenson, P.R.F., Fong, R., MacGillivray, H.T., 1984. *Mon. Not. R. Astron. Soc.*, **206**, 767.

Watson, F.G., 1988. *'Fiber Optics in Astronomy'*, A.S.P. Conference Series 3, p. 125, ed. S.Barden.

THE DURHAM/UKST GALAXY REDSHIFT SURVEY

A. Broadbent [1], D. Hale-Sutton [1], T. Shanks [1], R. Fong [1], A.P. Oates [2], F.G. Watson [3,4], C.A. Collins [4], H.T. MacGillivray [4], Q.A. Parker [4] and R.C. Nichol [5]

[1] Department of Physics, University of Durham, South Road, Durham DH1 3LE, U.K.
[2] Royal Greenwich Observatory, Madingley Road, Cambridge CB3 0HA, U.K.
[3] Anglo-Australian Observatory, Coonabarabran, NSW 2357, Australia.
[4] Royal Observatory, Blackford Hill, Edinburgh EH9 3HJ, U.K.
[5] Department of Astronomy, University of Edinburgh, Blackford Hill, Edinburgh EJ9 3HJ, U.K.

Abstract

We are currently engaged in a long term project to make a redshift survey of ~ 4000 galaxies with $b_j \leq 16^m.75$. When complete, we will have mapped a continuous volume of $\sim 4 \times 10^6$ Mpc^3 and to a depth $\sim 300h^{-1}$ Mpc in a region around the South Galactic Pole.

1 Description and Aims of the Survey

The Durham/UKST Galaxy Redshift Survey is a long term project which aims to extend the 2-D information in the Edinburgh/Durham Southern Galaxy Catalogue (Collins et al. 1988) to 3-D by measuring ~ 4000 redshifts of galaxies included in the catalogue. The Edinburgh/Durham Southern Galaxy Catalogue is a COSMOS survey of galaxies with blue magnitude, $b_j < 20^m.5$ over 60 UK Schmidt Telescope (UKST) fields, centred around the South Galactic Pole region.

We obtain the redshifts again using the UKST at Siding Spring in Australia in conjunction with FLAIR, a fibre coupled spectroscopy system. Optical fibre ends are positioned on the images of the target objects on a glass plate of the field being observed which is then mounted into the telescope. Two 35 fibre bundles are each fed from the telescope to an external spectrograph, with incoming data being recorded on a CCD camera system.

H. T. MacGillivray and E. B. Thomson (eds.), Digitised Optical Sky Surveys 389–395.
© 1992 Kluwer Academic Publishers.

Despite the fact that the UKST is only a 1.2m telescope, we are able to obtain blue spectra for 60 galaxies with $b_j \lesssim 17^m$ simultaneously in a minimum of 4×3000 s exposures, to an accuracy of $\pm 150 \mathrm{kms}^{-1}$. This is the equivalent of taking $3\frac{1}{3}$ minutes per galaxy and requires only one half of a night's observing time. By comparison, using a 2m telescope to obtain galaxy spectra individually to a similar resolution and limiting magnitude it would take typically 10–15 minutes per galaxy. Thus a whole night of observation would be required to obtain a comparable number of redshifts.

Also, adequate spectra can be obtained when seeing conditions are not good enough for taking photographic survey plates, which means that multiple fibre spectroscopy on the UKST is a very efficient use of telescope time.

The aim of the project is to map the 3-D structure of galaxies over the 60 fields by making a 1-in-3 random sampled redshift survey of galaxies of magnitude brighter than $b_j = 16^m.75$. At this depth there are typically 6 galaxies per square degree and so we need to position on each of the photographic survey plate centres only once in order to obtain the redshifts required for the whole of the large ($\sim 6° \times 6°$) field of view.

When complete, we will have mapped a continuous volume of $4 \times 10^6 \mathrm{Mpc}^3$ out to a distance of roughly $300 h^{-1}$ Mpc and the high sampling rate will enable us to investigate the distribution of galaxies over a wide range of scales ($1 < r < 300 h^{-1}$ Mpc). Thus we will be able to look for the structures claimed to be present in other redshift surveys such as walls, voids, sheets, filaments etc.

The IRAS redshift survey of Efstathiou et al. (1990) had a sampling rate of 1-in-6 and extends to a depth of $140 h^{-1}$ Mpc. The galaxies included in the IRAS survey are predominantly late types which are concentrated in regions of low galaxy number density. In our optically selected survey, all types of galaxies are included from both high and low number density regions.

2 Preliminary Results

At present we have completed a strip of 7 UKST fields between R.A. $21^h 39^m 30^s$ and $00^h 11^m 30^s$ at $\delta = -30°$ and have obtained over 300 galaxy redshifts. For each field redshifts have been determined typically for $\sim 80\%$ of the galaxies with $b_j < 16^m.75$ to an accuracy of $\sim \pm 150 \mathrm{kms}^{-1}$.

Figure 1 shows a few of the better examples of reduced galaxy spectra. The wavelength range is approximately 4000–5000Å. Most redshifts are obtained by cross-correlation of absorption features with template galaxies, but a few galaxies have hydrogen and oxygen emission lines. The wavelength range covered has the virtue of having very few sky emission features but quite a large number of absorption and emission features in the galaxy spectra. Red data have been taken in the 5800–7000Å range as well as blue data for some of the fields but, as illustrated in Fig. 2, there are strong sky emission lines which makes sky subtraction from galaxy spectra difficult. Also, there tend to be fewer features in the galaxy spectra and we have found that very few redshifts are deduced from red data alone.

Figure 3 shows preliminary results of the redshift survey. The R.A. slice covers the 7 fields already observed and the declination direction is projected onto the

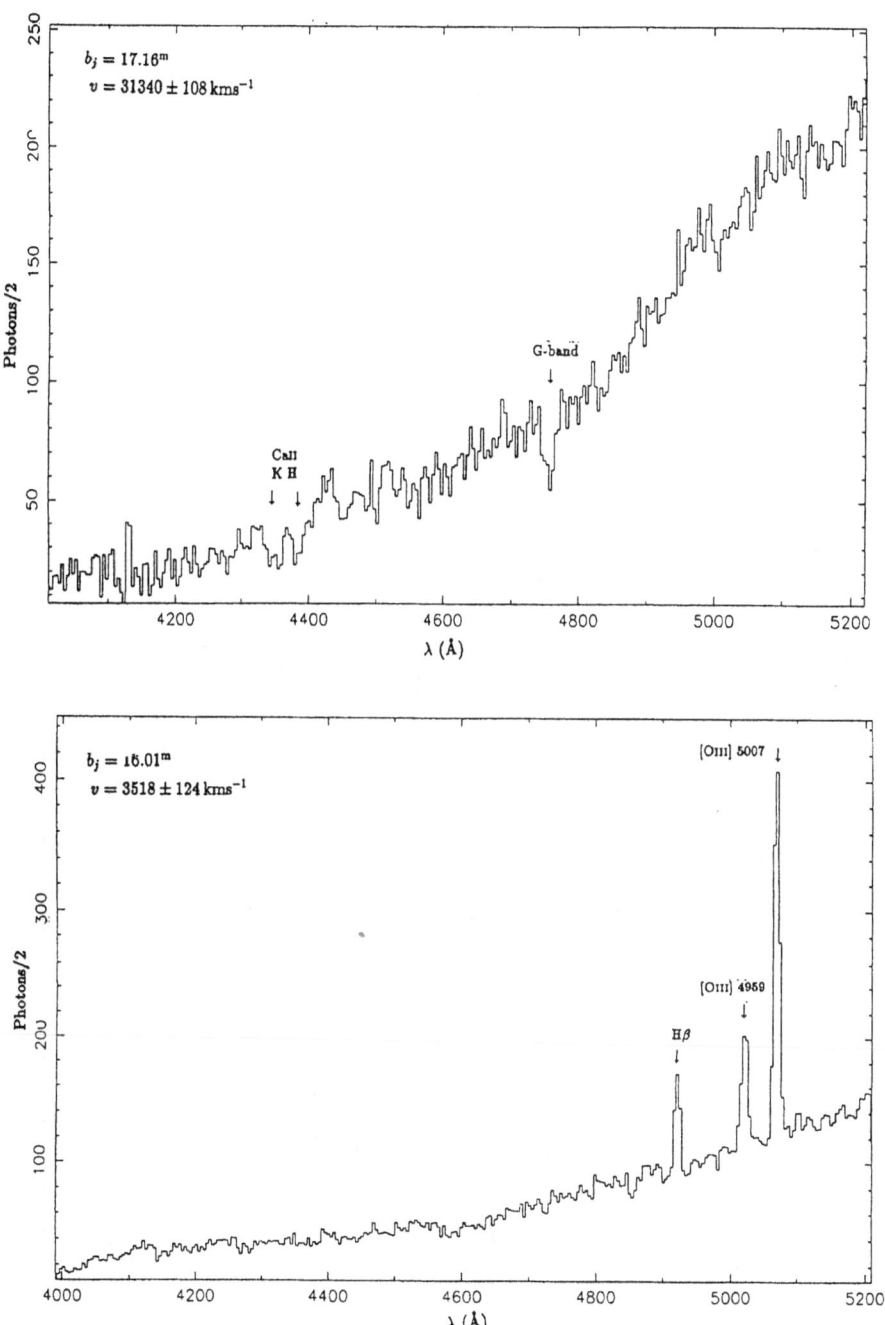

Fig. 1. On this and the following page are presented four examples of reduced galaxy spectra from the Durham/UKST Redshift Survey. The majority of our spectra cover the ~ 4000–5200 Å range although the response of the CCD chips extends up to ~ 9000 Å. Spectral features which are found commonly are marked on these examples.

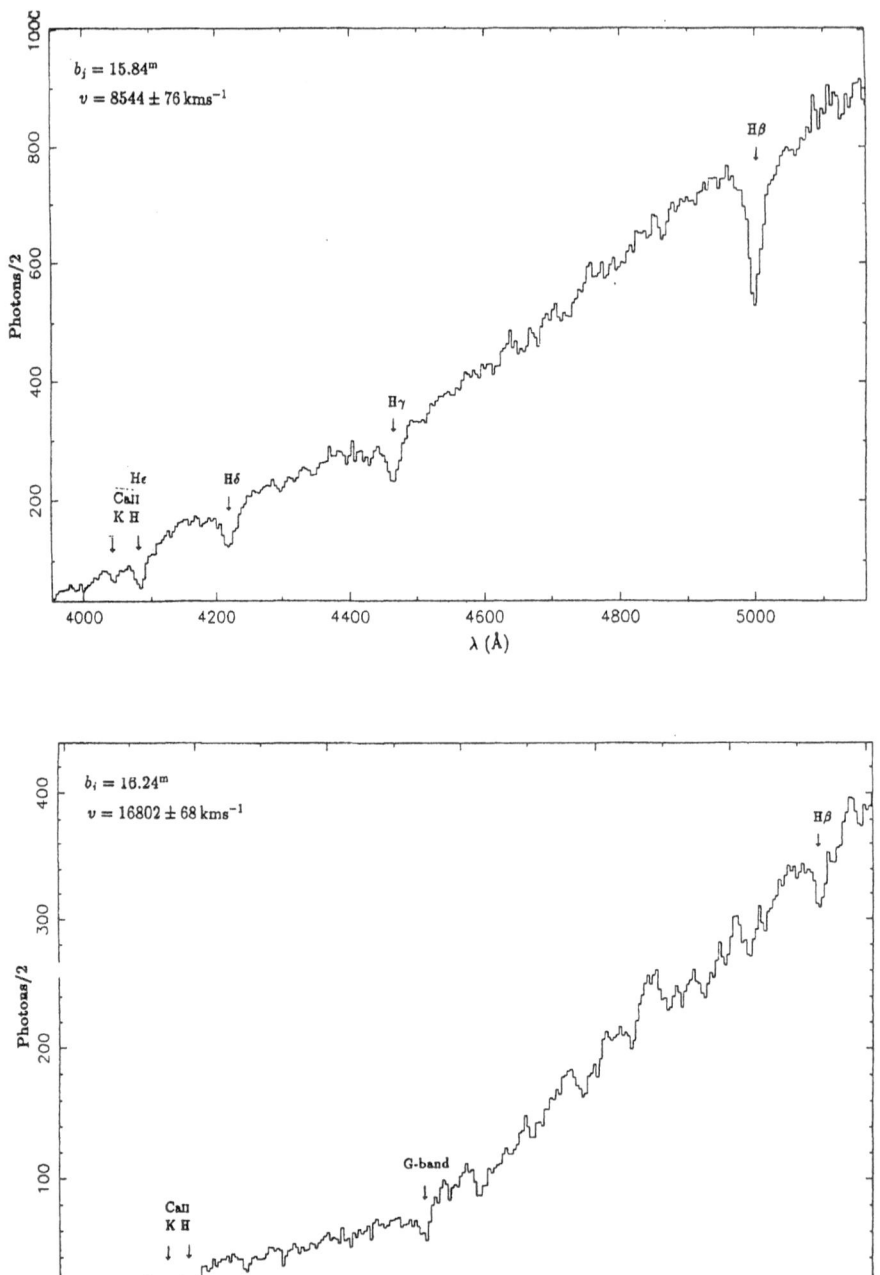

Fig. 1. Continued.

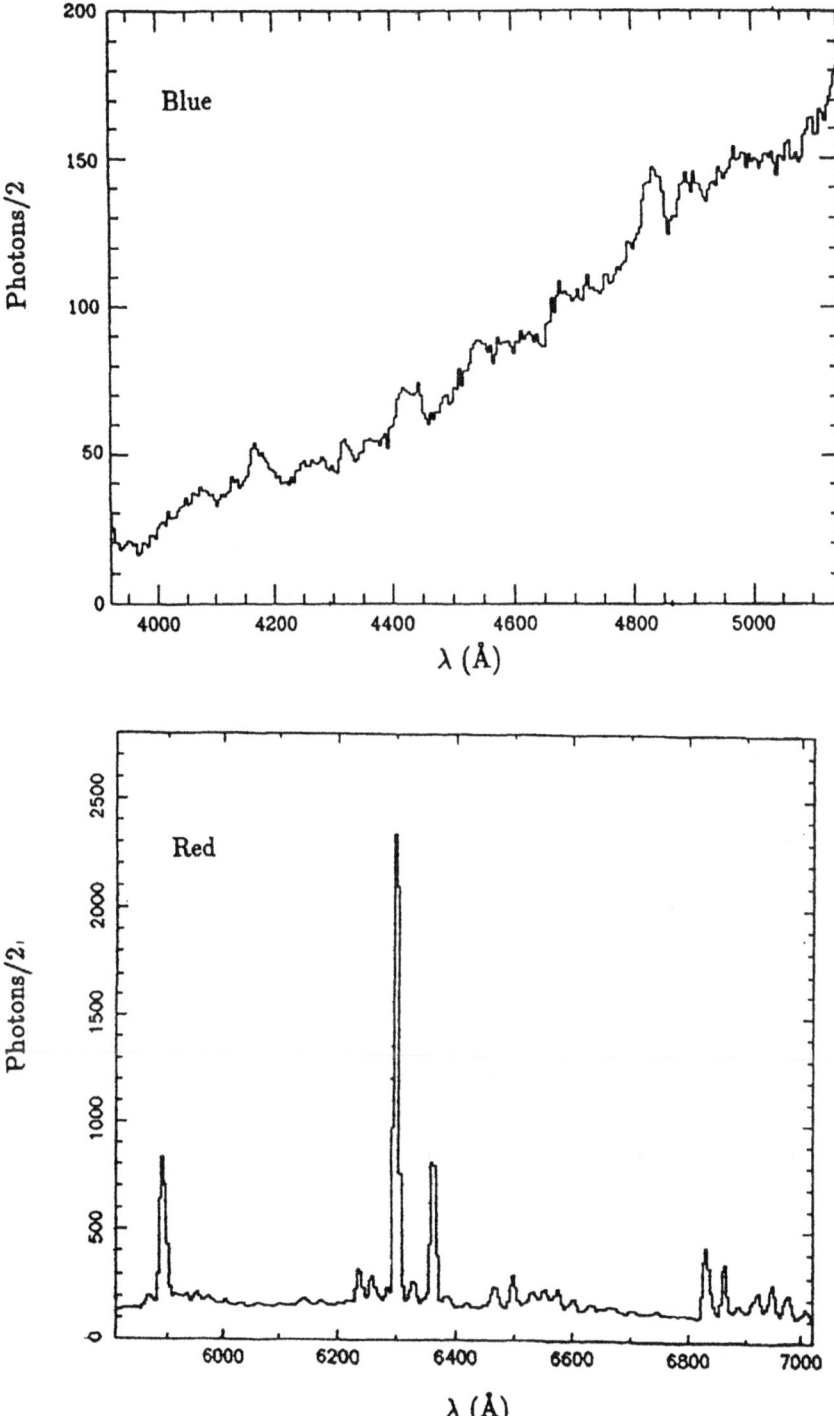

Fig. 2. Typical sky spectra in the blue (∼ 4000–5200 Å) and red (∼ 5800–7000 Å) wavelength ranges.

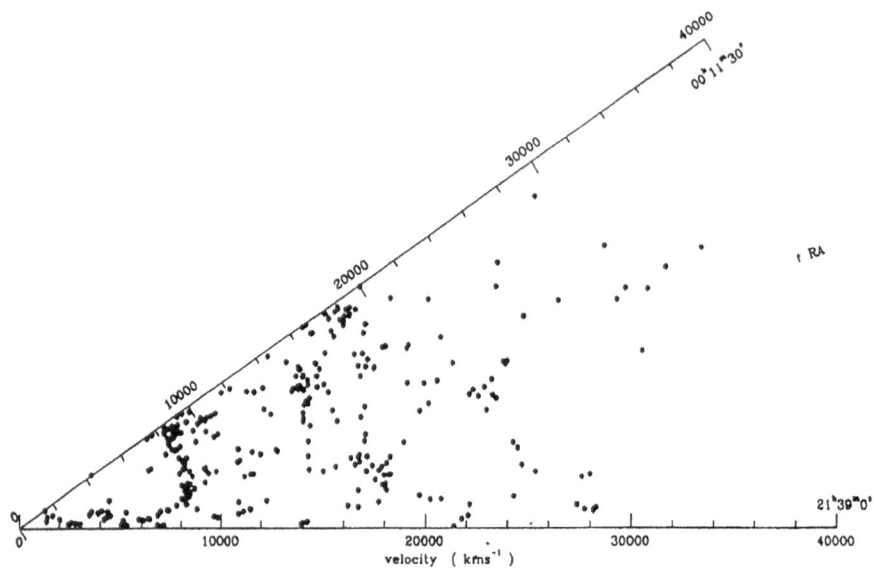

Fig. 3. A cone plot of the galaxy distribution between R.A. $21^h39^m30^s$ and $00^h11^m30^s$. There are 313 galaxies represented in this plot which all lie in the δ range $-27.5°$ to $-32.5°$.

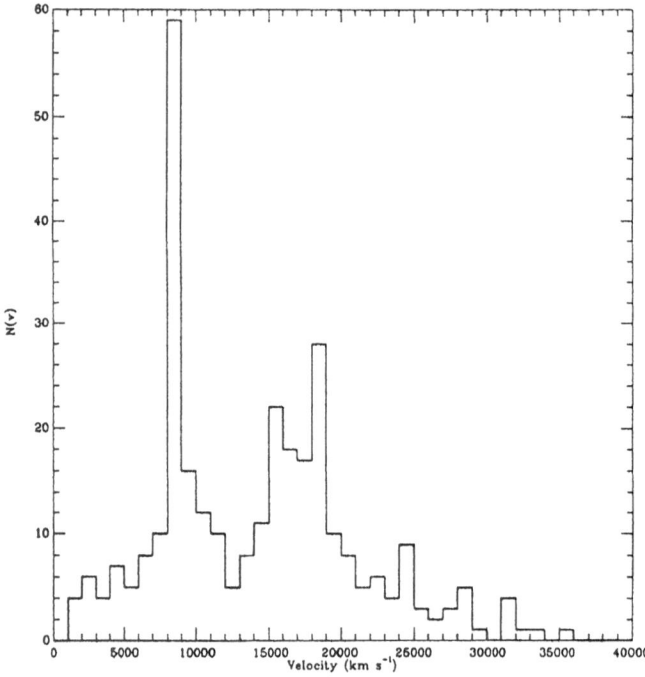

Fig. 4. A histogram plot showing the distribution of velocities measured from spectra of the survey galaxies lying in the 7 fields observed to date.

paper. Each dot represents the location of a galaxy deduced from its spectrum. Already there are interesting features appearing, most notably the strong feature at $\sim 8000\,\mathrm{kms^{-1}}$. As expected, this strong concentration of galaxies is also a prominent feature of the n(z) plot in Fig. 4 and will obviously affect the results obtained from a statistical analysis of the data, which we hope to carry out in the near future.

3 Future Developments of FLAIR

Currently the fibre system is being updated and improved with the new FISCH spectrograph which has better blue through-put and better resolution. Also, there will be more fibres per spectrograph. The fibre-positioning and plate-mounting procedure is being automated, which should reduce considerably the time required for preparation between observation of different fields.

References

Collins, C.A., Heydon-Dumbleton, N., MacGillivray, H.T., Shanks, T., 1988. In *'Large Scale Structures in the Universe'*, *IAU Symp.*, No. 130, p. 125, ed. Audouze, J. et al., Reidel.

Efstathiou, G.P., Kaiser, N., Saunders, W., Lawrence, A., Rowan-Robinson, M., Ellis, R.S., Frenk, C.S., 1990. *Mon. Not. R. Astron. Soc.*, **247**, 10P.

EXTENDED STRUCTURE IN THE 3-D GALAXY DISTRIBUTION AT THE GALACTIC POLES

T.J. Broadhurst [1], R.S. Ellis [2], D.C. Koo [3] and A.S. Szalay [4,5]

[1] Royal Observatory, Blackford Hill, Edinburgh EH9 3HJ, U.K.
[2] Physics Department, University of Durham, South Road, Durham, U.K.
[3] University of California Observatories, Lick Observatory, Santa Cruz, California, U.S.A.
[4] Department of Physics and Astronomy, The Johns Hopkins University, Baltimore, Maryland, U.S.A.
[5] Department of Physics, Eötvös University, Budapest, Hungary.

Abstract

The claim of Broadhurst et al. (1990 – BEKS) of a large-scale, regular structure seen in a deep galaxy redshift survey along the Galactic polar axis is strengthened by new off-axis redshifts. The transverse extent of this new spatial information is comparable with the scale 128Mpc * identified by BEKS, and allows us to demonstrate that the original pattern is due to genuine large scale, $\approx 50h^{-1}$Mpc, features lying at the poles, and is not the chance alignment of small clumps. The southern polar distribution has been independently mapped using rich clusters (Nichol et al., these proceedings) revealing structure in very good agreement with our extended galaxy distribution.

1 Introduction

The spatial structure of the galaxy distribution is more inhomogeneous and complex than has been hitherto imagined. Redshift surveys are well on the way to mapping out the local universe ($r < 300$Mpc), but it is evident from the gross structures revealed in these surveys that the depth is not yet great enough to contain a 'fair'

---

* $H_o = 100$ kms sec^{-1} Mpc^{-1} and $q_o = 0.5$.

H. T. MacGillivray and E. B. Thomson (eds.), Digitised Optical Sky Surveys 397–404.

sample of the galaxy distribution. For example, the CFA slices have running through them a large sheet like structure — the 'Great Wall' — which is coherent over $\approx 100 \times 50 \text{Mpc}$ at a depth of 70Mpc, close to the limiting depth of that survey (Geller & Huchra 1989). A less fully-sampled 3-D survey based on all sky IRAS photometry (Lawrence et al. 1991) extends deeper, to a radius of 250Mpc all round the sky, from which it is possible to obtain an estimate of density fluctuations on intermediate scales ($10 - 50 \text{Mpc}$). The observed variance of the number density distribution is surprisingly large compared with the standard biased Cold Dark Matter (CDM) model (Efstathiou et al. 1990; Saunders et al. 1991). This conclusion is also supported by the 3-D power spectrum of a number of large area samples of depth $z < 0.15$ (Peacock 1991, preprint), for which the implied density fluctuations are 30% at the largest adequately sampled scale of 100Mpc.

It is clear from such surveys that deeper samples extending to several hundred Mpc will be required to examine the statistical behaviour of such large structures. In the range $r > 100 \text{Mpc}$, homogeneity must be found for consistency with the highly isotropic microwave background. It is also on these large scales where any initial irregularities are able to survive unspoilt by subsequent gravitational clustering, and hence the regime for best discrimination between competing cosmogonies.

Pencil beams offer a different approach to sampling structure by sacrificing areal coverage for depth; 1-D structural information on scales of order the Horizon can be examined. This information comes as a free bonus from studies of galaxy evolution, for which the deepest pencil beams are primarily motivated [Broadhurst et al. 1988 (BES); Koo & Kron 1987].

To date, the deepest 1-D galaxy probe is the combined polar sample of BEKS, which stretches out to $z = 0.4$ both North and South, representing 30% of the Horizon (for $\Omega = 1.0$), with a width of only $\approx 5 \text{Mpc}$. The distribution of galaxies in this sample is unexpectedly regular, displaying at least 11 sharp 'spikes' separated by underdense regions. A characteristic scale of 128Mpc was reported by BEKS, with a probability of only 2.0×10^{-4} of arising from the known small-scale clustering of the galaxy 2 point correlation function.

We have now acquired substantially more data, off-angle to the original beams. These data allow a determination of the transverse extent of the BEKS redshift distribution and hence the reality of our claimed large-scale power.

2 Sample Selection for Extended Polar Redshift Survey

The sample selection was made from the COSMOS/UKST Southern Sky Object Catalogue (see Yentis et al., these proceedings). This data base provides reliable star/galaxy separation to $b_j = 21$. Four fields centered on the SGP were selected for which the photographic calibration is very accurate due to the presence of CCD sequences in this region of the catalogue.

These fields were then observed using AUTOFIB on the AAT. A total of 65 fibres are available for simultaneous spectroscopy and in 3 clear nights over 300 new redshifts were secured with 85% completeness in the range $17 < b_j < 20.5$. Figure 1a shows a coneplot of this new data together with the SGP sample of BEKS.

By selecting bright and faint limited samples and spacing the brighter fields over a larger area than the fainter ones, the resultant selection is approximately cylindrical of transverse radial extent $r_t < 50\mathrm{Mpc}$ in the redshift range $0.05 < z < 0.25$ (see Fig. 1a).

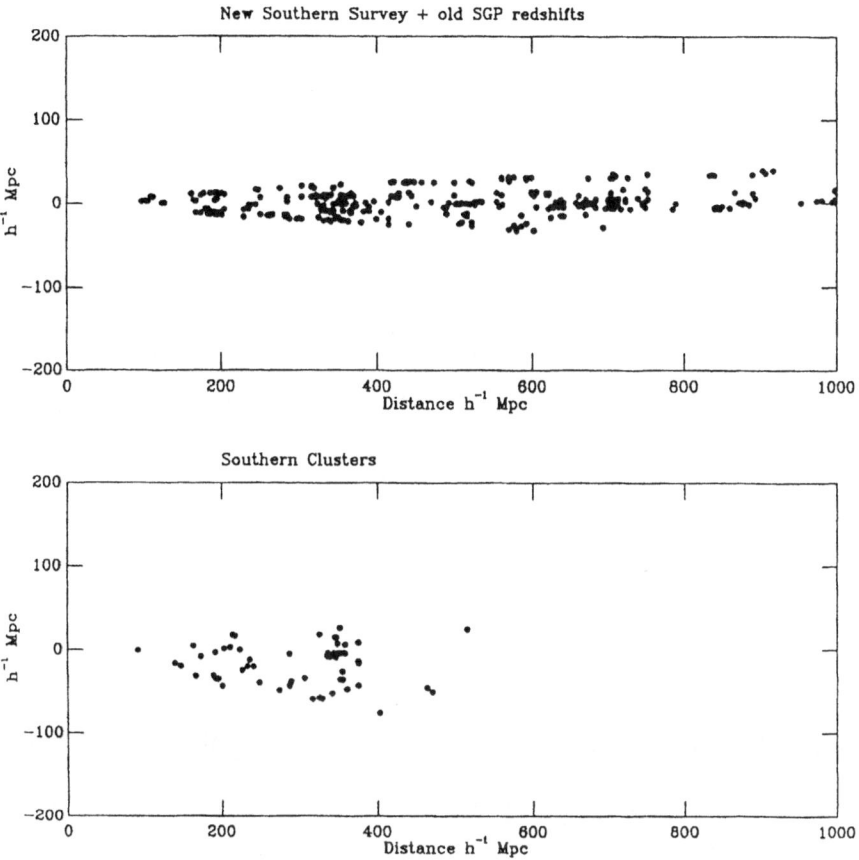

Fig. 1. a) Coneplot of the Southern Sample of Broadhurst & Ellis (1991, in preparation) centred on the SGP. **b)** Coneplot of the cluster distribution of Nichol et al. (these proceedings) centred on the SGP.

Similar new data has been obtained in the North (Ellman et al. 1991, in preparation) and the combined histogram of these data together with all available pencil beam data (1024 galaxies), in a cylinder of radius $50h^{-1}\mathrm{Mpc}$ along the Galactic polar axis, is plotted in Fig. 2a. The regular pattern observed in the original narrow beams of BEKS is clearly visible in this extended data set. The sharp peak at 70Mpc in the north is due to the well sampled 'Great Wall'. The 1-D power spectrum of this distribution reveals a sharp peak at 128Mpc (Fig. 2d), with relatively little power on all other scales. To assess the internal significance of this power, the cumulative distribution of power spectrum amplitudes can be constructed. The distribution has a large shoulder due to the single amplitude peak at 128Mpc which has a probability of 7×10^{-7} against the small amplitude white noise. Figure 2g is a periodogram of this sample. The phase distribution (assessed from the reduced χ^2 against a uniform phase distribution with errors assumed to be $\sqrt{n}$ of the number of objects in each phase bin) shows a high degree of coherence centered on ≈ 130Mpc, picking out very clearly the highly regular large scale structure.

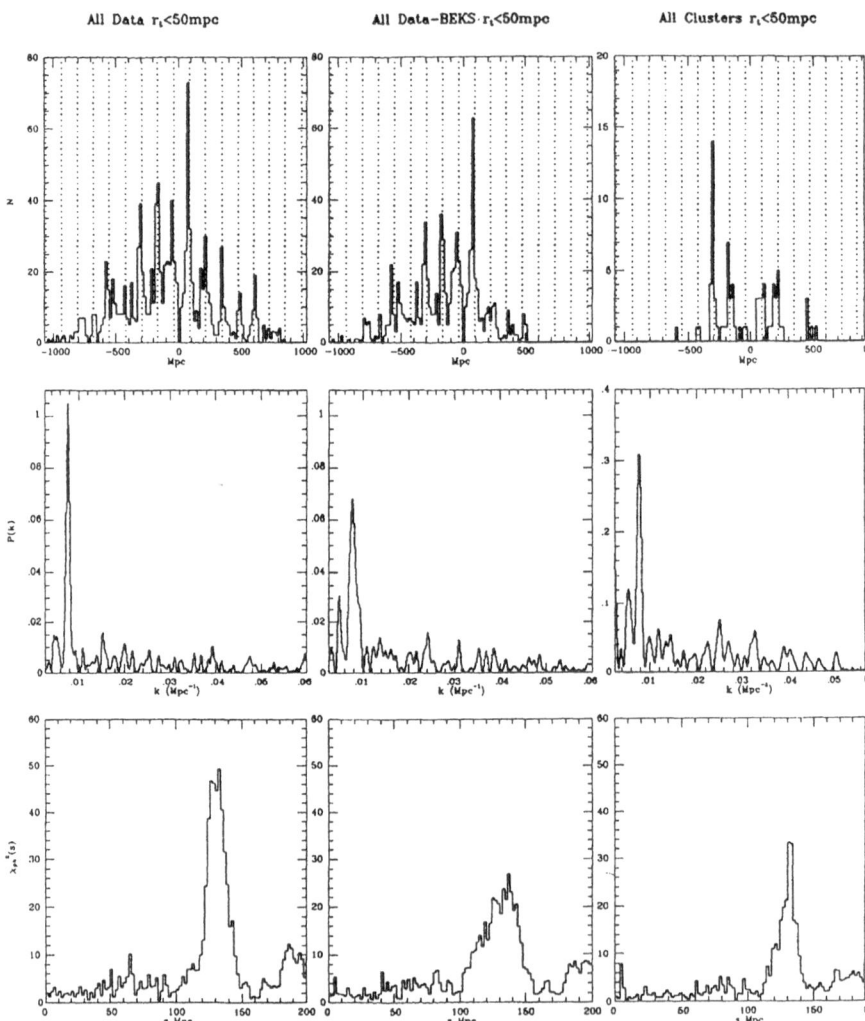

Fig. 2. The comoving distances (top row), 1-D power spectra (middle row) and periodogram (bottom row), for the full polar data set (lefthand column), without the original BEKS sample (middle column) and for all available rich clusters with redshift information (righthand column). All objects within $r_t < 50$Mpc of the Galactic Polar Axis are selected. Figure labels (a – i) referred to in the text proceed from top left to bottom right.

In order to test the reality of the original result, we can remove all the original BEKS data leaving a total of 754 redshifts. These remaining galaxies are spread evenly over $r_t = 50$Mpc from the polar axis and are more concentrated in a smaller redshift range than BEKS. Nevertheless the nearest few peaks, both North and South, are still very prominent (Fig. 2b) and this is reflected in the excess power at 128Mpc (Fig. 2e), which is as significant as that of the BES data, 1.0×10^{-4}, assessed from the internal noise.

Given the large increase in redshift information at both poles, it is now possible to assess the reality of the large scale power at each pole independently. The power spectrum of galaxies lying within a 50Mpc radius of the direction of the poles both show independently a sharp peak (Figs. 3a, b) at low frequency corresponding to 125 ± 10Mpc and 136 ± 10Mpc in the South and North respectively. This detection of consistency in these two directions is strong evidence of a more general coherence at this large scale.

3 Transverse Structure

The cone plot of Fig. 1a for the Southern data set (Broadhurst & Ellis 1991, in preparation) shows a number of structures which are extended in the direction transverse to the line of sight most notably at $z = 0.07$, 0.11 and 0.22. It is also evident that some galaxies are clustered off the main structures so that the general pattern is more complex than simply a series of 'Great Walls'. To quantify the extent of the transverse coherence, the phase distribution is calculated for period=128Mpc as a function of radial distance from the the the polar axis, r_t. Figure 4, column 1, shows the distribution of phases in annuli of increasing diameter about the polar axis. It is evident from column 1 that phase coherence is present at 128Mpc out to a diameter of 50Mpc. Above 50Mpc the coherence is lost. For comparison, three other radial phase patterns are also shown in Fig. 4 (columns 2, 3 and 4) at unrelated scales 30, 50 and 70Mpc where nothing of any interest is seen.

It is clear that the structures which contribute to the power at 128Mpc are not simply chance small-scale fluctuations as predicted by some (e.g. Kaiser & Peacock 1991), but are substantial structures typically ≈ 50Mpc in extent.

4 Polar Distribution of Rich Clusters

An independent assessment of the large scale power observed in the galaxy distribution is provided by the recent improvement in the Southern sky coverage of rich cluster redshift measurements. Figure 1b shows the spatial distribution of southern clusters (Nichol et al. these proceedings) for comparison with our extended polar galaxy distribution. The clusters have a very similar spatial distribution with two clear concentrations at $z = 0.065$ and $z = 0.11$. Combining the cluster survey with the compilation of Abell et al. (1989) and Postman et al. (1991) allows a reasonably uniform sample of clusters to be constructed in the polar region. Figure 2c shows the histogram of the comoving distance in the polar direction for a sample with $r_t < 50$Mpc (80 clusters). The clusters extend to $z < 0.15$ and most of the peaks of

Southern Data r$_t$<50mpc   Northern Data r$_t$<50mpc

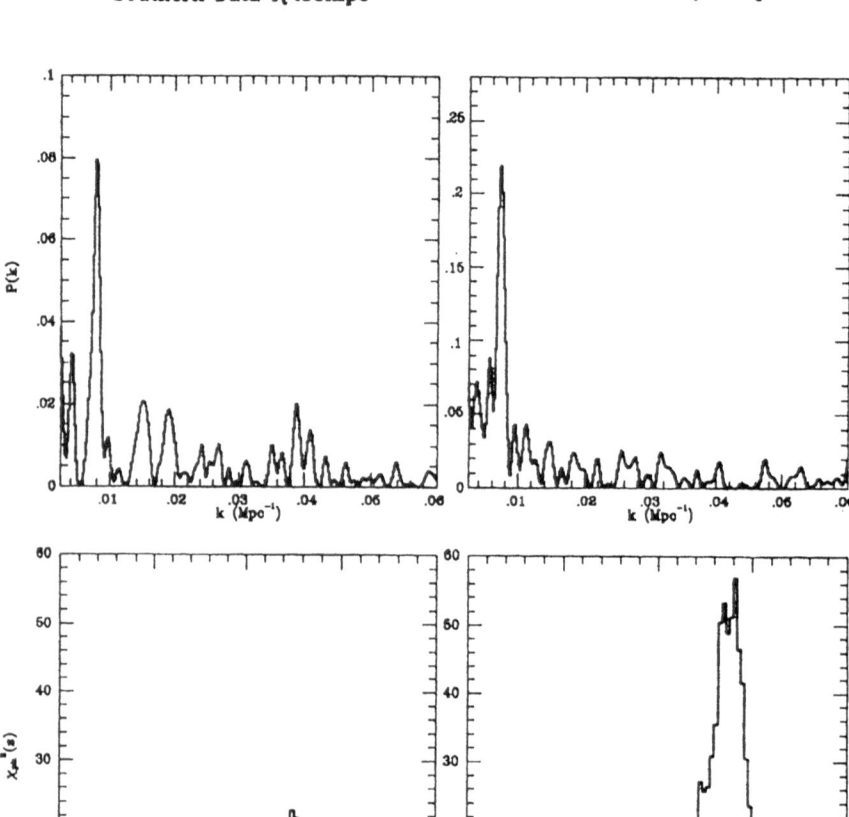

Fig. 3. Power spectra (top row) and periodogram (bottom row) for Southern (lefthand column) and Northern (righthand column) combined galaxy pencil beam sample.

galaxy distribution are picked out in this redshift range (see also Bahcall 1991 for comparison with superclusters) . The power spectrum and periodogram (Figs. 2f and 2i) show very clearly a large excess of power at 128Mpc in excellent agreement with the galaxy distributions presented here.

5 Conclusions

Using a new extended galaxy redshift survey close to North and South galactic poles, we have demonstrated the reality of the claimed 128Mpc scale identified in the much narrower polar sample of Broadhurst et al. (1990). The over-densities responsible

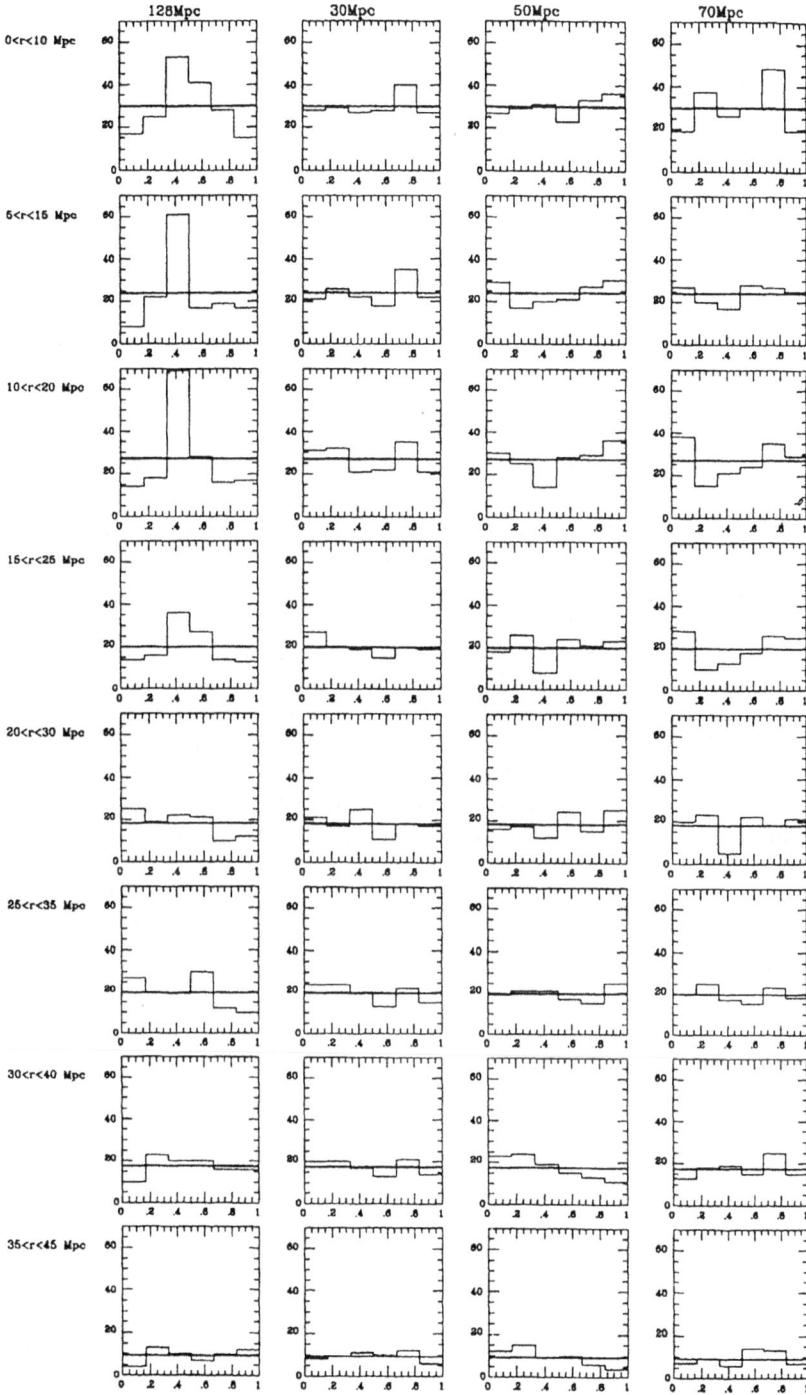

Fig. 4. Phase distribution of the southern galaxy sample as a function of radial distance from the polar axis. The leftmost column refers to the scale 128Mpc and the other columns to scales 30, 50 and 70Mpc respectively.

for the excess power at this scale are **not** simply small clumps intersected by chance in the original narrow beams, but are substantial features extending over 50Mpc. Both the regularity and spatial extent of the structures found in the polar galaxy distribution are independently supported by the polar distribution of rich clusters. This finding may indicate that only $\approx$ 50 significant independent structures will be found over the whole sky to $z = 0.1$, and hence to determine the statistical properties of the general galaxy distribution, a well sampled redshift survey of at least 500Mpc^3 will be required.

References

Abell, G.O., Corwin H.G., Olowin, R.P., 1989. *Astrophys. J. Suppl.*, **70**, 1.

Bahcall, N.A., 1991. *Astrophys. J.*, **376**, 43.

Broadhurst, T.J., Ellis, R.S., Shanks, T., 1988. *Mon. Not. R. Astron. Soc.*, **235**, 827.

Broadhurst, T.J., Ellis, R.S., Koo, D.C., Szalay, A.S., 1990. *Nature*, **343**, 726.

Efstathiou, G., Kaiser N., Saunders, W., Lawrence, A., Rowan-Robinson, M.R., Ellis, R.S., Frenk, C.S., 1990. *Mon. Not. R. Astron. Soc.*, **247**, 10.

Geller, M., Huchra, J.P., 1989. *Science* , **246**, 897.

Kaiser, N., Peacock J.A., 1991. *Astrophys. J.* In press.

Koo, D.C., Kron, R., 1987. In *IAU No. 124, Observational Cosmology*, p383. Eds. Hewitt, A., Burbidge, G.R., Fang, L.Z., D Reidel, Dordrecht.

Lawrence, A., Rowan-Robinson, M., Ellis, R.S., Frenk, C.S., Parry, I.R., Efstathiou, G., Kaiser, N., Xia, X-Y. 1991. *Mon. Not. R. Astron. Soc.* In preparation.

Postman, M., Huchra, J.P., Geller, M.J., 1991. Preprint.

Saunders, W., Frenk, C.S., Rowan-Robinson, M., Efstathiou, G., Lawrence, A., Kaiser, N., Ellis, R.S., Crawford, J., Xia, X-Y., Parry, I.R., 1991. *Nature*, **349**, 32.

A HOMOGENEOUS BRIGHT QUASAR SURVEY

F. La Franca [1,2,3], S. Cristiani [1], P. Andreani [1], A. Gemmo [1], R. Vio [1], C.
Barbieri [1], M. Lazzarin [1], M. Sanvico [1], L. Miller [3], H.T. MacGillivray [3], R.G.
Clowes [3], C. Goldschmidt [4], C. Gouiffes [5], A. Iovino [6] and A. Savage [7]

[1] Dipartimento de Astronomia, Vicolo dell'Osservatorio 5, I-35122 Padova, Italy.
[2] Instituto de Radioastronomia, Via Irnerio 46, I-40126 Bologna, Italy.
[3] Royal Observatory, Blackford Hill, Edinburgh EH9 3HJ, U.K.
[4] Department of Astronomy, University of Edinburgh, Blackford Hill, Edinburgh
EH9 3HJ, U.K.
[5] European Southern Observatory, D-8046 Garching, Germany.
[6] Osservatorio Astronomico de Brera, Via Brera 28, I-20121 Milano, Italy.
[7] UK Schmidt Telescope Unit, Coonabarabran, New South Wales 2357, Australia.

1 The Aim and the Problems

In undertaking a large shallow survey for quasars, we considered especially two issues:

(I) On the one hand, the shape of the Luminosity Function and the determination of the form of evolution as $L \propto (1+z)^k$ or $L \propto e^{T/\tau}$ are uncertain. Trends of this form, potentially telling about the mechanism to fuel the central engine (in principle they can reveal if the QSO phenomenon is driven by the surrounding environment or is determined by its nuclear conditions only), can be spuriously favoured by fits that overlook the observational biases. To probe the real trend, not only the database at faint magnitudes and higher redshifts has to be enlarged, but also the incompleteness at bright magnitudes should be bound or removed with better samples allowing an adequately sophisticated analysis.

The Palomar-Green (Schmidt & Green 1983, hereafter referred to as PG) bright quasar survey at the bright end of the Log N - Log S diagram is especially affected because of the shape of the luminosity function and of the low ratio ($\simeq 2.5$) F/σ, where F is the flux and σ^2 is the variance of the magnitudes, compared to other existing samples. A true increase of information about LF shape and evolution is obtained only combining rich 'homogeneous' samples, i.e. with sensibly matched

H. T. MacGillivray and E. B. Thomson (eds.), Digitised Optical Sky Surveys 405–409.
© 1992 *Kluwer Academic Publishers*.

signal-to-noise ratios: F/σ ought to be larger for the brighter samples with a magnitude limit crossing the steep branch of the LF.

From the present situation of quasar counts for quasars with $z < 2.2$ (see Fig. 1), it is clear that a new survey ($F/\sigma > 4$) is required for the bright part of the Log N - Log S diagram, where the PG sample is incomplete and the information from other surveys is still scanty.

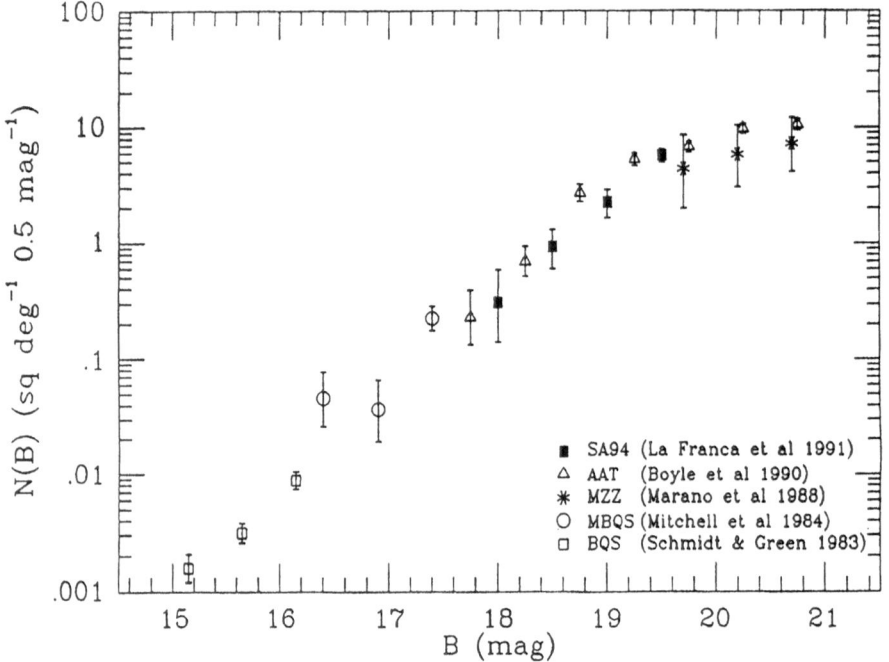

Fig. 1. The number-magnitude relation for QSOs with $z \leq 2.2$ and $M_B < -23.0$ ($H_0 = 50$, $q_0 = 0$) from surveys with complete spectroscopic identifications.

At those magnitudes several hundred square degrees of sky must be searched in order to derive a meaningful number of objects. Beyond the obvious problems, such as the large amount of observing time and the length of the study, intrinsic difficulties (non-uniform photometry and variable extinction over the surveyed area, etc.) call for a carefully coordinated search capable of insuring homogeneous criteria over the required area.

(II) On the other hand, the need is evident (see for example Andreani et al. 1991; Hartwick & Schade 1990) for a large connected volume of space to establish, in a way statistically significant, the different properties of the clustering. For example, the investigation of the nature of the scaling law observed at small length-scale can be achieved by applying algorithms, such as the estimators of the generalized dimensions of the system, whose significance is strictly related to the space volume sampled by the data.

We have undertaken such a difficult task in the framework of an ESO Key

Programme, which has obtained considerable support in the form of telescope and machine time at several institutions (Barbieri et al. 1989).

2 Characteristics of the Survey

- planned area $\sim$ 1500 deg^2 at high galactic latitudes $|b| > 50°$;
- limiting magnitudes $16 < B < 19$;
- $z < 2.2$ (in some areas the search has been extended up to $z = 3.2$);
- two Schmidt plates for each bandpass UBVRI (or UJVRI) are obtained at UKSTU or at ESO, possibly within a few months interval, in order to minimize the variability effects;
- all plates are scanned on the COSMOS machine at ROE;
- 8 fields have also objective prism data and have been analysed with AQD (Clowes 1986a, 1986b) plus a template matching technique (Cristiani et al. 1991), producing candidates in the $1.8 < z < 3.2$ redshift interval.

The Key Programme started in 1989; since then every summer and fall two spectroscopic runs with the ESO 2.2m and 1.5m telescopes, and two photometric runs with the ESO 2.2m or the Danish 1.5m and with the 1m telescopes have been allocated. The expected duration of the programme is 5 years.

3 Present Status

Up to the present time, UBVRI (or UJVRI) plates have been obtained for 25 fields of 25 deg^2 each and these have been scanned on COSMOS. 14 fields have been photometrically calibrated by means of our own CCD photometric standards. The faint limit of the photometry varies from field to field from B=19.5 to B=21.0; the accuracy is 0.1 mag both in mag and colour index. UVx QSO candidates have been selected in all these fields. Higher redshift candidates have been selected in all the objective-prism fields.

The follow-up spectroscopy has been completed for 10 UVx fields and 3 slitless fields.

Of the several hundreds of candidates so identified, some 600 have already been examined spectroscopically. The success rate is about 35%. As expected, most quasars turn out to have $z < 2.2$, although a fair number of higher redshift quasars is also obtained. Adding the quasars already known in the literature, the total number of quasars presently known in the area of our homogeneous bright quasar survey with $B < 19$ exceeds 500.

The data of this survey can be enlarged by the results of a similar search by L. Miller and C. Goldschmidt on 13 fields ($\sim$ 325 sq. deg.) in the north galactic cap (see Goldschmidt et al., these proceedings). The candidates in 4 fields (100 sq. deg.) have been already observed in the framework of a joint project and included in our computations.

In Fig. 2 the space distribution of the fields and of the confirmed quasars in our programme are shown.

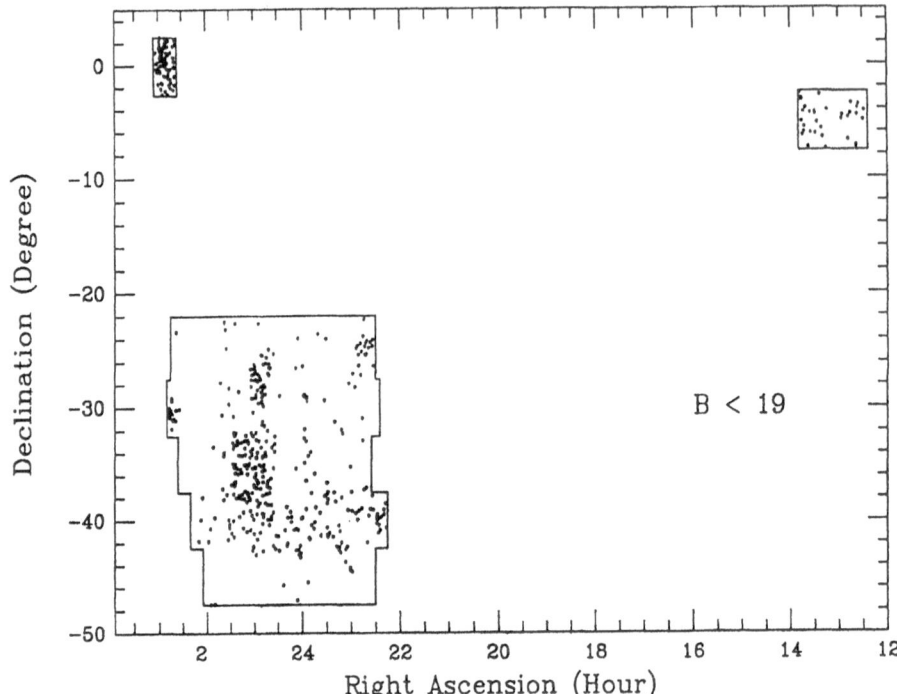

Fig. 2. The space distribution of the fields covered by this survey. The dots correspond
to the known quasars.

Up to now we have spent a significant amount of time on the candidates in the
fainter apparent luminosity bins, for which an area much smaller than 1500 deg^2 will
provide significant statistics. Therefore, the subsequent stages will be considerably
speeded up.

4 Other Goals

A better understanding of the quasar luminosity function and of the large-scale
clustering (at $z < 2.2$) will not be the only outcomes of this programme:

a) We expect to find hundreds of active galaxies for which a fine-tuning of the
   selection techniques will allow us to determine complete samples.
b) A study of the variability of the objects (by comparison with the plates of the
   original ESO blue survey) will allow us to study the LF of the BL Lac objects
   (and to check the completeness of the quasar survey), for which only scanty
   information is available at present.
c) The multi-colour technique will be exploited to select high redshift ($z > 2.2$)
   candidates.
d) Precise statistical information will be available for millions of objects, which
   can be used for stellar counts, galaxy studies, cross-correlation with catalogues
   at different wavebands etc.

References

Andreani, P., Cristiani S., La Franca F., 1991. *Mon. Not. R. Astron. Soc.* Submitted.

Barbieri, C., Cristiani, S., Andreani, P., Clowes, R.G., Gemmo, A., Gouiffes, C., Iovino, A., La Franca, F., Savage, A., Vio, R., 1989. *ESO Messenger* **58**, 22.

Clowes, R.G., 1986a. *Mon. Not. R. Astron. Soc.*, **218**, 139.

Clowes, R.G., 1986b. *Mitteilungen der Astron. Ges.* **67**, 174.

Cristiani, S., La Franca, F., Barbieri, C., Clowes, R.G., Iovino, A., 1991. *Mon. Not. R. Astron. Soc.* In press.

Hartwick F.D.A., Schade D., 1990. *Ann. Rev. Astron. Astrophys.*, **28**, 437.

La Franca, F., Cristiani, S., Barbieri, C., 1991. *Astron. J.* Submitted.

Schmidt, M., Green, R.F., 1983. *Astrophys. J.*, **269**, 352.

THE EDINBURGH MULTI-COLOUR SURVEY FOR QUASARS

C. Goldschmidt [1], P.S. Mitchell [1], L. Miller [2], R.S. Stobie [2], B.J. Boyle [3], F. La Franca [2,4], S. Cristiani [4], R.D. Cannon [5], W.K. Griffiths [6], S.J. Lilly [7] and R.M. Prestage [8]

[1] Department of Astronomy, University of Edinburgh, Blackford Hill, Edinburgh EH9 3HJ, U.K.
[2] Royal Observatory, Blackford Hill, Edinburgh EH9 3HJ, U.K.
[3] Institute of Astronomy, Madingley Road, Cambridge CB3 0HA, U.K.
[4] Dipartimento de Astronomia, Universita di Padova, Vicolo dell'Osservatorio 5, I-35122 Padova, Italy.
[5] Anglo-Australian Observatory, P.O. Box 296, Epping, NSW 2121, Australia.
[6] Department of Physics, University of Leeds, Woodhouse Lane, Leeds LS2, U.K.
[7] Department of Astronomy, University of Toronto, Toronto, Ontario, Canada.
[8] Joint Astronomy Centre, Hilo, Hawaii, U.S.A.

1 The Survey

The Edinburgh Multi-colour Survey is a large-area, accurately-calibrated UBVRI survey at intermediate magnitudes ($R \leq 18.5$). It consists of 2 plates in each of U,B,V,R and I in 13 UKST fields, covering a contiguous area of 330 deg^2. The plates were taken close together in time to minimise contamination and incompleteness due to variability, and have been scanned on the COSMOS machine at the Royal Observatory, Edinburgh.

1.1 Astronomical Motivations

This survey is being used to acquire complete and accurately calibrated samples of quasars in the following redshift and magnitude ranges:

a) $0.3 \leq z \leq 2.2$ and $15 \leq B \leq 18.5$.
b) $3.4 \leq z \leq 4.7$ and $17 \leq R \leq 18.5$.

H. T. MacGillivray and E. B. Thomson (eds.), Digitised Optical Sky Surveys 411–416.
© 1992 *Kluwer Academic Publishers.*

The main motivations are to define the luminosity function for luminous UVX and high redshift quasars better than previous surveys have done, to compare space densities of quasars over a large range of redshifts in order to determine the form of evolution and to look for evidence of clustering (in particular on large-scales) in order to provide constraints on galaxy formation theories.

1.2 Method of Calibration

There are two main problems that must be dealt with when calibrating photographic plates:

a) the relationship between COSMOS-measured magnitudes and 'true' magnitudes is non-linear. This relationship must be defined using photometric data of standard stars.

b) the COSMOS-measured magnitudes and errors in these measured magnitudes vary as a function of position on the plate (due to vignetting and emulsion desensitisation); these 'field effects' are especially important because the log(number)-magnitude relation for quasars is very steep, so a small change in the measured magnitudes will mean a large change in the number of quasars detected. This effect will appear as spurious clustering.

The plates were measured on COSMOS in image analysis mode. It was a requirement that each object should be present on each of the 2 plates in each waveband (this acts as a powerful discriminant against spurious objects with peculiar colours which would otherwise dominate the candidate lists); but not necessarily present in all 5 wavebands. Within each waveband, the positions of the objects were matched by using a transformation derived from the 40 000 brightest objects and then applying half that transformation to each plate.

Each waveband was separately calibrated using photoelectric and CCD stellar sequences. A spline fit to the COSMOS-true magnitude relation is adopted with the constraint that the observed log(number)-magnitude relation be a low order polynomial. This constraint ensures that the calibration curve is well-behaved, even for images which are fainter than the faintest calibration star.

Once we have calibrated catalogues in each waveband we can carry out the morphological analysis; this is useful not only to perform star/galaxy separation but also to look at the extent of field effects, since these affect the morphological parameters of images as well as their magnitudes. Morphological analysis is carried out on each plate using 3 COSMOS parameters; the axial ratio parameter b/a (where b is the minor axis and a is the major axis of each image), the area and the maximum intensity of each image, binned as a function of position and total intensity. The axial ratio is used to distinguish between stars and galaxies because for stars $(b/a) \to 1$, (although we have to take into account the fact that as images get fainter they get noisy and the axial ratio parameter deviates from 1). For a given intensity, the area of a galaxy will be larger than a star and the maximum intensity of a galaxy will be smaller than that of a star.

After each waveband has been separately calibrated and morphologically analysed, the wavebands are combined into one catalogue and the position of each object is transformed onto its coordinates derived in the R band. Then, analysis on position-dependent and magnitude-dependent errors is carried out by transforming number-magnitude histograms in the R band (the waveband suffering least from field effects) binned as a function of plate position onto the central histogram on the best plate in each field. For the other wavebands, we transform colour histograms (i.e. $U - R, J - R, V - R$, and $R - I$) onto the central histogram on the best plate in each field.

1.3 The Spectroscopic Survey

Quasar candidates are chosen by virtue of their non-stellar colours. UVX candidates are required to be blue in $U - B$ relative to the main sequence and red in $B - R$ to distinguish them from hot blue stars. For the high-redshift candidates, strict morphological analysis is carried to minimise the numbers of red galaxies which would otherwise swamp the candidate lists. Then the candidates are required to lie away from the main-sequence locus in 4-D colour space. It has been shown (Mitchell et al. 1990) that this multi-colour selection technique is successful at selecting 50% of quasars in the redshift range $3.4 \leq z \leq 4.7$. To date, spectroscopic confirmation of the UVX sample is nearly completed up to $B = 18$. In the future, the UVX survey will be extended to $B \leq 18.5$ using the FLAIR fibre multi-object system at the UK Schmidt.

1.4 Recent Results

For luminous quasars, the luminosity function is a straight power law. Therefore one cannot distinguish between luminosity evolution and density evolution. Hence, if we have a sample of quasars with the same luminosity at different redshifts then we can characterise the evolution in terms of pure density evolution; i.e. changes in the space density as a function of redshift. Previous results from this survey (Mitchell et al. 1990) have shown that the space density of quasars with $M_B = -28$ at $z \approx 3.5$ is equal to that of quasars of the same luminosity at $z \approx 2$. This implies that quasar evolution slows down at $z \geq 2$. However the luminous quasars at lower redshifts were taken from the Palomar-Green (PG) survey (Schmidt & Green 1983) which is thought to be significantly incomplete (Wampler & Ponz 1985). Thus, in order to carry out a proper comparison between the data at different redshifts, we compared the data at high redshifts with a complete sample of luminous UVX quasars taken from the Edinburgh survey. We compared the space densities of these UVX quasars to those of the PG survey, MBQS survey and the AAT survey; see Fig. 1.

Within the same magnitude range, all our data lie above all the PG data, implying that that survey is indeed incomplete. Our faintest data point agrees well with the brightest data point from the AAT and our data also agree well with MBQS. Boyle et al. (1988) showed that the best parameterisation of the evolution of UVX quasars was pure luminosity evolution with a simple dependence on $(1+z)$;

$$L(z) = L_0(1 + z)^k \qquad (1)$$

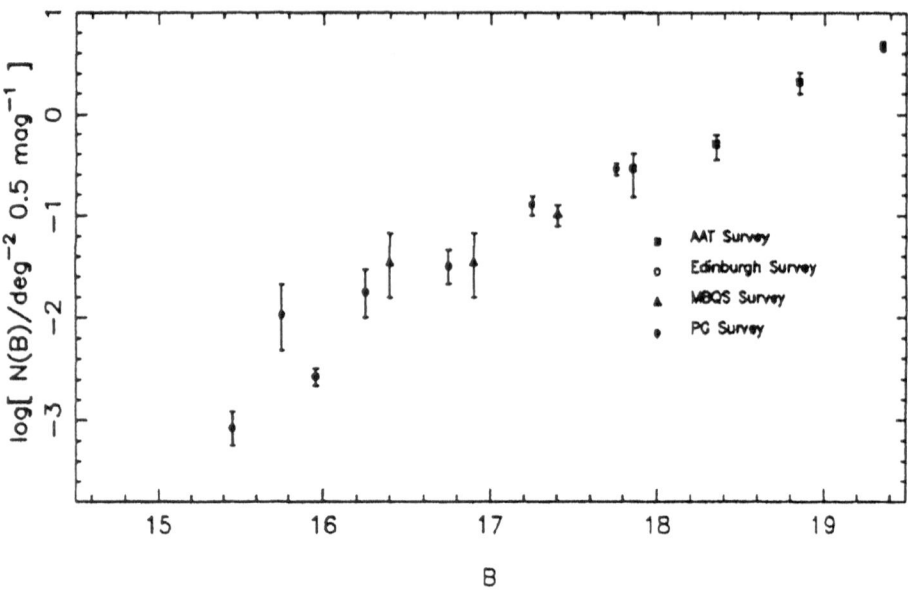

Fig. 1. Differential number-magnitude diagram comparing the surface densities of quasars found in the Edinburgh, PG, AAT and MBQS surveys.

which, as explained above, for luminous quasars is equivalent to $\rho(z) = \rho_0(1 + z)^f$ where L is the luminosity, ρ is the space density, and k and f are constants. Thus, a graph of $\log(\rho)$ against $\log(1 + z)$ should be a straight line.

As can be seen from Fig. 2, which shows $\log(\rho)$ against $\log(1 + z)$ for the Edinburgh and PG surveys, the data from the Edinburgh survey lies in a straight line up to $z \leq 2$ and then flattens off. This implies that a global change took place in the host and surrounding galaxies at $z \sim 2$ which triggered the quasar evolution. Just how the change in galaxy evolution influenced quasar evolution depends on whether quasars are long-lived or short-lived. More data is needed on luminous quasars, especially at intermediate redshifts, to determine the form of the evolution.

2 Acknowledgements

We acknowledge the UK Schmidt Telescope Unit of the Royal Observatory, Edinburgh for obtaining the plate material and the COSMOS unit for the measurements on which the survey is based. We acknowledge allocations at: ESO, the Steward Observatory of the University of Arizona, the University of Hawaii, the Isaac Newton Group, the Anglo-Australian Observatory, and the Cerro Tololo Inter-American Observatory. Data reduction was carried out on STARLINK. P. Goldschmidt acknowledges receipt of a SERC studentship.

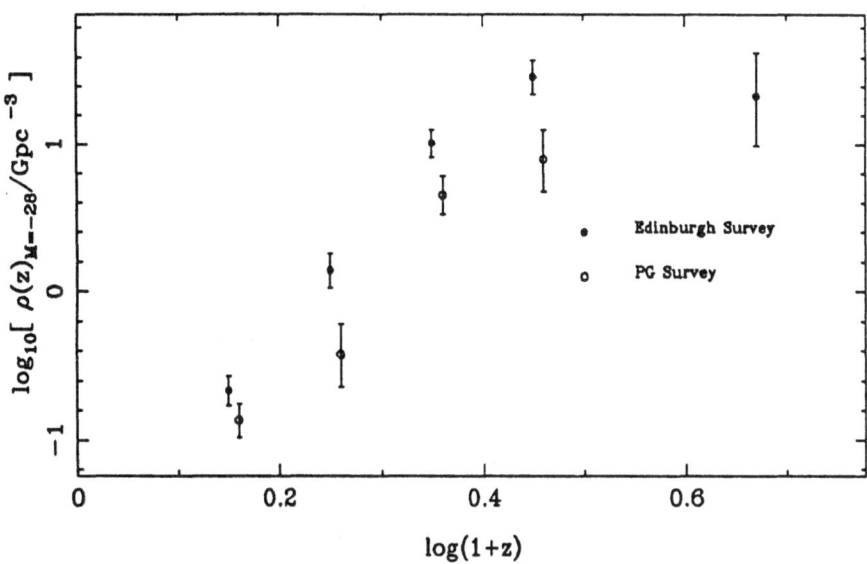

Fig. 2. Diagram of $\log(\rho(z))$ against $\log(1+z)$ for quasars from the Edinburgh and PG surveys with absolute magnitude $M_B = -28$.

References

Boyle, B.J., Shanks, T., Peterson, B.A., 1988. *Mon. Not. R. Astron. Soc.*, **235**, 935.
Miller, L., Mitchell, P.S., 1988. *Astron. Soc. Pacific Conf. Ser.*, **2**, 114.
Mitchell, K., Warnock, A., Usher, P.D., 1984. *Astrophys. J.*, **287**, L3.
Mitchell, P.S., Miller, L., Boyle, B.J., 1990. *Mon. Not. R. Astron. Soc.*, **244**, 1.
Schmidt, M., Green, R.F., 1983. *Astrophys. J.*, **269**, 352.
Wampler, E.J., Ponz, D., 1985. *Astrophys. J.*, **298**, 448.

Discussion

Schachter :
(1) What is the range of Galactic latitude?
(2) How many predicted quasars in each of the redshift bands do you expect?
(3) How clearly can you discriminate host galaxy starlight from bright nuclei in the Y-dimensional colour–colour diagrams?

Goldschmidt :
(1) The survey is at $b \approx 60°$.
(2) We expect about 160 UVX quasars and 6–30 quasars in the range $3.4 \leq z \leq 4.7$.
(3) For $z > 0.3$ we do not pick up the contribution from the host galaxy on the photographic plate because the host galaxies are intrinsically too faint; and the multi-colour method of selecting quasar candidates selects $\sim 98\%$ of

all UVX quasars. We also use morphological criteria to choose candidates, so this would discriminate against quasars whose host galaxies showed up on the photographic plates.

Odewahn :

For calibrating the faint end of your survey, you have made assumptions about the form of the stellar luminosity function. Is this method used independently in each bandpass, and what effect might this have in your colours for faint objects?

Goldschmidt :

In each waveband we have photometric and CCD sequences that go very faint and it is these data that defines the COSMOS–true magnitude relationship. The added requirement, that the n(m) relation should be 'smooth', is not a very stringent requirement and should therefore not affect the colours of any objects.

Clowes :

Do you find any evidence for either small or large–scale quasar clustering?

Goldschmidt :

We have not done any analysis to look for clustering yet, but we expect to be analysing the survey in the next few months.

Parker :

What are the relative success rates for quasar detection for the UVX and high Z QSOs from your data and have you any plans for the presumably astrophysically interesting non-quasars observed?

Goldschmidt :

The success rates are (a) $\sim$ 50% for UVX candidates (this is averaged for the whole magnitude range) and (b) $\sim$ 5% for high Z candidates. The completeness rates are $\sim$ 95% for UVX quasars and 50% for high redshift ones. To date, we have not used the rest of the survey ourselves, although people from Queen's University, Belfast, have used the survey to select main sequence B stars out of the Galactic plane.

APM SURVEYS FOR HIGH REDSHIFT QUASARS

R.G. McMahon [1] and M.J. Irwin [2]

[1] Institute of Astronomy, Madingley Road, Cambridge CB3 0HA, U.K.
[2] Royal Greenwich Observatory, Madingley Road, Cambridge ÇB3 0EZ, U.K.

Abstract

Two related surveys for high redshift quasars are described:
(i) The APM BRI survey for z > 4 quasars. This survey currently covers
~2000 deg^2 to a nominal limiting magnitude of R = 19 and has resulted
in eighteen new spectroscopically confirmed quasars with z > 4. No
decline in the space density of luminous quasars is observed over the
redshift range z = 2 to z = 4.5.
(ii) The APM has been used to identify radio sources from the MIT–
Greenbank (MG) 5Ghz radio survey on the POSS-I O and E plates.
A subset have been selected for follow-up on the basis of their extreme
red colours. Four new radio loud quasars with z > 3.5 have been discovered including one with z = 3.9.

1 Introduction

Over the last 5 years, the APM machine at the Institute of Astronomy, Cambridge
has been used for a large number of optical surveys for quasars based on direct
and objective prism plates taken with the UK Schmidt Telescope and also direct
and grens/grism plates from numerous 4m class telescopes. The results of many of
these studies are described in the proceedings of two conferences; 'Proceedings of a
workshop on optical surveys for quasars' (Osmer et al. 1988) or 'Space distribution
of quasars' (Crampton et al. 1991). Most of the quasars in these surveys have either
been discovered on the basis of their emission lines or blue continuum colours.
However, a small but significant number of the APM quasars have been discovered
primarily on the basis of their red colours. These are the highest redshift quasars.
In this paper, I describe two projects which exploit the red colours of high redshift

H. T. MacGillivray and E. B. Thomson (eds.), Digitised Optical Sky Surveys 417–424.
© 1992 *Kluwer Academic Publishers.*

quasars. In the first, the APM has been used to discover optically selected high redshift quasars purely on the basis of their extreme red colours on UKST B, R and I direct plates. In the second project, optical colours derived from Palomar Observatory Sky Survey (POSS) blue (O) and red (E) plates have been used to select a subset of the red stellar radio identifications for spectroscopic follow-up.

2 The APM BRI Survey for Optically Selected QSOs

The APM BRI colour survey is designed to look for bright (R < 19), high redshift quasars (4 < z < 5), by covering a large area, ~1000 deg^2 of sky. Quasar candidates are selected using APM measurements of UK Schmidt Telescope (UKST) B$_J$, R and I plates. Most of the spectroscopic follow-up of the candidates has been done using the Faint Object Spectrograph on the 2.5m Isaac Newton Telescope on La Palma using exposures of typically 600s. The survey limit is effectively defined by a combination of the depth of the B$_J$ band ($\sim$ 22.5) and the morphological discrimination possible on the R plates.

The rationale behind our survey is quite simple. The majority of redshift z > 3.9 quasars have a B$_J$ $-$ R, R $-$ I colour different from the locus of points occupied by normal stars. This is illustrated in Fig. 1, which shows a composite two colour (B$_J$ $-$ R, R $-$ I) diagram for stellar objects derived from APM measurements of standard UKST survey plates of one of the fields in our survey. Quasars with z < 3 lie in the region indicated by the ellipse centred on B$_J$ $-$ R $\sim$ 0.4, R $-$ I $\sim$ 0.2. The colours of a representative sample of confirmed quasars with 3 < z < 4 are indicated by heavy circles. Of these, the quasars with 3.0 < z < 3.3 were discovered using UKST IIIa-J objective prism plates, whilst the 3 quasars with 3.5 < z < 3.9 were discovered on UKST IIIa-F objective prism plates. The other symbols denote the location of some of the quasars with z > 4 found using the multi-colour method. Since each UKST field contains some 250 000 objects, minimising the contamination is a vital part of the survey. The main 'contaminants' in this region of colour space are compact red galaxies at a redshift of z $\sim$ 0.3. An accurate star/galaxy classification algorithm is therefore also an essential requirement and as noted earlier defines the effective R survey limit. Interestingly, the only other significant 'contaminants' that we have found are high latitude distant carbon stars and very low luminosity M dwarfs.

Important features of the colours of the quasars are: (i) the well-defined range in R $-$ I of z < 3 quasars due to the effects of continuum slope and emission lines; (ii) for the quasars with z > 3 there is a strong trend in B$_J$ $-$ R with increasing redshift because of increasing amounts of absorption due to the Lyα forest within the B$_J$ band; (iii) by a redshift z $\sim$ 4 the quasars are well separated from the foreground stars and there are **extremely few** outlying points in this region; (iv) above a redshift of z $\sim$ 4.2 the quasars lie so far away from the stellar locus that a simple B$_J$ $-$ R cut suffices to select them.

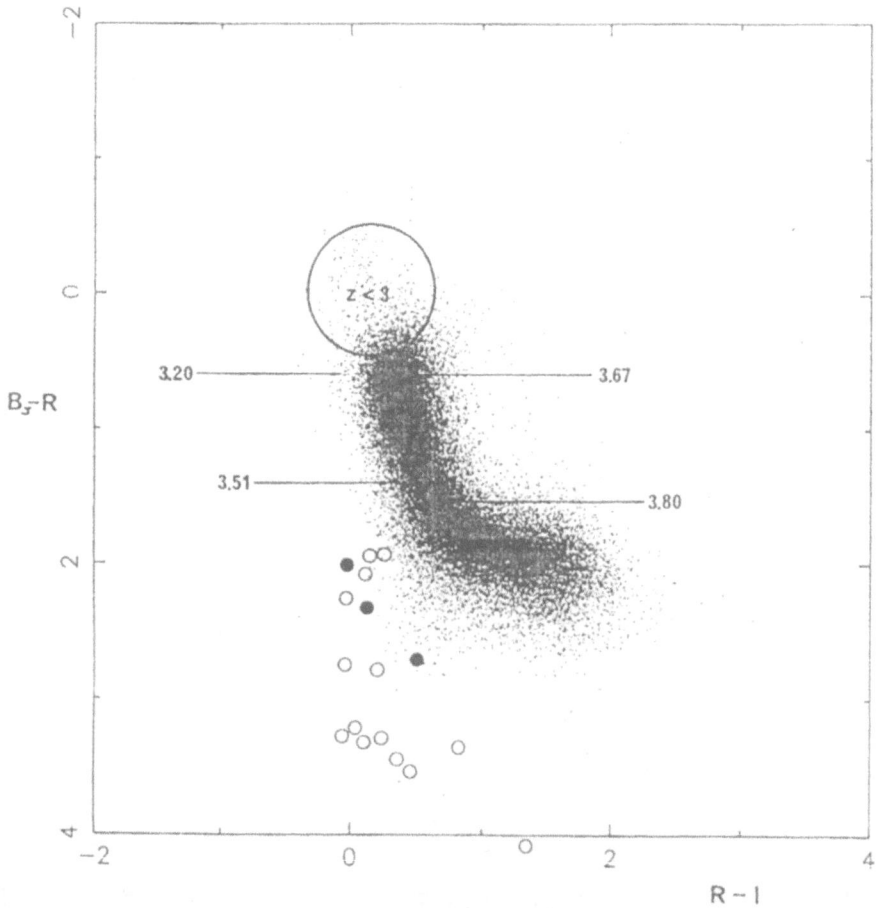

Fig. 1. A two–colour $B_J - R$, $R - I$ plot of stellar objects in a high galactic latitude UKST standard survey field. The region occupied by the majority of quasars with $z < 3$ is shown by the large ellipse. A number of quasars selected using UKST IIIa-F objective prism plates are shown together with the $z > 4$ quasars found during the APM survey (open circles) and the three $z > 4$ quasars found in the SGP (filled circles).

2.1 The APM B–R Survey

We had initially assumed that the quasar locus in a B_J, R, I two-colour diagram would be much closer to the stellar main sequence and I band plates would be necessary to help discriminate against foreground stellar interlopers. However, after completing the first 1000 deg^2 of the survey and examining quasar candidates much closer to the stellar locus (and finding no more than those we had found further out) it became apparent that a simple extension of the survey using just B_J and R plates was possible. Consequently a second 1000 deg^2 was surveyed in these two passbands only (a BRX survey for $z > 4.2$ cf. UVX for $z < 2.2$). Figure 2 shows a stellar ($B_J - R$, R) colour-magnitude diagram for a typical high galactic latitude field showing the location of the $z > 4$ quasars. Figure 3 shows how the $B_J - R$ colour varies with redshift with the highest redshift objects lying furthest from the main sequence. We are currently extending the $B_J - R$ survey over another 1000 deg^2 and over the next few years we hope to cover up to 10 000 deg^2 as the 2nd Epoch UKST R survey progresses. We may extend the survey to the Northern Hemisphere when POSS-II B and R plates become available in large numbers.

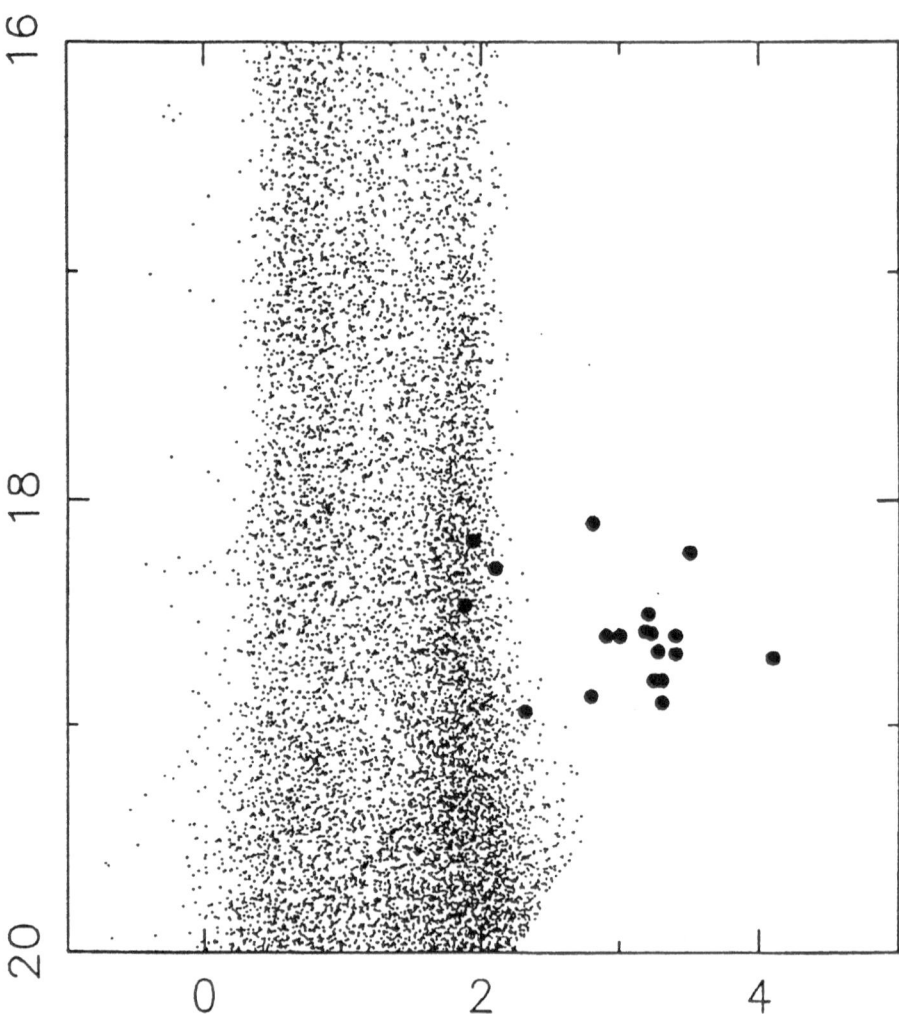

Fig. 2. $b_j - R$ versus R colour magnitude diagram based on APM measurements of UKST plates for a typical 10 deg^2 region at high galactic latitude ($|b| > 30°$). The colours of $z > 4$ quasars found during the APM survey are shown by the filled open circles.

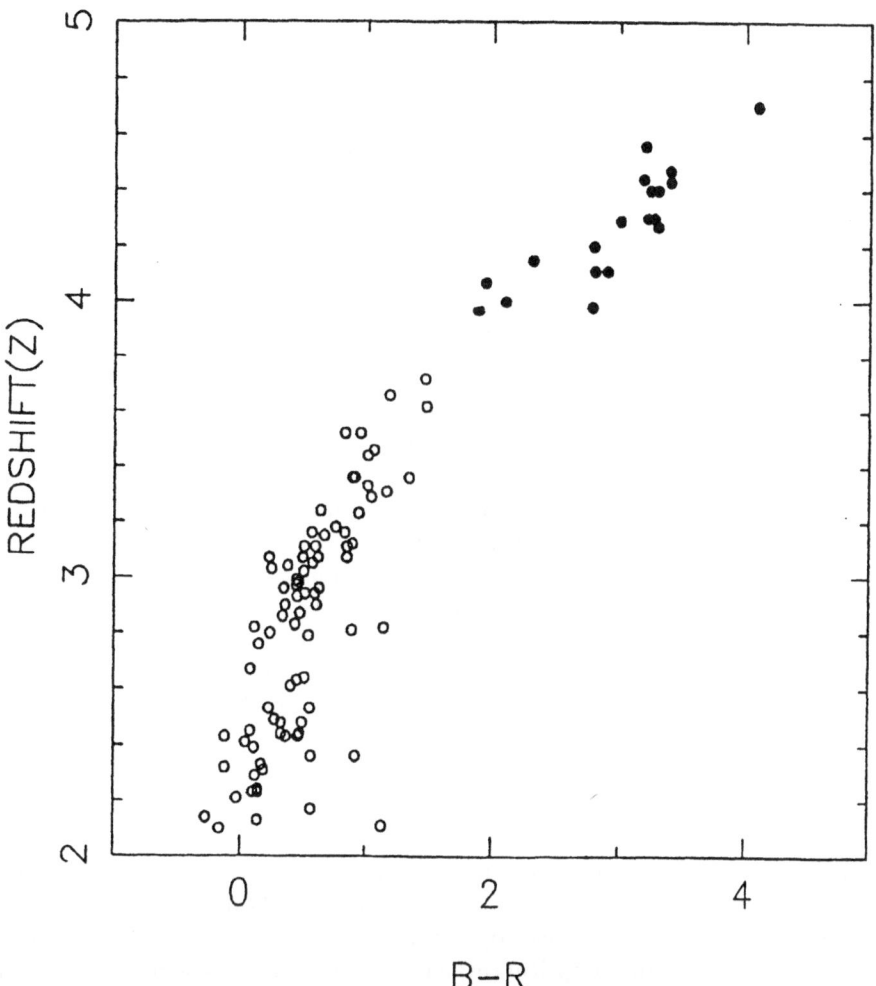

Fig. 3. b_j − R colour as a function of redshift for quasars from the APM z > 4 quasar survey (filled circles) and a selection of lower redshift quasars (open circles).

3 The MG/APM Survey for Radio Loud QSOs

Whilst the first quasars with redshifts in excess of 3 were discovered in the early 1970s using radio techniques, there has been little progress in determining how the space density of radio-loud quasars evolves, beyond even a redshift of 2, except to demonstrate that the rapid evolution observed at lower redshifts ceases at z ∼ 2. The main reason for this is that in a purely flux-limited sample of radio sources, the proportion of z > 3 quasars is low (∼5%) even when one restricts the sample to flat spectrum sources (e.g. Savage et al. 1988). The most recent analysis of all the available data on statistically complete samples by Dunlop & Peacock (1990) contains only two quasars with z > 2.5. Therefore, the prerequisite for any programme aimed at probing the evolution of the space density of radio-loud quasars at high redshift is a large number of sources (>1000) and a technique to remove the majority of the much commoner low redshift quasars from those that must be followed up spectroscopically.

3.1 The MIT–Green Bank (MG I) 5Ghz Radio Survey

The first MIT–Green Bank (MG I) 5GHz radio (Bennett et al. 1987) survey carried out with the now defunct NRAO 300 ft telescope covers 1.9 sr of sky in the declination band $-00°30' < \delta + 19°30'$ and contains $\sim$ 5000 radio sources with $S_{5GHz} >$100mJy. About 4000 of these sources have been reobserved with the VLA by B. Burke and his colleagues at MIT, and hence accurate radio positions ($\sim 1''$) and radio maps exist for these sources. Up until now, the primary use for this survey has been in a search for radio-loud gravitational lenses where a small sub-sample of the radio sources have been selected for optical follow-up based on their radio morphology.

The MG I survey will also provide a large sample of radio-loud quasars. It is expected that at least 1000 of the sources can be identified with stellar objects above the POSS O and E plate limits. Such a large sample is ideally suitable for many projects such as directly probing the evolution of the radio-loud quasar luminosity function above z = 3. Before any progress can be made, the radio sources must be identified in a systematic manner on the basis of positional coincidence alone over the whole region. Such a project using the APM and POSS O and E plates is now nearing completion.

3.2 APM Identification Procedure

Over the last 2 years, the APM machine has been used to digitise the high Galactic latitude ($|b| > 30$) POSS O and E plates at $1.0''$ resolution ($0.5''$ pixels) producing an astrometric and photometric catalogue accurate to $\sim 0.2''$ and 0.05mag respectively. This resultant optical catalogue has been used to identify, classify and determine the colours and magnitudes of all those MG sources that are visible on either of the POSS plates. Figure 4 shows a (O–E, E) colour-magnitude diagram (with an arbitrary colour zeropoint) for a complete sample of $\sim$ 300 optical identifications. This diagram is based on objects selected from $\sim$ 40 POSS fields. The photometry for each field has been zero-pointed by assuming the E plate limit is 20 and by assuming a universal magnitude independent position for the stellar ridge. The colours of the stellar identifications are characterised by a sharp peak and tail off to the red. The width (FWHM) of the peak is ~ 0.8 magnitudes ($\sigma = 0.35$mag). This demonstrates how good the internal magnitude system is, since a large fraction of this spread is due to the intrinsic colour distribution of the quasars themselves.

4 Spectroscopic Results

Rather than carry out an indiscriminate redshift campaign of all the radio-loud quasars in the resultant sample, we have initially concentrated on obtaining optical spectroscopy of the reddest optical identifications independent of optical morphology. Low resolution spectroscopy has been obtained mainly at the Shane 3.0m

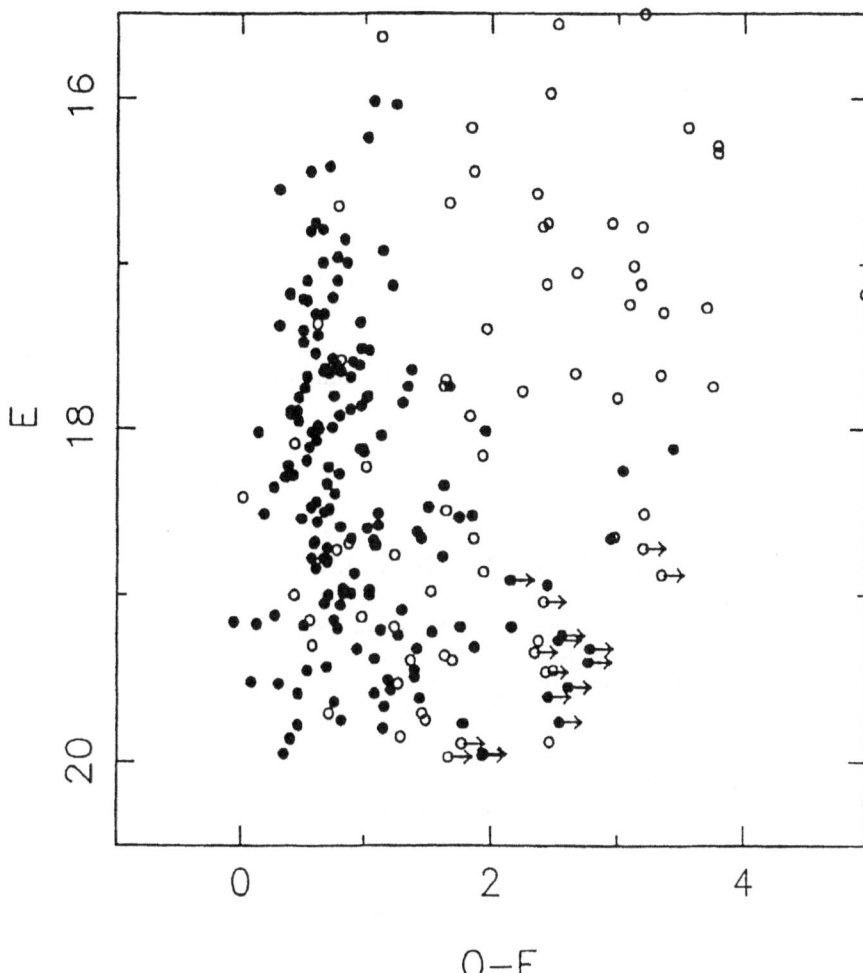

Fig. 4. The O–E v E colour magnitude diagram for the optical identifications of MG sources. Filled circles and open circles denote objects classified as stellar and non-stellar respectively on the E plate. Rightward pointing arrows denote the absence of an image on the O plate, i.e. a very red object.

telescope of Lick Observatory in collaboration with A. Wolfe (UCSD). A total of ∼ 40 spectra have been obtained for the programme, with the discovery of 7 new quasars with z > 3 to add to the 4 already known in the MG I region under investigation. 4 of the quasars have z > 3.5 including one with z = 3.9. Two of these z > 3.5 quasars had been previously classified as faint galaxies, one of which (MG0201+1120) is a ∼1Jy VLA calibrator. The remaining identifications were either lower redshift quasars, low redshift compact narrow emission line radio galaxies or featureless spectra. The featureless objects are either low redshift BL Lac objects or low redshift galaxies.

Over the next 12 months, the spectroscopic phase of the MG-I/APM programme will be completed. The identification programme is currently being extended over a larger fraction of the Northern Hemisphere aimed at identifying sources in the 6C, 7C, 8C, Texas and 87GB radio surveys.

5 Acknowledgements

The APM colour survey has been carried out in collaboration with Cyril Hazard. This work would have been impossible without the steady supply of high quality UKST plates. In particular the plate library at the UKSTU in ROE has been invaluable, since the BRI survey has involved mostly archival plate material. The MG/APM work has been primarily carried out in collaboration with Bernie Burke and Sam Connor at MIT.

References

Bennett, C.L., Lawrence, Garcia-Barreto, J.A., Hewitt, J.N., Burke, B.F., 1987. *Astrophys. J. Suppl.*, **61**, 1.

Crampton, D., et al. 1991. In *'Proceedings of a Workshop on the Space Distribution of Quasars', ASP Conference Series.*

Dunlop, J.S., Peacock, J.A., 1990. *Mon. Not. R. Astron. Soc.*, **247**, 19.

Osmer, P., et al. 1988. In *'Proceedings of a Workshop on Optical Surveys for Quasars', ASP Conference Series* No. 2.

Savage, A.J., et al. 1988. In *'Proceedings of a Workshop on Optical Surveys for Quasars', ASP Conference Series* No. 2, p. 204, eds. Osmer et al.

LARGE GROUPS OF QUASARS

R.G. Clowes [1] *and L.E. Campusano* [2]

[1] Royal Observatory, Blackford Hill, Edinburgh EH9 3HJ, U.K.
[2] Universidad de Chile, Casilla 36-D, Santiago, Chile.

Abstract

A short summary of the large groups of quasars is given, with particular emphasis on the most recent discovery — a group with a long dimension of $\sim 100\text{-}200h^{-1}$ Mpc. This group was discovered with an automated (AQD) survey for quasars, using COSMOS data of an objective-prism plate from the UK Schmidt Telescope.

1 Large Groups of Quasars

Some recent results on quasars are suggesting that their clustering properties are not so simple as was previously thought. The first description of quasar clustering was certainly very simple: with the possible exception of an occasional 'anomaly', quasars seemed to be unclustered — just scattered uniformly and randomly in space (e.g. Osmer 1981). Deeper samples and a clear detection of clustering (Shanks et al. 1988) then led to a revision, in which quasars are weakly clustered on small scales $< 10h^{-1}$ Mpc but are still scattered uniformly and randomly at larger scales. A complication (evolution) emerged with some tentative evidence for the existence of this small-scale clustering only at redshifts less than ~ 1.5 (Iovino & Shaver 1988). Now, the recent discoveries and recognition of large groups of quasars might lead to a further, quite drastic, revision. Large groups, also known as isolated groups, appear like relatively dense groupings embedded in the general distribution. They really are very large, having dimensions from $\sim 50h^{-1}$ to $\sim 200h^{-1}$ Mpc. It seems possible — consistent with existing data — that all clustering of quasars, even that on the small scales, $< 10h^{-1}$ Mpc, is localised in these groups, and that all other quasars, the majority, are distributed uniformly and randomly.

Large groups seem to be very rare: only three examples are known at present. The most recent discovery, Clowes & Campusano (1991), illustrated in Fig. 1, could be the largest structure so far observed in the universe. It has a long dimension of

H. T. MacGillivray and E. B. Thomson (eds.), Digitised Optical Sky Surveys 425–428.

$\sim 100\text{-}200h^{-1}$ Mpc, and its redshift of $z \sim 1.3$ implies that it had formed when the universe was only about one-third of its present age. The short dimension on the sky is $35h^{-1}$ Mpc and the depth in the redshift direction is $\sim 100h^{-1}$ Mpc. (Distances are comoving, for the present epoch.) The group is clearly elongated, but otherwise its morphology seems to be like a clump, with no evidence so far for gradients in the redshifts, sheets and bubbles, although there could be some sub-grouping on scales $\sim 20h^{-1}$ Mpc. Ten members of the group were identified from our AQD survey. A further three quasars, known from other sources, seem likely to be members also, and there are probably still more members that have not yet been observed. AQD (Clowes 1986; Clowes et al. 1984) is a computerised method for finding quasars, which uses COSMOS measurements of objective-prism plates from the UK Schmidt Telescope.

Webster (1982) found the first large group of quasars. This was a group of four with size $\sim 75h^{-1}$ Mpc at $z \sim 0.37$, which was in the survey of Osmer & Smith (1980). The second was found by Crampton et al. (1987; 1989), and is a group of 23 quasars at $z \sim 1.1$, with size of at least $60h^{-1}$ Mpc. In this case there is also sub-grouping at $\sim 15h^{-1}$ Mpc. This second group is now known to extend beyond the original survey limits (Cowley 1991, private communication). Note that all of these three large groups were found in surveys by slitless spectrocopy (i.e. objective-prism, grens, grism), which might be particularly suitable because of the combination of wide-angle coverage and a good success-rate for confirming candidates.

The sizes of these groups are similar to those of the large structures of galaxies that have been reported recently. Tully (1987) identified five supercluster complexes, each of size $\sim 225h^{-1}$ Mpc and containing ~ 50 rich clusters of galaxies, although they are disputed by Postman et al. (1989). Da Costa et al. (1989) found two structures of galaxies, one with size $60\text{--}150h^{-1}$ Mpc and one of at least $100h^{-1}$ Mpc, while Geller & Huchra (1989) found the 'great wall' of galaxies with size of at least $170h^{-1}$ Mpc. Broadhurst et al. (1990) showed that galaxies in the directions of both galactic poles are structured in a repetitive pattern with a spatial cycle of $128h^{-1}$ Mpc, extending over at least $1000h^{-1}$ Mpc.

Are the large groups of quasars related to the large structures of galaxies? A strong association might seem unlikely because the structures of galaxies seem to be emerging as commonplace features of the universe, whereas the groups of quasars seem to be rare. Of course, the groups of quasars are occurring much earlier in the history of the universe. There is some evidence that is consistent with an association of the two: Longo (1991) found that the 'great wall' is also detectable in the distribution of low-redshift AGNs, and Phillipps (1990, private communication) found from Schmidt plates that each of two quasars in Webster's (1982) group of four has an uncommonly rich, faint cluster of galaxies within $\sim 2'$. Deep CCD imaging for galaxy counts and absorption-line spectroscopy of background quasars should help to decide this question. Note that West (1991) finds evidence for filamentary superclustering at $z \sim 1$ in the alignment of radio emission from quasars and galaxies.

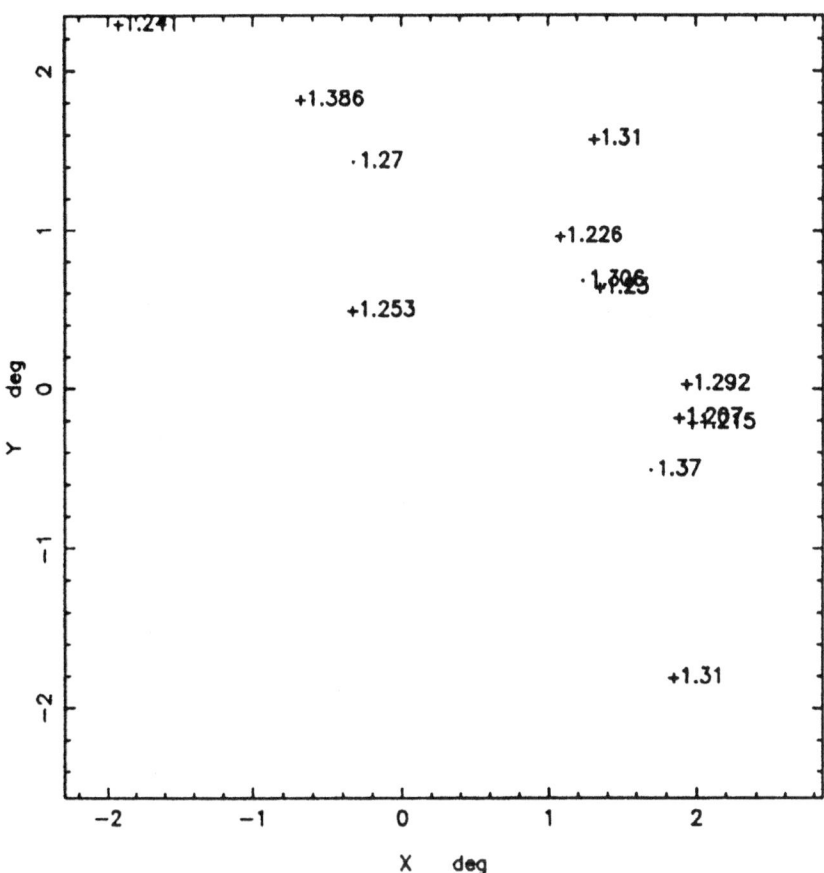

Fig. 1. The new large group of quasars, from Clowes & Campusano (1991). This figure shows a plot of the x, y coordinates in degrees (with respect to the field centre) of those quasars from the Clowes & Campusano (1991) survey that have $1.2 \leq z < 1.4$ (+symbols). Also shown are the further three quasars from other sources that have $1.2 \leq z < 1.4$ (• symbols). The redshifts are shown with the plotted points. The boundaries of the plot correspond to the area of the survey. Note that redshifts ending in zeros are truncated — all values are specified to three decimal places except that at 1.37.

References

Broadhurst, T.J., Ellis, R.S., Koo, D.C., Szalay, A.S., 1990. *Nature*, **343**, 726.

Clowes, R.G., 1986. *Mitteilungen der Astr. Ges.*, **67**, 174.

Clowes, R.G, Campusano, L.E., 1991. *Mon. Not. R. Astron. Soc.*, **249**, 218.

Clowes, R.G., Cooke, J.A, Beard, S.M., 1984. *Mon. Not. R. Astron. Soc.*, **207**, 99.

Crampton, D., Cowley, A.P, Hartwick, F.D.A., 1987. *Astrophys. J.*, **314**, 129.

Crampton, D., Cowley, A.P, Hartwick, F.D.A., 1989. *Astrophys. J.*, **345**, 59.

da Costa, L.N., Pellegrini, P.S., Willmer, C, Latham, D.W., 1989. *Astrophys. J.*, **344**, 20.

Geller, M.J., Huchra, J.P., 1989. *Science*, **246**, 897.

Iovino, A., Shaver, P.A., 1988. *Astrophys. J.*, **330**, L13.

Longo, M.J., 1991. *Astrophys. J.*, **372**, L59.

Osmer, P.S., 1981. *Astrophys. J.*, **247**, 762.

Osmer, P.S., Smith, M.G., 1980. *Astrophys. J. Suppl.*, **42**, 333.

Postman, M., Spergel, D.N., Sutin, B., Juszkiewicz, R., 1989. *Astrophys. J.*, **346**, 588.

Shanks, T., Boyle, B.J., Peterson, B.A., 1988. In *'Proceedings of a workshop on optical surveys for quasars'*, *Astronomical Society of the Pacific Conference Series* Volume 2, p. 244.

Tully, R.B., 1987. *Astrophys. J.*, **323**, 1.

Webster, A., 1982. *Mon. Not. R. Astron. Soc.*, **199**, 683.

West, M.J., 1991. Preprint.

DIGITISED OPTICAL SKY SURVEYS

Part Five:

OPTICAL IDENTIFICATION PROGRAMMES USING DIGITISED SURVEYS

THE IDENTIFICATION OF POTENTIAL COUNTERPARTS TO X–RAY BINARIES USING COSMOS

P. Roche and A. Norton

High Energy Astrophysics Group, Physics Department, Southampton University, Southampton SO9 5NH, U.K.

Abstract

We discuss a possible method for identifying the optical counterparts to X-ray binary systems using COSMOS data. This is based on the anomalous colours which they are expected to exhibit when compared with isolated stars in the same field. Using colour – colour plots for all objects in the area in and around the X-ray error boxes, it is shown that several candidates for further investigation become apparent in each case.

1 The Problem — X-ray Error Boxes

Attempting optical identifications of X-ray sources, especially from archive data, often presents astronomers with a bewildering number of candidates, mainly due to the 'accuracy' of the X-ray position. X-ray positions can be determined with sub-arcminute accuracy using imaging X-ray telescopes (e.g. Einstein), but such observations have not been made for the majority of known X-ray binary sources. Even when this level of accuracy is available, locations in the Galactic plane may still result in several optical candidates being present in or near the error box.

The presence of large databases of X-ray observations, containing archival material from earlier satellite experiments, has prompted us to attempt identification of certain types of source (namely transient X-ray binaries and low-mass X-ray binaries) based on the positional data. However, the accuracy of these older experiments is such that there are often several tens of stars within the error box, and others that lie close enough to the edges of the box to be considered 'possibles'. Typical error box sizes range from several tens of arcminutes (i.e. GINGA, UHURU, Ariel

H. T. MacGillivray and E. B. Thomson (eds.), Digitised Optical Sky Surveys 431–440.
© 1992 *Kluwer Academic Publishers.*

V), to several arcminutes (i.e. SAS-3, HEAO-1 A-3) down to the X-ray telescope error circles of sub-arcminute sizes.

An associated problem is the degree of trust which can be placed in the position of the X-ray error box. The majority of published X-ray positions give 90% confidence error boxes — obviously implying that there is still a chance that the counterpart lies outside the box (although probably within a few tens of arcseconds of the edge). Besides this, the actual positions of the error boxes themselves are often 'revised' after further analysis of the data, so that the search area now becomes much larger than simply the published error box. All in all, the job of optically identifying such sources is not an easy one.

When a source has been detected more than once, there will be a number of error boxes determined for it. This would appear to be a solution to the problem of the very large error boxes, as the source should lie in or near to the areas of overlap of the boxes. However, it is seldom the case that all the boxes cover the same area of sky, and the presence of more than one box can often complicate the problem (see Fig. 1).

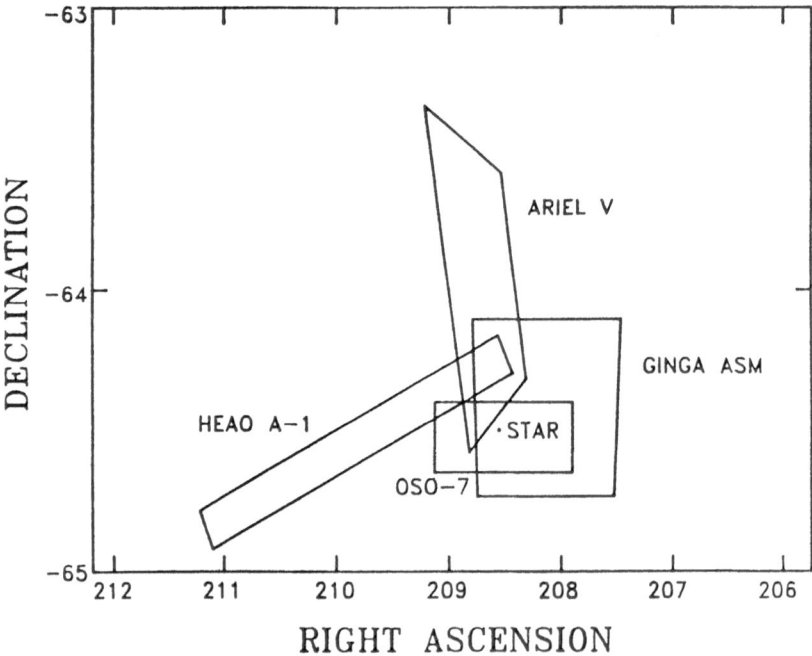

Fig. 1. The overlapping error boxes for a recently identified Be-star/X-ray source. Note that the counterpart lies within only 2 of the 4 error boxes.

2 The Targets — X-ray Binaries

The general class of accretion-powered X-ray binaries can be broadly sub-divided into three distinct types, namely high mass X-ray binaries (HMXBs), low mass X-ray binaries (LMXBs) and cataclysmic variables (CVs). See White (1989) and Osborne (1987) for reviews of their X-ray properties. So far we have applied the optical counterpart search technique to one possible LMXB and to several transient HMXBs. A similar approach to optically identifying X-ray selected CVs has recently been carried out by Hertz et al. (1990).

HMXBs : In these systems the primary is a massive ($\geq 10M_\odot$), early (O or B-type) star, which loses material via Roche-lobe overflow or stellar-wind processes. These sources can be further sub-divided into the supergiant systems (primary $\geq 20M_\odot$, close to filling its Roche lobe) and the transient systems, which contain an emission line star (i.e. a Be or Oe-type, rapidly rotating star).

Figure 2 (from van den Heuvel & Rappaport 1987) shows a simplified model of a Be/Neutron star system. The rapidly rotating Be star is surrounded by a shell or disk of material, which is continually replenished by the slow stellar wind. The neutron star, often in a highly eccentric orbit, accretes from this material and produces X-ray emission at its magnetic poles. The systems appear as X-ray pulsars, with the pulse representing the spin period of the neutron star. X-ray flares can occur at or near periastron passage, as the neutron star passes through the disk and begins to accrete the material. As the neutron star leaves the disk, the X-ray emission tails away until the source 'turns off'. Measurements of successive flares can lead to orbital period determination for these systems. However, the disk itself appears to be unstable, and can either disappear or undergo a 'shell event', when a large outburst of material results in a huge X-ray and optical flare. Therefore, flares can occur at times other than periastron, or simply not appear at all. Because of the inherent on-off nature of these sources, and the often long orbital periods (e.g. A0535+26 111 days, GX304-1 133 days), they are often not detected in X-ray surveys, and when they are it is usually by 'all sky monitors' with low spatial resolution rather than by imaging X-ray telescopes. Thus, there are many transient X-ray binaries with unknown orbital periods and poorly determined positions. It is these targets that we have attempted to identify.

LMXBs : In these systems, the primary is now a low mass ($\leq 2M_\odot$), late-type star and often intrinsically faint. Mass transfer is thought to occur via Roche-lobe overflow only and forms an accretion disc around the neutron star companion. Figure 3 shows a schematic diagram of such a source (taken from Epstein et al. 1986). Note also the surrounding corona which is believed to scatter X-rays from the central source. As shown in Fig. 3, two types of X-ray variability, namely dips and eclipses, are possible with this geometry, depending on the line of sight angle to the system. Other identifying features seen in the X-ray behaviour of some LMXBs are: bursts, thought to be caused by accretion instabilities or by thermonuclear flashes on the neutron star surface; and quasi-periodic oscillations with frequencies of a few Hz, whose exact cause is unknown.

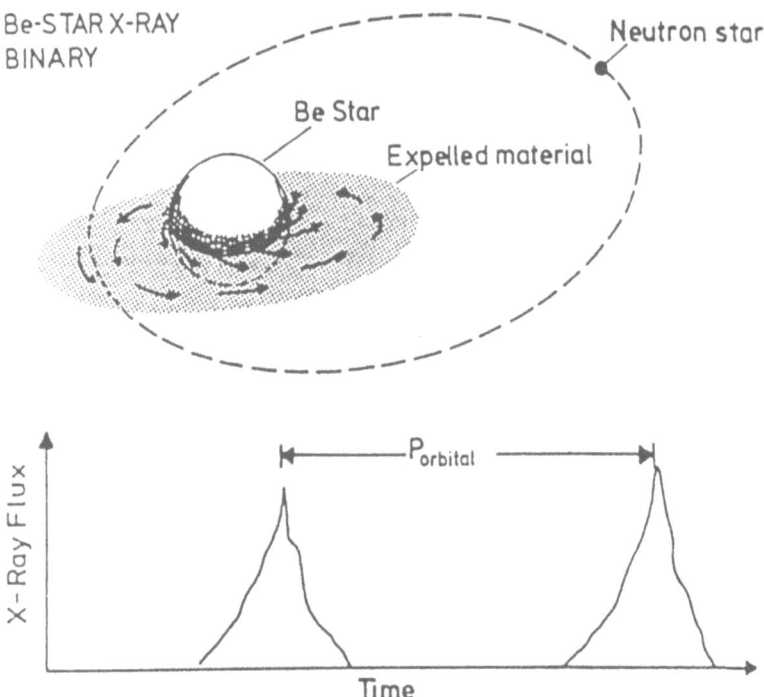

Fig. 2. Schematic diagram of a Be/Neutron star system. Large flares often occur around periastron passage.

3 The Search — What are We Looking For?

Once we have established with reasonable confidence that an X-ray source is either a transient HMXB or a low-mass system, we must then determine what optical features we are looking for in such an object. If we have some idea of the distance to the source, we can try to model the reddening expected, and predict a rough search area in our colour – colour diagrams. We are basically looking for anomalously coloured objects within or near the X-ray error circle. In addition, sources showing variability between two optical plates are also of interest — both transient systems and LMXBs are known to exhibit variability to some degree, and this may be a useful pointer when searching for such sources. If a star is the 'right' colour, within the error box, and/or variable, it becomes a potential candidate.

HMXB Transients : the circumstellar dust shell surrounding the star results in heavy reddening of the intrinsically bright O or B star. These stars are often found to lie above the main sequence and we are therefore looking for a bright, anomalously red object.

LMXBs : Many of the previously identified counterparts to LMXBs were discovered as a result of their ultraviolet excess relative to normal stars (Bradt & McClintock 1983). Hence, the counterparts we are looking for will appear anomalously blue in comparison with other objects in the surrounding area of sky.

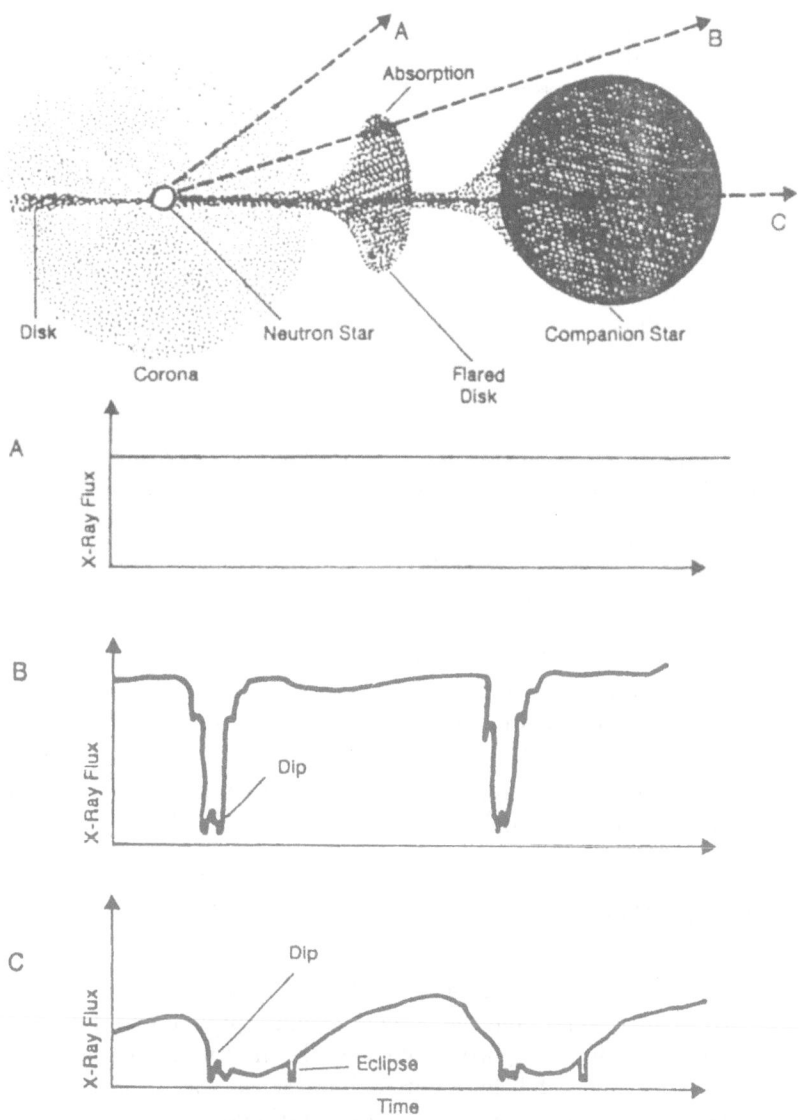

Fig. 3 Schematic diagram of a LMXB.

4 The Method — COSMOS Colour – Colour Plots

Once a suitable source has been found (optically un-identified, transient or suspected LMXB, with a reasonably small error box), we select the appropriate plates from the UK Schmidt plate library, have the relevant area scanned by COSMOS, and begin the search.

The sources we have looked at so far have X-ray error boxes in the region of a few arcminutes or less in size. Ideally, the small error circles of the EXOSAT

LE, Einstein HRI or ROSAT XRT are the easiest to search, but of course these are the ones most likely to have been optically identified already! However, there are some still left without identification, suggesting that the candidates are either very faint or lie outside the search area. The Galactic plane location of many X-ray binaries means that counterparts are likely to be heavily reddened, and in the case of the intrinsically red emission line stars this can lead to extremely faint objects (e.g. EXO2030+375 — a Galactic plane source at a distance of approx. 5kpc has a Be-star counterpart with V=19.7; see Coe et al. 1988). The use of UK Schmidt plates allows us to search for these faint, red objects.

Plate selection is via the UK Schmidt telescope unit on-line service, UKSCAT, then further consultation with the plate librarians. We have tried to obtain three good quality, long exposure plates in each of the b_j, R and I bands. This means that we can obtain well defined 2-colour information on each object within the search field and also perform a variability check between plates in a given band.

COSMOS initially scans an area of each plate which is two degrees square (to set the coordinate transforms for the plate), and then smaller 'target areas' centred on the expected source location are extracted from this. The actual size of the target area depends on how much confidence is placed in the positional accuracy of the error box (and how big the 'potential counterpart' zone is thought to be). As an example, we have used 5 arcminute square areas for our 2 or 3 arcminute square error boxes. Bearing in mind the factors that can increase the potential search area (Sect. 1), we must ensure that we search the error box and a reasonable-sized area around it. Thus, if the box is later moved as a result of re-analysis of the X-ray data, we should have already searched the new position. We initially scan only within the error box, then the search area is gradually increased until we are satisfied that there are no further realistic candidates present.

5 Results So Far

We will now discuss the two transient HMXBs and the potential LMXB which have been investigated so far, and the potential counterparts found. It appears that colour – colour and colour – magnitude plots are the most useful methods for selecting anomalous objects in each field. Figure 5 shows an example of such a scatter plot for each of the three targets, as well as a representation of the main sequence. Note that the panels in Fig. 5 show colours calculated as the difference between plate magnitudes. Since the plates vary (e.g. in exposure), the absolute colours calculated in each case should not be compared between one field and another. However, the relative colours of objects in each target field can be compared with the predicted main sequence and so give an indication of which objects are anomalous.

H1833-077 (Scutum X-1):

A hard X-ray transient discovered by a sounding rocket experiment (Hill et al. 1974), Scutum X-1 remains optically unidentified due to the existence of several possible error diamonds from the HEAO-1 A-3 experiment (Reid et al. 1980). Each diamond is approx. 2' by 0.5', and contains about 70-100 stars on the ESO Quick Blue survey plates (see Fig. 4), and this seemed an ideal candidate for testing

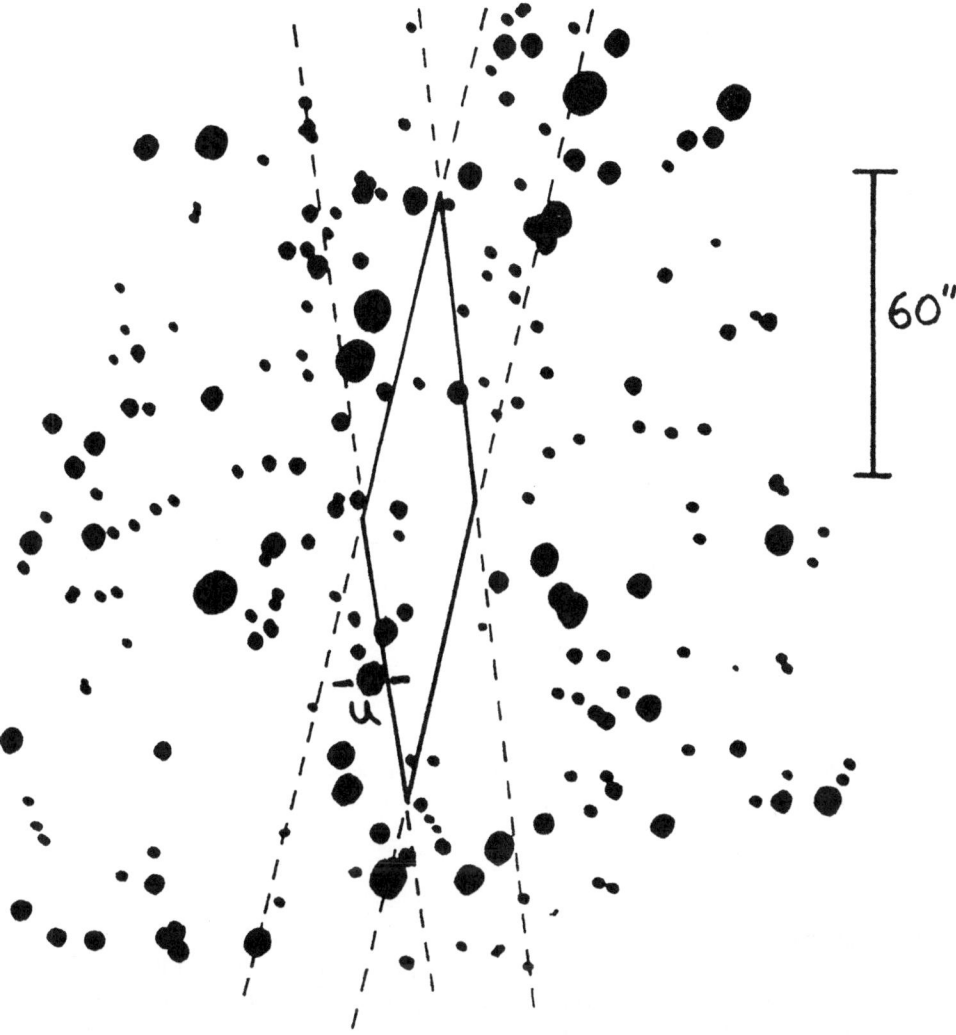

Fig. 4. One of the 5 HEAO-1 A-3 error diamonds in which the source Scutum X-1 is thought to lie (this is the 'most likely' of the 5). Anything within about 20″ of the edge has to be considered a possible candidate — there are thus about 60 candidates in this diamond alone.

the technique. Unfortunately, due to video tape degradation, the plate data for diamonds 3, 4 and 5 has been lost, and only the two 'most likely' diamonds have been searched so far. However, it does seem that star 'U' (Fig. 5b) is certainly a potential candidate, as it appears in an anomalous position, to the right of the early-type stars in Fig. 5a. The only other anomalously coloured star in or near the diamond is star 'M', which has turned out to be an M-type dwarf star (as its position in the colour – colour plot may suggest). If star U is not a Be-type star, then the plates for diamonds 3,4 and 5 will be rescanned and searched.

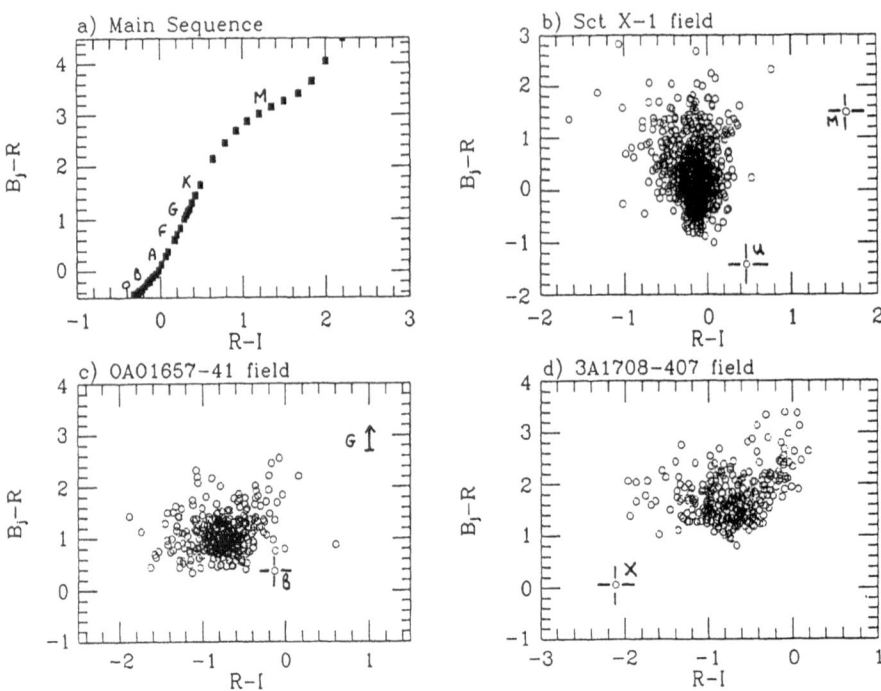

Fig. 5. The colour – colour plots showing the main sequence (**a**), and the distribution of stars in the 5 arcmin square fields searched for each source. Possible candidates are marked by a cross.

OAO1657-41:

Discovered by the Copernicus satellite in 1978 (Polidan et al. 1978), this hard X-ray transient source was initially (wrongly) associated with the variable binary star V861 Sco, leading to it briefly being considered a strong black-hole candidate. Later Einstein observations (which produced a 30″ error circle) revealed that the source was in fact a 38 second pulsar, and not coincident with V861 Sco (Parmar et al. 1980). It remains optically unidentified. An earlier search by Parmar down to V=16 revealed no candidates, so we have extended the depth of the search down to approximately V=21 using the UK Schmidt plates. There now appears to be a potential counterpart lying to the right of the early-type stars (see Fig. 5c, star B), as is the case for star 'U' in the Scutum X-1 field. An extremely red star which lies within the error circle and shows variability was initially the prime candidate. As a result of an AAT service observation, this has subsequently turned out to be a heavily reddened G-type star (star G in Fig. 5c). Nonetheless, this cannot be ruled out as a counterpart, since one theory suggests that this system is in fact an unusual pulsing LMXB (Nagase et al. 1984). However, should star B turn out to be a Be-type star, then this will probably be the correct identification, as there are several problems with the LMXB model.

3A1708-407:

This source has been observed by all the major X-ray surveys (e.g. Uhuru, Ariel V, HEAO-1) but the only pointed observations of it to date were made by EXOSAT and the TTM experiment on the Kvant module of the Mir space station. All these observations show a variable, strong X-ray flux and the source has been bright whenever it was looked at, so is unlikely to be a transient. The EXOSAT observation in particular was of interest since the 17 hour light-curve showed a broad, smooth dip lasting 6 hours. This was reminiscent of the absorption dips seen in certain LMXBs, as was the X-ray spectrum, and so such an identification was suggested (Norton 1988). However, since the observation was the last ever made by EXOSAT, there is some doubt about the reliability of the X-ray modulation, which may in fact be due to pointing errors of the satellite. The TTM observation made in 1989 provided a more accurate position than was previously available, with an error radius of < 1 arcmin (Patterson 1990). This meant that a COSMOS search for the counterpart was feasible. We had nine Schmidt plates scanned, and performed a search within a 5 arcmin radius of the TTM position.

The COSMOS colour – colour plot for this field shows only one object which is clearly separated from the majority of the stars in the surrounding area (see Fig. 5d, candidate marked X). Moreover, it is in the 'blue' region of the plot where we would expect LMXB counterparts to lie. The star itself is, however, about 1 arcmin outside the error circle given by Patterson (1990), but this position is expected to be revised to a more precise value in the near future (Skinner, private communication). We have obtained a spectrum of the candidate star from a service observation at the AAT which shows what appears to be a normal O-type star, with no N III or He II emission lines, which are usually seen in LMXB systems. Hence, if this star is the counterpart to the X-ray source it may be an unusually strong emitter of coronal X-rays and the X-ray object is not a LMXB after all. Otherwise, we must look again to try to find the 'real' counterpart to the LMXB lurking somewhere else on the sky survey plates.

6 Acknowledgements

We would like to thank Alan Wood for help with SCAR; Quentin Parker of the COSMOS unit; Mike Read of the UKSTU; and the staff of the AAT for all their help with this project. Thanks also to Laurence Jones for many useful discussions. The analysis presented here was carried out on the Southampton node of Starlink, which is funded by the UK SERC, primarily using the SCAR package.

References

Bradt, H.V.D., McClintock, J.E., 1983. *Ann. Rev. Astron. and Astrophys.*, **21**.

Coe, M.J., Longmore, A., Payne, B.J., Hanson, C.G., 1988. *Mon. Not. R. Astron. Soc.*, **232**, 865.

Epstein, R.I., Lamb, F.K., Priedhorsky, W.C., 1986. *Los Alamos Science, 13*, 2.

Hertz, P., Bailyn, C.D., Grindlay, J.E., Garcia, M.R., Cohn, H., Lugger, P.M., 1990. *Astrophys. J.*, **364**, 251.

Hill, R., Burginyon, G., Grader, R., Toor, A., Stoering, J., Seward, F., 1974. *Astrophys. J.*, **189**, L69.

Nagase, F., Hayakawa, S., Kii, T., Sato, N., Ikegami, T., Kawai, N., Makishima, K., Matsuoka, M., Mitani, K., Murakami, T., Oda, M., Ohashi, T., Tanaka, Y., Kitamoto, S., 1984. *Publ. Astron. Soc. Japan,* **36**, 667.

Norton, A.J., 1988. Ph.D. Thesis, Leicester University.

Osborne, J.P., 1987. *Astrophys. & Sp. Sci.*, **130**, 207.

Parmar, A.N., Branduardi-Raymont, G., Pollard, G.S.G., Sanford, P.W., Fabian, A.C., Stewart, G.C., Schreier, E.J., Polidan, R.S., Oegerle, W.R., Locke, M., 1980. *Mon. Not. R. Astron. Soc.*, **193**, 49p.

Patterson, T.G., 1990. Ph.D. Thesis, Birmingham University.

Polidan, R.S., Pollard, G.S.G., Sanford, P.W., Locke, M.C., 1978. *Nature*, **275**, 296.

Reid, C.A., Johnston, M.D., Bradt, H.V., Doxsey, R.E., Griffiths, R.E., Schwartz, D.A., 1980. *Astron. J.*, **85**, no.8, 1062.

van den Heuvel, E.P.J., Rappaport, S., 1987. *'Physics of Be Stars'*, Proc. IAU Colloq. No. 92, p. 361, eds. Slettebak, A. & Snow, C.P., CUP.

White, N.E., 1989. *Astron. & Astrophys. Rev.*, **1**, 85.

Discussion

Schachter :

Perhaps an all-sky survey with ASTRO-D would turn up some of these hard X-ray transients?

Roche :

Certainly an all-sky survey at higher energies than the ROSAT XRT, with similar positional accuracy, would be very useful for identifying further hard X-ray transient sources.

THE EINSTEIN SLEW SURVEY CATALOG

J.F. Schachter, M. Elvis, D. Plummer, G. Fabbiano and J. Huchra

Harvard-Smithsonian Center for Astrophysics, 60 Garden Street, Cambridge MA 02138, U.S.A.

Abstract

The Slew Survey catalog contains 1075 bright X-ray sources, including 557 objects with no previous X-ray detection. Two-thirds of the survey has been identified with counterparts of known optical type. Source samples to date provide insight on low-luminosity AGN and clusters. The remainder, which contains many uncatalogued BL Lacs and clusters, will be identified by using digitized photographic plates.

1 Introduction

All-sky surveys can help to address key problems in X-ray astronomy. Among these problems are (1) low-luminosity active galactic nuclei (LLAGN) and the soft X-ray background, and (2) the cluster X-ray luminosity function. We briefly summarize each of these problems below.

1.1 LLAGN and the Soft X-Ray Background

The steep luminosity function of AGN (meaning emission-line objects only) means that low luminosity AGN (e.g. Seyfert nuclei) are likely to provide a major part of the AGN contribution to the diffuse X-ray background (Schmidt & Green 1986). Yet optical color-selected samples (e.g. the Palomar Bright Quasar Survey; Schmidt & Green 1983) are incomplete at low luminosities ($M_V \geq -23$) because of dilution by host galaxy starlight. New X-ray–selected samples can be far more complete down to significantly lower luminosities ($M_V \sim -18$).

H. T. MacGillivray and E. B. Thomson (eds.), Digitised Optical Sky Surveys 441–451.
© 1992 *Kluwer Academic Publishers.*

1.2 The Cluster Luminosity Function

Edge et al. (1990) constructed a flux-limited sample $[f_x(2 - 10 \text{ keV }) > 1.7 \times 10^{11}$ ergs cm^{-2} s$^{-1}]$ of 55 clusters selected from the *HEAO-1* and *Ariel V* surveys. They found a statistically significant deficit of high-luminosity ($L_x \geq 5 \times 10^{44}$ ergs cm^{-2} s^{-1}) clusters at $z > 0.1$. If this effect is real, and not the result of a selection bias or their choice of a flux limit, then evolution must have occurred in $z > 0.1$ clusters. (But the Slew Survey clusters appear to be both high redshift and *over*luminous; see below).

1.3 The ROSAT All-Sky Survey

Ultimately, the ROSAT all-sky survey will provide a wealth of information to bear on these important problems. But the ROSAT survey, which will contain up to 100 000 sources, cannot plausibly be ready before the end of 1991. Even then, the identification effort required for the ROSAT medium and deep surveys is immense. To identify ~ 800 sources with $18 < V < 22$ will take, optimistically, 50 clear dark-time nights on 4-meter class telescopes, i.e. almost 100% of a 4-meter's dark time for one year. The current MPE plan (Trümper 1991, private communication) is to identify ~ 2000 'medium survey' sources in a selected ~ 600 square degrees of the sky during the first three years (i.e. up to the end of 1993-94). This is only a factor ~ 2 more than the existing Extended Einstein Medium Survey (Gioia et al. 1990).

For a decade now, X-ray *logN-logS* studies have been dominated by results from the Einstein IPC. The Einstein Deep and Medium surveys have effective *upper* limits to their flux range due to their limited sky coverage. As a result, we have only limited knowledge of the bright X-ray sky at low (IPC) energies. The energy range of ROSAT is significantly lower than that of Einstein, so that obscuration, both Galactic and intrinsic, is even more significant; thus, the population of sources ROSAT will detect will be biased toward softer spectra. These difficulties will enlarge the ambiguities in explaining the diffuse X-ray background, since its spectrum is only well determined at energies significantly above the ROSAT energy range.

2 The Einstein Slew Survey

We have constructed an all-sky, soft, X-ray survey from ~ 3000 individual slewing (i.e. travelling between pointings) observations of the IPC (Elvis et al. 1991). The Slew Survey covers 50% of sky at 6s exposure, and has a total effective exposure of $\sim 1/2 \times 10^6$ s (e.g. Fig. 1 of Plummer et al. 1991). Compared to the Medium Survey, the Slew Survey has a flux limit ~ 10 times higher (3×10^{-12} erg s^{-1}cm^{-2}, 0.2–4.0 keV) and has 30 times the area.

All the photon data of the Slew Survey, useful lists, and software tools are available either on CD–ROM (from the Einstein Data Products Office at CfA; email: edpo@cfa.harvard.edu) or via *einline*. A paper with updated counterpart identifications, and containing multiwavelength data for the Slew Survey (radio, IR, optical, soft and hard X-ray), is in preparation (Schachter et al. 1991).

Sources were detected by a percolation algorithm, which is more efficient than a sliding box method for a spatially sparse data set. We accepted sources as real if the Poisson probability of the observed counts relative to the local background is greater than 3.95×10^{-4} and the total number of counts in the source is ≥ 3. This yields a catalog of 1075 objects, the largest bright X-ray catalog to date, including 557 new X-ray sources.

A positional accuracy of $1.2'$ (90% confidence; $3'$ at 95% confidence) was achieved. The principal limitation on source localization is the slew aspect solution (details in Elvis et al. 1991).

3 Counterparts of Slew Survey Sources

3.1 Sources with Known Optical Types

We have performed an exhaustive search of standard catalogs of stars, galaxies, AGN and BL Lacs, clusters of galaxies, cataclysmic variables and X-ray binaries (see §6 of Elvis et al. 1991). In addition, we have searched the Simbad and NED databases. This program yields counterparts of known optical type (e.g. AGN, CV) for $\frac{2}{3}$ of the survey. A bar graph (Fig. 1) shows the relative numbers of sources with known optical types; also indicated are the fractions of each type that are new X-ray detections. Clearly, we have objects of every known type of X-ray source. Examples of new samples of sources are discussed below.

Identified Slew Survey AGN to date – There are 130 Slew Survey sources with counterparts in catalogs of AGN, of which 10 are new X-ray sources. One of the new X-ray sources is S5 0836+710, which has $z = 2.2$. To the list of identified AGN, we will probably add many of the new X-ray galaxies (21 to date) found in galaxy catalogs, but not currently known to possess active nuclei.

The identified Slew Survey AGN to date are compared with the Palomar Bright Quasar Survey sample in Fig. 2. Clearly, the Slew Survey sample can detect sources at least 3 magnitudes fainter in M_V, showing the efficiency of detecting LLAGN. Plotting the Slew Survey and the Medium Survey AGN on a similar graph (not shown) shows that the two samples can be readily combined to sample a large range of luminosity-redshift space.

Identified Clusters – We find 80 cluster counterparts, of which 11 are new X-ray detections. The new X-ray clusters are listed in Table 1, with richness class (R), distance class (D), and redshift for each. IPC count rates have been converted to X-ray luminosities for the sources with known or estimated (Huchra et al. 1991) redshifts, using $H_0 = 50$ km s^{-1} Mpc^{-1}. Note that the clusters are all high z (or large D), i.e. $z \gtrsim 0.1$, and also high L_X. This would tend to go against the Edge et al. (1990) result (§1).

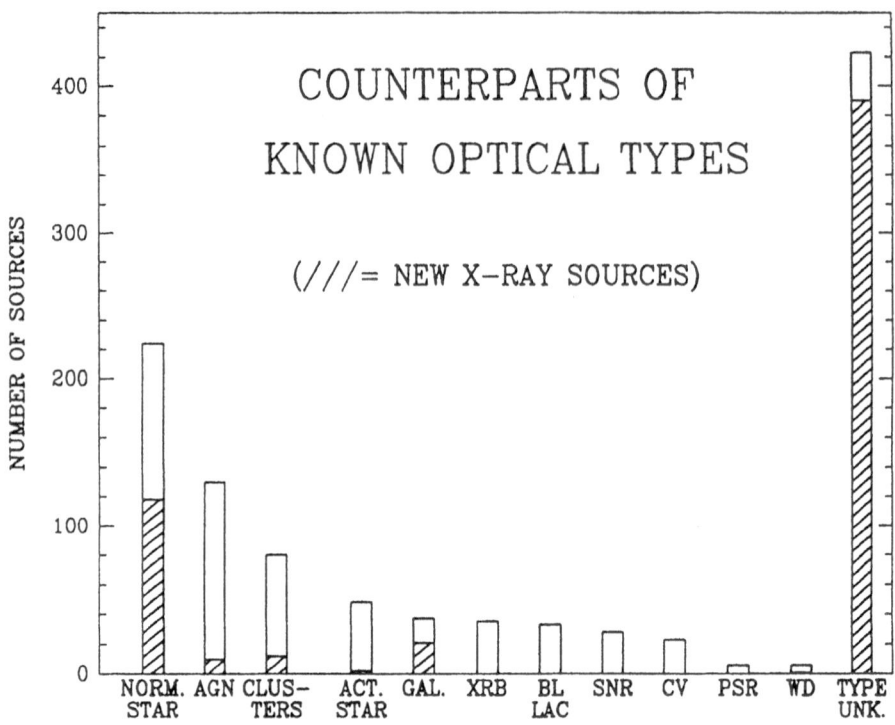

Fig. 1. Bar chart showing breakdown of Slew Survey sources possessing counterparts of known optical types. For each type, the hatched region represents the proportion of new X-ray sources. Abbreviations used are NORM. and ACT. for normal and active.

Table 1. New Identified X-ray Clusters

Name	Slew Name	R	D	z	IPC Ct. Rate	log L_X
A193	1ES0122+084B	1	4	0.048	$0.21^{+0.07}_{-0.06}$	$45.81^{+0.13}_{-0.14}$
A773	1ES0914+519	0	6	~0.20	$0.22^{+0.10}_{-0.08}$	~47.06
A1602	1ES1241+275	0	6	~0.24	$0.32^{+0.15}_{-0.12}$	~45.39
A1651	1ES1256−039	1	4	0.083	$0.51^{+0.20}_{-0.17}$	$44.67^{+0.14}_{-0.17}$
A1664	1ES1301−239	2	6	...	$0.20^{+0.07}_{-0.06}$	...
A2495	1ES2247+106	0	5	...	$0.76^{+0.33}_{-0.26}$	...
A3404	1ES0644−541	1	5	...	$0.27^{+0.09}_{-0.08}$	...
A3866	1ES2217−354	0	5	...	$0.40^{+0.21}_{-0.16}$	...
S724	1ES1310−327	0	4	...	$1.78^{+0.42}_{-0.36}$	...
S1158	1ES2349−561	0	5	...	$0.88^{+0.43}_{-0.33}$	...
ZW 314	1ES0058+345	...	...	...	$1.14^{+0.71}_{-0.51}$	...

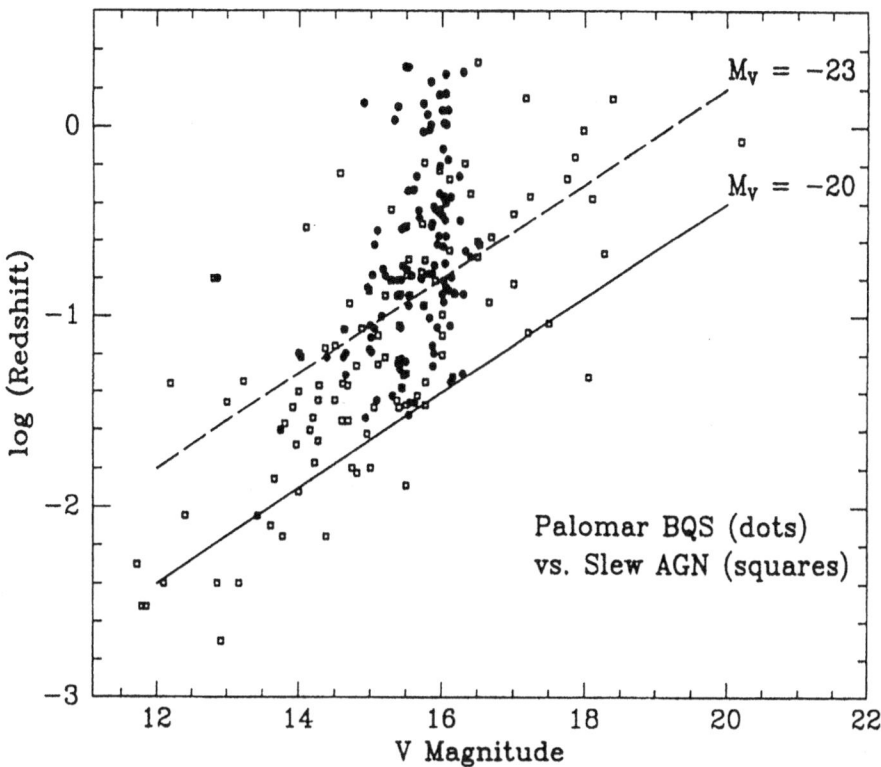

Fig. 2. Redshift-V magnitude distribution for the Palomar Bright Quasar Survey (BQS; solid dots), and for identified Slew Survey AGN to date (squares). The broader distribution in V magnitude for the Slew Survey AGN is a consequence of the X-ray selection. The Slew Survey AGN are seen to be intrinsically 3 magnitudes fainter in M_V.

Identified Active Stars – Some 48 stars in the Slew Survey were previously known to be active (e.g. FK Com, RS CVn, AD Leo). Of the stars *not* known to be active, but identified in normal stellar catalogs, there are 151 late-type ($> F$) counterparts, including 91 new X-ray sources. The spectral types of these new X-ray sources often suggest coronal activity (e.g. FIII). Many cases (> 30) have been confirmed in ongoing observations by R. Remillard of MIT at the MDM 1.3 m, and S. Saar of CfA with the echelle spectrograph at the McMath 1.6 m.

3.2 Counterparts without Known Optical Types

Predicted Optical Properties – There are 502 Slew Survey sources without known optical magnitudes, of which 79 have known optical types (mainly supernova remnants and low Galactic-latitude X-ray binaries), and the remaining 423 lack counterparts of known type. For this last group, we can estimate the distribution of V-magnitudes. This distribution is a useful diagnostic of the amount of optical observing required for complete identification. We use values of the X-ray–to–optical flux ratio (f_X/f_o) from Stocke et al. (1991), and the relation 1.0 IPC cts s^{-1} = 3.26×10^{-11} erg cm^{-2} s^{-1} (0.2–4.0 keV), appropriate to a power-law energy index of 0.5 and $N_H = 2.0 \times 10^{20}$ cm^{-2} (Gioia et al. 1984).

The estimated V-magnitude histogram is given in Fig. 3, where the typical uncertainty in determination of V is 1 mag. The bright end of the distribution ($V \leq 13$) is dominated by stars, where the lower limit depends on the completeness of stellar catalogs already searched. AGN account for the prominent ~2 mag wide feature centered at $V = 15$, while clusters and BL Lacs dominate the faint end (to $V = 19 - 20$). There are some 24 objects expected to be fainter than $V = 18$.

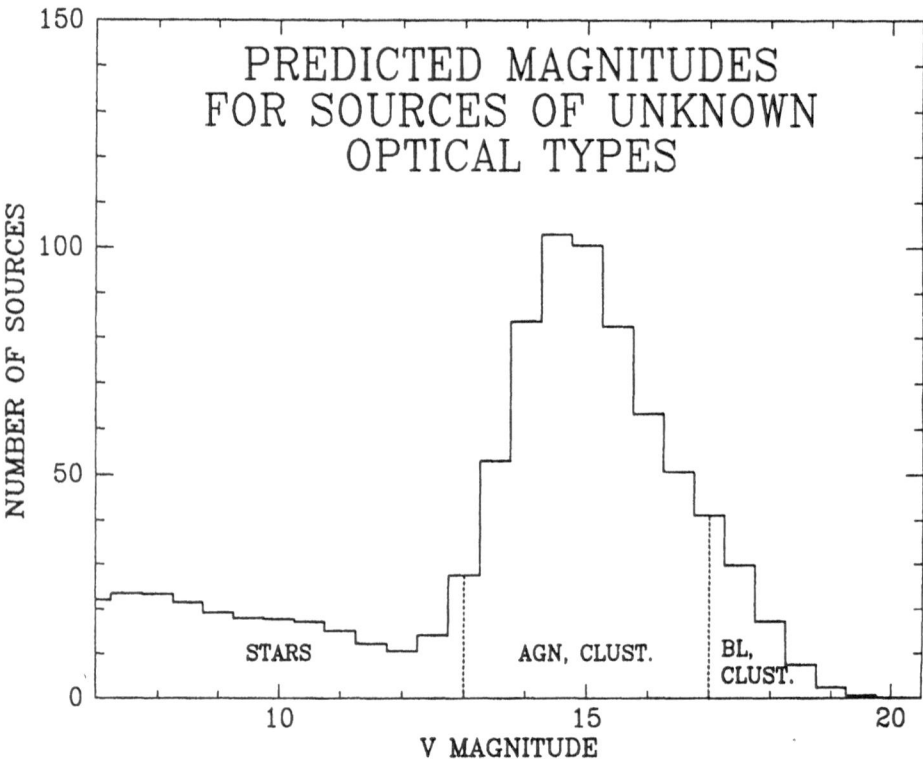

Fig. 3. Histogram of predicted V magnitudes for the 423 sources in the Slew Survey currently lacking optical types. The distribution has been divided into three regions, containing bright, relatively bright, and faint objects. The main and secondary contributors to each of these three regions are listed.

We see that all the sources are expected to have readily detectable counterparts in either the Palomar Optical Sky Survey or the UK Schmidt plates. A sizable fraction (66%) are far enough from the Galactic plane ($|b| \geq 20°$) so that confusion is not a severe problem. The large number of unidentified sources and the large error radius suggest that we need other discriminants for a true counterpart determination.

Radio and IR Counterparts – We have searched the Becker et al. (1991) catalog compilation of the the Green Bank 5 GHz survey of the northern sky [0° ≤ |b| ≤ 75°; Condon et al. 1989 (CBS)] to find counterparts to Slew Survey sources. Radio spectral indices (5 GHz – 0.365 Mhz) have also been tabulated by Becker et al. Figure 4 is an example of a diagnostic diagram to identify CBS counterparts. We show all the Slew Survey CBS counterparts known to be clusters, AGN, or BL Lacs; we also show the three CBS sources lacking optical counterparts (*A*, *B*, and *C*). On the basis of Fig. 4, sources *B* and *C* are probably BL Lacs.

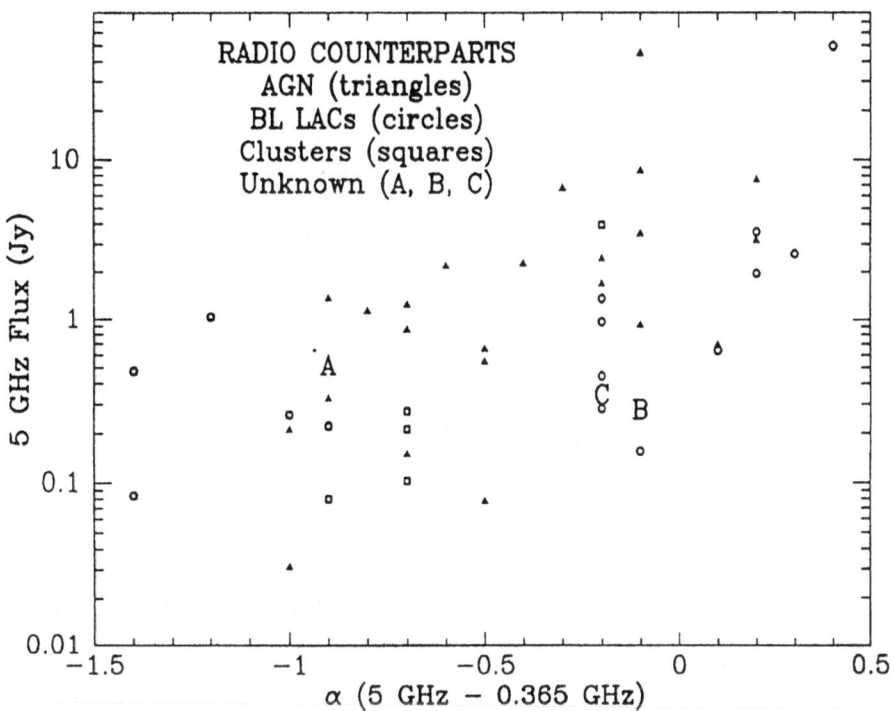

Fig. 4. Technique for determining optical types for Slew Survey sources with radio counterparts. Known AGN, BL Lacs, and clusters of galaxies in the Slew Survey are indicated with triangles, circles, and squares. The letters *A*, *B*, and *C* indicate counterparts of unknown optical type.

We will obtain 1400 Mhz fluxes for all the northern objects using the sky maps of Condon & Broderick (1985; 1986). Existing southern sky surveys (PKSCAT90; Condon et al. 1991) are not nearly as deep as the CBS survey. Therefore, we plan to use the PMN survey, which is the southern complement of the CBS survey (at 5 GHz and 843 MHz; described elsewhere in these proceedings).

IRAS Point Source Catalog colors suggest two new candidate T Tauri stars, one new molecular cloud core, BL Lacs, AGN, and starburst galaxies (color-color diagnostics from Emerson 1987 and Soifer et al. 1987).

Hubble Guide Star Catalog Counterparts – For the 502 Slew Survey sources lacking optical magnitudes, we find 364 possible identification in the Hubble Guide Star Catalog (GSC). This is consistent with the predicted magnitude distribution of the Slew Survey (Fig. 3) and the GSC flux limit of $V \sim 16$. There are an average of 3.7 possible counterparts per Slew Survey source, although one of these is probably just a chance superposition. The GSC has typically 100 sources per square degree (Lasker, these proceedings).

4 Digitized Resources

4.1 Overview

Using the results of the previous section, we find 45 sources (or only 4% of the survey) lacking any sort of counterpart. That is, these Slew Survey sources are (1) absent from standard catalogs, (2) absent from IRAS and radio surveys, and (3) absent from the GSC. These sources are expected to have the most extreme values of f_x/f_o and hence are mainly uncatalogued BL Lacs and clusters.

This again shows the advantage of X-ray selection. For example, the BL Lacs will be part of the $\sim$100 total BL Lac objects in the Slew Survey. This can be compared with only 87 in the *entire* Hewitt & Burbidge (1986) catalog. The BL Lac sample of the Slew Survey will be uniform and X-ray selected, which is the most efficient way to detect these sources (Stocke et al. 1989).

4.2 Digitized Searches

We have used the COSMOS-digitized UK Schmidt plates (see article by Yentis et al., in these proceedings), via the database at the Naval Research Laboratory (NRL). This work is in collaboration with H. Gursky and colleagues (principally B. Stuart, J. Wallin, and D. Yentis). Our initial interest was to obtain spectra of all objects in fields of sources favorable to the MMT ($\delta > -27°$). There are typically 25-50 objects within the conservative 3' (95% confidence) radius down to the plate limit ($b_j \sim 23$). The brightest objects in our fields have $b_j \sim 16 - 17$, e.g. 1ES2343$-$151 (Fig. 5). We find typically 3 to 6 sources with $b_j \leq 19$. The faintest Slew Survey sources should lie near the upper limit.

In a recent run at the MMT Blue Channel (covering 3200-7000 Å), we observed three Slew Survey fields (1ES1355$-$086, 1ES2248$-$163, and 1ES2343$-$151). We suspect that object number 19 in the field of 1ES2343$-$151 is the optical counterpart, a faint galaxy with $\lambda 4686$ absorption (Fig. 6). This may be a member of a faint cluster.

In the near future, we will query the COSMOS/UKST database for all the southern sources. Our collaborators at the University of Minnesota are providing digitized magnitudes for the 4% of the Slew Survey with no counterparts from the the Palomar plates (Pennington et al., these proceedings). The usefulness of having two-color information for X-ray source identification is apparent.

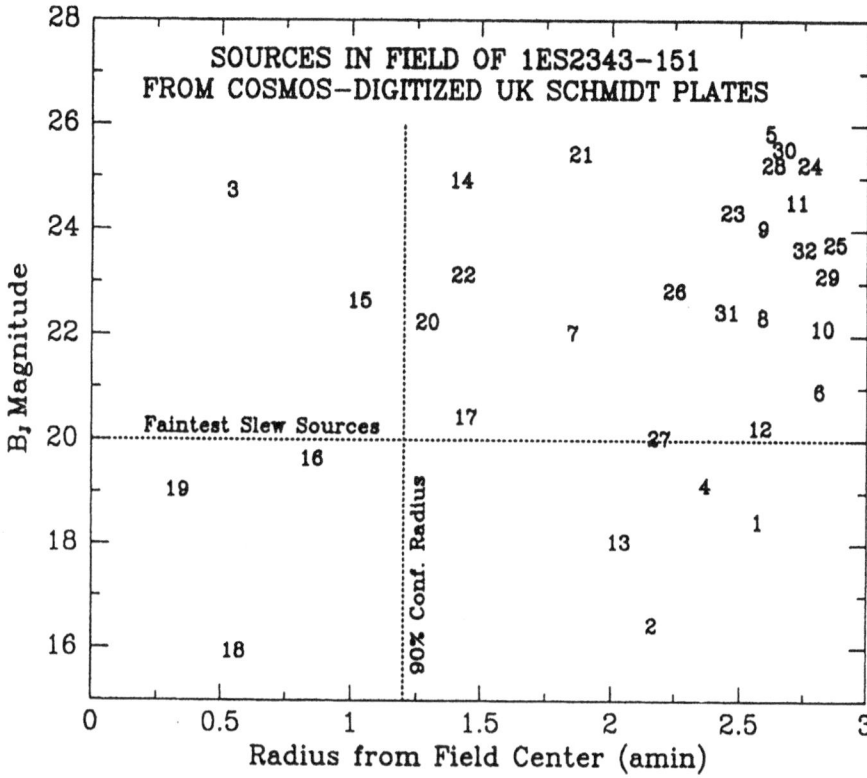

Fig. 5. Magnitude distribution of sources in the field of the Slew Survey source 1ES2343−151 as a function of central radius. All sources detected in the COSMOS-digitized UK Schmidt plate are listed by number. The most likely Slew Survey counterparts lie in the box at the lower left.

We are committed to complete identification of the Slew Survey. We have on-going observing programs to verify the new X-ray identifications, via optical spectroscopy at the Mt. Hopkins 60″, the MMT, and southern telescopes, in addition to existing collaborations on the Kitt Peak telescopes described in an earlier section. Radio positions will be obtained with the aid of the VLA, which significantly reduces the number of possible counterparts in a given field. This can be used to pick out BL Lacs.

All of the Slew Survey sources are expected to be strongly detected (> 100 counts) in the ROSAT all-sky survey. In a collaboration with Dr. J. Trümper of MPE, we will search the ROSAT survey at the Slew Survey positions. Slew Survey sources constitute a ROSAT survey subsample of manageable size with some of its most interesting objects.

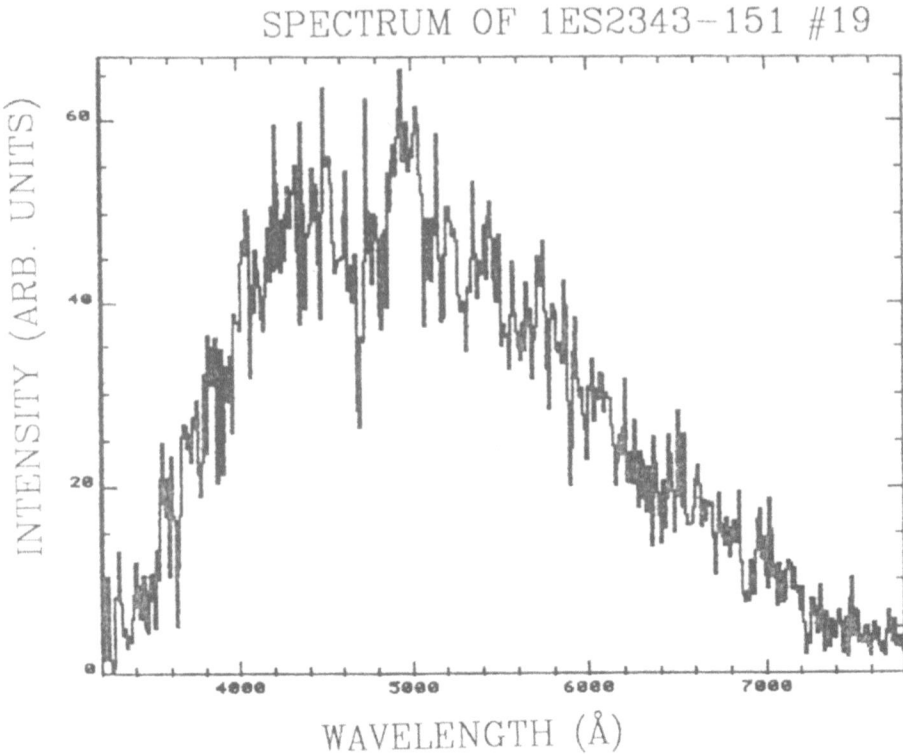

Fig. 6. MMT Blue Channel spectrum of object no. 19 in the field of the Slew Survey source 1ES2343−151. He II λ4686 absorption is seen, suggesting that the counterpart is a member of a faint cluster.

5 Acknowledgements

This research has made use of the Simbad database, operated at CDS, Strasbourg, France; and also the NASA/IPAC Extragalactic Database (NED) which is operated by the Jet Propulsion Laboratory, California Institute of Technology, under contract with NASA. Our work was supported by NASA grant NAG5-1201 (ADP).

References

Becker, R.H., White, R.L., Edwards, A.L., 1991. *Astrophys. J. Suppl.*, **75**, 1.
Condon, J.J., Broderick, J.J., 1985. *Astron. J.*, **90**, 2540.
Condon, J.J., Broderick, J.J., 1986. *Astron. J.*, **91**, 1051.
Condon, J.J., Broderick, J.J., Seielstad, G.A., 1989. *Astron. J.*, **97**, 1064.
Condon, J.J., Broderick, J.J., Seielstad, G.A., 1991. NRAO preprint.
Edge, A.C., Stewart, G.C., Fabian, A.C., Arnaud, K.A., 1990. *Mon. Not. R. Astron. Soc.*, **245**, 559.
Elvis, M., Plummer, D., Schachter, J., Fabbiano, G., 1991. *Astrophys. J. Suppl.* Accepted.

Emerson, J.P. 1987. In *'Star Forming Regions'*, *IAU Symp.*, No. 115, p. 16, eds. M. Peimbert, J. Jugaku, Reidel, Dordrecht.

Gioia, I.M., Maccacaro, T., Schild, R.E., Stocke, J.T., Liebert, J.W., Danziger, I.J., Kunth, D., Lub, J., 1984. *Astrophys. J.*, **283**, 495.

Gioia, I.M., Maccacaro, T., Schild, R.E., Wolter, A., Stocke, J.T., Morris, S.L., Henry, J.P., 1990. *Astrophys. J. Suppl.*, **72**, 567.

Hewitt, A., Burbidge, G., 1986. *Astrophys. J. Suppl.*, **65**, 603.

Huchra et al. 1991. In preparation.

Plummer, D., Schachter, J., Garcia, M., Elvis, M., 1991. CD-ROM issued by Smithsonian Astrophysical Observatory, Cambridge, MA.

Schachter, J., Elvis, M., Plummer, D., Fabbiano, G., 1991. In preparation.

Schmidt, M., Green, R.F., 1986. *Astrophys. J.*, **305**, 68.

Schmidt, M., Green, R.F., 1983. *Astrophys. J.*, **269**, 352.

Soifer, B.T., Houck, J.R., Neugebauer, G., 1987. *Ann. Rev. Astr. Astrophys.*, **25**, 187.

Stocke, J.T., Morris, S.L., Gioia, I.M., Maccacaro, T., Schild, R.E., Wolter, A., 1989. In *BL Lac Objects*, p. 242, eds L. Maraschi, T. Maccacaro, M.-H. Ulrich, Springer-Verlag, Berlin.

Stocke, J.T., Morris, S.L., Gioia, I.M., Maccacaro, T., Schild, R., Wolter, A., Fleming, T.A., Henry, J.P. 1991. *Astrophys. J. Suppl.* In press.

Discussion

Miller :

Can you rule out the Edge et al. (1990) result? If so, do you have an explanation for what they see?

Schachter :

Our new identified X-ray clusters are interesting, I think, because they are not what you would expect from the Edge et al. work. But we really need to have all the Slew Survey clusters identified (all ~200, rather than just 80) before we can make a detailed quantitative comparison with other samples.

THE ROSAT ALL-SKY SURVEY AND FIRST IDENTIFICATIONS OF X-RAY SOURCES USING DIGITISED OPTICAL PLATES

W. Voges

Max-Planck-Institut für Extraterrestrische Physik, Karl-Schwarzschild-Str. 1, D-8046 Garching, Germany.

Abstract

The ROSAT (Roentgensatellit) X-ray astronomy satellite has completed the first all-sky X-ray and XUV survey with imaging telescopes. About 50 000 new X-ray and 1 500 new XUV sources were detected. The identification of these sources using digitised optical sky surveys has begun.

1 Introduction

ROSAT was launched on June 1, 1990 into a circular orbit with an inclination of 53 degrees and an altitude of 580 km. After an initial 2-month period comprising the switch-on, calibration and verification phases, ROSAT began the first all-sky survey in the soft X-ray (0.07 – 2.4 keV; 100 – 5 A) and the extreme ultraviolet (0.025 – 0.2 keV; 500 – 60 A) bands using imaging telescopes. At the end of the six-month long survey, in February 1991, a series of pointed observations, proposed by scientists from the three countries involved in this program, Germany, the UK and the USA, was begun. We describe here the work in progress on the analysis of the survey data and the identification of the ROSAT X-ray sources.

2 The X-Ray Observations

The two instruments onboard ROSAT are an X-ray telescope, consisting of four nested Wolter type I mirrors (see Truemper 1983; Aschenbach 1988; Pfeffermann

H. T. MacGillivray and E. B. Thomson (eds.), Digitised Optical Sky Surveys 453–463.
© 1992 *Kluwer Academic Publishers.*

et al. 1988) and the EUV Wide Field Camera, consisting of three nested Wolter
type I mirrors (see Kent et al. 1990; Wells et al. 1990).

Some of the properties of the focal-plane detectors used for the survey are summarized in Table 1:

Table 1.

	XRT Position Sensitive Proportional Counter (PSPC)	WFC Microchannel Plate Detector (MCP)
Field of view	2 degrees	5 degrees
Energy band	0.07 – 2.4 keV	0.02 – 0.2 keV
Wavelength band	200 A – 6 A	600 A – 60 A
Energy resolution	0.4 at 1.0 keV	
Angular resolution	30 arcsec	

During the all-sky survey the sky was scanned by ROSAT in great circles in a
plane perpendicular to the solar direction. Therefore by following the sun the entire
sky was covered in half a year. Due to the method of scanning the highest exposure
time, approximately 50 000 sec, was reached at the ecliptic poles. The exposure time
decreased gradually with latitude, reaching a value of about 400 sec at the ecliptic
equator. During penetration of the radiation belts around the magnetic poles and in
the South Atlantic Anomaly the X-ray detector had to be switched off, decreasing
the exposure over parts of the sky.

For each photon event the time of arrival, the position in detector coordinates
and the pulse height information are recorded onboard. The Standard Analysis
Software System (SASS) developed at MPE for automatically processing all the
survey data calculates the sky coordinates and energy of each photon by applying
the aspect solution and calibration data to each X-ray event. Data are screened
according to various quality checks made using the housekeeping and attitude data.
The photons are finally collected in great circle strips 2 degrees wide passing through
the ecliptic poles.

During the survey one strip of data was completed every two days. Figure 1
shows an example of one strip quadrant (2×90 degrees), in which the pixel size is
96×96 arcsec and the photons lie in the energy band 0.4 – 2.4 kev. This strip, number 40, crosses the ecliptic equator at a longitude of 80 degrees. It contains about
150 X-ray sources, among which are two prominent supernova remnants, each filling
the strip width. The upper is HB9, and the lower is a soft remnant which remained
undiscovered until detected by ROSAT. Most of the events shown in Fig. 1 are
produced by cosmic X-ray background events, and only a small fraction are caused
by high-energy particles interacting with the detector. The data are analysed using

various source detection algorithms, comprising two sliding window techniques (differing only in how the background is derived) and a maximum-likelihood method. For each X-ray source found the following properties of the source are calculated: the detection likelihood, the source position in celestial coordinates, the source and local background count-rates, the exposure time, spectral fit parameters such as the power law index, temperature and column density, two hardness ratios, the source extent and corresponding likelihood, and finally temporal variability flags produced by a light-curve analysis (χ^2 test), a Kolmogorov-Smirnov test, and a Fast Fourier analysis. Figure 2 shows samples of the results, in the form of histograms of the x- and y-position errors as defined in detector co-ordinates by the maximum-likelihood analysis, the exposure time in seconds, the source count-rate ($\log N / \log S$), the extent, and the hardness ratio. Here the hardness ratio is defined as the difference between the hard (0.4 – 2.4 keV) and the soft counts (0.07 – 0.4 keV), divided by the total counts (0.07 – 2.4 keV). In these distributions contributions of double, sometimes multiple entries, where the same source is detected in two or more overlapping strips, have not yet been removed.

3 The Identification Process

In Fig. 3 we show the distribution in galactic coordinates of X-ray sources detected by ROSAT during the all-sky survey. As in Fig. 2, multiple entries have not not yet been removed from this plot. More than 90% of the survey data have been analysed so far. The missing part of the sky, the empty slices on the plot, has been observed, but the aspect solution is not yet available due to the use of a special satellite mode during that part of the survey.

As the number of new X-ray sources found is at least an order of magnitude higher than the number of previously known sources (about 5000), a high priority is given to source identification. A version of the SIMBAD catalog, containing 640 000 entries and compiled especially for the ROSAT correlations, provides a possible identification in roughly 30% of all cases. In addition, other catalogs of known sources in the radio, UV, infrared, X-ray, and gamma-ray domains, compiled by members of the institute for the ROSAT data analysis, are cross-correlated with the X-ray source positions.

An important role in the identification of ROSAT sources is played by digitised optical sky surveys, for example the UK Schmidt survey of the southern hemisphere. The blue (IIIa-J) plates of this survey have been scanned and digitised at the Royal Observatory in Edinburgh (ROE) using the COSMOS machine and associated analysis software, and in a collaborative effort between ROE and the U.S. Naval Research Laboratory (NRL) the data have been made available to the ROSAT group at MPE. This is a very powerful tool for selecting the most interesting and best candidates for identification purposes, and further in selecting targets for follow-up optical observations. In this respect the accuracy of the determination of the X-ray source position plays an important rôle, both in locating the error circle correctly and in minimising its radius. In order to assess the accuracy of the X-ray positions, we have correlated ROSAT positions with those of bright stars in the SIMBAD catalog, of which the large majority are X-ray sources. In

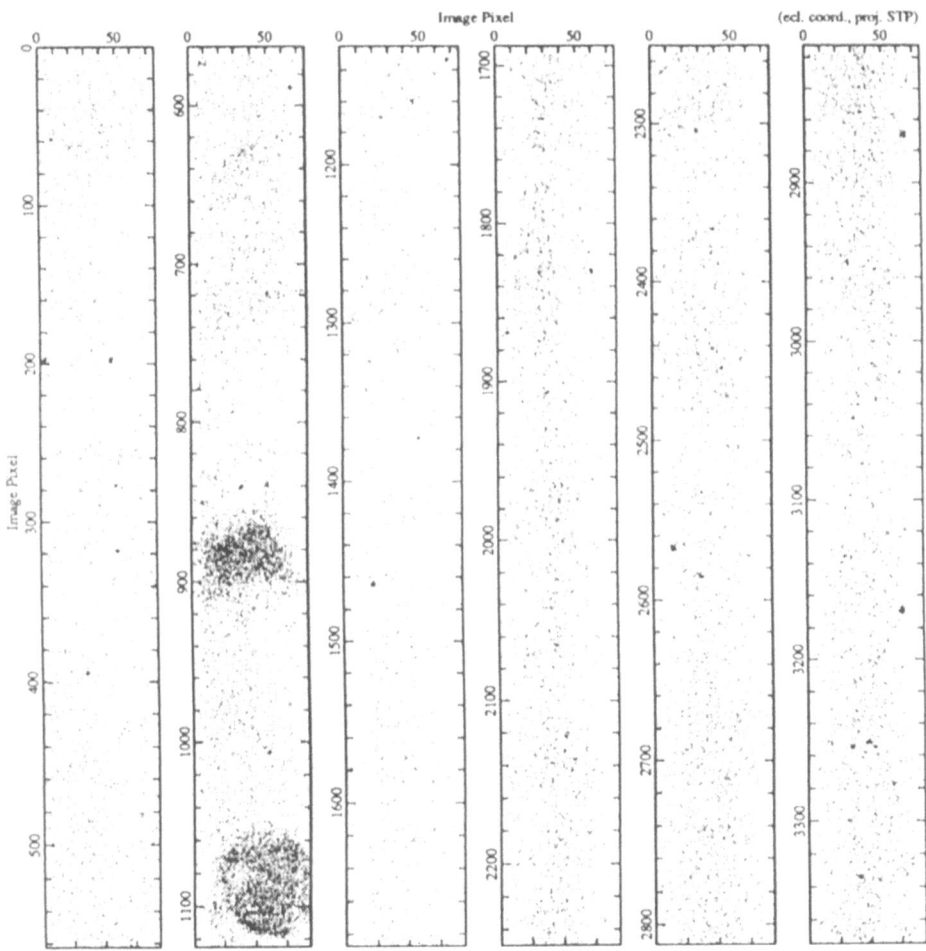

Fig. 1. The six segments, each 2° × 15°, comprise one quadrant of a ROSAT survey strip. Each 2° wide strip follows a great circle through the ecliptic poles, and this particular strip (strip 40) crosses the ecliptic equator at a longitude of 80°. The quadrant contains 150 sources, of which two are supernova remnants covering almost the full strip width. The upper one is HB9 and the lower a newly discovered remnant. The events on this map are produced mostly by the cosmic X-ray background, and only a small fraction is due to particle interactions in the detector. The energy band of this map is 0.4 – 2.4 keV.

Fig. 4, which shows this correlation for SIMBAD stars found within 2 arcmin of the ROSAT position, 90% of the ROSAT X-ray sources are found within 30.5 arc sec of those SIMBAD stars which have a position accuracy of better than 1.5 arc sec. Of the 30% of the X-ray sources having a counterpart in the SIMBAD catalog 75% are stars, 10% galaxies and 15% clusters of galaxies, AGNs, quasars and other miscellaneous objects. An analysis has shown that about 25% of these associations are coincidental. This leaves about 75% of the ROSAT sources which have yet to

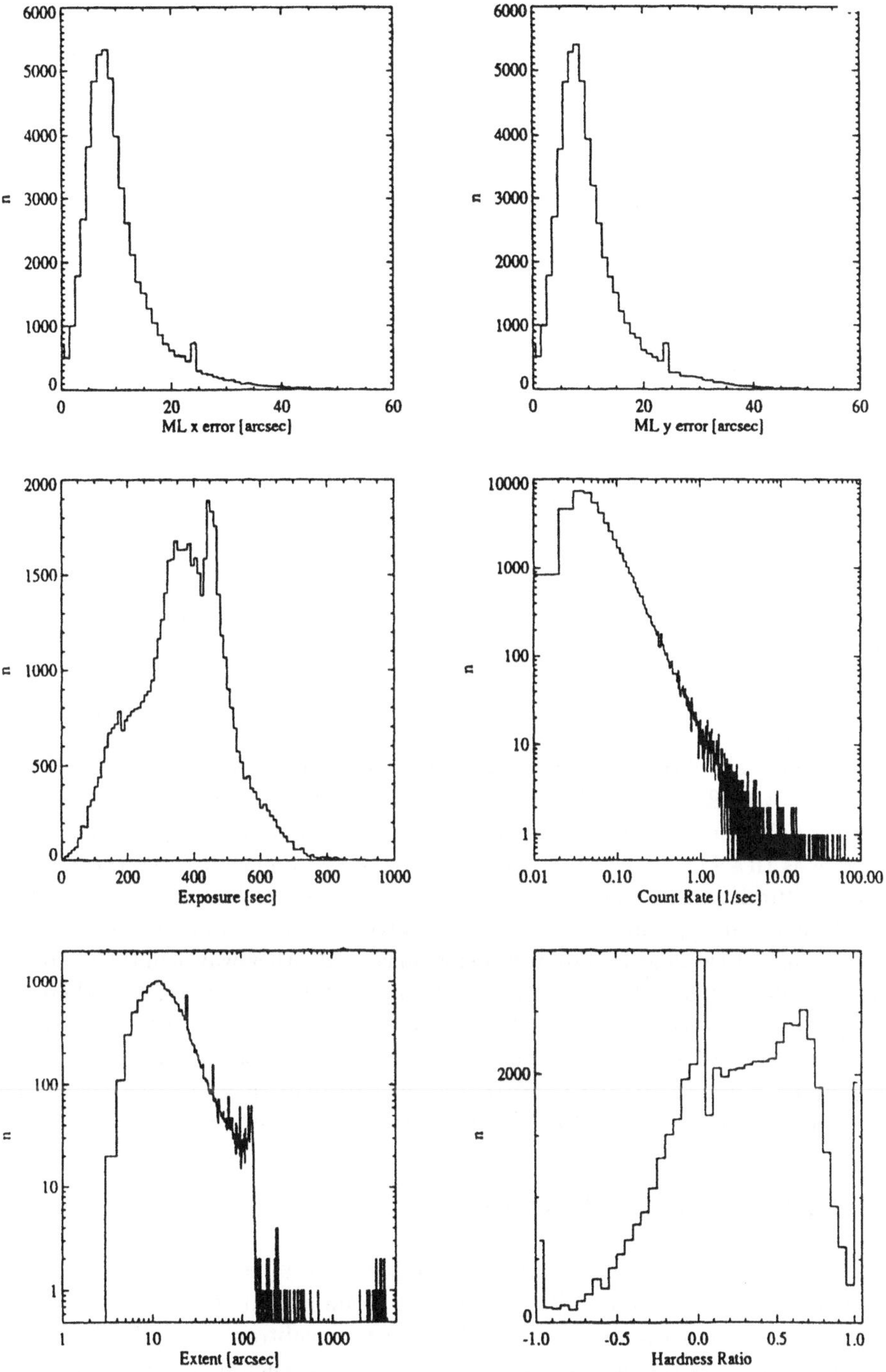

Fig. 2. Six histograms summarising some results of the analysis of a large ensemble of survey X-ray sources by the ROSAT Standard Analysis System (SASS). The upper two show the distributions of the source position errors, as deduced by a maximum likelihood analysis. The second row shows the distributions of the exposure (observation) time and the source count-rate ($\log N / \log S$ curve), and at the bottom are the distributions of the source angular extent, as deduced by the maximum likelihood analysis, and the spectral hardness ratio.

Galactic II Coordinate System

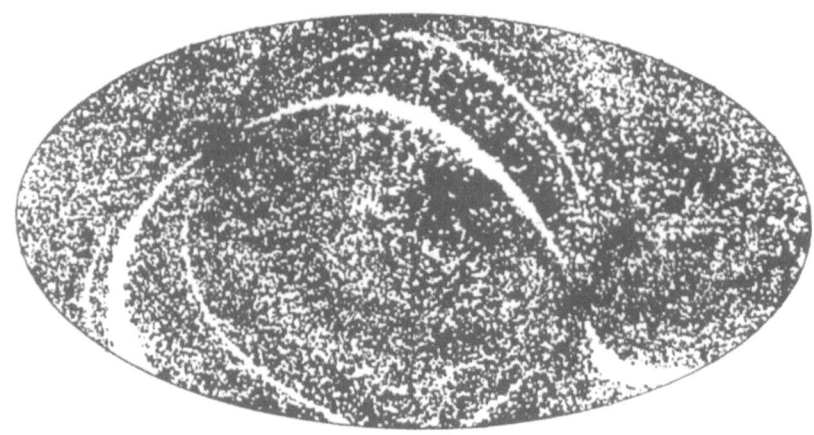

Fig. 3. An all-sky map in galactic coordinates showing the distribution of the ROSAT survey X-ray sources.

be identified.

With the aid of the SIMBAD correlations we estimate that the number of QSO's identified with ROSAT sources is of the order of 500. Using the IRAS point source catalog (250 000 sources) we estimate that the total number of X-ray sources identifiable as IRAS sources to be approximately 2000. Similar investigations with the Condon-Brodend-Seiestad 6cm radio survey catalog (140 000 sources) lead to the conclusion that appromately 2000 X-ray sources should be radio sources.

The use of catalogs and digitised sky surveys at other wavelengths is not completely sufficient for identification and study of ROSAT sources, and therefore a comprehensive program of optical follow-up studies has been organized, with emphasis on selected regions of the sky. For example, extensive campaigns are under way at La Silla (ESO), the Landessternwarte Heidelberg in Mexico, at the large telescope on Mauna Kea in Hawaii, and at the Anglo-Australian Telescope (AAT).

4 Digitised Optical Surveys

The 894 fields of the UK Schmidt Southern Sky survey, each about 6.5 deg square, have been scanned at ROE with a resolution of about 1 arc sec, yielding about 0.6 gigabytes of data per plate. The COSMOS software subtracts the plate background and then analyses the characteristics of each object, such as position, size, shape and magnitude. The software is capable of resolving confused regions and of distinguishing between stars and galaxies. The resulting 'object catalog' contains about 500 million objects (see the contribution by Yentis et al. in these proceedings), and for use at MPE the data base, stored on optical disc, has been compressed to about 8 gigabytes. Installed also at MPE is software which can extract and manipulate

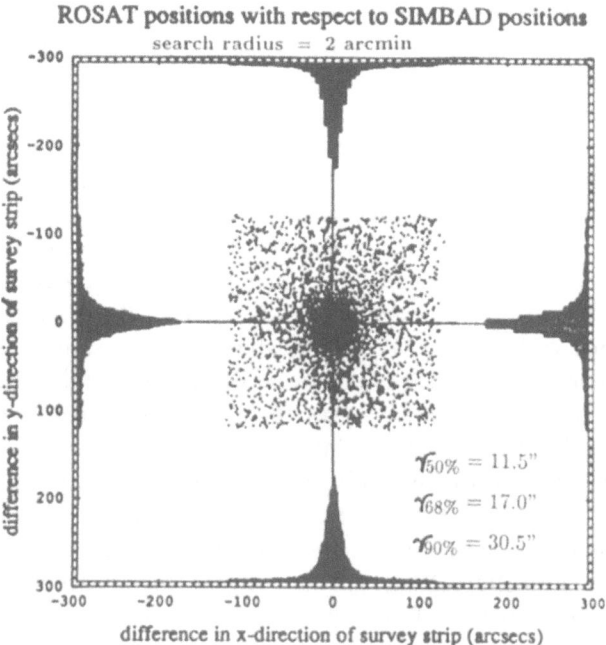

Fig. 4. The result of a calibration of the pointing system accuracy. Nearly all bright stars are detected by ROSAT as X-ray sources. The plot shows the distribution of the offsets of the ROSAT-measured position from the positions listed in the SIMBAD catalog, which have an error of less than 1.5 arc sec.

the data for the purposes of source identification. One of the tasks of this software is the production of 'finding charts' for each ROSAT source, which show the optical objects around the X-ray position.

Two examples are shown in Fig. 5, where the ellipses represent galaxies, the crosses stars and the dots objects too faint to permit identification as a star or a galaxy. The magnitudes, positions and separations from the ROSAT position are listed at right, and the dashed circle is the ROSAT 90% confidence error circle. In Fig. 5(a) the error circle of the ROSAT source RX J0518.1-5835 contains two galaxies and a faint object, so that the identification is not immediately clear. However optical follow-up observations at La Silla by T. Fleming (MPE) and L. Pascini (ESO) have shown that the three shaded galaxies have similar redshifts, about 0.1, and therefore may belong to a cluster. It is not yet clear whether the X-ray source is associated with the cluster which is identifiable clearly in the top right quarter of the map. The finding chart for the source RX J2235.8-2603, shown in Fig. 5(b), reveals an unambiguous identification with a bright, nearby spiral galaxy, which is listed in the SIMBAD catalog as the Sc galaxy NGC 7314.

In the northern hemisphere source identification is being assisted by direct and objective-prism plates obtained with the Calar Alto Schmidt telescope in Spain. The availability of spectral and image data facilitates the classification of sources, and the Hamburg Sternwarte has undertaken the scanning and digitisation of 225 fields with moderate resolution using a 1010G PDS microdensitometer. The resulting images and spectra, stored on optical disks, have been correlated with the positions of a sample of some 1800 ROSAT X-ray sources in 54 of these fields. In 550 cases no conclusion could be reached because the error circle contained too many candidates, and in 127 cases the error cicle was found to be empty. Of the remaining sources, 520 were identified with AGNs, 139 with clusters of galaxies, and 475 with stars. The objective-prism spectra were found to be especially useful in identifying M-stars as X-ray sources, for in many cases these were not listed in the SIMBAD catalog. Although the objective-prism plates do not go as deep as the optical images accessible to ROSAT, they are especially useful in the identification of AGNs (Bade et al. 1991), which constitute the majority of the extragalactic ROSAT sources.

At present there does not exist a single, computer-based process for the systematic identification of all ROSAT survey sources. More experience with the identification procedures is required before such a process can be designed. However, the elements of such a procedure are being accumulated in both a computer archive, and a printed archive allowing rapid access to individual results. The latter comprises a series of binders, referred to as 'books', each devoted to one survey strip and containing the following information:

1. Summaries of cross-correlation of ROSAT source positions with catalogs of known sources, such as the Simbad catalog and the Abell and Zwicky catalogs of clusters of galaxies.
2. The results of the SASS analysis of the X-ray data.
3. The X-ray-to-optical flux ratios of the ROSAT optical candidates. This has proven useful in the Einstein Medium Sensitivity Survey for discrimination between stellar and extragalactic sources. In Fig. 6 we show a plot of this ratio against the optical magnitude for a sample of ROSAT sources identified firmly using the SIMBAD catalog. The distinction between, for example, the majority of stars and extragalactic objects is clear in this diagram.
4. A finding chart for each ROSAT source.

5 Identifying Clusters of Galaxies

The identification of clusters of galaxies among the ROSAT survey sources requires special techniques, because no single optical object within the X-ray error circle will reveal the presence of a cluster, and instead we must examine the distribution of galaxies around the ROSAT position. In addition to correlating ROSAT source lists

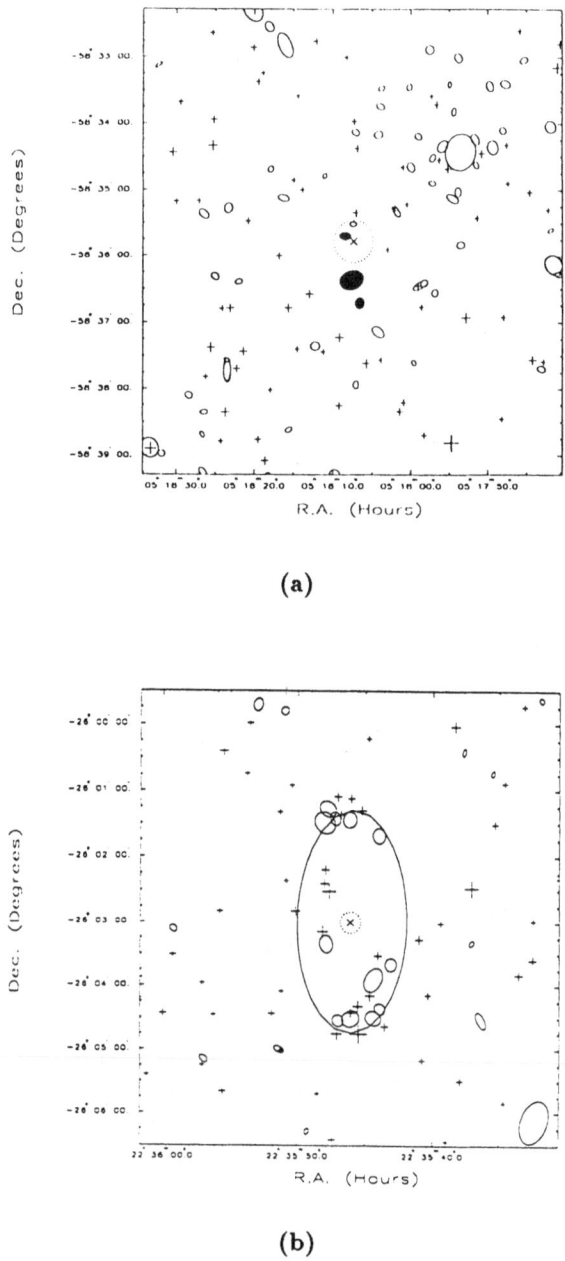

Fig. 5. The finding charts of two ROSAT sources, produced using the ROE/NRL object catalog. This catalog describes the characteristics of some 500 million objects in the southern sky, as defined by the scanning of the UK Schmidt survey plates, and the subsequent digitisation and analysis of the data. The crosses are stars, the ellipses galaxies, and the dots objects too faint to allow a determination of their nature. The dashed line defines the 90% confidence ROSAT error circle. The spectra of the three galaxies which have been shaded in the upper figure (a) have been measured subsequently, and all lie close to 0.1, so that possibly the X-ray source is a cluster of galaxies. In (b) the X-ray source is seen to lie near the center of a bright galaxy, subsequently identified as the Sc spiral galaxy NGC 7314.

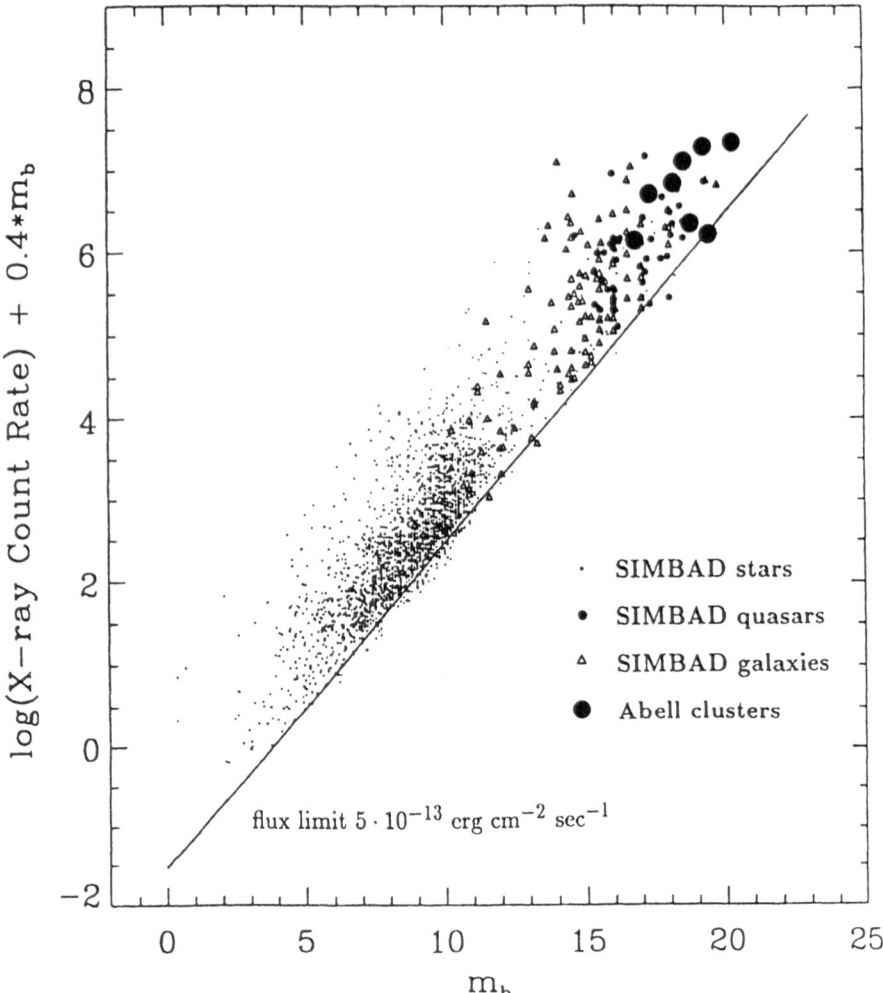

Fig. 6. The ratio of the X-ray-to-optical flux is a useful discriminant of the nature of the X-ray source, in particular in discriminating between stars and extragalactic objects. The plot shows a correlation of the flux ratio with the optical magnitude for a sample of ROSAT sources already identified using the SIMBAD catalog. Such a plot is used to set thresholds for this ratio.

with published catalogs, principally the Abell and Zwicky catalogs, two techniques have been developed for use with Schmidt telescope surveys. In the first technique, developed at ROE and NRL, the UK Schmidt object catalog has been searched systematically for significant groupings of galaxies using objective, computer-driven algorithms. These algorithms include both a search for significant regions of high galaxy surface density, and a percolation or 'nearest neighbour' search procedure. Some 70 000 groupings fall out of this analysis, a large fraction of which are small clumps of less than 10 galaxies. Each of these groupings is then subjected to a series of analyses to test its significance and measure its properties. For example Abell's analysis is applied to a large sub-sample comprising the richer groupings. The result is a computer-derived catalog of clusters in the southern sky, which is then correlated with the ROSAT source lists.

The second technique developed at ROE and NRL analyses directly the distribution of galaxies around the ROSAT source position, as illustrated in Fig. 4(a). The galaxies are counted in five concentric circles of radius 1,3, 5,7.5 and 10 arc minutes, and the results are then tested by comparison with the counts obtained in 1000 random samplings made on each Schmidt plate. In both these techniques quantities describing the completeness of the extracted sample of clusters, and the estimated contamination by chance associations, are calculated. Thresholds for these parameters must be set before the ROSAT data base can be searched for clusters, and some compromise between completeness and contamination must be struck. Other parameters, such as X-ray extent and hardness ratio, and X-ray-to-optical flux ratio, can be invoked to improve the correlations.

The result of the searches made to date, in an area of about 3000 square degrees around the south galactic pole, is a list of close to 200 clusters identified firmly as ROSAT X-ray sources, and in addition a list of 200 less secure identifications which are under investigation. Extrapolating these results to the whole survey, we derive a conservative lower limit to the total number of ROSAT clusters of about 2000.

6 Acknowledgements

The author would like to thank the many people who have contributed to the work described above, first the ROSAT team at MPE, in particular H. Ebeling, K. Dennerl, H. Boehringer, and J. Englhauser, second H.T. MacGillivray and C.A. Collins at ROE, and finally R.G. Cruddace, H. Gursky, B.V. Stuart, J. Wallin and D.J. Yentis at NRL.

References

Aschenbach, B., 1988. *"Design, Construction, and Performance of the ROSAT High Resolution X-Ray Mirror Assembly"*, Appl. Optics, **27**, 1404-1413.

Bade, N., Engels, D., Fink, H., Hagen, H.-J., Reimers, D., Voges, W., Wisotzki, L., 1991. *"ROSAT AGN Identified on Objective Prism Plates"*, Astron. Astrophys. Submitted.

Kent, B.J., Reading, D.H., Swinyard, B.M., Graper, E.B., Spurett, P.H., 1990. *"EUV Band Pass Filters for the ROSAT Wide Field Camera"*, Proc. SPIE, **1344**, 255-266.

Pfeffermann, E., Briel, U.G., Hippmann, H., Kettenring, G., Metzner, G., Predehl, P., Reger, G., Stephan, K.H., Zombeck, M.V., Chappell, J., Murray, S.S., 1988. *"The Focal Plane Instrumentation of the ROSAT Telescope"*, Proc. SPIE, **733**, 519-532.

Truemper, J., 1983. *"The Rosat Mission"*, Adv. Space Res., **2**, No. 4, 241-249.

Wells, A., Abbey, A.F., Barstow, M.A., Cole, R.E., Pye, J.P., Sims, M.R., Spragg, J.E., Watson, D.J., et al., 1990. *"First In-Orbit Measurements with the ROSAT XUV Wide Field Camera"*, Proc. SPIE, **1344**, 230-243.

OPTICAL IDENTIFICATION OF X-RAY SOURCES FROM DIGITISED SKY SURVEY PLATES

B.J. McLean and R. Burg

Space Telescope Science Institute, 3700 San Martin Drive, Baltimore MD 21218, U.S.A.

Abstract

In this paper, we describe the procedure that has been developed to identify the optical counterparts of the ROSAT X-ray sources in the northern
hemisphere. This is performed by the identification and categorisation of
all optical objects contained within the X-ray error circle using the digitised sky survey images that are available in the GSSS archive at STScI.
Additional ground based observations can then be scheduled on particular
objects of interest or to resolve any ambiguous identifications.

1 Introduction

The Guide Star Selection System (GSSS) group has built up an archive of all the
Schmidt plates that were scanned with the two PDS 2020G measuring machines
in order to support the creation of the Guide Star Catalogue (Lasker et al. 1990)
and the operation of the Hubble Space Telescope. Over 2000 plates, digitised at a
resolution of 1.7 arcsec per pixel, have been stored on WORM optical discs. These
include the SERC-J survey and Equatorial extension, the Palomar 'Quick-V' survey
and the Palomar–National Geographic E plates. These images, or subsets of them,
can be retrieved from optical disc to examine any field of interest and are done so
routinely to support HST operations and staff research.

There is a collaboration between the Max Plank Institute (MPE) which operates
ROSAT and STScI to identify the optical candidates for the northern hemisphere
X-ray sources. A similar collaboration exists between MPE, ROE and NRL for the
southern hemisphere.

H.T. MacGillivray and E.B. Thomson (eds.), Digitised Optical Sky Surveys 465–469.

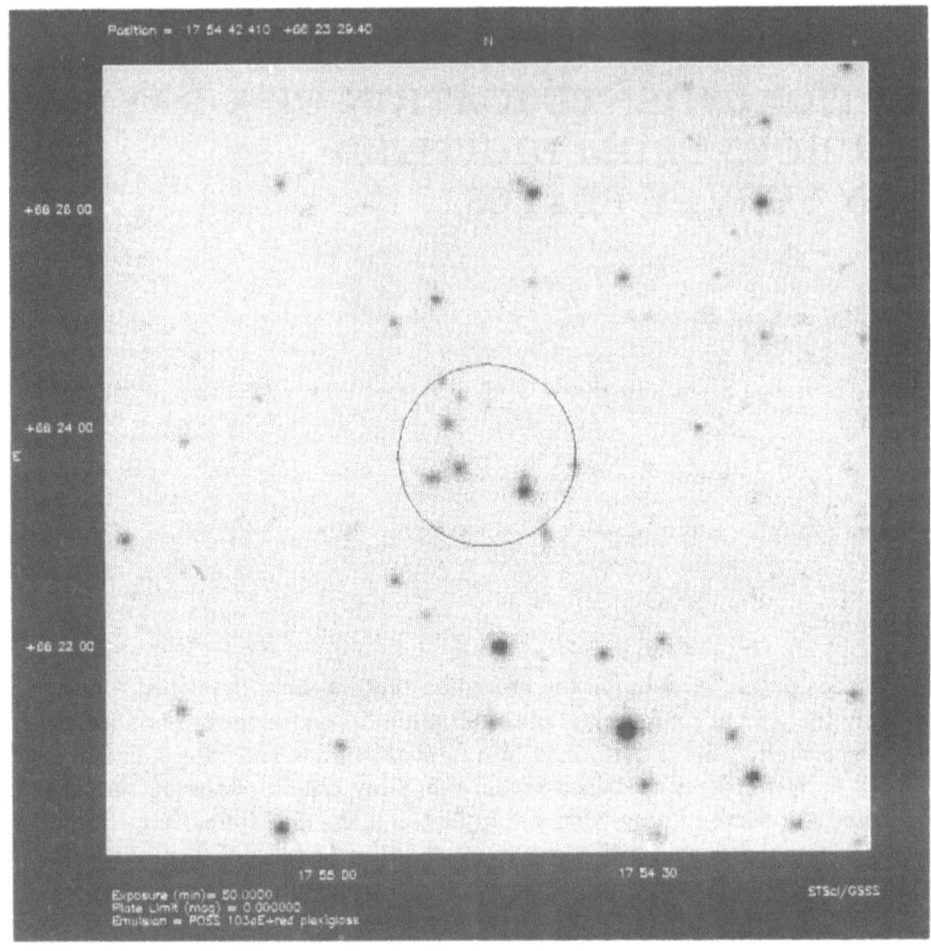

On these and following pages are shown images retrieved from the GSSS processing system.

Initially, we shall use the POSS-I E plates to obtain our candidates lists as they are longer exposure and more likely to detect the optical counterparts of some of the more distant extragalactic sources even though they have more cosmetic defects. In the second phase of this project, the Quick-V plates will be used to provide colours and proper motions for the brighter objects.

2 Plate Processing

Each of the digitised plates is processed to obtain an object catalogue with positions, magnitudes and classifications. One of the major problems is that the POSS-I plates are uncalibrated, so we have implemented the areal profile technique used by the APM group (Bunclark 1982) to determine the stellar profile and D-I curve in order to perform the photometric calibration using the Guide Star Photometric

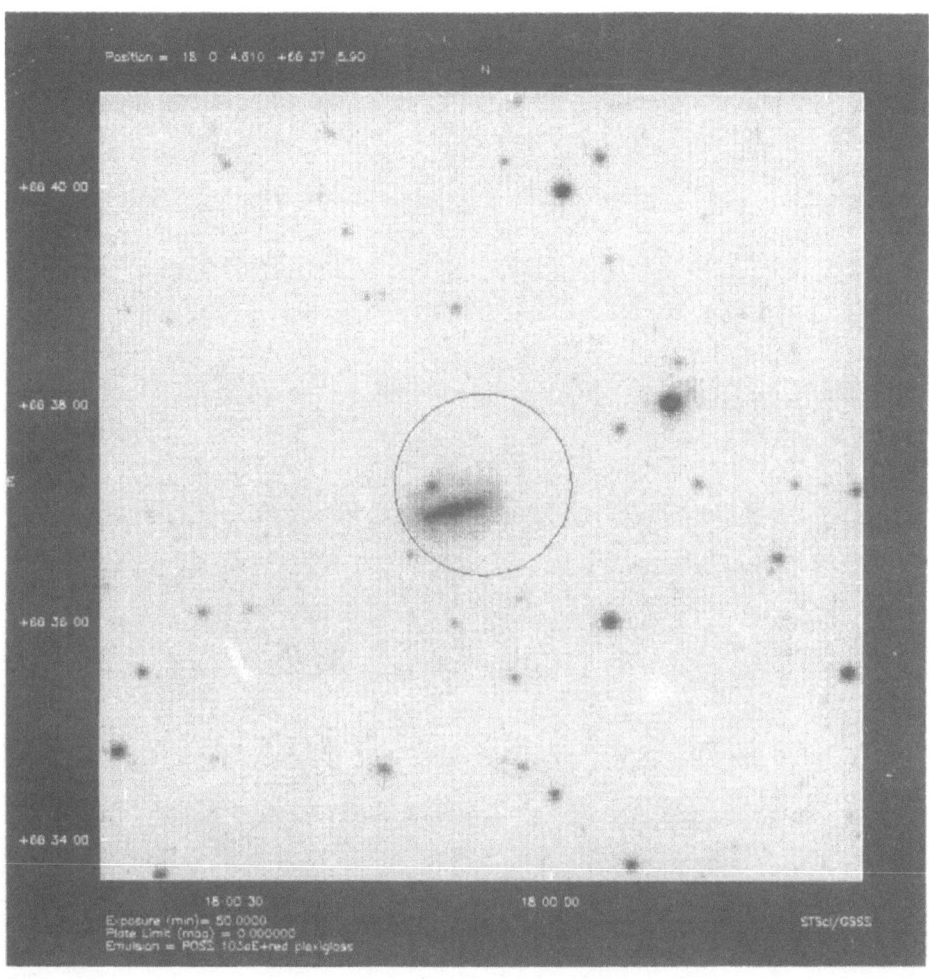

Catalogue (Lasker 1988) as standards for each plate. We have also implemented the object deblending algorithm developed by the COSMOS group (Beard et al. 1990). Astrometry is obtained using the same plate model and reference catalogues as used for the GSC (Russell et al. 1990).

In addition to providing an object catalogue for identifying optical counterparts, we are creating maps of both the stellar and galaxy distributions for off-line analysis in other projects, in particular to detect clusters of galaxies which will appear as extended X-ray sources and may not be in the initial ROSAT source list.

3 Candidate Identification

The positions of the X-ray sources detected in the ROSAT scan strips will be sent to STScI on a regular basis once they have been completely processed through the ground system and X-ray analysis phase. These strip detections will be entered

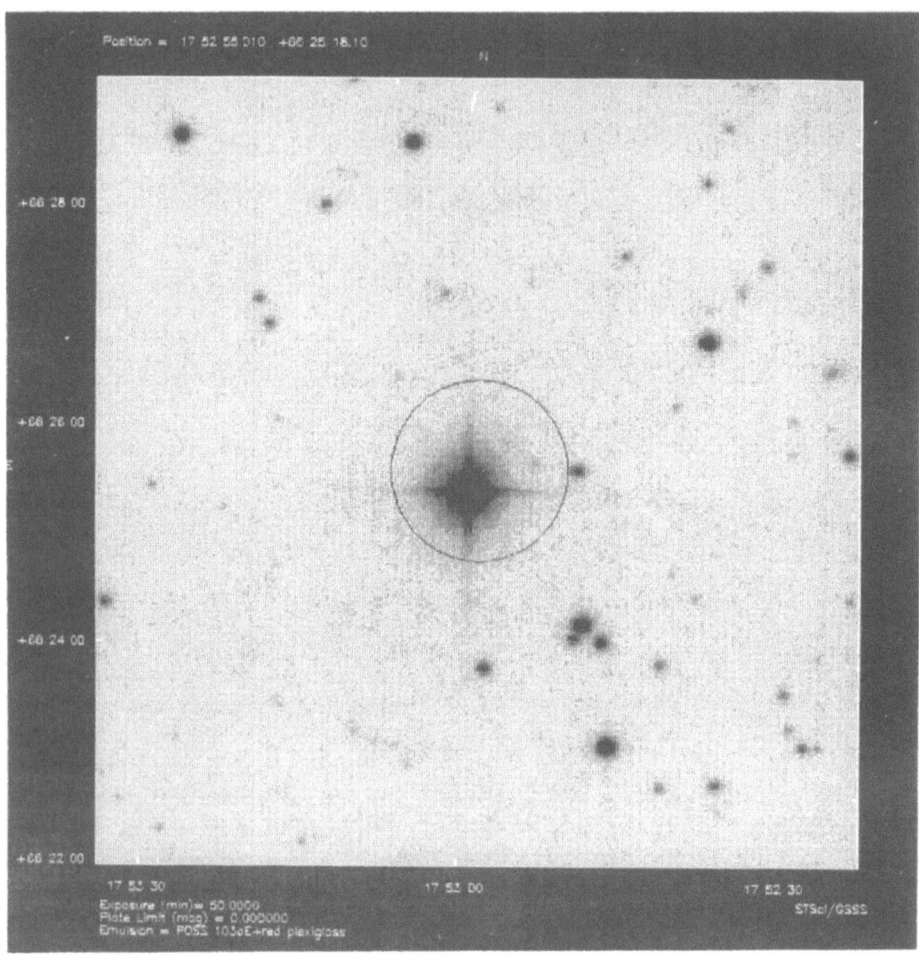

into a position database and the plate on which they will be found is identified. Whenever all sources that will appear on a given plate have been received, that plate will be retrieved from the archive and the object catalogue obtained. (see Fig. 1)

The catalogue of optical objects will be correlated with the X-ray positions to obtain the lists of optical candidates. These lists are returned to MPE where the optical identification is being coordinated. In addition, we will create finding charts in order to perform follow-up ground based observations (see plates).

The current status of the project is that only the North Ecliptic Pole and HZ43 regions have been processed as these were supplied as test fields. The first ROSAT strips were received at STScI in late May 1991 and full-scale processing is just beginning. Follow-up ground based observations on the NEP are currently in progress.

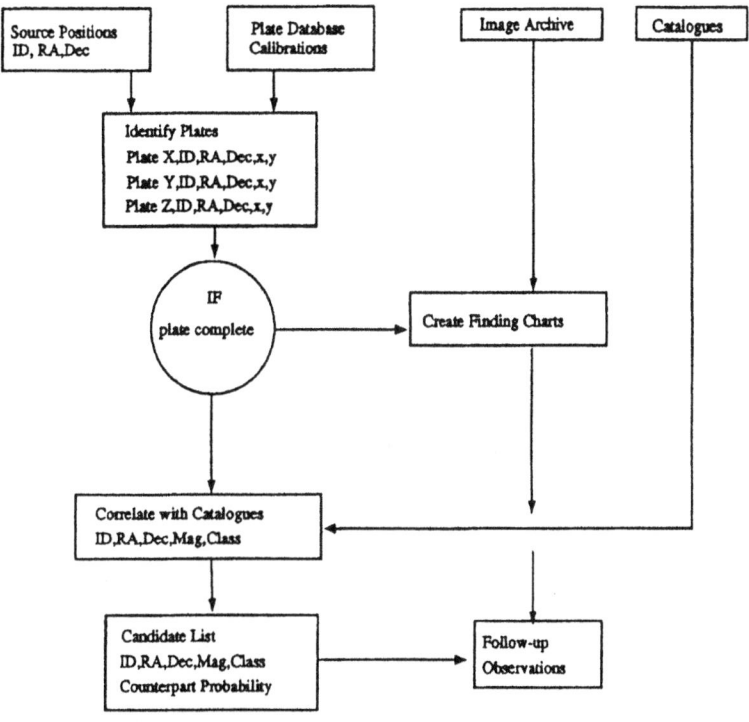

Fig. 1. Data flow of identification pipeline.

References

Beard, S.M., MacGillivray, H.T., Thanisch, P.F. 1990. *Mon. Not. R. Astron. Soc.*, **247**, 311.

Bunclark, P.S., 1982. *Proceedings of the Workshop on Astronomical Measuring Machines* (Edinburgh), eds. Stobie and McInnes.

Lasker, B.M. et al., 1988. *Astrophys. J. Supp.*, **68**, 1.

Lasker, B.M. et al., 1990. *Astron. J.*, **99**, 2019.

Russell, J.L. et al., 1990. *Astron. J.*, **99**, 2059.

OPTICAL IDENTIFICATION OF SOURCES IN THE IRAS FAINT SOURCE DATABASE

R.D. Wolstencroft [1], H.T. MacGillivray [1], C.J. Lonsdale [2], T. Conrow [2], D.J. Yentis [3], J.F. Wallin [3] and G. Hau [4]

[1] Royal Observatory, Blackford Hill, Edinburgh EH9 3HJ, U.K.
[2] Infrared Processing and Analysis Center, California Institute of Technology, Pasadena, California, U.S.A.
[3] Naval Research Laboratory, 4555 Overlook Ave. SW., Washington DC 20375, U.S.A.
[4] St John's College, Oxford University, Oxford, U.K.

1 Introduction

The need to establish the optical counterparts of the sources in any major non-optical sky survey is self evident. When the original IRAS Point Source Catalogue (PSC) was published, only about 28% of the $\sim 250\,000$ sources had positional associations with objects in astronomical catalogues. Although these associations are not equivalent to identifications, in the case of IRAS sources with optically bright counterparts the two are in general synonymous. Nevertheless, despite the extensive study of IRAS sources both with and without associations over the past 8 years, it is clear that the optical counterparts are unknown for more than half of the PSC sources. With the availability of the IRAS Faint Source Database, which contains $\sim 750\,000$ sources (including the PSC) down to a considerably fainter limit, the need for a systematic identification program is even more clear-cut if the potential of the IRAS mission is to be fully realised. Furthermore, the launch of the non-survey ISO mission in 1993 adds to the urgency of this program.

With the recent completion of the COSMOS/UKST optical catalogue of the southern sky (produced in a collaboration between ROE and NRL, see paper by Yentis et al. 1991, in this volume) we now have a tool for carrying out the identification of all southern sources in the IRAS Faint Source Database (FSDB). In

H. T. MacGillivray and E. B. Thomson (eds.), Digitised Optical Sky Surveys 471–483.
© 1992 *Kluwer Academic Publishers.*

anticipation of the completion of the ROE/NRL catalogue, we set up a collaboration between ROE, IPAC and NRL to get this project underway, and in this paper
we describe its present status: we discuss the properties of the FSDB (section 2),
outline the identification method (section 3) and present the results of a pilot study
carried out at high galactic latitudes (section 4).

2 The IRAS Faint Source Survey Database

This database (FSDB) contains all IRAS point sources with a signal-to-noise ratio
(SNR) of at least 3 in any band. It comprises two parts, namely:

The Faint Source Catalogue	(FSC)	173,044	sources
The Faint Source "Reject" File	(FSR)	593,516	
Total	(FSDB)	766,560	

The FSC contains the highest reliability sources with a reliability $> 90\%$ at
12μm and 25μm and $> 80\%$ at 60μm which are essentially equivalent to a SNR > 6
at 12μm, 25μm and 60μm and minimum flux densities of 180, 290 and 260 mJy;
whereas the FSR contains all FSDB sources (SNR > 3) not in the FSC , which
generally means sources with $3 <$ SNR < 6. However since the FSC is restricted
to the unconfused parts of the sky ($|b| > 10$ deg at 12μm and 25μm; $|b| > 20$
deg at 60μm), the FSR does contain a small number of high SNR sources at low
galactic latitudes. At 100μm the presence of cirrus complicates the picture in the
FSR (there are no 100μm only sources in the FSC): typical limits at 100μm are
SNR > 9 and 90% completeness at 1.1Jy. Overall the FSC is 2.5 to 3 times more
sensitive than the PSC and contains $\sim 100\,000$ sources that are not in the PSC.

Some estimate of the nature of an IRAS source (e.g. star versus galaxy) may
be made when the flux in at least 2 bands is known. However, even in the cases
of the PSC and FSC, half the sources are single band sources so that their nature
can only be guessed (without a proper optical identification) from a study of their
distribution over the sky. The reason why the star versus galaxy question is important in the identification process will be discussed in section 3. For the FSR,
86% of the sources are in the single band category, which illustrates the difficulty
of identifying the faintest sources in the FSDB. The spectral classifications of the
FSDB, FSC and PSC sources are listed in Table 1: note that for example (1000)
and (0100) indicate 12μm only and 25μm only sources respectively.

Of the numerically large classes it is likely, based on colour and associations
with catalogued objects, that the (1000) and (1100) classes are dominated by stars,
viz. 359324 (46.9% of FSDB); the (0010) and (0011) classes are mostly galaxies,
viz. 196403 (25.6% of FSDB); and 210833 (27.5%) are 'other', of which the largest
is the 106923 (0100) class.

The Galactic latitude distribution of the FSDB single band sources is illustrated
in Fig. 1. The star versus galaxy classification depends naively on stars being concentrated to the Galactic plane and the galaxies having a uniform number density

Table 1. Spectral classification of the FSDB.

Band	PSC 2		FSC 2		FSDB	
Comb.	N	f (%)	N	f (%)	N	f (%)
1000	67332	27.39	61232	35.39	260955	34.04
1100	67015	27.26	40339	23.31	98369	12.83
1110	13233	5.38	6907	3.99	13744	1.79
1111	6343	2.58	4671	2.70	7495	0.98
0100	3936	1.60	367	0.21	106923	13.95
0110	3642	1.48	2382	1.38	6840	0.89
0111	3873	1.58	4621	2.67	6021	0.79
0010	19264	7.84	26913	15.55	164105	21.41
0011	22702	9.23	23556	13.61	32298	4.21
0001	33146	13.48	0	0	65766	8.58
1101	2025	0.82	103	0.06	224	0.03
1010	1100	0.45	715	0.41	1779	0.23
1011	520	0.21	1017	0.59	1551	0.20
1001	1207	0.49	215	0.12	424	0.06
0101	501	0.20	6	0.00	66	0.01
Total	245839		173044		766560	

over the sky. If the majority of single band sources in a given band are either stars or galaxies, then the 12μm sources are mostly stars, the 60μm sources are mostly galaxies and the 100μm only sources are cirrus (rather than galaxies). In the case of the 25μm sources, their spatial distribution suggests that they are galaxies: however, there are relatively few galaxies with fluxes at 12μm, 25μm and 60μm which when placed at a greater distance would be detected at 25μm only. Another possibility is that many of the (0100) class are (1100) class objects (stars) that have lost their band 1 (12μm) fluxes for some reason, perhaps due e.g. to confusion at low Galactic latitude. However, such objects are unlikely to define the major component because of the uniform number-density distribution. A further possibility is that there is a significant fraction of spurious sources in the (0100) class.

The probability that a FSDB source corresponds to a true astronomical source (the reliability) can be estimated by analysing the dependence of source counts on SNR and by looking at the optical identifications for a small sample of sources. This allows a rough estimate to be made of the number of real FSDB sources: for the FSR sources (SNR between 3 and 6 in the main) the appropriate reliability — allowing for variations in survey coverage — is 58%, 48%, 40% and 40% (at 12μm, 25μm, 60μm, 100μm) which, for the single band sources, yields:

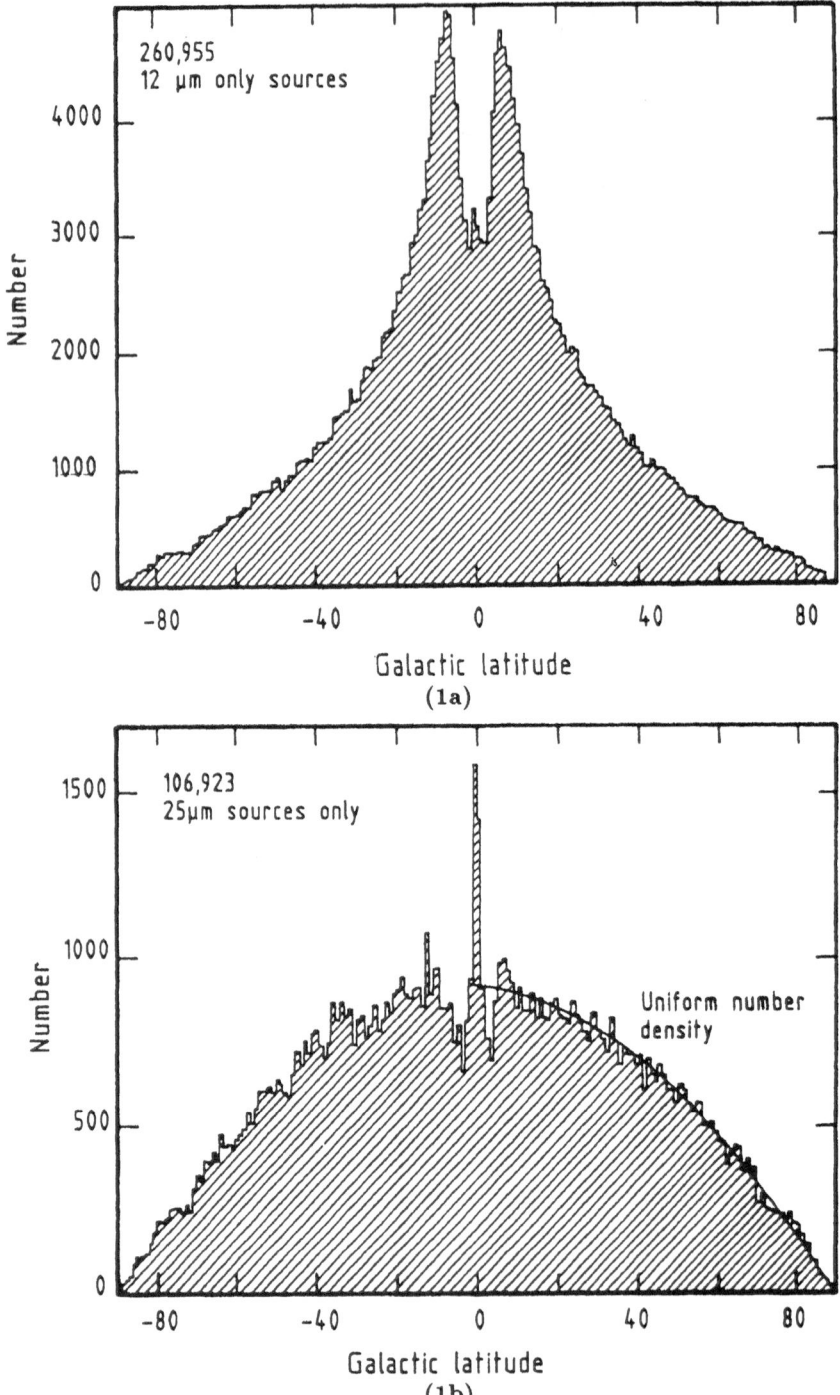

Fig. 1. Number of single band sources in the FSDB at each galactic latitude (1 degree steps) for **(a)** 12μm, **(b)** 25μm, **(c)** 60μm and **(d)** 100μm only sources. The curve for a uniform density population is shown on the 25μm only figure. The large dips very close to the plane are an artifact of the source extraction algorithm.

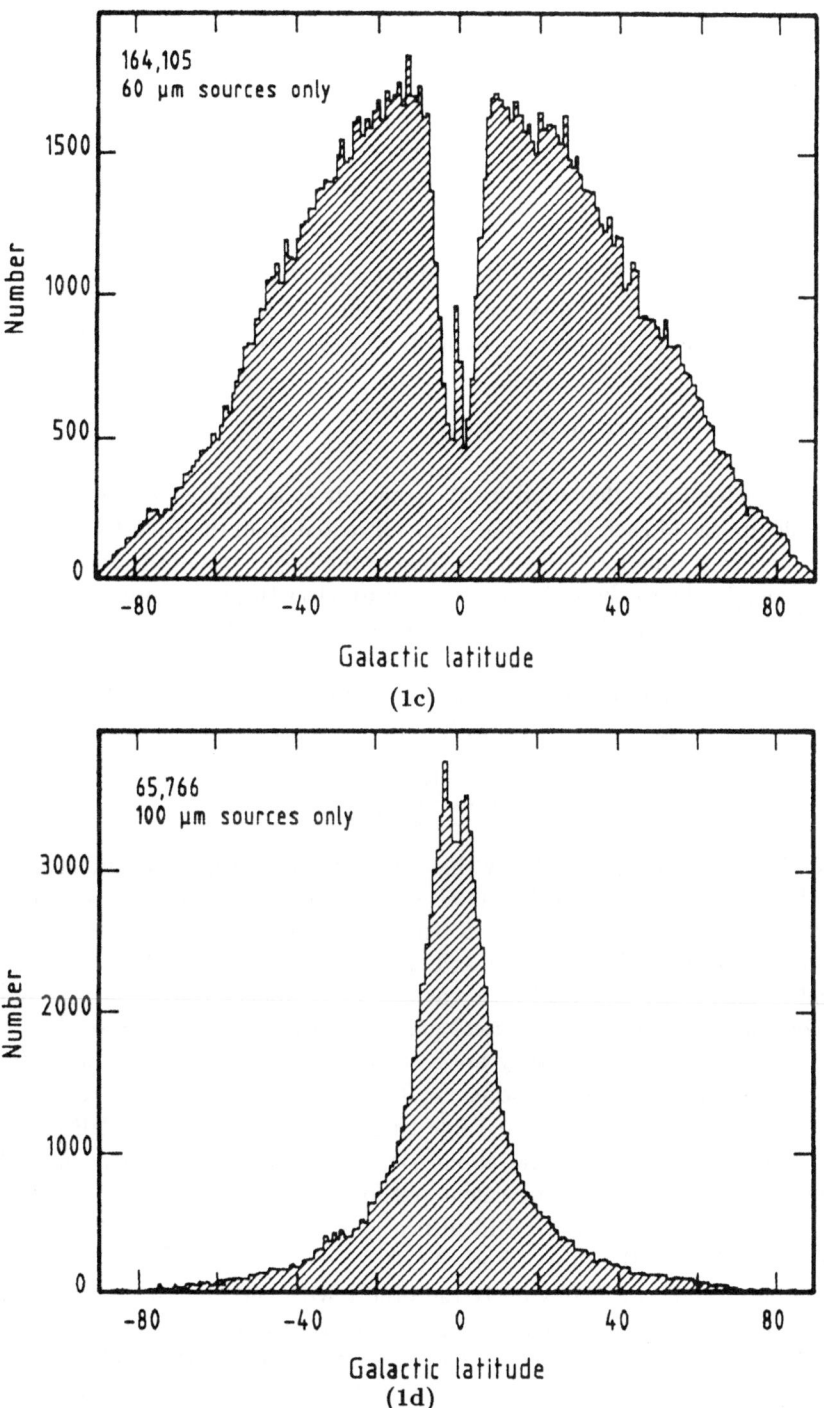

164,105
60 μm sources only

Number

Galactic latitude

(1c)

65,766
100 μm sources only

Number

Galactic latitude
(1d)

| | FSDB | 'Real' |
	Sources	Sources
12 μm only	199273	116000
25	106556	51000
60	137192	55000
100	65766	26000

This suggests that 248 000 of the 510 000 single band FSR sources may be real. Adding in the higher reliability sources from the rest of the FSDB using the appropriate factors results in a best guess of about 470 000 real FSDB sources.

3 The Identification Method

Our approach follows that used by Wolstencroft et al. (1986). Optical candidates within 90 arc sec of each IRAS source are extracted from the ROE/NRL database. The 90 arc sec is chosen to be well in excess of the major axis of the 95% probability error ellipse for the weakest sources. In order to select the best candidate, we use the likelihood ratio (LR) criterion: LR is largest when the IRAS error ellipse is small (bright IRAS source), the optical candidate is bright (reduced probability that the candidate is an unrelated object lying close to the IRAS source by chance) and the optical candidate lies close to the IRAS source position. The likelihood ratio is

$$LR = \frac{Q\exp(-r^2/2)}{2AN(B)} \tag{1}$$

where Q is the a priori probability that the IRAS source has a counterpart in the optical catalogue, r is the positional offset between the IRAS and optical positions, A is the area of the one sigma IRAS error ellipse and $N(B)$ is the surface density of optical objects brighter than or as bright as the candidate. The objects in the optical catalogue are classified into star or galaxy images so that in principle one can, for example, select only optical galaxy candidates for IRAS sources whose colours clearly indicate that they are galaxies: thus in this case the identification may be much easier, especially when $N(B)$ is much less for galaxies than for stars (see Table 2). It can be seen from Table 1 that about 26% of the FSDB are mostly galaxies [(0010) and (0011)] and about 47% are mostly stars [(1000) and (1100)], so that in principle LR may be calculated for galaxy $N(B)$ only, or stellar $N(B)$ only, in the majority of the trial identifications. For the remaining 27% LR must be determined using the sum of both $N(B)$s.

The best candidate is obviously the object with the highest value of LR and this candidate will almost certainly be the true optical counterpart provided that the LR value is (1) above an acceptable threshold value, and (2) significantly above the LR value of the second best candidate. The threshold value is dictated by a balance between completeness (low LR) and reliability (high LR): estimates in the literature range between about 2 and 10. Since the completeness and reliability

Table 2. Number of galaxies and stars per sq. deg. as bright or brighter than b$_j$ at the South Galactic Pole.

b$_j$	N[b$_j$] galaxies	N[b$_j$] stars	N[b$_j$] galaxies + stars
21	851	931	1782
20	331	632	963
19	115	469	584
18	42	295	337
17	11	191	202

depend on factors such as the Galactic latitude, IRAS colours, star or galaxy, and so on, so also does the *LR* threshold. A useful way to evaluate the threshold is to compare the *LR* distribution of the best candidates for a reasonably large sample of IRAS sources of a given class (say stars based on IRAS spectral class) with the same sample where the IRAS positions have been randomly shifted to produce a false match distribution. Two examples for sources with stellar colours at the South Galactic Pole and at $b = -30$ deg are shown in Fig. 2: it appears that a higher threshold is needed at lower latitude (*LR* $\sim$ 10 at $b = -30$ deg) than at the pole (*LR* $\sim$ 3) as expected in a region of higher star density.

4 Pilot Study of Galaxies at the South Galactic Pole

We selected FSDB sources with 60μm detections and SNR (60μm) > 4 and applied the condition $F(25\mu\mathrm{m})/F(60\mu\mathrm{m}) < 1.6$ to strongly discriminate against stars: this included 60μm only sources with 25μm upper limits satisfying this constraint. The sample area was defined by the Edinburgh-Durham Southern Galaxy Catalogue (EDSGC) (Heydon-Dumbleton et al. 1989) which is based on a subset of the ROE/NRL catalogue: the RA and DEC limits, defined by 60 UKST fields, cover an area $\sim$ 1500 sq. deg near the SGP. There were 4124 FSDB sources selected in the area. Optical galaxy candidates from the EDSGC down to b$_j$ = 20.5 were selected within a distance of 2 arc min from each source: note that a selection distance of 90 arc sec is considered adequate for the full identification program now in progress. The number of candidates for each source ranged from 0 to 5 with a median of 2. Following Wolstencroft et al. (1986), we adopted a trial threshold of 3 for this pilot study. The identification statistics were as follows:

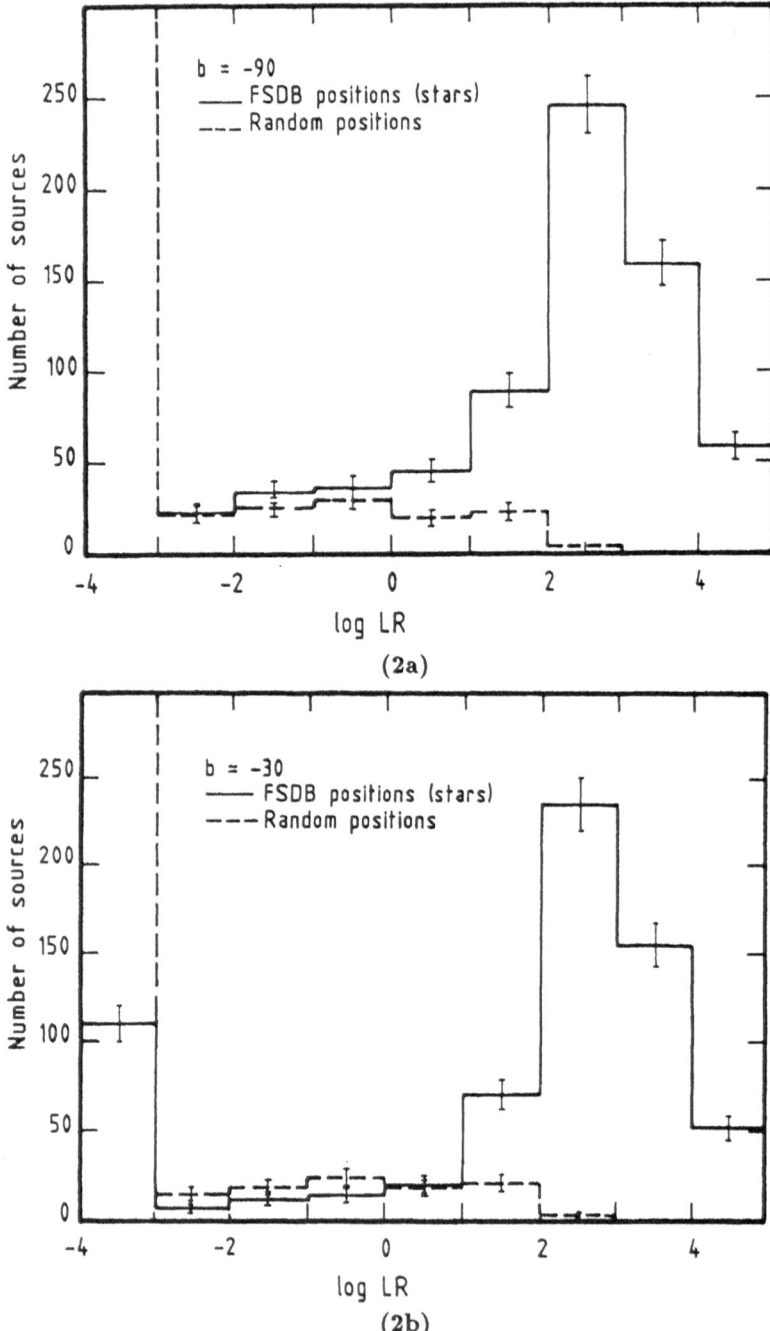

Fig. 2. Histogram of the Likelihood Ratio (LR) for the best matched (highest LR) optical candidates for each source where (**a**) the actual IRAS positions are used (solid line) and (**b**) the IRAS positions have been randomly shifted (dotted line) to produce a false match distribution. Histograms are shown for both $b = -90$ and $b = -30$ degrees.

No. of candidates with $LR > 3$	No. of sources
2 or more	734
1	2436
0	954
Total	4124

This implies that the fraction of IRAS galaxies with optical counterparts having $b_j < 20.5$ is 77%. Of the 954 without counterparts, 242 had no candidates within 2 arc min and the remaining 712 had one or more candidates below the threshold. As expected, the unidentified sources are preferentially the weaker sources (see Fig. 3) with 50% of the unidentified sources having $F(60\mu m) < 0.20 Jy$. A more instructive way of looking at this is shown in Fig. 4 where the dependence of the fraction identified, f, on SNR ($60\mu m$) is plotted: f rises rapidly from $67.5(\pm 2.2)\%$ at SNR = 4 to 5 to $86.0(\pm 5.5)\%$, where the errors are based on Poisson statistics. The asymptotic value of f at high SNR (say > 20) is 254/284 i.e. $89.4(\pm 5.6)\%$: until we have a much larger sample size and have decided on the 'correct' LR threshold, we consider an attempt to interpret the apparent absence of counterparts for $\sim 10\%$ of the brightest sources to be premature. As to the interpretation of the 23% unidentified in the sample overall, apart from the proper choice of the LR threshold there are at least 3 possible factors involved: (1) the misclassification of a small fraction of the optical images in the ROE/NRL southern sky database, either galaxies being classified as stars or vice-versa — this is unlikely to be a problem except at $b_j > 19$ where $\sim 5\%$ may be misclassified: however even if the steep fall-off in the magnitude histogram for counterparts at $b_j > 19.5$ (Fig. 5) were due to this effect, it is clear that a smooth extrapolation of this curve from 19.0 to 20.5 would contribute very little to the 23%; (2) there may be a small number of IRAS galaxies with optical counterparts fainter than 20.5 — e.g. galaxies that are very heavily reddened: such galaxies will have been detected by IRAS if $\log[L(IR)/L(b_j)] > 0.4 b_j - 6.5$, i.e. if $L(IR)/L(b_j) > 50$ or 200 at $b_j = 20.5$ or 22 respectively. Galaxies with these values of $L(IR)/L(b_j)$ exist in our sample (see Fig. 6) but it seems unlikely that such galaxies are a major component of the unidentified sources; and (3) the $F(25\mu m)/F(60\mu m)$ criterion may admit a few objects that are not galaxies such as T Tauri stars, planetary nebulae and stars associated with reflection nebulae and HII regions (see Walker et al. 1989); however the above are mostly confined to low Galactic latitudes, so that contamination is likely to be minimal. The relative importance of these and other factors should become apparent as the new identification program develops. One final point of interest coming out of the pilot study is that there appears to be very little difference in the IRAS colours of the identified and unidentified sources: for example the histograms of $F(60\mu m)/F(100\mu m)$ (Fig. 7) are essentially identical apart from the identified sources being slightly hotter (shifted by 0.05).

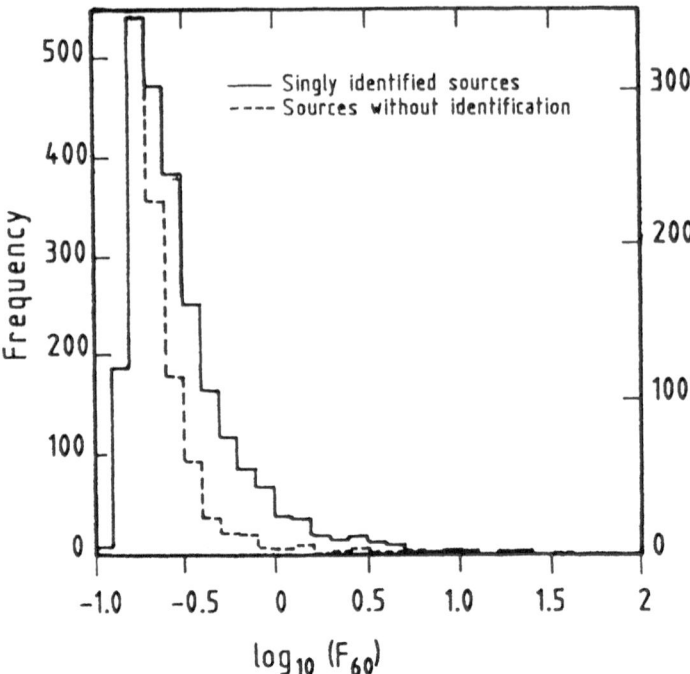

Fig. 3. Histogram of $F(60\mu m)$ for the 2436 sources with a single counterpart (solid line) and for the 954 sources with no identification (dashed line). The unidentified sources are preferentially weaker.

5 Conclusions

Less than about 20% of the $\sim 750\,000$ IRAS FSDB sources have either identifications or associations and it is very clear that a program of optical identifications is needed. We have described the identification methods and results of a pilot study of high Galactic latitude sources which illustrate the main problems encountered: these include the need to establish the dependence of the likelihood ratio (LR) threshold on Galactic coordinates and IRAS band combination, and the importance of using IRAS colours where possible to determine whether a source is likely to be a star or a galaxy. We have recently embarked on a fully automated identification of the southern FSDB sources using the ROE/NRL database of the southern sky. The final catalogue, which will also be made available to the astronomical community in the form of an on-line database, will list all candidates above the LR threshold, although we anticipate that the majority of sources at high and intermediate Galactic latitude will have one counterpart only.

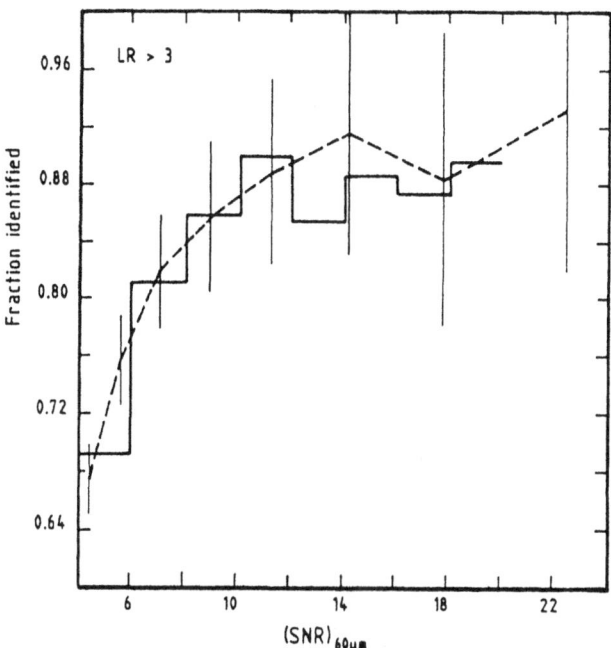

Fig. 4. Dependence of the fraction of FSDB sources identified (one or more candidates) on the 60μm SNR: 3170 out of 4124 sources (77%) are identified with $LR > 3$. The stepped curve/points with error bars refer to bins of equal/increasing width.

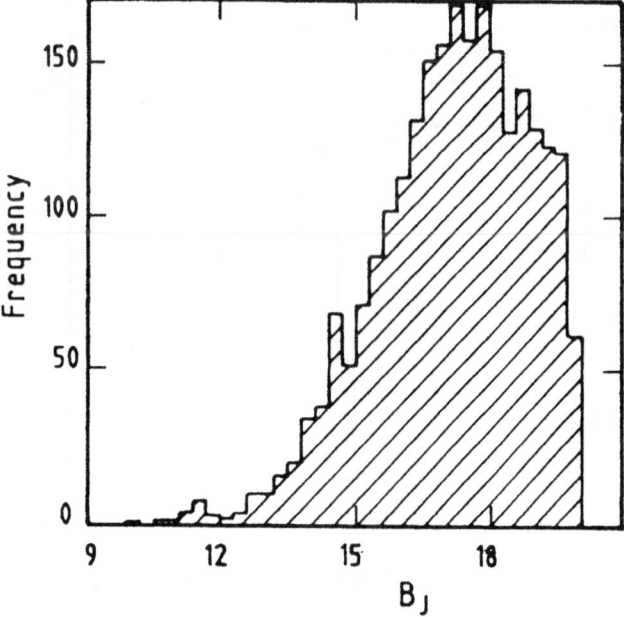

Fig. 5. Histogram of b$_j$ for the 2436 sources with a single counterpart (0.3 mag. bins).

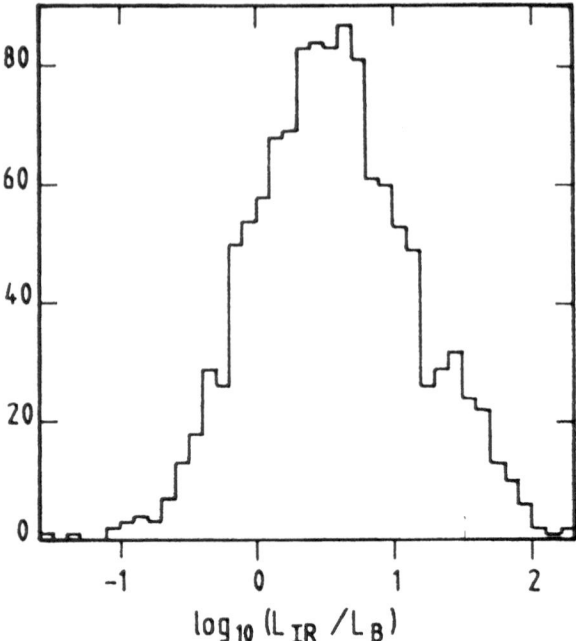

Fig. 6. Histogram of $\log[L(IR)/L(b_j)]$ for the 1214 single counterpart sources with a good quality flux density at both 60μm and 100μm.

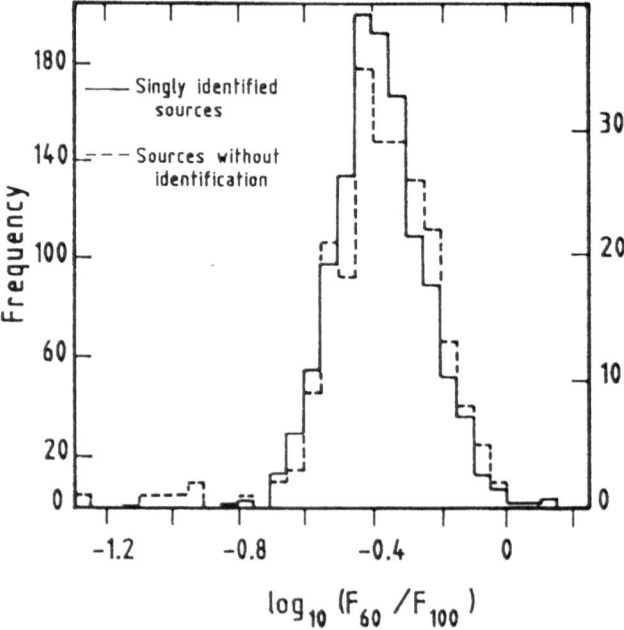

Fig. 7. Histograms of $\log[F(60\mu m)/F(100\mu m)]$ for the 1214 single counterpart sources and 229 unidentified sources that have a good quality flux density at both 60μm and 100μm.

References

Heydon-Dumbleton, N.H., Collins, C.A., MacGillivray, H.T., 1989. *Mon. Not. R. Astron. Soc.*, **238**, 379.

Walker, H.J., Cohen, M., Volk, K., Wainscoat, R.J., Schwartz, D.E., 1989. *Astron. J.*, **98**, 2163.

Wolstencroft, R.D., Savage, A., Clowes, R.G., MacGillivray, H.T., Leggett, S.K., Kalafi, M., 1986. *Mon. Not. R. Astron. Soc.*, **223**, 279.

OPTICAL IDENTIFICATION OF THE NEW PARKES/MIT/NRAO RADIO SOURCES FROM DIGITISED SKY SURVEY DATA

A. Savage [1] *and A.E. Wright* [2]

[1] Schmidt Telescope, Anglo-Australian Observatory, Coonabarabran, Australia.
[2] Australia Telescope National Facility, Parkes, Australia.

Abstract

In 1990 the southern sky was surveyed at Parkes using a multi-beam receiver at frequencies of 843 and 4850MHz. The surveys cover the Declination range between +10 and -90 degrees and are complete in Right Ascension, an area of 7.30 steradians. Preliminary analysis of the 5Ghz data indicates a flux limit of about 30mJy, and we estimate we will find some 90 000 new sources above this limit, or some 90 radio sources per Schmidt field. Such source densities, combined with accurate positions from the Australia Telescope, means that this survey will be ideally suited to an automated identification procedure using the Digitised Sky survey data from the UK Measuring Machines and Schmidt Telescope plate material.

1 Introduction

The PARKES/MIT/NRAO (PMN – Griffith et al. 1991) survey of the southern sky is a collaboration between institutions in Australia and the United States of America, using Australian telescope facilities. The primary aim of the project was to survey the whole sky south of −36 degrees declination at a frequency of 4850MHz to a five sigma flux limit of 30mJy. Secondary aims of this project were to extend the survey from −36 degrees declination as far north as time permitted and to make a concurrent 843MHz survey using a separate, off-axis receiver. In the event, the southern survey was carried up to +10 degrees. By doing this we complemented the northern survey of Condon et al. (1989) at the same frequency. The northern survey was made with the NRAO Greenbank 91m radio telescope in October 1987.

485

H. T. MacGillivray and E. B. Thomson (eds.), Digitised Optical Sky Surveys 485–489.

It covered 6.0 steradians of sky in the declination range $0°$ to $+75°$ at the same frequency of 4850MHz, with a resolution of 3.7 arcmin by 3.3 arcmin. Their rms map noise is about 5mJy and about five times greater than the rms confusion level. There are 10 000 sources per steradian stronger than 25mJy and their rms position uncertainties in each coordinate range from 10 arcsec for the strongest sources to 30 arcsec for the weakest sources observed near the sun. Preliminary automated identification procedures using Palomar Schmidt plate material are already being used on these data in collaboration with the APM group to attempt to find gravitational lens candidates (B.F. Burke, private communication). A southern extension to -40 degrees declination is in progress using the NRAO 43m radio telescope also at Greenbank. This overlap with the Parkes survey will enable reliable completeness estimates to be made on the two independent northern and southern surveys.

The only existing large-area, high frequency radio survey in the south is that of Bolton and collaborators (Bolton et al. 1979, and references therein) which contains around 8200 sources to a 2700MHz flux limit corresponding to about 160mJy at 5000MHz for sources of average radio spectral index. This survey has a source density of only 8 sources per Schmidt field and a high chance coincidence rate in that there is always one object within the radio position error box, making it not suitable for automated identification procedures. Only the brighter galaxies were claimed as identifications and an ultra-violet excess discriminant was used to select candidate quasar identifications. Arcsecond radio positions were required to find high-redshift quasar candidates (Peterson et al. 1982) and identify galaxies at the limit of the UK Schmidt plate material (Downes et al. 1986).

2 The Southern Radio Survey

The total reduction, evaluation and display of the data from the PMN 4850MHz survey will be completed by the end of 1991 and will be made available in the form of a Point Source Catalogue. However, we can indicate a few of the preliminary results. We find a standard error on the peak flux of a source at this frequency of between 5 and 8mJy and an overall positional accuracy of about ±25arcsec in Right Ascension and about ±20 arcsec in Declination. A test area, covering 0.11 steradians in the RA range 22 to 23 hours and Declination range -36 to -88 degrees (epoch 1990.45), contains 1367 sources with fluxes above 25mJy giving a source density of 12 400 per steradian, comparable with the northern survey result and giving about 100 radio sources per Schmidt field. Such radio source densities are ideally suited to an automated optical identification procedure using digitised sky survey data from Schmidt Telescope plate material, providing the errors in these radio positions can be reduced to a few arcseconds by mapping with the VLA for the northern sources or positioning with the Australia Telescope for the southern sources. However, to obtain accurate positions for all the southern sources would take some 400 days of telescope time.

3 Automated Optical Identification

a) Small Area Programmes

Six deep Parkes selected regions (Downes et al. 1986) with a limiting flux of 100mJy at 2700MHz (corresponding to about 60mJy at 5000Mhz for sources of an average spectral index) cover a similar area of 0.075sr and contain 178 sources, giving about 30 sources per field. Accurate positions were obtained for these sources from VLA maps of the regions surrounding each source, giving arcsecond positions and additional information on source structure. The search area for candidate identifications was determined by the radio structure for unresolved and slightly resolved sources, so the identifications are limited only by the accuracy with which the radio and optical frames can be matched. In the case of this survey, and to allow for any unrecognised systematic effects, we adopted a pessimistic search radius of 2.5 arcseconds. Thus an automated procedure could be effectively used for the identification of these sources from both Palomar Observatory Sky Survey plates and from UK Schmidt Telescope plates. The initial search was carried out by scanning the POSS 'O' plates with the COSMOS measuring machine (see MacGillivray & Stobie 1984) working in image analysis mode. This allowed semi-automatic measurement of the positions of the identification candidates. Even so, a few radio source fields were processed by direct measurement at ROE with a Packmann x-y machine because in these cases a very faint candidate lay below the COSMOS threshold. The accurate radio data defined the structure and positions sufficiently well for an examination of existing Schmidt plate material to yield an identification content of 56 per cent of the sample down to V~21. Magnitudes of all the candidate identifications had to be obtained individually using B and R CCD frames from the AAT because no external magnitude calibration existed for the optical material.

b) Full-Sky Programmes

Since this preliminary small area attempt at identifications using digitised data, a wealth of experience has been gained by using these techniques to identify the optical counterparts of the all sky IRAS surveys. Wolstencroft and collaborators (Wolstencroft et al. 1986) identified the optical counterparts of IRAS point source catalogue sources (Beichman et al. 1985) in a 300 square degree area centred on the South Galactic Pole. The positions and COSMOS magnitudes of identification candidates were obtained using Schmidt Telescope plates scanned by COSMOS. These magnitudes have an internal accuracy of about 0.2mag although zero-points have not been externally calibrated for every plate. Subsequently, this automated procedure was extended to cover 3000 square degrees (Savage et al. 1987). Sutherland and collaborators (Sutherland et al. 1991) have used the APM galaxy survey based on scans with the APM measuring machine (Kibblewhite et al. 1984) of 4600 square degrees to automatically identify nearly 5000 sources in the IRAS Faint Source Catalogue (Moshir et al. 1989). An automated image classification algorithm was applied and the plates have been matched to a uniform magnitude system utilising plate overlap zones and external CCD calibration (Maddox et al. 1990) . However, the external photometric calibration is sparse being limited to only 70 fields (some containing only bright objects), with less than six calibrated galaxies in each of these fields, again giving rise to possible zero-point errors throughout the complete catalogue.

One perceived problem with this automated procedure for IRAS sources was the large positional errors associated with the IRAS positions, necessitating searches for optical counterparts within one to two arcminutes and the very high probability of a chance coincidence with an unrelated optical object. For optical objects to $b_j=22.5$, there are 9 and 36 objects to this limit within such error circles (Jauncey et al. 1982), meaning that identifications with faint objects cannot be made unambiguously. However, fortunately most IRAS identifications are with rare objects, stars brighter than 11 magnitude or with galaxies brighter than 17 magnitude, and the error circle is an ellipse with a very small minor axis. Thus the spurious identification rate turned out to be very low.

Spurious identifications will not be a problem for those southern radio sources where we have Australia Telescope positions giving a search radius of only 2.5 arcseconds, even though most of the identifications will be faint galaxies near the plate limit. At least 10 days of Australia Telescope time will be required to position even 3000 sources, which should then give about 1700 identified sources to the limit of the digitised catalogues compiled from Schmidt plates.

4 Conclusion

It should be a fairly straightforward procedure to extend the knowledge gained from the automated full sky identification procedures for IRAS sources to a full-sky procedure for accurately positioned radio sources. The scientific benefits of such a procedure will be greatly enhanced if the digitised catalogues are accurately photometrically calibrated at the time when the respective point source catalogues are compared.

References

Beichman, C.A., Neugebauer, G., Habing, H.J., Clegg, P.E., Chester, T.J., 1985. *Explanatory Supplement to the IRAS Catalogues and Atlases*, US Government Printing Office, Washington, DC.

Bolton, J.G., Wright, A.E., Savage, A., 1979. *Aust. J. Phys. Astrophys. Suppl.*, **46**, 1.

Condon, J.J., Broderick, J.J., Seielstad, G.A., 1989. *Astron. J.*, **97**, 1064.

Downes, A.J.B., Peacock, J.A., Savage, A., Carrie, D.R., 1986. *Mon. Not. R. Astron. Soc.*, **218**, 31.

Griffith, M., Burke, B.F., Fletcher, A., Behrens, G., Chestnut, C., Wright, A.E., Hunt, A.J., Troup, E., et al., 1991. *Proc. A. S. A.* In press.

Jauncey, D.L., Batty, M.J., Gulkis, S., Savage, A., 1982. *Astron. J.*, **87**, 763.

Kibblewhite, E.J., Bridgeland, M.T., Bunclarke, P.S., Irwin, M.J., 1984. In *Proc. Astronomical Microdensitometry Conference*, NASA **2317**.

MacGillivray, H.T., Stobie, R.S., 1984. *Vistas Astr.*, **27**, 433.

Maddox, S.J., Efstathiou, G., Sutherland, W.J., 1990. *Mon. Not. R. Astron. Soc.*, **246,** 433.

Moshir, M., Kapan, G., Conrow, T., McCallon, H., Hacking, P., Gregorich, D., Melnyk, M., Rohrbach, G., et al., 1989. *Explanatory Supplement to the IRAS Faint Source Survey*, JPL, Pasadena.

Peterson, B.A., Savage, A., Jauncey, D.L., Wright, A.E., 1982. *Astrophys. J. Letts.*, **260,** L27.

Savage, A., Clowes, R.G., MacGillivray, H.T., Wolstencroft, R.D., Leggett, S.K., Puxley, P.J., 1987. In *NASA Conference Publication* **2466,** p. 537.

Sutherland, W.J., McMahon, R.G., Maddox, S.J., Loveday, J., Saunders, W., 1991. *Mon. Not. R. Astron. Soc.*, **248,** 483.

Wolstencroft, R.D., Savage, A., Clowes, R.G., MacGillivray, H.T., Leggett, S.K., Kalafi, M., 1986. *Mon. Not. R. Astron. Soc.*, **223,** 279.

DISCUSSION MEETING

MacGillivray:

In opening this discussion session, I would like to refer back to the discussions we had at the meeting in Geneva in 1989. You will recall that at that meeting, Professor Carlos Jaschek posed a number of questions. After some deliberation, a consensus of agreement was reached, and the result of our deliberations was summarised very lucidly by Professor Westerhout. I would like here to bring to your attention again the questions that were raised by Jaschek, together with the responses at that time by Westerhout, in order to examine how much progress has been made in the intervening 2 years or what, if anything, has happened over the past 2 years to change our attitudes towards some of the issues. Subsequently, I would like to raise some points of my own that I feel should be discussed at the present meeting.

First, the questions and responses from the 1989 meeting:-

1) What should the digitised surveys contain: just raw pixels or catalogues of objects?

It was decided that they should contain both! I don't think anyone would now dispute that in an ideal situation we would indeed like both. However, at the present time the cost of storing all-sky pixel information (several hundreds of Gbytes, or Terabytes) is still prohibitive for many groups. It must also be borne in mind that I suspect 95% of the science coming from the digitised surveys is based upon the final processed object catalogues. Only a small ($\sim$ 5) percentage really require the raw pixel information (such as e.g. searches for dwarf galaxies). The advent of digital film for the storage medium will certainly aid the transport of such data. In summary, I think that we will see the machine groups certainly storing the raw pixel data, but I think the main user demand will be in the processed object catalogues. Fast network links will probably allow the smaller pixel-oriented community to have access to the raw data.

2) Should everything be digitised at the highest possible scanning resolution?

We indeed foresaw in 1989 that we would want to digitise at as high a resolution as possible. It is interesting to see that the 'median' of the pixel size used by the various scanning groups is about 10 microns. This should indeed be sufficient to extract the information from the Schmidt surveys with an oversampling of 3 at 2 arcsecs seeing. We must also bear in mind, however, the possible use of fine-grain emulsions for future major sky surveys, and I note that the UKST group are

H. T. MacGillivray and E. B. Thomson (eds.), Digitised Optical Sky Surveys 491–498.
© 1992 *Kluwer Academic Publishers.*

pressing for 5 micron pixels to fully extract all the information from these films. Clearly, further work is advisable in examining the optimum pixel size for digitising the sky survey material, both existing and intended.

3) Should one archive the high-resolution scans and only distribute the lower resolution data?

It was agreed that we should certainly archive the data at the highest possible resolution, and it was considered acceptable to distribute lower resolution data. It depends, to a certain extent, on the capacity of the media available, and at the present time we are perhaps not quite ready for the routine transport of Terabytes of data. It depends also upon the typical requirements for the data. I suspect that most recipients would not be interested in the vast bulk of the information (e.g. the sky background data), and I personally think that we are going to see an increase in the use of 'compression' techniques applied to the raw data and much interest in the distribution of highly compressed versions. It was interesting to see the work of Baruch and collaborators and to see that compression factors of 350:1 do not lead to substantive loss in the 'visual' image. Such compressed frames would certainly be quite acceptable for, for example, the purposes of verification of correct object acquisition at large telescopes.

4) What should the physical medium be for data distribution?

There can be no doubt that, at present at least, CD-ROMs still appear to be the most suitable medium for many reasons, some of which were also expressed in 1989 (viz. cost of manufacture and widespread availability). A note of caution was expressed at the Geneva meeting, however, in that it was perhaps too early to 'fix' the distribution device yet and we should perhaps wait for new developments. There is certainly still some wisdom in this:– with limited resources at each group's disposal, it is always advisable to 'hold-off' for as long as possible and to 'go' with the most sensible device or medium available at the time. On the other hand, one could hold-off for ever! I suspect that the situation with the CD-ROMs will not change drastically over the coming few years, apart from the fact that the cost of reproduction will become more accessible to many groups, and we will therefore see the groups being more able to distribute their own data.

5) Should surveys be published by plate zones or sky coordinates?

Professor Westerhout was quite emphatic that they should be published according to plate zones. The reason is simply the ease at which one can 'home-in' on a particular area of sky. Certainly we have seen how effective this is with the COS-MOS/UKST Southern Sky Object catalogue, where Dr. Yentis has been able to show that the use of the data still stored in plate zones has enabled very fast access to any area of sky. I accept the merits of the plate zone methodology. However, I also think that subsets of the catalogues consisting of much smaller numbers of objects (for example a catalogue solely consisting of galaxies say down to $b_j = 20$) will probably be published rather in monolithic form. I suspect we will see many specialised sub-catalogues mushrooming in the future due to their smaller storage requirements.

6) Should provision be made for long-term ($\sim$ 100 years) archiving of plate scans and object catalogues?

The answer to this in 1989 was: most definitely! I agree entirely. I think we must digitise the major sky surveys once and for all, and have the data archived for future use. We are perhaps in the 'golden era' of photographic-plate digitising machines. There has been much talk at this conference of the arrival soon of digital sky surveys based on the use of large-format CCD arrays on dedicated telescopes, and it is indeed very exciting to see the progress in Japan described by Mr Doi at this conference. Several other groups around the world are also planning such digital sky surveys, and it has been suggested that POSS-III would be undertaken using CCDs rather than photographic plates. I think it is perfectly possible that photographic plates will be eclipsed in about 10-20 years as the main medium for undertaking future sky surveys. However, I also think that the difficulties of such digital sky surveys and performing them in an homogeneous way should not be underestimated. Certainly, anyway, such future digital sky surveys will have to rely on the digitised photographic sky surveys for a considerable time to come as the 'first epoch' material due to the long time baseline that these data provide. In my opinion, therefore, in view of the known deteriorations in long-term storage of photographic plates, it is very very important to be able to archive the digitised scans for 'posterity'.

7) Will each scanning institute also act as the distributor for its data?

In 1989 we agreed that the Data Centres should do it, if they are equipped. My own personal view is that the catalogues are now getting to be very large, and it should be the responsibility of the different groups to 'arrange' for the publication and distribution of their own data. However, I hope we can hear from a representative of one of the Data Centres with a more informed statement on the 'official' policy on this subject.

8) What should be done to ensure that the user can cross-identify objects on plates with other data existing in other data bases and/or archives?

In 1989 we thought that the data centres should be responsible, but I invite comment from a representative from such to clarify the rôle of the data centres with regard to this question as perceived by themselves.

9) Is it reasonable to distribute sky surveys on film, or should we now directly precognise digitised surveys?

My own opinion is as expressed in 1989: viz. that sky surveys will still be distributed on films for some years yet. The reason has been clearly expressed by Professor Reddish in his introductory talk: it is very convenient to just take the film out of its envelope, to hold it up to the light and to 'see' what is on it. We are perhaps not yet at the point of having full mastery over displaying in a suitable way the digitised data. Nonetheless I do see a growing demand for the distribution of digitised pixel data in the future. For the coming few years, I see the digitised copies of sky surveys complementing the film copies. On the other hand, there can be no doubt that on the longer term, once users become more familiar with handling digitised (and probably highly compressed) sky survey data, that your requested copy of whichever sky survey will arrive as a pack of CD-ROMs (or equivalent) rather than as a pack of films.

Now for some questions of my own that I would like to see form the basis for further discussion at the present meeting:

1) Should there be coordination or exchange with regard to calibration sequences?

We have seen from the talks over the past 2 days that different groups are in the process of obtaining CCD sequences at telescopes for the purpose of calibrating the scans. In view of the fact that telescope time is expensive and there is some difficulty in obtaining time for this type of purpose, I wonder if it is not better that the sky be 'split-up', the different groups begin allocated separate areas and the whole thing subsequently merged? Of course, there should be some overlap to verify results. This would obviously require coordination and a degree of good-will between the different groups, but I am convinced of the need to 'work together' in this area to avoid unnecessary duplication.

2) Should there be coordination of the various digitisation programmes?

There are a number of issues involved here: We need to identify what are the actual digitisation programmes of the different groups; are we all intent on digitising the same plates? What is behind the digitisation programmes; are they being driven by scientific objectives or are we just digitising the plates for the sake of digitising the plates? Should there be one group whose prime purpose is to undertake and distribute the definitive scan of all the sky survey material?

To put forward my answers to my own questions, I think it is clear that we are not all digitising exactly the same material, although there are areas in common. It is evident, for example, that some groups are only interested in digitising the IIIa-J plates of the POSS-II survey, while others are interested in digitising all three colours. I think we have here an indication that there is indeed sufficient of a difference between the aims of the different groups that I do not foresee too much of a clash. It is clear that the different groups are being driven by different scientific aims, whether those aims are to support space platforms, to study the structure of the Galaxy, to detect brown dwarfs or to investigate the large-scale structure of the Universe from analysis of the distribution of galaxies, clusters and quasars. I also think it is good, however, that there should be some overlap in the scientific aims. I consider it to be very powerful, for example, that both the

COSMOS and the APM galaxy catalogues provide evidence for more power in the angular correlation function on large angular scales than is predicted by the Cold Dark Matter (CDM) model for galaxy formation. The fact that the two results are in such close agreement when the scanning techniques and software are so different indicates to me that we can have strong confidence in the result. However, being the cautious person I am, I hope that another group will work on the problem to further substantiate the result! With regard to the final issue addressed above, I think that I personally would like to see a diversity of data available. It is quite possible that different machines are suited best for different types of plate or data needs.

3) Should there be a comparison of the different machines?

This was also suggested at the last meeting, but was not picked up at the time as a definite goal that we should pursue. My own opinion is that I feel it may be worthwhile, but I do not think it is essential: so long as we obtain comparable results in the science, then I feel that this is an inherent comparison of the machines. Also, it is an extremely time-consuming task to perform which should only be undertaken by a completely unbiased observer. My own experience is that we undertake a comparison of the machines in order to 'prove' our own pet machine! Such comparisons carry little weight. Nonetheless, it is something that needs to be done in order to put valuable comparative information 'on the record'.

4) Should there be a comparison of software techniques?

This is perhaps an item which is much easier to undertake, and the outcome of which is more clear-cut. I know that Dr. Okamura has informed words to say on this subject, and so I leave the job of persuasion to him, although I must add for myself that I do think this is an area that we should indeed follow-up.

5) Should we merge with the IAU Working Group on 'Astronomical Photography'?

This is an issue that was raised at the last meeting also but there was a great deal of indecision. Events have happened in the intervening time, although there has been no formal contact between our two groups. From informal discussions between those present here, I gauge the opinion that such a merger would lead to too big a group, and too many topics for discussion. It is better to have smaller focussed meetings as this one.

6) Where should we hold the next meeting, planned for 1993?

I invite suggestions from the audience. Personally I think the next meeting should be in the U.S.A. There are three very active machine groups in the U.S.A. (Minnesota, Flagstaff and the STScI), and I think one of these sites would be a highly desirable venue for the next meeting.

R. Pennington :

As regards the various measuring machine groups and what each is saving from their scanning, I would like to point out that this is a matter of economics. Each group is saving as much as they can afford to save.

F. Bonnarel and F. Ochsenbein :

On behalf of the Strasbourg Stellar Data Centre, we should like to provide some answers to MacGillivray's comments:

What should be the rôle of Data Centres?

1. Should they give an access to the data? Yes ...but how many 'digitised surveys' should they provide and which ones?

2. The typical help provided by Data Centres can be the following: they can do their own identifications with known objects, because they have some knowledge and some tools to do that. They can provide on-line a view of the neighbourhood of known objects. They can distribute sets of standard sequences.

3. What are the ongoing activities and projects in Strasbourg? The cross-identifications of IRAS, IUE and GSC with the Simbad database are in progress. A project for an interactive sky atlas is being studied.

S. Okamura :

As a starting point of the discussion, I would like to repeat what I said in my short article in the last 'Digitised Optical Sky Surveys' Newsletter (1990, ed. H.T. MacGillivray).

From the 1990s onward, we will be definitely living in the era of digitised sky surveys. At present, the major data source is Schmidt plates digitised with scanning machines. This will remain so for the coming 10(?) years. However, wide-field CCD cameras attached to the next-generation 8-10m telescopes as well as to the existing telescopes including Schmidts will gradually take over from the plates. Another important aspect in the 1990s is the advent of non-optical sky surveys. In the near infrared, mosaics of 1000 x 1000 (PtSi) or 256 x 256 (HgCdTe) arrays will be available quite soon. In the far infrared, COBE has already mapped the sky and ISO will produce some survey data. In the X-ray, ROSAT is taking deep images of the sky and a CCD imaging spectrometer on board the next Japanese satellite ASTRO-D will also bring us a lot of X-ray images. Surveys have been carried out in the radio as well.

In the era of digitised sky surveys, the importance of the software, which does the jobs such as image detection, de-blending, star/galaxy separation, morphological classification, and correction for confusion, cannot be emphasised too strongly. Several examples of such software were reported in this conference and I believe that the techniques are most advanced in the optical regime. Such software will soon be of far wider use than today because it will be an indispensible means of a wide variety of research. There is a need here for standard data sets.

The need is, I suppose, two-fold. The standard data sets are necessary for the intercomparison of the results:

1. by different programs applied to the identical data and
2. by the same program applied to different data from different telescopes / instruments.

The former is intended to check the performance of the program and help to identify items requiring possible improvements and to examine fine-tuning necessary between photographic and CCD data. The latter is intended to check the reliability of new instruments including next-generation telescopes. I think that the former has the first priority at the moment. So I propose, in fact, I ask you, to establish standard data sets (both photographic and CCD) for several 'selected areas'.

I have been working mainly in the field of detailed surface photometry of galaxies. I feel that the present situation of the digital sky survey business is somewhat similar to that of galaxy surface photometry in the early 1970s. At that time, there were only a few groups which were actively producing the data in the field. At the time of the data compilation for RC2 (early 1970s), detailed surface photometry was available for only 118 galaxies. It was a very wise decision that the Working Group on Galaxy Photometry and Spectrophotometry of Commission 28 of IAU chose, as early as the 1960s, several galaxies as standard galaxies based on the proposal by Professor G. de Vaucouleurs. Newcomers in the field were gently encouraged to measure the standard galaxies first to check the reliability of their new system. This arrangement gave a great contribution to the subsequent tremendous increase in activities in the field. Surface photometry of galaxies is now a technique of daily use by almost anyone who wants to do it. I believe that the standard data sets in digitised sky surveys will have the same effect in the field.

N. Weir :

To what extent are 'thresholded map' data useful to retain as an intermediate form between (and in addition to) images and pixels? Are they a more useful compact form of the pixel data than the compressed forms discussed by Dr. Baruch et al.? I would point out that my work on the use of FOCAS-type image classification algorithms could be performed on the thresholded pixel data, hence vitiating the need for the full pixel information.

R. Pennington :

We have all talked about scanning and archiving this data. I feel it necessary to point out that we will probably only get one opportunity to perform this scanning. We should treat this as the unique opportunity that it is and make every effort to preserve the data. These data will be used in the future for many programmes because they represent the first generation of digital data.

Lasker and Maddox :

Noting that many groups are engaged in extensive programs of CCD photometry for the calibration of Schmidt plates, and that the current levels of coordination among groups appear inadequate to avoid unnecessary duplication, we suggest the following proposal:

1. That each group prepares a text file containing detailed descriptions of observations planned and observations completed, with the understanding that the files will be updated sufficiently frequently to be a useful source of information to other groups.

2. That each calibration team puts its file in a public place on one of its computers that is network accessible, with the expectations that interested

colleagues will copy it as required and that the file may be mailed freely (without editing!) to others for whom network copying may present technical difficulties.

The advantages of this scheme are that each group is responsible for the dissemination of its own plans, so that no editor or news manager is required.

The usefulness of this procedure depends on adequate dissemination of the location of the files. It is suggested that each calibration group e-mail the full path of its files to each of the other groups. To get this process started, the list of e-mail addresses posted at the meeting will be distributed to all groups.

When pre-publication sharing of calibration data appears advantageous, it can be worked out scientist-to-scientist, following the normal conventions of collaboration, thus avoiding all the complexities, caveats, and other delicacies associated with the distribution of preliminary catalogs.

CONFERENCE OVERVIEW AND CONCLUDING REMARKS

R.D. Cannon

Anglo-Australian Observatory, P.O. Box 296, Epping, NSW 2121, Australia.

1 Introduction

First, I have to say how appropriate it was that Vincent Reddish was here on the first day to give the introductory talk. He masterminded both the COSMOS measuring machine (and its predecessor, GALAXY) and the UK 1.2m Schmidt Telescope, first as project scientist and then as manager, and ultimately as Director of the ROE. So, in a very real sense he laid the foundations for a great deal of what we have been discussing here this week. It was interesting to realise that only now, almost 20 years on, we are close to achieving his vision of having readily available objective, quantitative data on all optical objects, covering the whole sky to very faint magnitudes. It was also salutory for me to realise that it is now 18 years since I joined the UK Schmidt Telescope Unit, with a remit to carry out a complete southern sky survey in about three years! In the event that task, like the digitisation of the sky surveys, took a bit longer than anyone expected.

At this meeting, we have had two days mainly devoted to the measuring machines and the practicalities of handling digitised data, followed by two days of presentation of the latest scientific results. Both main topics have been extremely interesting, with lots of new developments. As well as attempting to give an overview of the meeting, I propose to comment on what I see as some of the most exciting developments, and on the most important issues to be tackled in the future.

2 Measuring Machine Developments

As a general point, I think there is now widespread realisation that the technical problem is not simply to build faster or more accurate machines, although those are needed. Instead, we have to look at the complete system between the telescopes and astronomical research. There is a large 'black box', into which photographic plates are fed and out of which (eventually) emerge scientific papers. Inside the box are

H. T. MacGillivray and E. B. Thomson (eds.), Digitised Optical Sky Surveys 499–508.
© 1992 *Kluwer Academic Publishers.*

three layers: the measuring machines, the computers and software for analysing the data, and the astronomers (usually graduate students) who do the hard work. There are complex interactions between these three layers, somewhat like the neural networks discussed by Odewahn. Making the digitised data available to astronomers, and providing the tools to enable astronomers to exploit the data, are every bit as important as getting good data in the first place.

One of the most striking developments is the number of new machines which have recently, or will soon, come into service. For many years the digitisation of large areas of the sky seemed to be almost a British preserve, with COSMOS and the APM the two fastest large plate measuring machines in the world. Now there are ten to a dozen machines with similar capabilities in use or soon to be commissioned. Nearly all are designed for measuring large Schmidt telescope plates, although some are optimised for specific types of work, while at least one, at Hamburg, is aimed rather at increasing the accuracy and speed with which the smaller plates from astrographic telescopes can be measured. We heard here about the machines at the Space Telescope Science Institute (Lasker), the US Naval Observatory (Monet) and Minnesota (Pennington) in the USA; at ESO (Grøsbøl), Paris (Guibert), Leiden (Deul), Hamburg (de Vegt and Winter) and Münster (Seitter) in Europe; and at Tokyo (Doi) in Japan. The fact that all these machines exist perhaps says something slightly negative about the quality or accessibility of most of the digitised data generated so far, but also something positive about the long-term prospects for photography, and I shall return to both these points later.

The older machines are not static, however. We have heard about, and seen something of, the new SuperCOSMOS machine at ROE (Miller and Paterson), which will have dramatically enhanced performance in terms of both speed and accuracy. The APM machine, now part of the new RGO in Cambridge, has also been significantly upgraded, and Lasker described how the PDS 2020 machines at the STScI are being enhanced before embarking on the measurement of the northern and southern Second Epoch Surveys from the Oschin Schmidt at Palomar and the UK Schmidt Telescope in Australia, the latter now operated by the AAO.

Several of the descriptions of new measuring machines included discussion of how the output data should be handled, and techniques for image analysis. Odewahn presented a novel neural network approach to star/galaxy separation, while Baruch compared different data compression techniques which make storage problems more tractable and data retrieval faster.

3 Why Remeasure the Sky Survey Plates?

This question is equivalent to one of the key questions raised at the previous DOSS meeting in Geneva in 1989: should we store all the raw pixel data from plate scans, or just sets of parameterised images? This is related in turn to the question: why build more measuring machines?

The answer has to be that up to now, none of the measuring machines has been able to extract *all* of the information from photographic plates, particularly the large, high resolution Schmidt survey plates. The best photographs from the large Schmidts have stellar images about 1 arcsecond or $15\mu m$ in diameter (FWHM), and record densities over a range from near zero to more than four (ANSI diffuse); the intrinsic resolution of current emulsion types means that astrometric accuracies of better than 0.1 arcsec are readily attained. To match the data stored on sky survey photographs, an ideal measuring machine must have a positional accuracy of 1μm or better and be able to measure densities as high as five with an accuracy of say one percent in a spot of diameter 10μm. In fact the real requirement is even more demanding than this and is likely to tighten further if emulsion resolution improves, as seems probable. Although Nyquist sampling theory says that a 10μm spot should be adequate for measuring positions and distinguishing stars from galaxies, there is a third class of objects on sky survey photographs: artifacts and blemishes. These too have to be recognised by the measuring machines.

It is well known that the first reaction of an astronomer faced with an unexpected image or anomalous result in some digitised data, is to go back and look at the original photograph to see what is 'really' there. Small sharp specks of dirt are then immediately distinguishable from true star images, and even the notorious microspots ('gold spot disease') are often recognisably different from astronomical images. What this means is that the machines and the image analysis software are not extracting all the information from the plates, information that is readily available to the human eye and brain combination using a microscope. Clearly, scanning spot sizes of 20μm or larger, which were the best available for large area scans until recently, will make many blemishes indistinguishable from faint stars or galaxies.

Spot sizes of around 10μm (SuperCOSMOS) to 15μm (STScI PDS) should indeed be adequate for obtaining accurate positions and photometry, and for star / galaxy separation, but they may still not be good enough to eliminate some artifacts. In terms of astronomical applications, this should mean that projects which depend on large samples, such as the cosmological distribution of galaxies or the structure of the Galaxy, should be done perfectly well with the best machines soon to be available. However, programmes which involve looking for very rare objects, of which perhaps the most extreme example is the project to detect gravitational microlensing, which was described here by Moniez, will still have problems with contamination by spurious objects. I suspect that to really extract everything from Schmidt photographs, spot and step sizes as small as 5μm will be required. This will be even more true if the spectacular initial results with the Kodak 4415 emulsion that Russell presented here are confirmed, and if that film comes into regular astronomical use.

As a consequence of this incomplete information extraction, it has always been necessary up to now to remeasure the same original plates for different applications, or as the performance of the machines has improved. Obviously, it would not have been sensible to try to archive and distribute all the raw pixel data from earlier scans. Soon, however, the machines should be able to extract virtually all of the information from the plates, in which case archiving of the scan data will presumably become worthwhile. Archiving has the unlooked for advantage that it may be the best way to preserve the original data from the depredations of microspots which, as Morgan pointed out in a review of survey material from the UK Schmidt, still seem to be rampant despite our best efforts to keep the 'disease' under control.

Archiving is also of course rapidly becoming more practicable and affordable as computer technology advances.

4 Calibration

Adequate photometric calibration of machine data from photographic plates is still a major task. The photographic process is intrinsically non-linear, and large Schmidt plates are liable to be non-uniform due to a variety of causes such as vignetting in the telescope, deficiencies in hypersensitising or processing techniques, and variations in image structure due to differential atmospheric refraction. Star images also have density variations over very small length-scales and are not easily calibrated using the large-area grey-scale spots currently printed on many Schmidt plates. The minimum requirement for stellar photometry from a single plate is to fix the zero point of the magnitudes; a photoelectric sequence of faint stars will serve to check the linearity of the machine magnitude scale, while to fully calibrate a plate requires several such sequences distributed over the field.

For the large scale galaxy surveys involving mosaics of many Schmidt plates, such as the Durham/Edinburgh and Oxford/APM surveys described here by Nichol and Maddox respectively, remarkably good results have been obtained by carefully matching the magnitudes in the overlap regions of adjacent fields. Nevertheless, the accuracy of the magnitude scales and the possibility of systematic errors remains the limiting factor in interpreting the counts of faint galaxies, quasars and stars. Two important initiatives to try to eliminate this problem were described here. The STScI has already established a photoelectric sequence near the centre of every sky survey field to calibrate the data for the HST Guide Star Catalog. Postman and colleagues are now setting out to obtain CCD frames centred on the faintest photoelectric standard star in each of these sequences. These CCD sequences are being obtained with relatively small telescopes and so will not go as faint as the data from sky-limited Schmidt plates. Maddox described the plans by a large consortium of UK and Australian astronomers to obtain deeper CCD frames in all high latitude fields, using the 3.9m AAT. Clearly it will be useful if these two groups, and other astronomers working in smaller numbers of fields, can coordinate their efforts to provide all-sky faint photometric calibration.

While on the subject of photometric calibration, let me try to correct one common error for which I am partially responsible. We should *not* be using J to denote magnitudes from plates taken using Kodak IIIa-J emulsion. J is one of the standard photometric bands in the near infrared, at around 1.2μm, and we are going to see a lot more use of this band now that infrared array detectors are coming into common use. Having two totally different J bands will cause hopeless confusion. For the sky surveys, the IIIa-J plates typically cover from 390nm to 540nm, which amounts to the standard B band plus about a third of V. The best solution seems to be to use B_J or b_J. I now regret that I was responsible for introducing the use of the J prefix for UKST plate numbers back in 1973 when we embarked on the first IIIa-J sky survey.

Another aspect of calibration concerns the use of Schmidt plates for astrometry. The conference produced an interesting apparent divergence of views on the astrometric accuracy achievable on Schmidt plates, with Taff showing that there are asymmetrical shifts of order a second of arc over distances of a few tens of arcminutes, while Bienaymé and others at this meeting, and previously, have been able to obtain astrometric results accurate to tens of milliarcseconds. The answer seems to be that Taff's results apply to absolute astrometry, while much higher accuracy can be attained locally for relative astrometry if large numbers of reference stars can be used. It appears that the bending of Schmidt photographic plates to conform to the spherical focal surface introduces non-radial, non-linear distortions, but probably even these can be modelled if sufficient data are available.

5 Photography Versus CCDs for Future Sky Surveys

Several speakers, including Monet and Irwin, spoke strongly in favour of using CCDs directly for all-sky surveys, while Doi described an impressive and ambitious plan to equip the Kiso Schmidt telescope with an array of 64 large format CCDs. Given the problems of calibrating photographic data, the much higher quantum efficiency of CCDs compared with the photographic process, and the tremendous recent advances in the size of CCDs and the power of computers available for on-line data processing, this suggestion has to be taken very seriously. On a simple photon counting basis the two techniques have already become comparable, in that an array of a few large CCDs may have a twentieth the area of a large photographic plate, but about twenty times the sensitivity. It seems entirely reasonable to suppose that larger arrays of larger CCDs will soon be available and should be the fastest way to survey large areas of sky; indeed, for some projects, such as searching for faint supernovae, CCDs are already the 'best buy'.

CCDs do however have their own limitations, and it is not clear that they will easily do everything which can currently be done photographically. Most CCDs in practice turn out to be neither strictly linear detectors nor cosmetically perfect, although both these effects can usually be corrected for. They also suffer from cosmic rays generating spurious events, although usually these too can be readily removed. More significantly, the photographic plate and the CCD should not be compared on the basis of area but on the basis of number of pixels; most current CCDs have pixel sizes in the range 19μm to 25μm, equivalent to 1.5 arcsec in a large Schmidt, whereas the IIIa- series emulsions resolve features at the 7μm or 0.5 arcsec level. Thus, although the photographic plate may be only 50 times larger in area than a big CCD, it has about 1000 times as many pixels. In other words, since the CCD has about 20 times the quantum efficiency, to survey the sky to the same depth and resolution will still take about 50 times longer or require an array of 50 CCDs each of size 2000 × 2000 pixels. Handling the data flow from such an array of CCDs is not a trivial matter either; the photographic plate still constitutes a rather compact and convenient way to record, store and display large amounts of astronomical data.

However, perhaps the most serious problems with using CCDs for sky surveys will come in the fields of large-scale astrometry and photometry. Accurate astrometry will require a grid of positional reference stars fine enough to cover all CCD fields, while small changes in atmospheric transparency or 'seeing' tend to produce striations in sky maps built up by raster scanning. Perhaps these difficulties will be overcome best by putting an array of CCDs in a wide field Schmidt telescope, as planned at Kiso.

Another point I must make is that while CCDs look much superior to photographic plates in theory, in practice it is proving surprisingly difficult to do substantially better when studying faint extended features. For example, to determine the colours of some of the faint jets and arcs around galaxies, found on sky survey photographs by Malin and others using photographic amplification techniques, is currently at the limit of what can be achieved with a CCD on a 4m telescope. Finally, the exciting results presented by Russell using the new Kodak 4415 emulsion on film show that photography too has not yet reached its limits, and may well be capable of quantum efficiencies as high as ten percent.

Despite all these reservations, it is obvious that CCDs have already taken over most of the tasks thought to be the preserve of photography only a decade or so ago, and it seems very likely that wide-field deep imaging will go the same way. The only uncertainty seems to be the timescale. My guess is that we will be doing photography in Schmidt telescopes for another decade, but perhaps not for much longer than that.

6 The Distribution of Sky Survey Data

Digitising the sky surveys is only half the battle: the second step, equally important, is to make sure that the data are available to astronomers everywhere. Several very encouraging developments have been reported at this meeting. The STScI has already issued the Guide Star Catalog on two CD ROMs, and now plans to make the full set of image data for the whole sky available, on sets of about 100 CDs. Yentis demonstrated the COSMOS/UKST catalog of the southern sky, set up at NRL, MPE and ROE primarily to assist with the identification of ROSAT sources, but clearly applicable to many other programmes. Thus at last we are close to realising the dream whereby any astronomer, anywhere in the world, will be able simply to dial up the coordinates of his or her favourite bit of the sky and have a digitised map of the region instantly displayed on a computer graphics terminal. These astronomers include those of us working in Australia, where at the moment taking photographs for the sky surveys is very like most other Australian industries: we are primary producers and export useful raw materials to the rest of the world, but we are not taking enough part in the later 'value added' activities which make use of the primary produce.

I think that astronomers have made a mistake in thinking that the cost and manpower effort involved in reproducing and distributing the digitised sky surveys will be too much for the community to bear. These digitised surveys will form the Palomar and ESO/SERC Sky Atlases of the future, and will lead to a similar revolution in how we do astronomy. Bearing in mind that the production of the earlier atlases required teams of photographers working in specialised laboratories for many years, it is surely not too much to invest several person–years of effort into organising, copying and distributing the digitised surveys, and to producing software tools to make it easy for astronomers to exploit these databases.

7 Other General Issues

The matter of archiving of machine data was one of several questions raised at the first DOSS meeting in Geneva in 1989. That list of questions was reviewed here by MacGillivray, and it seems that the answers to most of them are already clear and unambiguous. One matter still outstanding is what physical medium to use to store and distribute digitised data. At present, the CD ROM seems the best buy, but the technology is still evolving rapidly and the answer may be different in a few years. Another not unrelated issue is whether film copies should continue to be distributed in future, or just the digitised data. For the time being it seems that the photographic copies are still necessary, but this may not be true for much longer.

MacGillivray also raised a number of new questions for the 1991 meeting, including how the digitisation and related calibration programmes should be coordinated, and how the various measuring machines and their associated software can be compared and tested. These were not formally resolved, although clearly collaboration is very desirable. Okamura pointed out the need for standard data sets for making comparisons, and suggested that some 'selected areas' should be set up.

8 Astronomical Highlights

The second half of the conference consisted of about two dozen reports on recent scientific results obtained using data from measuring machines. These covered a great variety of astronomical fields and involved many people at different institutions around the world, and clearly demonstrated the vitality of this branch of astronomical research. It would be invidious to try to make a personal selection of highlights, but it is perhaps worth noting that there is now a class of projects which has only become feasible since the advent of fast measuring machines, and which ultimately must be the main justification for the continued investment of money and effort in the development of such machines. These projects split into three classes: large-scale statistical studies, searches for rare objects, and the identification of sources detected at non-optical wavelengths.

In the first category we have studies of the structure of our Galaxy and of the Magellanic Clouds, represented here in the papers by Yamagata and Gardiner respectively. By obtaining multi-colour photometry for samples of stars orders of magnitude larger than were previously available, fundamentally new results are being derived on the structure and stellar content, and hence evolution, of these Local Group galaxies.

The large scale distribution of galaxies in the Universe is one of the most exciting applications of the digitised surveys, leading as it has done to a revision of the current 'conventional' type of Cold Dark Matter cosmological model. Recent results from the COSMOS and APM surveys were presented by Nichol, Raychaudhury, Hale-Sutton and Maddox, and included spectacular 2-dimensional maps of the galaxy distribution covering substantial fractions of the sky. Other speakers including Broadbent and Broadhurst described the start of the next step, which is to obtain redshifts for large samples of galaxies in order to probe the true 3-dimensional structure, for example by using multi-object fibre optic systems. It is even possible to determine redshifts from low dispersion objective prism spectra, as described by Seitter, although others have found it difficult to achieve sufficiently high accuracy by this technique.

Further cosmological data come from the distribution of quasars, which in the measuring machines context means finding rare objects but doing so systematically over large areas of sky. The results to date can perhaps best be described as intriguing, in that some structure is seen but the samples or areas covered are still too small to yield results of high statistical significance; reports were given here by Goldschmidt, McMahon and Clowes. However, this is clearly an area with enormous potential, since the quasars offer a unique opportunity to map out the 3-dimensional structure of the Universe at redshifts between one and five. Digitised sky survey data offer by far the best way of finding quasars in the first place, using one or more of several different techniques: multi-colour surveys, variability, from objective prism spectra and as radio or X-ray sources. Several hundred quasars can be identified in a single Schmidt sky survey field, corresponding to rates of order one object in a thousand.

Searches for much rarer classes of objects, such as brown dwarfs (which are rare at least in magnitude-limited samples) were mentioned by Hawkins, while Russell described the comparatively easy searches for asteroids, which can and have been carried out without using measuring machines, but which are much more efficiently and completely done by machine. But perhaps the most extreme example of a rare object search was that described by Moniez, to try to catch gravitational microlensing events caused by low-mass black holes in the halo of our Galaxy moving in front of stars in the Magellanic Clouds. Since these events are expected to be rare, shortlived and unrepeatable, they will be very hard to identify unambiguously.

The third class of machine-dependent sky survey activity is the identification of sources detected at other wavelengths, and six papers were presented here involving surveys covering the whole electromagnetic spectrum. Once such surveys include more than a few thousand sources, it is impracticable to attempt wholesale identifications without using machine data. If the source positions are accurate to one or two arcseconds, the entire procedure can be automated, but this is rarely practicable. More often, recourse must be had to secondary identification criteria, which is where the capability to display all the images in an area of the sky on a computer terminal comes into its own. Gursky, Roche, Schachter and Voges all discussed the identification of X-ray sources, where the main problem is that the error boxes are still usually quite large, sometimes multiple, and often in the crowded Galactic plane. Additional information, such as the colours or variability of candidate stars, has to be used to narrow down the search. Wolstencroft described one large-scale programme to identify infrared sources found by IRAS, where the problem was made a bit easier despite the large (around 20 arcsec) error boxes by the fact that candidate identifications were mostly expected to be relatively bright stars and galaxies. The same is not true for radio source catalogues, where many of the identifications will be with very faint galaxies and quasars, and a significant fraction will be below the optical detection limit. In such cases the only reliable procedure is to follow up the initial radio survey, usually carried out by a single-dish radio telescope, with an accurate position measurement for every source using an interferometer, as proposed by Savage here for the new PMN radio survey. Only then is it possible to make almost-certain optical identifications, based on positional coincidence alone.

Finally, there were several presentations of results which, while they could have been obtained before the era of digitised sky surveys, were certainly obtained much more easily and reliably using machine data. These include the search for faint dwarf stars in the Pleiades by Hambly, and for low surface brightness galaxies in the Fornax cluster by Phillipps.

9 The Future of Digitised Sky Surveys

Any selection of major future trends is bound to be subjective and personal, and almost certain to be wrong in some crucial respects. Nevertheless, it is important to emphasise for the benefit of other astronomers, and of those who sit on peer review committees and funding bodies, that the era of digitised sky surveys is only just beginning. It has taken a decade or more for astronomers, engineers and software gurus working together to make the wholesale digitisation of Schmidt sky survey photographs a realisable dream. While detailed prediction of where this will lead is impossible, there are some areas which we can be sure will feature. These include:

1. Readily accessible all-sky digitised optical data, in four or five colour bands, to the limit of current large Schmidt telescopes (around $U = 19.5, B = 21.5, R = 20$ and $I = 18.5$, as described herein by Hawkins). These data will include proper motions for the nearer stars, using the Second Epoch Survey currently under way in the south on the UK Schmidt, and the Palomar II surveys in the north.

2. In-depth studies using many plates of selected areas, for example to obtain accurate proper motions of faint stars to study Galactic kinematics as well as parallaxes of faint nearby stars, and to undertake systematic searches for variable stars.

3. Fuller exploitation of objective prism plates, which contain a wealth of largely untapped spectroscopic information, for example in the identification of X-ray sources.

4. Use of digital plate addition techniques, to attain much higher signal-to-noise ratios than can be achieved with a single plate, or to push to fainter magnitude limits.

Finally, there are related developments which are crucial for the proper exploitation of the digitised surveys, specifically:

1. The provision of adequate follow-up facilities to study large samples of faint objects found in the sky surveys; for example, multi-object fibre systems such as FLAIR which permits simultaneous spectroscopy of up to 90 targets over the 6.5 degree square field of the UK 1.2m Schmidt Telescope, or the 2dF facility currently under construction for the 3.9m AAT, which will enable observation of up to 400 objects over a two degree field.

2. Improved performance from the measuring machines, in terms of speed and both astrometric and photometric accuracy. SuperCOSMOS is one machine which should be a major step forward in these respects, although even there it is not clear that the minimum spot size and density range will really permit extraction of all the data from survey photographs, particularly if the latest film emulsions fulfil their early promise.

3. The use of large CCDs to calibrate photographic data, and to push studies to even fainter limits in selected areas. While the CCDs do present a challenge to Schmidt photographic surveys, it is not clear how soon or how completely they will supplant photography, and in any event there is now a vast database of astronomical survey photographs (some 15 000 to date from the UK Schmidt, for example) which alone justifies the continued development of digitising machines.

In conclusion, it seems to me that we have had a most interesting and successful meeting, in that a lot of ideas have been exchanged and many new contacts made. The digitisation of sky surveys is a field of astronomy which is very much alive and well, and probably about to enter a new era where several major areas of research will be revolutionised through the availability of vast amounts of quantitative data. I add my thanks to those of others to the many people involved in the organisation of this Second DOSS Conference, and look forward enthusiastically to the next one in a couple of years' time.

Discussion

Westerhout :

I beg to differ with the speaker's opinion that all-sky CCD work is far in the future. The technology is all there. Granted, very fast computers would be needed at the telescope, but that's almost there. Do we invest in thousands of photographic plates, or one mosaic of CCDs? Our telescopes will be more versatile, as they will be able to survey and/or go deeper in magnitude. I predict that POSS II and other currently running big surveys are the last done photographically. We will have CCD surveys within the next 5 years. Measuring machines will take another 10 years to cover the backlog of plates; thereafter they will be obsolete!

AUTHOR INDEX

SUBJECT INDEX

OBJECT INDEX